Quantum Chemistry

Quantum Chemistry

Raymond Daudel
Centre National de la Recherche Scientifique,
La Sorbonne, Paris
Georges Leroy, Daniel Peeters and Michel Sana
Laboratoire de Chime Quantique,
Université Catholique de Louvain

A Wiley–Interscience Publication

JOHN WILEY & SONS
Chichester · New York · Brisbane · Toronto · Singapore

***Library of Congress Cataloging in Publication Data*:**

Main entry under title

Quantum chemistry.
'A Wiley–Interscience publication.'
Includes index.
1. Quantum chemistry. I. Daudel, Raymond.
QD462.Q344 1983 541.2′8 82-23688

ISBN 0 471 90135 0

***British Library Cataloguing in Publication Data*:**

Quantum chemistry.
1. Quantum chemistry.
I. Daudel, Raymond
541.2′8 QD462

ISBN 0 471 90135 0

Typeset and printed in Northern Ireland at The Universities Press (Belfast) Ltd.
Bound at the Pitman Press Ltd., Bath, Avon

Contents

Preface

In order to realize the achievement constituted by the presentation, in 1983, of the basic principles, general theory, methodology and a number of essential results of quantum chemistry in a volume of about 550 pages, it suffices to remark that this volume has practically exactly the same dimension as the *Quantum Chemistry*, published in 1959 by R. Daudel, R. Lefebvre and C. Moser of which it is a largely transformed second edition. The last 15 years have witnessed such a development, sophistication and diversification of the methods of quantum chemistry, the extension of their application to so many new directions of exploration, an abundance of results in so many distinct fields that such an achievement seemed *a priori* improbable. But here it is. Drs Daudel, Leroy, Peeters and Sana have accomplished a remarkable synthesis of today's knowledge of the concepts, ideas and methods of computation which preside over the study of the structure of atoms and molecules in their static and dynamic behaviour. Having these data summed up in a clear presentation in one handy volume, together with the fundamental references to original papers, represents a working tool of great commodity and usefulness.

I hope that this book will be helpful to all chemists in their endeavour to understand what molecules are and that it will incite young students to join the exploration.

Bernard Pullman

General Physical Constants

Speed of light in vacuum	c	2.997925_1	$\times 10^{8}$ m s^{-1}
Gravitational constant	G	6.670_5	$\times 10^{-11}$ N m^2 kg^{-2}
Elementary charge	e	1.60210_2	$\times 10^{-19}$ C
		4.80298_7	$\times 10^{-10}$ e.s.u.
Avogadro number	N_A	6.02252_9	$\times 10^{23}$ mol^{-1}
Mass unit	u	1.6604_2	$\times 10^{-27}$ kg
Electron rest mass	m_e	9.10908_{13}	$\times 10^{-31}$ kg
Proton rest mass	m_p	1.67252_3	$\times 10^{-27}$ kg
Neutron rest mass	m_n	1.67482_3	$\times 10^{-27}$ kg
Faraday constant	F	9.64870_5	$\times 10^{4}$ C mol^{-1}
Planck constant	h	6.62559_{16}	$\times 10^{-34}$ J s
	$\hbar = h/2\pi$	1.054494_{25}	$\times 10^{-34}$ J s
Fine structure constant	$\alpha = 2\pi e^3/hc$	7.29720_3	$\times 10^{-3}$
Charge to mass ratio for electron	e/m_e	1.758796_6	$\times 10^{11}$ C kg^{-1}
Quantum of magnetic flux	h/e	4.13556_4	$\times 10^{-15}$ Wb
Rydberg constant	R_∞	1.0973731_1	$\times 10^{7}$ m^{-1}
Bohr radius	a_0	5.29167_2	$\times 10^{-11}$ m
Compton wavelength of electron	$h/m_e c$	2.42621_2	$\times 10^{-12}$ m
Electron radius	$r_e = e^2/m_e c_2$	2.81777_4	$\times 10^{-15}$ m
Thomson cross-section	$8\pi r_e^3/3$	6.6516_2	$\times 10^{-29}$ m^2
Compton wavelength of proton	$\lambda_{C,p}$	1.321398_{13}	$\times 10^{-15}$ m
Gyromagnetic ratio of proton	γ_p	2.675192_7	$\times 10^{8}$ rad s^{-1} T^{-1}
Bohr magneton	μ_B	9.2732_2	$\times 10^{-24}$ J T^{-1}
Nuclear magneton	μ_N	5.05050_{13}	$\times 10^{-27}$ J T^{-1}
Proton moment	μ_p	1.41049_4	$\times 10^{-26}$ J T^{-1}
Gas constant	R	8.31434_{35}	$\times$J K^{-1} mol^{-1}
Boltzmann constant	k	1.38054_6	$\times 10^{-23}$ J K^{-1}
First radiation constant ($2\pi hc^3$)	c_1	3.74150_9	$\times 10^{-16}$ W m^2
Second radiation constant (he/k)	c_2	1.43879_6	$\times 10^{-2}$ m K
Stefan–Boltzmann constant	σ	5.6697_{10}	$\times 10^{-8}$ W m^{-2} K^{-4}
Permittivity constant ($1/c^2\mu_0$)	ε_0	8.854184_7	$\times 10^{-12}$ F m^{-1}
Permeability constant	μ_0	4π	$\times 10^{-7}$ H m^{-1}
Gravitational acceleration	g	9.80665 m s^{-2}	
Thermochemical calories	cal	4.1840 J	
Normal atmospheric pressure	P	1.01325	$\times 10^{5}$ N m^{-2}
Litre	l	1.000028	$\times 10^{-3}$ m^3
Zero degree Celsius	°C	273.16 K	

C = coulomb, e.s.u. = electrostatic units, F = farad, G = gauss, H = henry, Hz = hertz, J = joule, K = kelvin degree, kg = kilogram, m = meter, mol = mole, N = newton, rad = radian, s = second, T = tesla, W = watt, Wb = weber.

PART I

GENERAL QUANTUM CHEMISTRY

CHAPTER 1

Basic Ideas and Methods

1.1 ABOUT WAVE MECHANICS

1.1.1 First Insight into the Structure of Wave Mechanics

It is assumed that the reader already has a certain knowledge of wave mechanics. Therefore, time will not be spent recalling basic concepts.

Bridgman has said[1] that a physical concept is a set of operations. The quantitative idea of length, for example, is equivalent to the set of operations which permits the length to be measured. It is useful to distinguish between experimental operations and rational ones. Experimental operations are made in a laboratory with the help of various apparatuses. Rational operations can be made in our minds with the help of papers and pencils or electronic computers.

Wave mechanics appears to be an efficient procedure of associating rational operations with experimental ones. Let us assume that we have experimental evidence leading to the conclusion that there is an electron in a given region of a laboratory at a certain time t. If at that time a small electronic counter is placed at a given point M of that region there exists a certain probability that a 'top' corresponding to the penetration of the electron into the counter will be heard. Therefore it is useful to associate a *probability* wave function $\Psi(\mathrm{M}, t)$ with the presence of an electron in a given space. This is the de Broglie wave,[2] defined in such a way that the probability $\mathrm{d}p$ of finding the electron in an elementary volume $\mathrm{d}v$ surrounding the point M at the time t is given by the equation:

$$\mathrm{d}p = |\Psi(\mathrm{M}, t)|^2 \,\mathrm{d}v \tag{1.1}$$

As we assume that the electron is in the considered space the probability of finding the electron in the all space must be unity. Therefore:

$$\int_{\mathbb{R}^3} |\Psi(\mathrm{M}, t)|^2 \,\mathrm{d}v = 1 \tag{1.2}$$

if $\mathbb{R}^3$ denotes the corresponding three-dimensional space.

Therefore all wave functions have a finite norm and belong to the ensemble of functions which have a finite norm. A vectorial space can be built on that ensemble in such a way that any function can be represented by a vector in that space. The corresponding function space is called the Hilbert space $\mathbb{E}$.

Any operation of measure of an electronic property is associated by wave mechanics to an operation on the corresponding wave function. Table 1.1 contains a list of operators associated with classical properties.

Table 1.1 Correspondence between classical quantities and wave mechanical operators

Classical quantities		Operators
Position	x	$\times x$
(coordinates)	y	$\times y$
	z	$\times z$
Mass	m	$\times m$
Charge	e	$\times e$
Planck constant	h	$\times h$
Potential energy	$V(x, y, z)$	$\times V(x, y, z)$
Momentum	p_x	$\frac{h}{2\pi i}\frac{\partial}{\partial x}$
	p_y	$\frac{h}{2\pi i}\frac{\partial}{\partial y}$
	p_z	$\frac{h}{2\pi i}\frac{\partial}{\partial z}$
Total energy		$\frac{h}{2\pi i}\frac{\partial}{\partial t}$

We can derive other operators from Table 1.1 by using the following rules. If a classical quantity A is the sum of two quantities B and C, the operator A_{op} is the sum of the operators B_{op} and C_{op}. If a classical quantity A is the product of two quantities B and C, the operator A_{op} is the product of the operators B_{op} and C_{op}.

To be more rigorous it would be neccessary to introduce a more complete system of vectorial spaces because some of the operators of Table 1.1 (such as x, y, or z) have eigenvectors in a Hilbert space like $\mathbb{E}$. This system of space must be made of:

(a) A nuclear space N on which a family of scalar products is defined having specific relations between the products. The space of rapidly decreasing functions could be this space.
(b) The Hilbert space $\mathbb{E}$.
(c) The dual N^* of the space N. This dual could be the space of distributions. In this space an operator like x possesses eigenvectors which are Dirac measures.

Let us use Table 1.1 to find another expression of the operator associated with the experimental measure of the energy of an electron. The classical energy would be written as:

$$E = \tfrac{1}{2}mv^2 + V(x, y, z) \tag{1.3}$$

where v denotes the velocity of the particle and $V(x, y, z)$ the potential energy. We can also write:

$$E = \frac{p_x^2 + p_y^2 + p_z^2}{2m} + V(x, y, z) \tag{1.4}$$

The corresponding operator, called the *hamiltonian operator H*, is therefore:

$$H = -\frac{h^2}{8\pi^2 m}\left(\frac{\partial^2}{\partial x^2} + \frac{\partial^2}{\partial y^2} + \frac{\partial^2}{\partial z^2}\right) + V(x, y, z) \tag{1.5}$$

The theory will be consistent if (and only if) the effect of the operator H on a wave function is equivalent to the effect of the other energy operator—$(h/2\pi i)(\partial/\partial t)$; that is to say if:

$$H\Psi(\mathrm{M}, t) = -\frac{h}{2\pi i}\frac{\partial\Psi(\mathrm{M}, t)}{\partial t} \tag{1.6}$$

Equation (1.6) is the wave equation. Any wave function must satisfy equation (1.6).

In simple cases the result of the effect of an operator A_{op} on a wave function Ψ is another function belonging to the same space $\mathbb{E}$:

$$A_{op}\Psi = \Phi \tag{1.7}$$

If the resulting Φ is proportional to $\Psi(\Phi = \alpha\Psi)$ the function Ψ *is* said to be an *eigenfunction* of the operator and the proportionality constant α is an eigenvalue of that operator. Equation (1.7) becomes:

$$A_{op}\Psi = \alpha\Psi \tag{1.8}$$

At a given operator A_{op} there corresponds a set of eigenvalues $\alpha_1, \alpha_2, \ldots, \alpha_i, \ldots$ which will be symbolized as $\{\alpha_i\}$. This set of numbers is called the *spectrum of the operator.*

The *first principle* of wave mechanics states that *the set of possible measures of a given quantity A is the spectrum of the operator A_{op} associated with that quantity by following the rules previously stated.* For example, the possible measures of the energy of an electron described by the hamiltonian operator H is the set of eigenvalues E_i which satisfies the equation:

$$H\Psi_i = E_i\Psi_i \tag{1.9}$$

This is the Schrödinger equation.[3]

It appears that wave mechanics establishes a kind of isomorphism between an ensemble of sets of experimental operations and an ensemble of sets of rational operations.

Let us consider an electron moving in the field of a fixed electric charge $+e$. Such a system is a good model to represent an hydrogen atom. If r denotes the distance between that charge and a point M the potential energy which the electron would have if it were at that point is:

$$V(r) = -\frac{e^2}{r} \tag{1.10}$$

The Schrödinger equation can be explicited as:

$$-\frac{h^2}{8\pi^2 m}\Delta\Psi_i(\mathrm{M}, t) - \frac{e^2}{r}\Psi_i(\mathrm{M}, t) = E_i\Psi_i(\mathrm{M}, t) \tag{1.11}$$

where Δ (called the Laplacian operator) denotes the sum of the second derivatives:

$$\frac{\partial^2}{\partial x^2} + \frac{\partial^2}{\partial y^2} + \frac{\partial^2}{\partial y^2} \tag{1.12}$$

It can be mathematically shown that the only negative E_i values for which convenient solutions can be found are those which satisfy the equation:

$$E_i = \frac{-2\pi^2 m e^4}{n^2 h^2} \tag{1.13}$$

where n is an integer called the *total quantum number*. Therefore any measures of the energy of the hydrogen atom will give one of the numbers E_i which obey equation (1.13).

Let $\Psi(\mathrm{M}, t)$ be the wave function of an electron. Consider a property A of that electron. We are led to consider the corresponding operator A_{op} and the equation:

$$A_{\mathrm{op}} f_i(\mathrm{M}, t) = \alpha_i f_i(\mathrm{M}, t) \tag{1.14}$$

which defines the eigenfunctions f_i and the eigenvalues α_i which coincide with the possible measures of A (first principle). In simple cases $\Psi(\mathrm{M}, t)$ can be expanded as a linear combination of the various eigenfunctions:

$$\Psi(\mathrm{M}, t) = \sum_i C_i(t) f_i(\mathrm{M}, t) \tag{1.15}$$

The *second principle* states that *the probability p_i of finding the value α_i as a measure of A at the time t is*:

$$p_i = |C_i(t)|^2 \tag{1.16}$$

Therefore if the wave function $\Psi(\mathrm{M}, t)$ coincides with one of the eigenfunctions of the hamiltonian H we can write:

$$\Psi(\mathrm{M}, t) = \Psi_i(\mathrm{M}, t) \tag{1.17}$$

Therefore the probability of finding the energy E_i is unity. *As we are sure that the energy of the electron is* E_i, the electron is said to be in a *stationary state.*

1.1.2 A Very Simple Detailed Example of Application of Wave Mechanics

Let us imagine an electron constrained to move along a segment of a straight line of length l with origin at point O. In that one-dimensional space the Schrödinger equation reduces to:

$$-\frac{h^2}{8\pi^2 m}\frac{d^2\Psi_i(M, t)}{dx^2}+V(M)\Psi_i(M, t)=E_i\Psi_i(M, t) \tag{1.18}$$

for $0<x<l$, if x denotes the coordinate of M. It will be assumed that $V(M)$ is a constant in that interval. In order to secure the electron in the segment it is necessary to introduce a potential $V(M)$ which tends to infinity when x tends to 0 or l. As a potential is defined except for an additive constant, $V(M)$ can be taken as zero for $0<x<l$. Equation (1.18) becomes:

$$-\frac{h^2}{8\pi^2 m}\frac{d^2\Psi_i}{dx^2}=E_i\Psi_i \qquad \text{for } 0<x<l \tag{1.19}$$

The solution of that equation is:

$$\Psi_i(x, t)=k(t)\sin(a_i x+b) \tag{1.20}$$

Introducing that last expression in (1.12), the following equation can readily be obtained:

$$E_i=\frac{h^2 a_i^2}{8\pi^2 m} \tag{1.21}$$

Only certain functions defined by equation (1.20) are convenient if we take account of the behaviour of equation (1.18) when x tends to 0 and l. The second member of equation (1.18) is always finite. Therefore when x tends to 0 or l the first member *must* also remain finite. As the second derivative of Ψ_i remains finite and V tends to infinity, Ψ_i must tend to 0. Therefore it is necessary that:

$$a_i l=n_i \qquad \text{and} \qquad b=q\pi \tag{1.22}$$

n_i and q being integers. Equations (1.21) and (1.22) lead to:

$$E_i=\frac{n_i^2 h^2}{8ml} \tag{1.23}$$

and any measure of the energy of the electron belongs to the set of numbers E_i given by equation (1.23). That simple discussion clearly shows the origin of a quantum number like n_i. The shape of the eigenfunction being independent of the value of q, we shall select the value:

$$q=0$$

The eigenfunction is now:

$$\Psi_i(x, t) = k(t) \sin \frac{n_i \pi}{l} x \tag{1.24}$$

To obtain the shape of $k(t)$ let us consider a *stationary state of* the electron for which the wave function coincides with the eigenfunction Ψ_i. Thus Ψ_i must be a solution of the wave equation:

$$-\frac{h^2}{8\pi^2 m} \frac{\mathrm{d}^2}{\mathrm{d}x^2} \Psi = -\frac{h}{2\pi \mathrm{i}} \frac{\partial \Psi}{\partial t} \tag{1.25}$$

However:

$$-\frac{h}{2\pi i} \frac{\partial \Psi}{\partial t} = -\frac{h}{2\pi i} \frac{\partial k(t)}{\partial t} \sin \frac{n_i \pi}{l} x \tag{1.26}$$

and

$$-\frac{h^2}{8\pi^2 m} \frac{\mathrm{d}^2 \Psi_i}{\mathrm{d}x^2} = E_i k(t) \sin \frac{n_i \pi}{l} x \tag{1.27}$$

Therefore we must have:

$$E_i k(t) = -\frac{h}{2\pi i} \frac{\partial k(t)}{\partial t} \tag{1.28}$$

that is to say:

$$k(t) = N \mathrm{e}^{-(2\pi \mathrm{i}/h) E_i t} \tag{1.29}$$

where N is a constant. The wave function associated with an electron constrained to remain on the segment is finally:

$$\Psi_i(x, t) = N \sin \frac{n_i \pi}{l} x \mathrm{e}^{-(2\pi \mathrm{i}/h) E_i t} \tag{1.30}$$

The probability of finding the electron between x and $x + \mathrm{d}x$ is given by the formula:

$$\mathrm{d}p = |\Psi_i|^2 \, \mathrm{d}x \tag{1.31}$$

and the probability density $\rho(x, t)$:

$$\rho(x, t) = \frac{\mathrm{d}p}{\mathrm{d}x} = |\Psi_i|^2 \tag{1.32}$$

As Ψ_i is a *complex* function the square of the modulus $|\Psi_i|^2$ can be written as:

$$|\Psi_i|^2 = \Psi_i \Psi_i^* \tag{1.33}$$

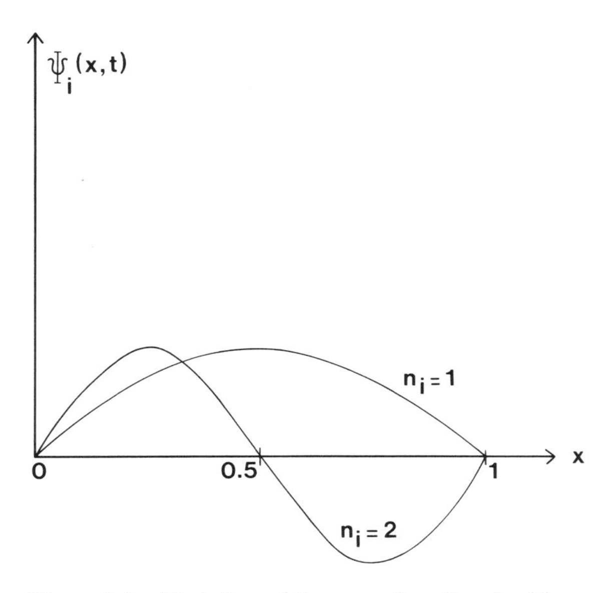

Figure 1.1 Variation of the wave function ψ with x

if Ψ_i^* denotes the complex conjugate of Ψ_i. Therefore:

$$\begin{aligned}\rho(x, t) = |\Psi_i|^2 &= N^2 \sin^2 \frac{n_i \pi}{l} x \mathrm{e}^{-(2\pi \mathrm{i}/h)E_i t} \mathrm{e}^{+(2\pi \mathrm{i}/h)E_i t} \\ &= N^2 \sin^2 \frac{n_i \pi}{l} x\end{aligned} \tag{1.34}$$

The electronic density for a stationary state does not depend on the time. Figure 1.1 shows how $\Psi_i(x, t)$ depends on x.

When the electron is in its ground state ($n_i = 1$) the electronic density ρ vanishes for $x = 0$ and $x = 1$ and reaches a maximum for $x = \frac{1}{2}$. For the excited state corresponding to $n_i = 2$ the density ρ is zero for $x = 0$, $x = \frac{1}{2}$ and $x = 1$. The number of modes of the wave function and therefore *the number of modes of the electronic density ρ increases with the energy of the electron.*

1.2 WAVE FUNCTIONS, NUCLEAR CONFORMATION AND ELECTRONIC STRUCTURE

1.2.1 The Hydrogen Atom

Before going into technical details of how to calculate molecular wave functions *basic ideas* concerning the electronic structure and nuclear conformation of molecules will be analysed. The simplest monoatomic molecule is the hydrogen atom! Some features concerning the electronic wave function associated with that atom are now discussed.

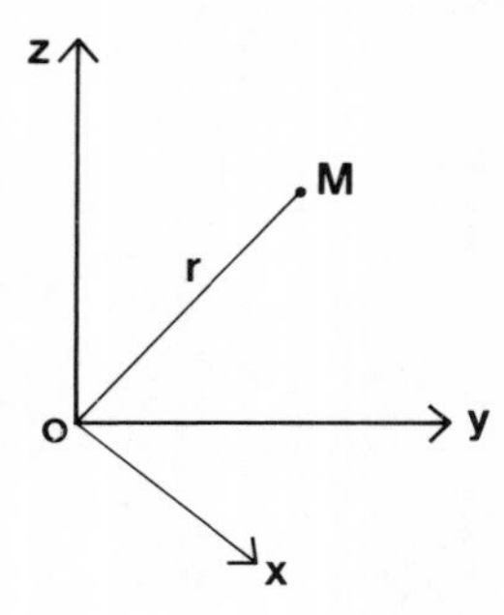

Figure 1.2 Coordinates of point M

In principle the hydrogen atom is a two-bodies problem, but happily for such a system the gravity centre theorem applies within the framework of wave mechanics.[4] Therefore the study of the movement of the two particles may be divided in two parts: (a) the study of the movement of their gravity centre and (b) the study of the movement of the electron with respect to the nucleus. Consider a classical coordinate system and a point M. Let r be the distance between that point and the origin of the coordinate system (Figure 1.2).

The Schrödinger equation for the relative movement is:

$$-\frac{h^2}{8\pi^2\mu}\Delta\Psi_i(\mathrm{M}, t)-\frac{e^2}{r}\Psi_i(\mathrm{M}, t)=E_i\Psi_i(\mathrm{M}, t) \tag{1.35}$$

where μ denotes the reduced mass:†

$$\mu=\frac{mM}{m+M} \tag{1.36}$$

Therefore the probability density of finding the nucleus at the origin *and* the electron at point M is:

$$\rho(\mathrm{M})=|\Psi(\mathrm{M}, t)|^2 \tag{1.37}$$

for a stationary state. As m is very small with respect to M it is possible to replace μ by m. The resulting equation:

$$-\frac{h^2}{8\pi^2 m}\Delta\Psi_i(\mathrm{M}, t)-\frac{e^2}{r}\Psi_i(\mathrm{M}, t)=E_i\Psi_i(\mathrm{M}, t) \tag{1.38}$$

is identical with the one representing an electron moving in the field of a fixed electric charge $+e$.

Consider a stationary state corresponding to a bonded electron; the energy E_i will be negative. It is well known that, in that case, the eigenvalues obey the equation:

$$E_n=-\frac{2\pi^2 me^4}{n^2h^2} \tag{1.39}$$

† m denotes the mass of the electron and M the mass of the nucleus.

where n is an integer, the *total quantum number*. Any measure of the energy of a hydrogen atom (in the bonded state) will give one of the E_n values.

If $n = 1, 2, 3, \ldots$, the electron is said to be in the K, L, M, N *shell* of the hydrogen atom. The concept of shell is equivalent to the concept of energy.

The eigenfunctions are:

$$\Psi_{n,l,m}(\mathrm{M}) = R_{n,l}(r)\Theta_{l,m}(\theta)\Phi_m(\varphi) \qquad (1.40)$$

where l, m are integers (or zero) which obey the relations:

$$0 \leqslant l < n \qquad \text{and} \qquad -l \leqslant m \leqslant l$$

The radial part of Ψ is related to Laguerre functions and the angular part to Legendre functions. They are analysed in many books.[5]

The corresponding wave function is therefore:

$$\Psi_{n,l,m}(\mathrm{M}, t) = R_{n,l}(r)\Theta_{l,m}(\theta)\Phi_m(\varphi)\mathrm{e}^{-(2\pi\mathrm{i}/h)E_n t} \qquad (1.41)$$

It is seen that for a given value of n, that is to say for a given value of the energy, there are various possible wave functions. *The energy is degenerate.* It is easy to see that the number of independent wave functions associated with a given energy E_n is n^2 (Stoner's rule). If $l = 0, 1, 2, 3, \ldots$, the electron is said to be in the state s, p, d, f, . . . If $m = 0, \pm 1, \pm 2, \pm 3$, the electron is said to be σ, π, δ, φ, ... In the framework of this language the function $\Psi_{2,1,0}$ is a L, p, σ function.

The K, s, σ function is simply:

$$\Psi_{1,0,0} = \frac{1}{\sqrt{\pi}a_0^{3/2}}\,\mathrm{e}^{-r/a_0} \qquad (1.42)$$

where:

$$a_0 = \frac{h^2}{4\pi^2 e^2 m}$$

Therefore the probability dp of finding an electron in a volume dv at a distance r of the nucleus is:

$$\mathrm{d}p = \frac{1}{\pi a_0^3}\,\mathrm{e}^{-2r/a_0}\,\mathrm{d}v \qquad (1.43)$$

Figure 1.3 shows the shape of that function. It can be seen that the probability of finding the electron at a given point in the space is at a maximum in the immediate vicinity of the nucleus. This result is in complete disagreement with Bohr's theory in which the electrons revolve in a circle of radius a_0 concentric about the nucleus.

The wave mechanical description of the hydrogen atom gave Yukawa and Sakata a very interesting idea. They[6] predicted that an electron of an atom could be captured by the nucleus due to a nuclear reaction like the following:

$$\text{Proton} + \text{electron} \rightarrow \text{neutron} + \text{neutrino}$$

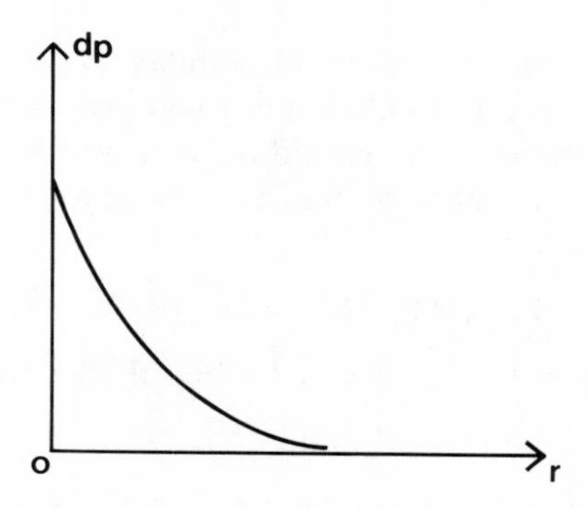

Figure 1.3 The probability dp as a function of the radius r

That phenomenon of electron capture was discovered, two years later, by Alvarez.[7] As the Fermi nuclear field responsible for the capture vanishes at a very short distance from the nucleus (10^{-12} cm), it can be concluded that in agreement with wave mechanical theory the electrons are able to come very near the nuclei. If Bohr's theory was correct electron capture would be impossible, as the Bohr radius a_0 is of the order of 0.5×10^{-8} cm, i.e. 10,000 times higher than the Fermi nuclear field range. We must add that for energetic reasons electron capture does not occur in the case of hydrogen. However, it is well known with the beryllium isotope $^7_4$Be, which disintegrates as follows:

$$^7_4\text{Be} \rightarrow {}^7_3\text{Li} + \nu$$

The half-life of this radioactive reaction is of the order of 50 days.

By using the function (1.41) it is possible to calculate the mean value $\bar{r}$ of the distance which separates an electron from the nucleus:

$$\bar{r} = n^2 a_0 \left\{ 1 + \tfrac{1}{2} \left[1 - \frac{l(l+1)}{n^2} \right] \right\} \tag{1.44}$$

It is found that $\bar{r}$ increases quickly with n but that for a given shell $\bar{r}$ decreases with l. It is also possible to show that:

$$\overline{\left(\frac{1}{r}\right)} = \frac{1}{a_0 n^2}$$

Therefore the mean value of the potential energy $\bar{V}_n$ of the electron is:

$$\bar{V}_n = \overline{\left(-\frac{e^2}{r}\right)} = -\frac{e^2}{a_0 n^2} = -\frac{4\pi^2 m e^4}{n^2 h^2} \tag{1.45}$$

It is seen that:

$$E_n = \tfrac{1}{2} \bar{V}_n \tag{1.46}$$

The *virial theorem* applies in wave mechanics. There is no mystery about the origin of the energy. It is simply half the mean value of the potential energy. The mean value $\bar{T}_n$ of the kinetic energy is therefore given by the expression:

$$\bar{T}_n = -E_n \tag{1.47}$$

In order to describe the distinction between localized and delocalized bonds it is useful to discuss the localizability of the electron of the hydrogen atom. We shall say[8] that a set of normalized functions $f_i(\mathrm{M})$ is localized with a precision at least equal to ε in an ensemble of volume V_i not having common points when:

$$\int_{V_i} |f_i|^2 \, \mathrm{d}v \geqslant 1-\varepsilon \tag{1.48}$$

for any i.† Consider the two functions $\Psi_{1,0,0}$ and $\Psi_{2,0,0}$. Let V_1 be the volume inside the sphere of radius 1.67 Å (angstrom units) and V_2 the remaining part of the space. It is easily shown that:

$$\int_{V_1} |\Psi_{1,0,0}|^2 \, \mathrm{d}v \geqslant 0.92 \tag{1.49}$$

$$\int_{V_2} |\Psi_{2,0,0}|^2 \, \mathrm{d}v \geqslant 0.92 \tag{1.50}$$

Therefore the two functions are localized with an accuracy of 8 per cent. in the two volumes V_1 and V_2.

A general study of the set of functions $\Psi_{n,l,m}$ shows that those belonging to different shells can be precisely localized in separate volumes, whereas those belonging to the same shell cannot be so.

To end this section we shall recall briefly how the electronic density depends on the angles θ and φ. For the small values of l and m the angular parts of the function $\Psi_{n,l,m}$ are:

$$\text{For an } s \text{ electron: } \Theta_{0,0}\Phi_0 = \frac{1}{2\sqrt{\pi}} \tag{1.51}$$

$$\text{For a } p \text{ electron: } \begin{cases} \Theta_{1,0}\Phi_0 = \dfrac{\sqrt{3}}{2\sqrt{\pi}} \cos\theta \\ \Theta_{1,\pm 1}\Phi_{\pm 1} = \dfrac{\sqrt{3}}{2\sqrt{2\pi}} \sin\theta \, \mathrm{e}^{\pm \mathrm{i}\varphi} \end{cases} \tag{1.52}$$

The electronic distribution for an s state does not depend on the angles; it has a *spherical symmetry*. The electronic distribution for a $p\sigma$ state has a maximum for $\theta = 0$ or $\theta = \pi$, that is to say along *the z axis*. There is a *certain*

† In a more mathematical language:

$$\exists\{V_i\}, \text{mes}\,(V_i \cap V_j) = k_i\delta_{ij}, \forall_i \Rightarrow \langle f_i \,|\, f_i \rangle_{V_i} \geqslant 1-\varepsilon$$

concentration of the electronic density in the proximity of that axis. Conversely, the electronic density for a $p\pi$ state appears to be *mainly located near the x, y plane* ($\theta = \pi/2$).

1.2.2 The Hydrogen Molecule Ion

The simplest diatomic molecule, the hydrogen molecule ion, is a three-bodies problem. Happily the Born–Oppenheimer approximation makes it possible to divide the calculation of the wave function into two parts. That approximation will be analysed in Section 2.1.2. It is based on the fact that the nuclear masses are much larger than the electron mass. It is possible to show that in many cases we can calculate an electronic wave function as if the nuclei were fixed and in a further step take account of the movement of the nuclei.

Let us apply that procedure for the hydrogen molecule ion H_2^+. During the first step of the calculation the nuclei are replaced by two electronic charges $+e$ fixed arbitrarily at points A and B. Let r_{AB}, r_A, r_B be respectively the distance between A and B, the distance between a point M and A, and the distance between M and B. The electronic equation for the stationary states are:

$$-\frac{h^2}{8\pi^2 m}\Delta\Psi_i(\mathrm{M}, t)+\left(\frac{e^2}{r_{AB}}-\frac{e^2}{r_A}-\frac{e^2}{r_B}\right)\Psi_i(\mathrm{M}, t)=U_i\Psi_i(\mathrm{M}, t) \qquad (1.54)$$

For any choice of points A and B we can solve this equation to obtain a set of eigenvalues U_i and a set of eigenfunctions Ψ_i. Therefore the U_i and Ψ_i values depend on r_{AB}. Figure 1.4 shows the variation of some functions $U_i(r_{AB})$ obtained by solving equation (1.54) for various values of r_{AB}.[9] It

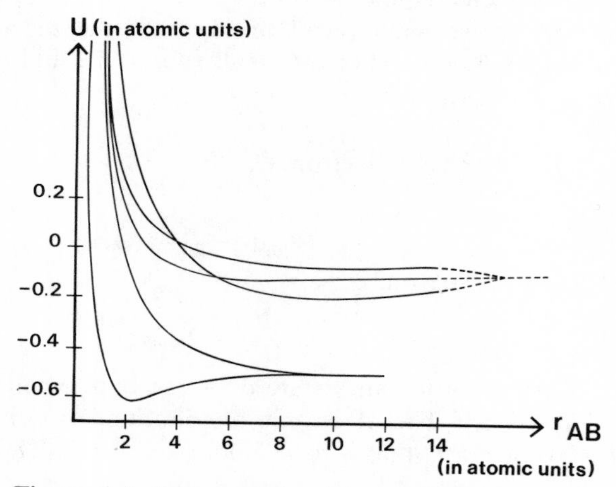

Figure 1.4 The energies U as a function of the interatomic distance r_{AB}. Reproduced by permission of Gautheir-Villars

appears that some curves possess a minimum for a certain value of r_{AB}. These correspond to stable electronic states. This is the case for the curve corresponding to $U_1(r_{AB})$ where the functions are classified in increasing order of the eigenvalues for an arbitrary value of r_{AB}. For that state the minimum electronic energy is reached when:

$$r_{AB} = 1.06 \text{ Å (2 atomic units)}$$

It is called the equilibrium distance r_e. *That value coincides exactly with the measurements obtained experimentally of the mean distance which separates the two nuclei of* H_2^+ *in the ground state. Therefore, wave mechanics can be used to determine the nuclear conformation of a molecule in a certain electronic state.* In principle, we only need to solve an equation like (1.54) and to find the positions of the nuclei minimizing the corresponding U_i.

When starting from an equilibrium distance r_e, the nuclei are separated to an infinite distance, the molecule being dissociated to form the system $H + H^+$. The corresponding energy difference:

$$D_e = U_i(\infty) - U_i(r_e) \tag{1.55}$$

is called the electronic dissociation energy. It is not equivalent to the experimental dissociation energy as the movement of the nuclei is not taken into account; however, it is possible to obtain D_e from experimental results. In the case of the ground state the theoretical calculations evocated by Figure 1.4 give:

$$D_e = U_i(\infty) - U_i(1.06) = 2.7773 \text{ eV } \dagger \tag{1.56}$$

This value agrees very well with the experimental results.

Figure 1.4 shows that for some electronic states the energy U_i has no minimum. The energy curve is a repulsive one. The corresponding state is unstable and the molecule dissociates spontaneously.

From the eigenfunctions Ψ_i it is easy to calculate the electronic density $\rho_i(M)$:

$$\rho_i(M) = |\Psi_i|^2 \tag{1.57}$$

Figure 1.5 shows the results of such a calculation for the ground state of the H_2^+ molecule ion when the internuclear distance coincides with the equilibrium distance r_e. The upper curve shows the value of the function along the line passing through the two nuclei. The lower figure shows contour lines for values 0.9, 0.8, . . . , 0.1 times the maximum value. It can be seen that (as for the hydrogen atom) the electronic density reaches a maximum at the nuclei. It can also be seen that the electron nearly always remains 'between' the nuclei in that small part of the space which has the shape of a cigar. *This is the simple picture of a chemical bond given by wave mechanics.* The electron is attracted by the nuclei. That attraction compensates for the repulsive

† Electronvolt.

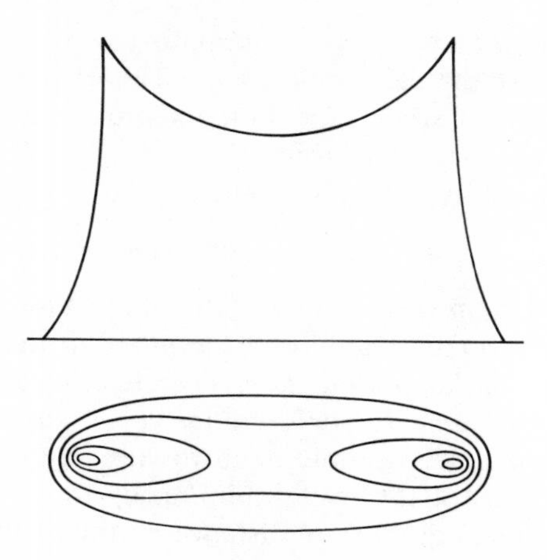

Figure 1.5 Electronic density along the line of the nuclei and in a plane containing this line. Reproduced by permission of Gauthier-Villars

forces which tend to separate the nuclei. The resulting movement of the electron is shown by Figure 1.5 to be when the nuclei are at a distance of 1.06 Å. The movement of the nuclei will be described further.

1.2.3 The Helium Atom

The helium atom can be treated as a two-bodies problem if it is replaced by a model in which the nucleus is replaced by a fixed charge $+2e$. That approximation is convenient because the mass of the nucleus is much larger than the mass of the electron. Let r_a, r_b, r_{ab} be respectively the distance between the charge $2e$ and a point M_a, the distance between the charge $2e$ and a point M_b, and the distance M_aM_b. The Schrödinger equation can be written:

$$-\frac{h^2}{8\pi^2 m}\Delta_a\Psi_i(M_a, M_b)-\frac{h^2}{8\pi^2 m}\Delta_b\Psi_i(M_a, M_b) + \left(\frac{e^2}{r_{ab}}-\frac{2e^2}{r_a}-\frac{2e^2}{r_b}\right)\Psi_i(M_a, M_b)=E_i\Psi_i(M_a, M_b) \quad (1.58)$$

If x_a, y_a, z_a denote the coordinates of point M_a and x_b, y_b, z_b the coordinates of M_b:

$$\Delta_a=\frac{\partial^2}{\partial x_a^2}+\frac{\partial^2}{\partial y_a^2}+\frac{\partial^2}{\partial z_a^2}$$

and (1.59)

$$\Delta_b=\frac{\partial^2}{\partial x_b^2}+\frac{\partial^2}{\partial y_b^2}+\frac{\partial^2}{\partial z_b^2}$$

The probability of simultaneously finding electron 1 in a volume dv_a surrounding point M_a *and* electron 2 in a volume dv_b surrounding M_b for a stationary state of the helium atom described by the wave function:

$$\Psi_{ti}(M_a, M_b, t) = \Psi_i(M_a, M_b)e^{-(2\pi i/h)E_i t} \tag{1.60}$$

is:

$$dp_{12} = |\Psi_i(M_a, M_b)|^2 \, dv_a \, dv_b \tag{1.61}$$

The probability dp_{21} of finding electron 2 in dv_a and electron 1 in dv_b is:

$$dp_{21} = |\Psi_i(M_b, M_a)|^2 \, dv_a \, dv_b \tag{1.62}$$

Note that, by convention, the first point in Ψ is the one where we are looking for the first electron and the second point the one where we are looking for the second electron.

As the hamiltonian operator is symmetrical with respect to the two points M_a and M_b, the function Ψ is symmetrical or antisymmetrical with respect to a permutation of that point depending on the value of E_i:†

$$\Psi(\overset{1}{M}_a, \overset{2}{M}_b) = \pm\Psi(\overset{1}{M}_b, \overset{2}{M}_a) \tag{1.63}$$

therefore:

$$dp_{12} = dp_{21} \tag{1.64}$$

in agreement with the indistinguishability principle. *All electrons of the system play the same role in the system.*

As a consequence of the symmetry properties of the Ψ_i functions, if we calculate, for example, the mean value $\overline{r_1}$ of the distance between electron 1 and the nucleus:

$$\overline{r_1} = \int \Psi^*(M_a, M_b) r_a \Psi(M_a, M_b) \, dv_a \, dv_b \tag{1.65}$$

and the mean value $\overline{r_2}$ of the corresponding distance for electron 2:

$$\overline{r_2} = \int \Psi^*(M_a, M_b) r_b \Psi(M_a, M_b) \, dv_a \, dv_b \tag{1.66}$$

we shall obtain:

$$\overline{r_1} = \overline{r_2} \tag{1.67}$$

This is a general result. *The mean value of the distance between an electron and the nucleus of an atom is the same for all the electrons of the atom.*

From the experimental viewpoint it is not possible to distinguish between the various electrons of an atom. *To distinguish between K electrons, L*

† It is assumed the δE_i is not degenerate from the space function viewpoint.

electrons, . . . for a polyelectronic atom and to distinguish between core electrons, valence electrons, σ electrons and π electrons in a molecule is an old custom based on Bohr's old theory but is nonsense in the framework of wave mechanics. We shall see, however, that when we use certain kinds of approximate wave functions it is useful to introduce K orbitals, L orbitals, σ orbitals and π orbitals. That is quite a different matter. It will be shown that it is very dangerous to establish a direct bridge between electrons and orbitals.

In the case of the helium atom it has been possible to compute very elaborate wave functions. It is therefore necessary to understand something of the electronic structure of a given state of helium atom without the help of orbitals. This can be done in the following way. Let us consider the first excited state which in Bohr's theory would correspond to an electron moving on the K orbit and an electron moving on the L orbit. Let $\Psi_2(\mathrm{M}_a, \mathrm{M}_b)$ be the eigenfunction solution of equation (1.58) associated with this state.

It is not possible to distinguish with the help of that function between a K electron and an L electron, but we can arbitrarily divide the space of the atom into two parts. The spherical symmetry of the considered state would suggest that a sphere with an arbitrary radius R centred at the nucleus could be taken as a frontier between the two parts. Imagine an experiment which would make it possible to measure the number of electrons located in the two parts of the space at a given time. Three events are possible:

(a) The two electrons are inside the sphere.
(b) The two electrons are outside the sphere.
(c) One electron is inside; the other is outside.

The knowledge of Ψ_2 permits the probability of each event to be calculated. We shall denote the probability of the ith event by $P_i(R)$, as obviously it is a function of R. For each value of R the set of probabilities P_1, P_2, P_3 gives a certain amount of information about the localizability of the electrons in the considered excited state of helium atom. *Information theory*[10] gives a quantitative measure of that amount of information and introduces the *missing information function I*, also called the *a priori* indetermination function:

$$I = \sum_i P_i \log_2 P_i^{-1} \qquad (1.68)$$

For the considered problem the function I will depend on the radius R. As we look for the partition of the space which gives the maximum information we are led to select the value of R which minimizes the function I. Figure 1.6 represents the variation of I with R obtained by using a convenient electronic wave function for the first excited state of the helium atom. The minimum of I corresponds to $R = 1.75a_0$. Then the values of the P_i values are:

$$P_1 = 0.052, \qquad P_2 = 0.028, \qquad P_3 = 0.920$$

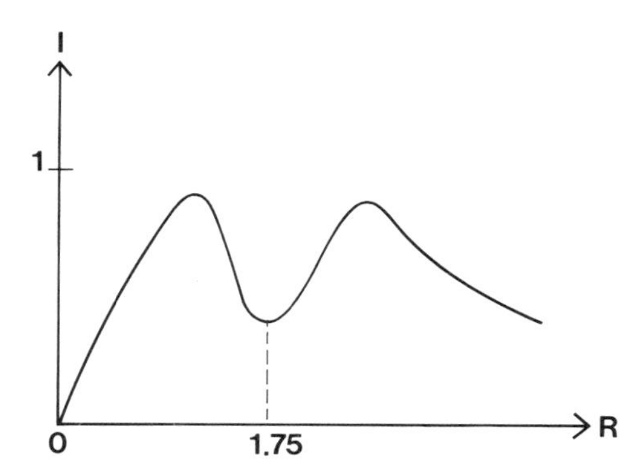

Figure 1.6 The missing information function as a function of R

The third event is clearly the leading one. *There is a high probability of finding one electron inside the sphere and one outside.* Therefore to establish a certain bridge with the old language we shall say that the sphere of radius $R = 1.75a_0$ corresponds to the best partition in *loges* of the space of the helium atom in its first excited state. The volume inside the sphere will be called the K loge and the volume outside the sphere the L loge. With such a language it can be said that the *leading event* (or most probable one) associated with that excited state corresponds to one electron in the K loge and one electron in the L loge. This language is not forbidden by the indistinguishability principle because now we do not distinguish between electrons but between regions of the space in which any electron can be located.

Note that when I reaches a minimum there is almost always a leading event, that is to say an event for which the probability is much higher than the probability of any other event. It must be added that the partitions of the space corresponding to $I = 0$ have no interest—they are trivial and do not correspond to a real division of the space (see Figure 1.6 for example). The loge concept was introduced by Daudel,[(11)] loge theory has been developed by various authors[(12)] and the bridge between loge theory and information theory has been established by Aslangul.[(13)]

1.2.4 The Hydrogen Molecule

As in the case of the hydrogen molecule ion the Born–Oppenheimer approximation will be used. The electronic Schrödinger equation is:

$$-\frac{h^2}{8\pi^2 m}\Delta_a \Psi_i(\mathrm{M}_a, \mathrm{M}_b) - \frac{h^2}{8\pi^2 m}\Delta_b \Psi_i(\mathrm{M}_a, \mathrm{M}_b) + \left(\frac{e^2}{r_{ab}} - \frac{e^2}{r_{a\mathrm{A}}} - \frac{e^2}{r_{a\mathrm{B}}} - \frac{e^2}{r_{b\mathrm{A}}} - \frac{e^2}{r_{b\mathrm{B}}} + \frac{e^2}{r_{\mathrm{AB}}}\right)\Psi_i(\mathrm{M}_a, \mathrm{M}_b) = U_i \Psi_i(\mathrm{M}_a, \mathrm{M}_b) \quad (1.69)$$

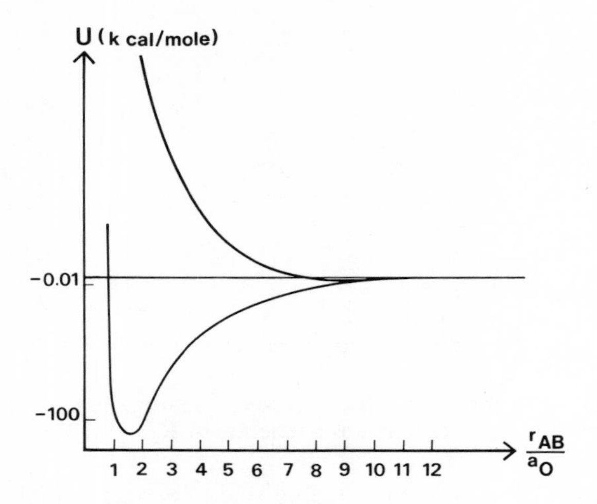

Figure 1.7 The energies U as a function of r_{AB}.
Reproduced by permission of Gauthier-Villars

In equation (1.69), M_a and M_b denote two arbitrary points of the space, A and B the positions of the fixed charge $+e$ replacing the nuclei, r_{ab} the distance between M_a and M_b, r_{AB} the distance between A and B, r_{aA}, r_{aB} the distances between M_a, A and B respectively and r_{bA}, r_{bB} the distances between M_b, A and B.

Figure 1.7 shows how the two smallest U_i depend on r_{AB}. It appears that the two first electronic states of H_2 are stable. The curve associated with the ground state reaches a minimum when:[14]

$$r_{AB} = 0.74127\ \text{Å}$$

The experimental mean value of the internuclear distance for that state is:

$$r_{AB} = 0.74116\ \text{Å}$$

The corresponding theoretical electronic dissociation energy is:

$$D_e = 4.7467\ \text{eV}$$

and the experimental measure is:

$$D_e = 4.7466 \pm 0.0007\ \text{eV}$$

The agreement is remarkably good, which shows how satisfactorily the principles of wave mechanics work.

The electronic density $\rho_1(M)$ has been computed. The figure showing the contour lines presents the same aspect as for the molecule ion but corresponds to a shorter and thicker 'cigar'. The upper curve of Figure 1.7 reaches its minimum when:

$$r_{AB} = 8.4a_0\ (\sim 4\ \text{Å})$$

that it to say for a much larger distance. The corresponding D_e is very small. To understand the origin of the difference of behaviour between U_1 and U_2 it must be said that the ground state eigenfunction $\Psi_1(M_a, M_b)$ is symmetrical with respect to the permutation of M_a and M_b. Conversely, the eigenfunction $\Psi_2(M_a, M_b)$ associated with the first electronic excited state is found to be antisymmetrical with respect to the same permutation. Therefore:

$$\Psi_2(M_b, M_a) = -\Psi_2(M_a, M_b) \tag{1.70}$$

If we now try to find the two electrons in that state at the same point M we shall have to consider the value of the eigenfunction when $M_a = M_b = M$. However, from equation (1.70) it can be seen that:

$$\Psi_2(M, M) = -\Psi_2(M, M) = 0 \tag{1.71}$$

Therefore the probability density of finding two electrons at the same point is zero in that state.

Further, it would appear that in that state the electrons possess a strong tendency to stay far away, which explains why the equilibrium internuclear distance is rather large. In the ground state, $\Psi_1(M_a, M_b)$ being symmetrical, the only reason for the repulsion between the electrons is the coulombic repulsion. That repulsion balances the attraction between the nuclei and electrons for a small internuclear distance. In the first excited state the symmetry property of the wave function adds another source of repulsion. The compensation between the two kinds of repulsion and the attraction between the nuclei and electrons is only reached at larger internuclear distances.

Let us go back to the ground state. Consider the two hydrogen atoms approaching from infinity to the equilibrium distance. It is obvious that during the approach there is an important change in the electronic distribution. This change can be measured by the *difference density function* $\delta(M)$ introduced by Roux, Besnainou and Daudel:[15]

$$\delta(M) = \rho(M) - \rho^f(M) \tag{1.72}$$

where $\rho(M)$ is the actual electronic density at point M and $\rho^f(M)$ is the density which would exist at that point if the two atoms were superposed without any change in their electronic density. Therefore, at point M where the formation of the chemical bond has led to an increase in the electronic density the function δ is positive. Where the formation of the bond has led to a decrease of the electronic density δ is negative. Figure 1.8 shows the variation of the difference density function along the nuclear line in the hydrogen molecule[16] (ground state). It can be seen that in agreement with chemical intuition $\delta(M)$ *is positive 'between' the nuclei.*

The chemical bond leads to an increase in the electronic density in that region. Many authors have used the difference density function to *study* the effect of chemical binding on the electronic density in molecules. We shall give various examples later on.[17]

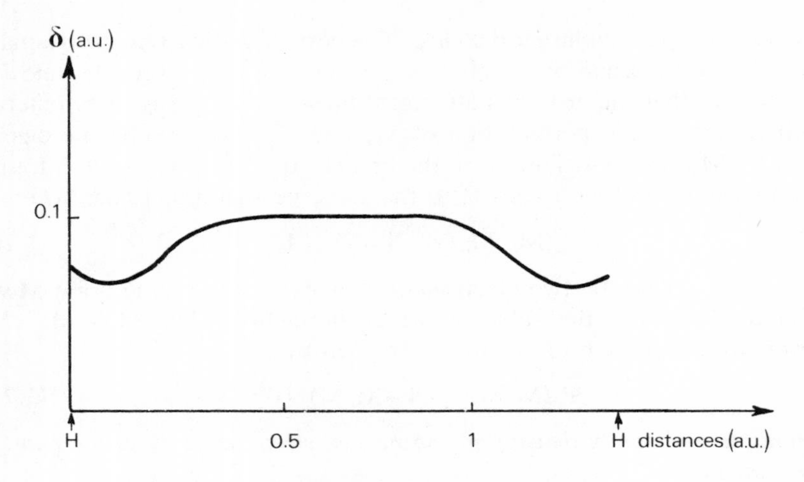

Figure 1.8 Difference density function of H_2

1.2.5 Structure of Polyelectronic Atoms

It is well known that in order to explain the complexity of atomic spectra and the surprising value of the gyromagnetic ratio Uhlenbeck and Goudsmit(18) introduced the concept of spin. They attributed an angular momentum and a magnetic moment to the electron and assumed that it had the spinning motion about an axis of an electrically charged body. Although this precise description may remain questionable the existence of the angular momentum does not. In the framework of wave mechanics the spin of an electron is represented by four operators: $\mathcal{S}^2$, $\mathcal{S}_x$, $\mathcal{S}_y$ and $\mathcal{S}_z$. The operator $\mathcal{S}^2$ is associated with the square of the length of the spin, while $\mathcal{S}_x$, $\mathcal{S}_y$ and $\mathcal{S}_z$ are concerned with the various projections on the three axes. The four operators are related by the equation:

$$\mathcal{S}^2 = \mathcal{S}_x^2 + \mathcal{S}_y^2 + \mathcal{S}_z^2 \tag{1.73}$$

The precise expression will be recalled later on. Now we shall simply say that $\mathcal{S}^2$ has only one eigenvalue—$\frac{1}{2}(\frac{1}{2}+1)h^2/4\pi^2$—and that $\mathcal{S}_x$, $\mathcal{S}_y$ and $\mathcal{S}_z$ have two eigenvalues:

$$\pm\tfrac{1}{2}\frac{h}{2\pi}$$

The only possible measure of the magnitude of the angular momentum is therefore $\sqrt{\frac{3}{4}}\, h/2\pi$ and a measure of one of the projections of the spin along one of the three axes is necessarily $\pm h/4\pi$. The simplest procedure to take account of the spin is now described.

Let us consider a unique electron. If we wish to find the electron somewhere with a projection of the spin $+h/4\pi$ on the z axis, it will be described with the wave function:

$$\Psi_T(\mathrm{M}, h/4\pi, t)$$

the probability density of finding that electron at point M at time t being:

$$\rho(\mathrm{M}, h/4\pi, t) = |\Psi_T(\mathrm{M}, h/4\pi, t)|^2$$

If, on the contrary, we wish to find the electron with a negative projection of the spin, it will be described with the function:

$$\Psi_T(\mathrm{M}, -h/4\pi, t)$$

If, now, we consider a two-electron system and if we wish to find simultaneously electron 1 at point M_a with a positive projection of the spin and electron 2 at point M_b with a negative projection of the spin we shall write the function.

$$\Psi(\overbrace{\mathrm{M}_a, +h/4\pi}^{1}, \overbrace{\mathrm{M}_b, -h/4\pi}^{2}, t)$$

For an n-electron system, if we are looking for electron 1 at M_a with a spin projection ω_a, electron 2 at M_b with a spin projection ω_b, and so on, we shall introduce the function:

$$\Psi(\mathrm{M}_a, \omega_a, \mathrm{M}_b, \omega_b, \ldots, \mathrm{M}_i, \omega_i, \ldots, \mathrm{M}_n, \omega_n, t)$$

Obviously we shall have:

$$\omega_i = \pm \frac{h}{4\pi} \tag{1.74}$$

Returning to a two-electron system, the probability density of finding simultaneously electron 1 at point M_a with the spin projection ω_a and electron 2 at point M_b with the spin projection ω_b *is*:

$$\rho_{ab} = |\Psi(\overbrace{\mathrm{M}_a, \omega_a}^{1}, \overbrace{\mathrm{M}_b, \omega_b}^{2}, t)|^2 \tag{1.75}$$

The probability density of finding electron 2 *at* point M_a with the spin projection ω_a and electron 1 at point M_b with the spin projection ω_b is:

$$\rho_{ba} = |\Psi(\overbrace{\mathrm{M}_b, \omega_b}^{1}, \overbrace{\mathrm{M}_a, \omega_a}^{2}, t)|^2 \tag{1.76}$$

The indistinguishability principle states that:

$$\rho_{ba} = \rho_{ab} \tag{1.77}$$

If (to simplify the discussion) we assume that the wave functions are real except for the exponential part which depends on the time having a square modulus equal to unity, we can conclude that:

$$\Psi(\mathrm{M}_b, \omega_b, \mathrm{M}_a, \omega_a, t) = \pm\Psi(\mathrm{M}_a, \omega_a, \mathrm{M}_b, \omega_b, t) \tag{1.78}$$

This introduces a new principle—the well-known *Pauli principle.* This states that taking the nature of the electrons into account the *minus* sign is the only convenient sign to use in equation (1.78). *More generally, the wave function associated with a system of electrons is antisymmetrical with respect to a permutation of two groups of space and spin coordinates*:

$$\begin{aligned} &P\Psi(\mathrm{M}_a, \omega_a, \ldots, \mathrm{M}_i, \omega_i, \mathrm{M}_j, \omega_j, \ldots, \mathrm{M}_n, \omega_n, t) = \\ &\mathrm{M}_i, \omega_i, \mathrm{M}_j, \omega_j \rightarrow \mathrm{M}_j, \omega_j, \mathrm{M}_i, \omega_i \\ &\Psi(\mathrm{M}_a, \omega_a, \ldots, \mathrm{M}_j, \omega_j, \mathrm{M}_i, \omega_i, \ldots, \mathrm{M}_n, \omega_n, t) \\ &\quad = -\Psi(\mathrm{M}_a, \omega_a, \ldots, \mathrm{M}_i, \omega_i, \mathrm{M}_j, \omega_j, \ldots, \mathrm{M}_n, \omega_n, t) \end{aligned} \tag{1.79}$$

A very important consequence follows. Let us assume that:

$$\mathrm{M}_i = \mathrm{M}_j = \mathrm{M} \quad \text{and} \quad \omega_i = \omega_j = \omega \tag{1.80}$$

Equation (1.79) becomes:

$$\begin{aligned} &\Psi(\mathrm{M}_a, \omega_a, \ldots, \mathrm{M}, \omega, \mathrm{M}, \omega, \ldots, \mathrm{M}_n, \omega_n, t) \\ &\qquad = -\Psi(\mathrm{M}_a, \omega_a, \ldots, \mathrm{M}, \omega, \mathrm{M}, \omega, \ldots, \mathrm{M}_n, \omega_n, t) \end{aligned} \tag{1.81}$$

As only zero is equal to its opposite we are led to the conclusion:

$$\Psi(\mathrm{M}_a, \omega_a, \ldots, \mathrm{M}, \omega, \mathrm{M}, \omega, \ldots, \mathrm{M}_n, \omega_n, t) = 0 \tag{1.82}$$

The probability density of finding two electrons at the same point M with the same spin projection ω is zero. As Ψ is a continuous function it can be stated that two electrons of identical spin projections do not like to occupy the same small region.

To simplify, we shall say that an electron has a negative or positive spin when the component of the spin along the z axis is negative or positive. Two electrons can be said to have the same spin if the components of their spins along the z axis have the same measure and to have opposite spins if the z components of their spins have opposite values.

The same method of using the concept of loge that helped to solve the problem concerning the helium atom can be used for more complex atoms. However, some new features will appear. Let us first consider the lithium atom. As it contains three electrons we can make a choice between two kinds of partitions of the space into loges. Obviously, we can consider the division into two loges and the division into three loges. As we have no *a priori* reason to choose one method in particular it is best to try both. Therefore we shall try to find the best division into two loges by minimizing the missing information function I. In the same way, we shall then look for the best division into three loges. Finally, we shall select the non-trivial division ($I \neq 0$) which corresponds to the smallest minimum.

With a complex atom this procedure becomes very tedious. As always, in wave mechanics it becomes impossible to solve the problem rigorously. Only approximate solutions can be obtained. We cannot reach the *best* partition of the space into loges but only *good* partitions into loges. Consider the fluorine ion F^- in its ground state. A good partition into loges is obtained by dividing the space into two parts: the region inside a sphere centered at the

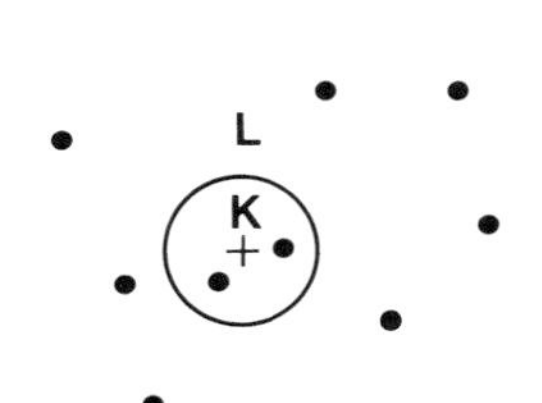

Figure 1.9 The leading electronic event for the ion F^-

nucleus with radius $r = 0.35a_0$ and the region outside that sphere. The leading event is symbolized by Figure 1.9. There is an 81 per cent. probability of finding two electrons of opposite spins in the sphere, the others being outside. Therefore, it can be said that the sphere corresponds to a K loge and the remaining part of the space to an L loge.

Usually, a good division into loges of the space associated with an atom in its ground state is obtained by cutting up that space into spherical rings, all concentric with the nucleus. Table 1.2 gives some values of the radii of the spheres corresponding to the various frontiers between loges, calculated from the Hartree–Fock wave functions.[19] They correspond approximately to the minimum probability density of finding in the atoms an electron at a certain distance from the nucleus. We can obtain an idea of the mean volume v which 'tends to occupy' an electron visiting a given loge by dividing the volume of that loge by the number of electrons found in it during the leading event. Also we can compute the mean value of the electric potential p which acts on a given electron visiting that loge. Thus:[20]

$$p^{3/2}v = \text{constant} \tag{1.83}$$

for all atoms and all loges. *A kind of Boyle–Mariotte low exists between the 'electric pressure' and the 'vital space' required by an electron.*

Table 1.2 Frontier of loges in atoms (radii in atomic units)

Loge	Be	F^-	Al^{2+}	Ca^{2+}	Rb^+
K	1.12	0.35	0.22	0.13	0.06
L				0.64	0.26
M					1.15

To complete the geometrical picture of atoms we shall introduce another concept: *the most probable electronic configuration.* Consider, for example, the helium atom characterized by the wave function $\Psi_T(M_a, \omega_a, M_b, \omega_b, t)$. For a stationary state the square of the modulus of that function does not depend on the time. It can be represented by an hypersurface in seven-dimensional space. Usually that surface shows various minima and maxima

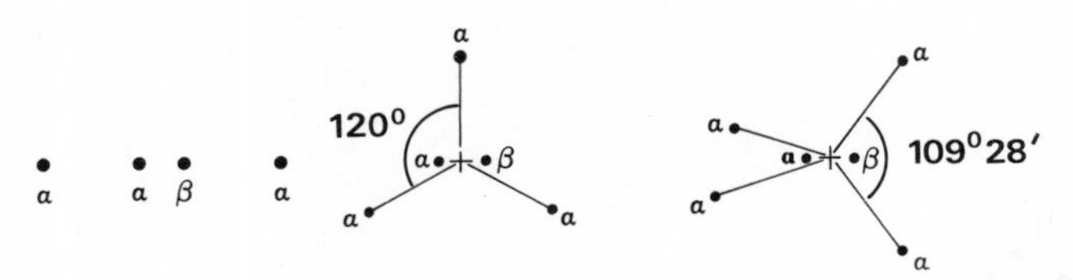

Figure 1.10 The most probable configuration for some small atoms. Reproduced by permission of Gauthier-Villars

and there is a maximum which is higher than all the others. The set of two points M_a and M_b which correspond to that maximum is the most probable electronic configuration of the atom. Effectively, from all possible groups of two points the considered set is that in which there is *the highest probability density of finding simultaneously an electron in* M_a *with a spin component* ω_a *along the z axis and another in* M_b *with a spin component* ω_b *along the same axis.*

Such language may be a little confusing because the word 'configuration' is used with a completely different connotation. It would be better to speak of *the most probable set of electronic positions* in the atom.

The calculations have been made for various states of small atoms. Figure 1.10 shows some results.[21] It can be seen that for all considered cases the most probable electronic configuration corresponds to a pair of electrons with opposite spins at the nucleus and therefore inside the *K* loge. On the figure the frontier of that loge is evoked by a circle. The other electrons are in the *L* loge. *They possess the same spin in the considered states and together form with the nucleus the largest possible angles* (180° for two electrons, 120° for three, 109°28′ for four). This shows that the 'Pauli repulsion' between electrons possessing the same spin is very significant. In fact, this *repulsion is the main origin of bond angles.*

1.2.6 The Concept of Bond; Localized and Delocalized Bonds

Following the discovery of the electron many attempts were made to build an electronic theory of the chemical bond. From a certain viewpoint the famous paper of Lewis[22] remains the basis of modern theories of chemical bonds. However, by coming within the framework of wave mechanics the Lewis concepts have undergone a severe transformation. Once more the indistinguishability principle and the resulting symmetry properties of electronic wave functions are responsible for that evolution, which, in fact, can be considered as a revolution.

Lewis has associated some specific electrons with the various bonds of a molecule. He localized some electrons along a bond. In fact, from the wave mechanical viewpoint all the electrons of a given molecule participate in the interaction between any pair of nuclei.

It is not possible to distinguish, *a priori*, from the electronic viewpoint between bonded atoms and non-bonded atoms. Furthermore, all electrons equally contribute to the 'binding energy' between two particular nuclei. Finally, a large gap appears between the old electronic theory of the chemical bond and the new one based on a purely wave mechanical treatment. It becomes difficult to establish a rigorous bridge between chemical intuition based on a considerable amount of empirical observation leading to a partially localized picture of molecules and wave mechanics which, at first sight, gives a blurred description of those molecules. The concept of orbital provided the first approach to establish such a bridge. However, as electronic computers now make it possible to calculate elaborate wave functions which have not a simple expression in terms of orbitals it is better to use the much more general concept of loge.

Consider, to begin the analysis, the case of a simple diatomic molecule such as the lithium molecule Li_2. A good partition into loges is obtained by dividing the space into three parts with the help of two spheres of equal radius R, one concentric about each nucleus.[23] The value of R corresponding to the minimum of the missing information function I is:

$$R = 1.53a_0$$

Figure 1.11 evokes the leading event. It corresponds to one pair of electrons with opposite spins in each spherical loge, the remaining one being outside these spheres. The probability of the leading event is very large (0.95). As the spherical loges have about the same radius as the K loge in the free lithium atom they are called *core loges.* The remaining part of the space can be called the *bond loge.* Using this terminology it can be said that there is a high probability of finding two electrons with opposite spins in the bond loge. *This is the modern translation of the Lewis idea.*

This kind of analysis can be used for larger molecules, but before doing so it is useful to introduce some topological concepts. As we cannot, *a priori*, distinguish between bonded and non-bonded atoms we need another criterion to define the *nearness* of two atoms. In other words, when do we say that two atoms (or more precisely two nuclei) are *adjacent* in a given molecule? A simple procedure is based on equilibrium internuclear distances which can be measured by using various techniques such as spectroscopic methods, electron and X-ray diffraction. Consider, as an example, a

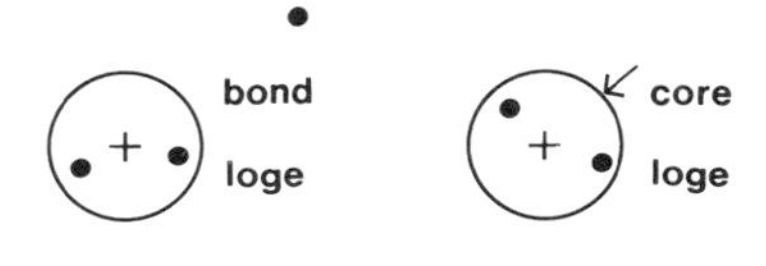

Figure 1.11 The leading electronic event for the lithium molecule

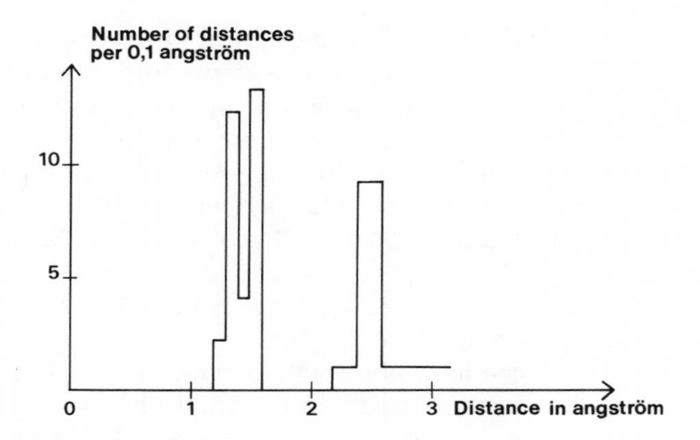

Figure 1.12 Variation of the probability of finding a CC bond with the length of the bond

large family of molecules containing carbon nuclei. Let us take into consideration *all* possible internuclear CC distances in that set of molecules. We can build an histogram by plotting the number of CC distances which is found in various CC distance intervals. Figure 1.12 shows the result of such a study based on approximately one hundred internuclear CC distances.[24] The graph shows the existence of a first region of possible distances between 1.2 and 1.5 Å as well as a forbidden region (from 1.6 to 2.2 Å). Beyond 2.2 Å all distances become possible.

The most important fact is the existence of an isolated interval of possible distances centered at 1.4 Å with a spread of 0.2 Å. *Two carbon nuclei can be said to be neighbours if they are separated by a distance inside that interval.* This procedure can be extended to other kinds of nuclei and therefore it becomes possible to define the *neighbourhood* of a given nucleus in a molecule by taking account only of experimental data.

If such an analysis is made for the methane molecule it is found that the carbon nucleus has four neighbouring hydrogen nuclei but that two hydrogen nuclei are not neighbours. This result agrees with the classical formula in which a bond exists between the carbon nucleus and each hydrogen nuclei but no bond between the hydrogen nuclei.

The loge localization analysis of the methane molecule has been made by Sanchez and Ludeña[25] with the help of electronic wave functions computed by Moccia[26] and by Bader and Preston.[27] A good partition of the space into loges is obtained by considering a spherical loge surrounding the carbon nucleus and four quadrant equivalent loges (Figure 1.13) bounded by spherical triangles at the surface of the spherical loge. The spherical loge has a radius of $0.56a_0$. Each quadrant loge contains one hydrogen nucleus. The leading event corresponds to the presence of two electrons of opposite spins in each loge. Therefore the spherical loge can be said to be a core loge or more precisely a K loge. The probability of finding two electrons with

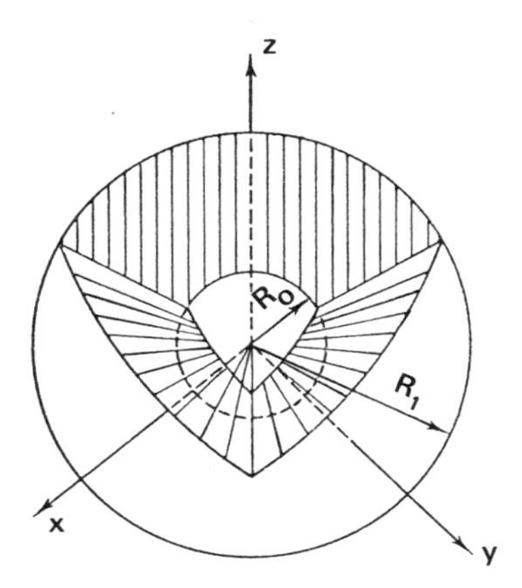

Figure 1.13 Schematic representation of quadrant loges (from Sanchez and Ludeña[25])

opposite spins in that loge is 89 per cent. Each quadrant loge is a bond loge or more precisely a *localized two-electron loge* as it is localized between the core of the carbon atom and a neighbouring hydrogen nucleus. Thus, each bond loge corresponds to a classical chemical simple bond—a localized two-electron bond.

The number N of electrons which are found during the leading event in the bond loges is said to be the number of *bonding electrons.* As it is easy to predict the number of electrons found in the core loges for a leading event it is very easy to predict the number of bonding electrons. Let us take the case of propane. It is obvious that a good partition into loges will contain a K loge surrounding each carbon nucleus. This suggests that two electrons will be found in each core for the leading event. As propane contains 26 electrons the number of bonding electrons is:

$$N = 26 - 3 \times 2 = 20$$

When the number of bonding electrons is twice the number P of pairs of neighbouring nuclei a good division into loges usually contains a localized two-electron loge between each pair of neighbouring nuclei. In such cases the molecules are said to be saturated. This occurs in the case of propane when $P = 10$. Figure 1.14 evokes a good partition into loges for that molecule and the corresponding leading event.

In certain cases it is not possible between only two nuclei (or cores) to localize a region where there is a high probability of finding a certain number of electrons with a precise organization of spins. This occurs when:

$$N < 2P \qquad \text{or} \qquad N > 2P$$

Consider diborane B_2H_6. This molecule contains 16 electrons. A good partition into loges will contain two boron core loges, each containing two

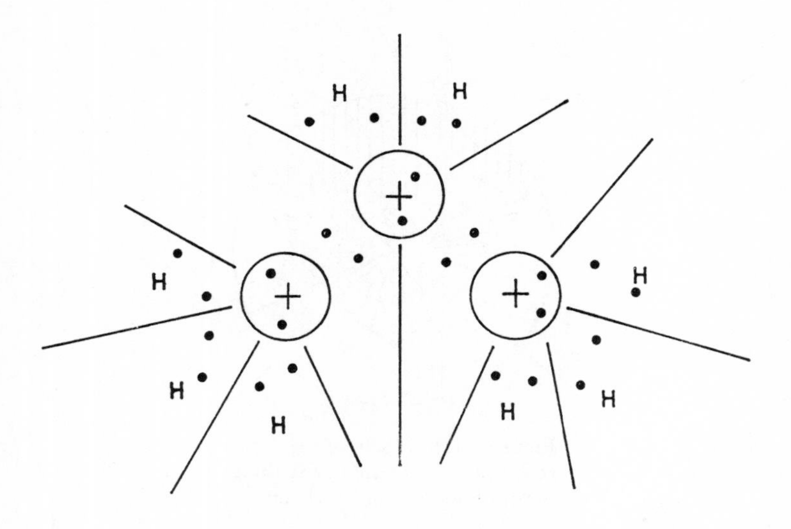

Figure 1.14 The leading electronic event for the propane molecule

electrons during the leading event. Therefore:

$$N = 16 - 2 \times 2 = 12$$

The geometrical arrangement of the nuclei of this molecule is given in Figure 1.15. The double arrows represent the pairs of neighbouring nuclei,

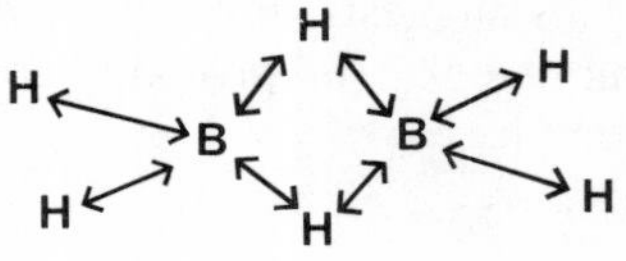

Figure 1.15 Diborane geometry

of which there are 8. Therefore:

$$N < 2P$$

It is not possible to associate a two-electron localized bond with each pair of neighbouring nuclei. From the experimental viewpoint the four outer BH bonds behave in the same way as normal localized two-electron bonds. This leads us to attempt a division into loges similar to the one described in Figure 1.16(a) by the assumed leading event; Figure 1.16(b) obviously corresponds to another event possessing the same probability p. Therefore:

$$2p \leqslant 1$$

There are two leading events (l.e.) in the missing information function

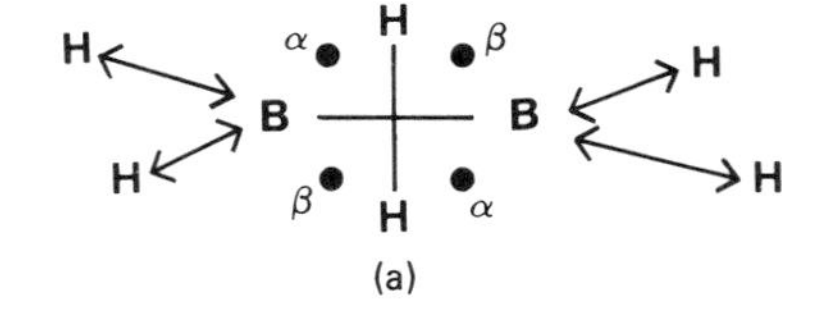

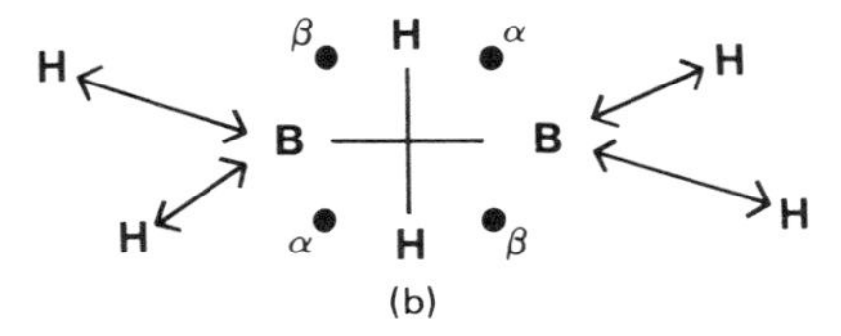

Figure 1.16 Two important electronic events for the diborane molecule

introducing the two terms:

$$I_{\text{l.e.}} = p \log_2 \frac{1}{p} + p \log_2 \frac{1}{p}$$

If $p \approx 0.5$ the contribution $I_{\text{l.e.}}$ is approximately unity.

If we now consider the division into loges evoked by Figure 1.17 it can be seen that the probability of the leading event will be at least equal to $2p$ and therefore approximately unity. Its contribution to the missing information therefore vanishes. It can be concluded that the missing information associated with the division into loges of Figure 1.17 will probably be smaller than the missing information corresponding to Figure 1.16. Therefore Figure 1.17 probably corresponds to a better division into loges. It contains two-electron loges extended over three cores (or nuclei). Such loges are said to be *delocalized* and by definition represent *delocalized bonds* (more precisely, two-electron bonds delocalized over three cores—B, H and B).

Figure 1.18 shows a typical good division of the space of a molecule into loges and the corresponding leading event. It contains four *core loges* A, B, C and D, a *localized two-electron loge* AB, a *four-electron loge delocalized* over B, C and D and a loge adjacent to only one core (the core A), in which there is a high probability of finding two electrons with opposite spins. The last loge is called a *lone pair loge.*

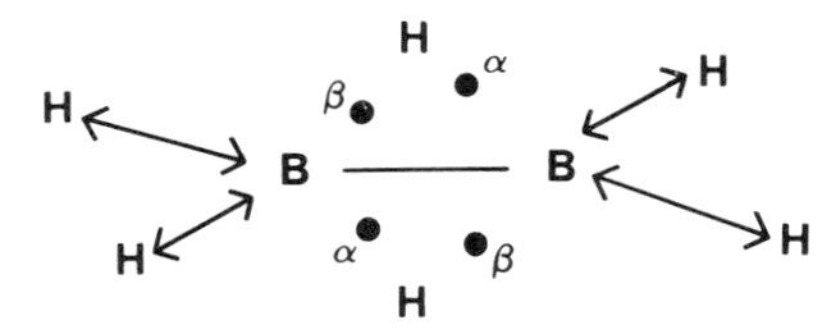

Figure 1.17 The leading electronic event for the diborane molecule

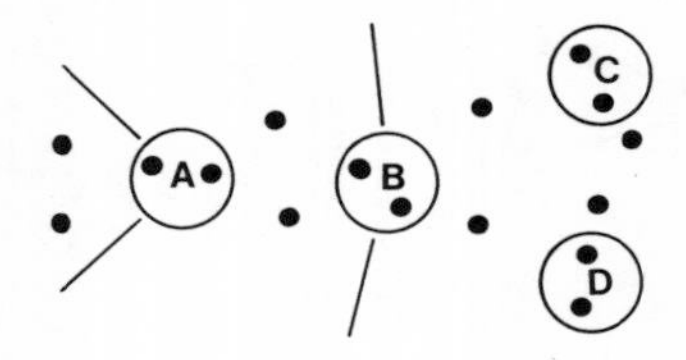

Figure 1.18 A typical partition into loges

It is anticipated (and will be shown later) that when it is possible to obtain a good partition into loges for a molecule any property A associated with the molecule can be expressed as:

$$A = \sum_i A_i + \sum_{i<j} A_{ij} \tag{1.84}$$

where A_i represents the contribution of the ith loge to the molecular property and A_{ij} the contribution associated with interaction terms coming from the ith and the jth loges. In fact, when the operator associated with the quantity A is a monoelectronic operator there are no interaction terms. Therefore:

$$A = \sum_i A_i \tag{1.85}$$

Equation (1.85) provides the basis of a clear understanding of the origin of empirical molecular additivity rules and helps in the interpretation of the additive properties of the Faraday effect.[28] In the case of quantities associated with the bielectronic operator the interaction terms do not vanish. They yield, for example, the isomerization energies.[29]

Let us go back to the methane molecule. The number N of bonding electrons is 8. In the leading event there are eight electrons (four with a positive spin, four with a negative one) surrounding the core loge of the carbon atom. We know that electrons possessing the same spin tend to form between them the largest possible angle with the nucleus. Therefore we anticipate that the four electrons with positive spins will tend to form tetrahedral angles. This is also true for the four electrons having negative spins. Because there is often one electron of positive spin (and also one of negative spin) in the vicinity of each C—H axis we must expect that the angle between the C—H axes will also be 109°28′. This explains why methane is a tetrahedral molecule.

Now, consider the molecule NH_3. It also contains 10 electrons. During the leading event (Figure 1.19) two will be found in the nitrogen K core and therefore eight will be outside (four of each spin). As in the case of methane, the angles between the electrons possessing the same spin tend to be of approximately 109°. Therefore the HNH angle will also have approximately that value. A lone pair loge appears, which is able to fix a proton, giving the tetrahedral NH_4^+ ion. In the water molecule the situation is very similar but two lone pair loges appear. In a molecule like $B(CH_3)_3$ it can be seen that for the leading event there are six electrons in the vicinity of the boron core

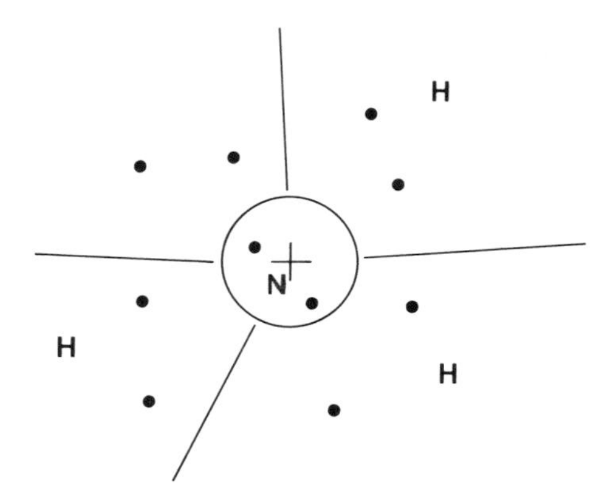

Figure 1.19 The leading electronic event for the NH_3 molecule

(three with positive spins and three with negative ones). This is why in that molecule the boron nucleus and the three carbon nuclei are coplanar and form angles of 120°. Many other examples of molecular geometry have been discussed in a similar way.[30]

Finally, it turns out that when the space of a molecule is divided into regions in order to obtain the maximum information about the localizability of its electrons, quasi-spherical volumes appear near the nuclei (the core loges), between which are other volumes closely corresponding to the classical concept of bond. For the leading event there is a certain number of electrons in each bond loge which corresponds to the number of electrons usually associated with the bond. *Filtered through the wave mechanical principles the empirical concept of bond becomes a volume of the molecular space surrounding some atomic cores in which there is a high probability of finding a certain number n of electrons with a certain organization of their spins.*

Furthermore, as a consequence of the Pauli principle (that is to say the antisymmetrical character of the wave functions), a strong 'repulsion' appears between electrons possessing the same projection of spin. This effect plays an essential role in the determination of bond angles.

REFERENCES

1. P. W. Bridgman, *The Logic of Modern Physics*, Macmillan, 1928.
2. L. de Broglie, *Ann. Phys.*, **3,** 22, (1925).
3. E. Schrödinger, *Ann. der Physik*, **79,** 361 (1926).
4. J. L. Destouches, *J. de Phys.*, **5,** 320 (1934); L. de Broglie and J. L. Destouches, *Compt. Rend. Acad. Sci.*, **201,** 369 (1935).
5. L. Pauling and E. B. Wilson, *Introduction to Quantum Mechanics*, McGraw-Hill, 1935.
6. H. Yukawa and S. Sakata, *Proc. Phys. Math. Soc. Japan*, **17,** 467 (1935).
7. L. W. Alvarez, *Phys. Rev.*, **52,** 134 (1937).

8. R. Daudel, *Compt. Rend. Acad. Sci.*, **237,** 601 (1953); *The Fundamentals of Theoretical Chemistry*, Pergamon, 1968, p. 30.
9. E. Teller, *Z.f. Phys.*, **61,** 458 (1930); E. A. Hylleraas, *Z.f. Phys.*, **71,** 739 (1931); G. Jaffe, *Z.f. Phys.*, **87,** 535 (1934); S. K. Chakravarty, *Phil. Mag.*, **28,** 423 (1939); D. R. Bates, K. Ledsham, and A. L. Stewart, *Phil. Trans. Roy. Soc.*, **A 246,** 215 (1953).
10. L. Brillouin, *La Science et la Théorie de l'Information*, Masson, 1959; N. Wiener, *Cybernetics, A Control and Communication in the Animal and the Machine*, Hermann, 1948.
11. R. Daudel, *Compt. Rend. Acad. Sci.*, **237,** 601 (1953); *Les Fondements de la Chimie Théorique*, Gauthier-Villars, 1956 (English ed., *The Fundamentals of Theoretical Chemistry*, Pergamon, 1968).
12. S. Odiot, *Cah. Phys.*, **81,** 1 (1957), **84,** 23 (1957); E. V. Ludeña and V. Amzel, *J. Chem. Phys.*, **52,** 5923 (1970); E. V. Ludeña, *Int. J. Quantum Chemistry*, **5,** 395 (1971); G. Sperber, *Int. J. Quantum Chemistry*, **5,** 189 (1971).
13. C. Aslangul, *Compt. Rend. Acad. Sci.*, **B 272,** 1 (1971). For a general review see C. Aslangul, R. Constanciel, R. Daudel and P. Kottis, *Advances in Quantum Chemistry*, **4,** 93 (1972).
14. W. Kolos and C. C. J. Roothaan, *Rev. Modern Phys.*, **52,** 219 (1960).
15. M. Roux, S. Besnainou and R. Daudel, *J. Chim. Phys.*, **53,** 218 (1956).
16. S. Bratos, R. Daudel, M. Roux and M. Allavena, *Rev. Modern Phys.*, **32,** 412 (1960).
17. R. F. W. Bader, W. H. Henneker and P. E. Cade, *J. Chem. Phys.*, **46,** 3341 (1967); A. Pullman, *Aspects de la Chimie Quantique Contemporaine*, CNRS, 1971, p. 254.
18. G. E. Uhlenbeck and S. Goudsmit, *Naturwiss*, **13,** 953 (1925), *Nature*, **117,** 264 (1926).
19. S. Odiot, *Compt. Rend. Acad. Sci.*, **237,** 1399 (1953).
20. S. Odiot and R. Daudel, *Compt. Rend. Acad. Sci.*, **238,** 1384 (1954).
21. J. W. Linnett and J. Pople, *Trans. Far. Soc.*, **47,** 1033 (1951).
22. G. N. Lewis, *J. Amer. Chem. Soc.*, **38,** 762 (1916).
23. H. Brion, R. Daudel and S. Odiot, *J. Chim. Phys.*, **51,** 553 (1954).
24. R. Daudel, *Electronic Structure of Molecules*, Pergamon, 1966, p. 60.
25. M. Sanchez and E. V. Ludeña, *Int. J. Quantum Chem.*, **6,** 1113 (1972).
26. R. Moccia, *J. Chem. Phys.*, **40,** 2164 (1964).
27. R. F. W. Bader and H. J. T. Preston, *Theoret. Chim. Acta*, **17,** 584 (1970).
28. R. Daudel, F. Gallais and P. Smet, *Int. J. Quantum Chemistry*, **1,** 873 (1967).
29. R. Daudel, *Compt. Rend. Acad. Sci.*, **270,** 929 (1970).
30. R. J. Gillespie, *Molecular Geometry*, Van Nostrand Reinhold, 1972.

CHAPTER 2

Calculating Wave Functions

2.1 METHODS WHICH DO NOT ASSUME ELECTRONIC LOCALIZATION

2.1.1 The Independent Electron Model for Atoms; Atomic Orbitals

From the practical viewpoint, solving the Schrödinger equation is a very difficult problem. Even in the case of the helium atom rigorous solutions have not been obtained. Only the proof of the existence of such solutions has been given.[1] Therefore all theoretical atomic and molecular physics is built on approximate wave functions. The basis of the most commonly used procedure for calculating electronic wave functions is the independent electron model.

This model will be presented here for the case of atoms. Let us go back to the helium atom represented by two electrons moving in the field of a fixed electric charge $+2e$ (Section 1.2.3). The Schrödinger equation is:

$$-\frac{h^2}{8\pi^2 m}\Delta_a \Psi_i(\mathrm{M}_a, \mathrm{M}_b) - \frac{h^2}{8\pi^2 m}\Delta_b \Psi_i(\mathrm{M}_a, \mathrm{M}_b) + \left(\frac{e^2}{r_{ab}} - \frac{2e^2}{r_a} - \frac{2e^2}{r_b}\right)\Psi_i(\mathrm{M}_a, \mathrm{M}_b) = E_i \Psi_i(\mathrm{M}_a, \mathrm{M}_b) \quad (2.1)$$

If the letter T is used to represent the kinetic operators and the letter F the potential ones, equation (2.1) becomes:

$$\begin{aligned} H\Psi_i(\mathrm{M}_a, \mathrm{M}_b) &= (T_a + T_b + F_{ab} + F_a + F_b)\Psi_i(\mathrm{M}_a, \mathrm{M}_b) \\ &= E_i \Psi_i(\mathrm{M}_a, \mathrm{M}_b) \end{aligned} \quad (2.2)$$

As it is not possible to obtain rigorous solutions of that equation we are led to consider a simpler equation for which exact solutions can be obtained. The independent electron model tends to neglect the coulombic interaction between the electrons. The hamiltonian H^0 of that model reduces to:

$$H^0 = H - \frac{e^2}{r_{ab}} = H - F_{ab} = T_a + F_a + T_b + F_b \quad (2.3)$$

and the Schrödinger equation becomes:

$$H^0\Phi_i(M_a, M_b) = (T_a + F_a + T_b + F_b)\Phi_i(M_a, M_b)$$
$$= E_i^0\Phi_i(M_a, M_b) \tag{2.4}$$

It is very easy to obtain exact solutions of equation (2.4). Let us consider the equation:

$$(T+F)\varphi_k(M) = \left(-\frac{h^2}{8\pi^2 m}\Delta - \frac{2e^2}{r_M}\right)\varphi_k(M) = \varepsilon_k\varphi_k(M) \tag{2.5}$$

Except for a factor 2 in the potential operator this is the equation of the hydrogen atom and has been solved rigorously.[2] The solutions are the well-known *hydrogenic functions.*

Let us consider a product such as:

$$\Upsilon(M_a, M_b) = \varphi_k(M_a)\varphi_{k'}(M_b) \tag{2.6}$$

We can easily see that:

$$H^0\Upsilon = (T_a + F_a + T_b + F_b)\varphi_k(M_a)\varphi_{k'}(M_b)$$
$$= \varphi_{k'}(M_b)(T_a + F_a)\varphi_k(M_a) + \varphi_k(M_a)(T_b + F_b)\varphi_{k'}(M_b) \tag{2.7}$$

Taking account of equation (2.5) it can readily be seen that:

$$H^0\Upsilon = \varphi_{k'}(M_b)\varepsilon_k\varphi_k(M_a) + \varphi_k(M_a)\varepsilon_{k'}\varphi_{k'}(M_b)$$
$$= (\varepsilon_k + \varepsilon_{k'})\Upsilon \tag{2.8}$$

It turns out that Υ is an eigenfunction of the hamiltonian H^0. Therefore it is a rigorous solution of the independent electron model associated with the helium atom. For that reason the functions φ_k are called *atomic orbitals.* It can be seen that a product of two such orbitals is an eigenfunction of H^0 which corresponds to an eigenvalue which is simply the sum of the energies ε_k associated with the considered atomic orbitals:

$$E^0 = \varepsilon_k + \varepsilon_{k'} \tag{2.9}$$

This result can be generalized for every atom. *A product of n atomic orbitals is a rigorous solution of the independent electron model for an atom containing n electrons. The corresponding eigenvalue is the sum of the energies associated with each orbital.*

When a simple product of atomic orbitals is used as a wave function to represent a state of an atom the electrons are really independent. Remember that two events are said to be independent when the probability p_{12} that the two events occur simultaneously is simply the product of the respective probabilities p_1 and p_2 of occurrence of each isolated event:

$$p_{12} = p_1 p_2 \tag{2.10}$$

Assuming that:

$$\Phi = \varphi_1(M_a)\varphi_2(M_b) \tag{2.11}$$

let us calculate the probability p_{12} of simultaneously finding electron 1 in dv_a and electron 2 in dv_b. We have:

$$p_{12} = |Y|^2 \, dv_a \, dv_b = |\varphi_1(M_a)|^2 \, dv_a \, |\varphi_2(M_b)|^2 \, dv_b \tag{2.12}$$

The probability p_1 of finding electron 1 in dv_a, whatever the position of electron 2, may be obtained by integrating p_{12} over all possible positions of M_b. Therefore:

$$\begin{aligned} p_1 &= \int p_{12} \, dv_b = |\varphi_1(M_a)|^2 \, dv_a \int |\varphi_2(M_b)|^2 \, dv_b \\ &= |\varphi_1(M_a)|^2 \, dv_a \end{aligned} \tag{2.13}$$

if $\varphi_2(M_b)$ is normalized. For analogous reasons:

$$p_2 = |\varphi_2(M_b)|^2 \, dv_b \tag{2.14}$$

Comparison of equations (2.12), (2.13) and (2.14) leads to the conclusion that:

$$p_{12} = p_1 p_2 \tag{2.15}$$

It has already been stated that the movements of the two electrons are independent: there is no correlation between them.

It would appear that orbital φ_1 was a guide for electron 1 and orbital φ_2 a guide for electron 2. This seems to be confirmed by the following facts. If the independent electron model is used the ground state of the helium atom can be represented by the function:

$$\begin{aligned} \Phi_1 &= \Psi_{1,0,0}(M_a)\Psi_{1,0,0}(M_b) \\ &= Ks(M_a)Ks(M_b) = N^2 e^{-2r_a/a_0} e^{-2r_b/a_0} \end{aligned} \tag{2.16}$$

where N denotes a normalizing constant. The energy of the ground state will be:

$$E_1^0 = 2\varepsilon_1 \tag{2.17}$$

where ε_1 represents the energy associated with the orbital $\Psi_{1,0,0}$.

As the energy ε_1 associated with a hydrogen function corresponding to a nucleus of charge Ze is:

$$\varepsilon_1 = -\frac{2\pi^2 m Z^2 e^4}{n^2 h^2} \tag{2.18}$$

the ground state energy of the helium atom can be written as:

$$E_1^0 = -\frac{16\pi^2 m e^4}{n^2 h^2} = -108 \text{ eV} \tag{2.19}$$

within the framework of the independent electron model. The experimental value being −78.6 eV, it is seen that neglecting the coulombic interaction between the electrons leads to an excess of bonding energy of approximately 30 eV.

Consider an excited state represented by:

$$\Phi_2 = Ks(\mathrm{M}_a)Ls(\mathrm{M}_b) \tag{2.20}$$

It will have the energy:

$$E_2^0 = \varepsilon_1 + \varepsilon_2 \tag{2.21}$$

where ε_2 denotes the energy associated with Ls (or $\psi_{2,0,0}$). Therefore the transition energy between the two states is:

$$E = E_2^0 - E_1^0 = \varepsilon_2 - \varepsilon_1 \tag{2.22}$$

It can be concluded that to go from the ground state to the considered excited state we have to excite an electron from orbital φ_1 to orbital φ_2.

Consider now the ionization of the considered excited state of helium. The ground state of the corresponding ion He^+ is represented by the function:

$$E_1^+ = \Psi_{1,0,0}(\mathrm{M}) = Ks(\mathrm{M}) \tag{2.23}$$

Its energy is simply ε_1, the ionization energy being:

$$I = E_2^0 - E_1^+ = \varepsilon_2 \tag{2.24}$$

We can say that to obtain the ion we have to extract electron 2 from the orbital Ls. This is the shell model language.

Such a description would be permissible if the spin did not exist and if the indistinguishability and Pauli principles were not admitted. However, the spin does exist and the indistinguishability and Pauli principles are admitted. *We shall now see that, as a consequence, all the shell model language is inconsistent.*

Before doing so, it is first necessary to recall some properties of spin operators. It has already been stated that the spin of an electron is represented by four operators: the three operators $\mathcal{S}_x$, $\mathcal{S}_y$ and $\mathcal{S}_z$ associated with the various projections of the spin and the operator $\mathcal{S}^2$ associated with the square of its length. The four operators are connected by the following relation:

$$\mathcal{S}^2 = \mathcal{S}_x^2 + \mathcal{S}_y^2 + \mathcal{S}_z^2 \tag{2.25}$$

Furthermore, the operators obey the same commutation rules as the ordinary angular momentum operators. Therefore:

$$\begin{aligned}
\mathcal{S}_x\mathcal{S}_y - \mathcal{S}_y\mathcal{S}_x &= -\frac{h}{2\pi \mathrm{i}}\mathcal{S}_z \\
\mathcal{S}_y\mathcal{S}_z - \mathcal{S}_z\mathcal{S}_y &= -\frac{h}{2\pi \mathrm{i}}\mathcal{S}_x \\
\mathcal{S}_z\mathcal{S}_x - \mathcal{S}_x\mathcal{S}_z &= -\frac{h}{2\pi \mathrm{i}}\mathcal{S}_y \\
\mathcal{S}^2\mathcal{S}_i - \mathcal{S}_i\mathcal{S}^2 &= 0
\end{aligned} \tag{2.26}$$

where $\mathcal{S}_i$ denotes one of the three operators $\mathcal{S}_x$, $\mathcal{S}_y$ or $\mathcal{S}_z$.

As a measure of a projection of the spin is only $\pm h/4\pi$ we need only to represent each $\mathscr{S}_i$ as an operator possessing these two numbers as eigenvalues. We can select the three following two-dimensional matrices:

$$S_x = \frac{h}{4\pi}\begin{Vmatrix} 0 & 1 \\ 1 & 0 \end{Vmatrix}, \qquad S_y = \frac{h}{4\pi}\begin{Vmatrix} 0 & -i \\ i & 0 \end{Vmatrix}, \qquad S_z = \frac{h}{4\pi}\begin{Vmatrix} 1 & 0 \\ 0 & -1 \end{Vmatrix} \tag{2.27}$$

which operate on a vectorial two-dimensional space (the so-called spin space) and possess eigenvectors. However, these vectors can be represented by functions using the following procedure. Consider the two functions $\alpha(\omega)$ and $\beta(\omega)$ (where ω can only take the two values $\pm h/4\pi$) which satisfy the relations:

$$\begin{aligned} \alpha\left(+\frac{h}{4\pi}\right) &= 1, \qquad \beta\left(+\frac{h}{4\pi}\right) = 0 \\ \alpha\left(-\frac{h}{4\pi}\right) &= 0, \qquad \beta\left(-\frac{h}{4\pi}\right) = 1 \end{aligned} \tag{2.28}$$

If now the effect of a matrix A on a function $\sigma(\omega)$ is defined by the expressions:

$$\sigma'(\omega) = A\sigma(\omega)$$

with

$$\begin{aligned} \sigma'\left(+\frac{h}{4\pi}\right) &= a_{11}\sigma\left(+\frac{h}{4\pi}\right) + a_{12}\sigma\left(-\frac{h}{4\pi}\right) \\ \sigma'\left(-\frac{h}{4\pi}\right) &= a_{12}\sigma\left(+\frac{h}{4\pi}\right) + a_{22}\sigma\left(-\frac{h}{4\pi}\right) \end{aligned} \tag{2.29}$$

it is readily seen that the functions $\alpha(\omega)$ and $\beta(\omega)$ are eigenfunctions of the operators $\mathscr{S}_i$. Let us calculate, for example:

$$\sigma'(\omega) = \mathscr{S}_z\alpha(\omega)$$

We obtain:

$$\begin{aligned} \sigma'\left(+\frac{h}{4\pi}\right) &= \frac{h}{4\pi}\,\alpha\left(+\frac{h}{4\pi}\right) \\ \sigma'\left(-\frac{h}{4\pi}\right) &= -\frac{h}{4\pi}\,\alpha\left(-\frac{h}{4\pi}\right) = 0 \end{aligned} \tag{2.30}$$

Therefore:

$$\mathscr{S}_z\alpha(\omega) = \frac{h}{4\pi}\,\alpha(\omega) \tag{2.31}$$

and $\alpha(\omega)$ appears to be the eigenfunction associated with the eigenvalue $+h/4\pi$ of the operator $\mathscr{S}_z$. It is easy to show that $\beta(\omega)$ corresponds to the eigenvalue $-h/4\pi$.

If an n-electron system is now considered, a set of spin coordinates ω_1,

$\omega_2, \ldots, \omega_j, \ldots, \omega_n$ is introduced and an operator $\mathscr{S}_i$ is defined as:

$$\mathscr{S}_i = \sum_j \mathscr{S}_i^{(j)} \tag{2.32}$$

each operator $\mathscr{S}_i^{(j)}$ being one of the matrices (2.27) acting on the coordinate ω_j. Let us go back to the helium case:

$$\mathscr{S}_z = \mathscr{S}_z^{(1)} + \mathscr{S}_z^{(2)} \tag{2.33}$$

Consider the effect of this operator on the function $\alpha(\omega_1)\alpha(\omega_2)$:

$$\begin{aligned} \mathscr{S}_z\alpha(\omega_1)\alpha(\omega_2) &= (\mathscr{S}_z^{(1)} + \mathscr{S}_z^{(2)})\alpha(\omega_1)\alpha(\omega_2) \\ &= \alpha(\omega_2)\mathscr{S}_z^{(1)}\alpha(\omega_1) + \alpha(\omega_1)\mathscr{S}_z^{(2)}\alpha(\omega_2) \\ &= \alpha(\omega_2)\alpha(\omega_1)\left(\frac{h}{4\pi} + \frac{h}{4\pi}\right) \end{aligned} \tag{2.34}$$

The function $\alpha(\omega_1)\alpha(\omega_2)$ is an eigenfunction of $\mathscr{S}_z$, the corresponding eigenfunction being $h/2\pi$.

Starting from these definitions it is possible to show that the operators $\mathscr{S}^2$ have the eigenvalues given by the expression:

$$S(S+1)\frac{h^2}{4\pi^2} \tag{2.35}$$

S being integer if n is even and half-integer if n is odd. For a given value S of that quantum number the eigenvalues corresponding to a projection of the spin are:

$$s\frac{h}{2\pi} \tag{2.36}$$

with:

$$-S \leqslant s \leqslant S$$

s being integer or half-integer in the same conditions as S.

In the case of a two-electron system it is easy to show that the function:

$$\alpha(\omega_1)\beta(\omega_2) - \beta(\omega_1)\alpha(\omega_2) \tag{2.37}$$

is an eigenfunction of $\mathscr{S}^2$ (eigenvalue 0) and also of $\mathscr{S}_z$ (eigenvalue 0). The three functions:

$$\begin{aligned} &\alpha(\omega_1)\beta(\omega_2) + \beta(\omega_1)\alpha(\omega_2) \\ &\alpha(\omega_1)\alpha(\omega_2) \\ &\beta(\omega_1)\beta(\omega_2) \end{aligned} \tag{2.38}$$

are eigenfunctions of $\mathscr{S}^2$ for the eigenvalue $2h^2/4\pi^2$. They are also eigenfunctions of $\mathscr{S}_z$ and correspond respectively to the eigenvalues 0, $h/2\pi$, $-h/2\pi$.

This is why it is said that the eigenvalue 0 of $\mathscr{S}^2$ describes a *singlet* state and the eigenvalue $h^2/2\pi^2$ a *triplet* state. Furthermore, it is seen that the singlet state spin function is antisymmetrical with respect to a permutation of the spin coordinate whereas the triplet state spin functions are symmetrical.

To satisfy the Pauli principle we must build a total wave function antisymmetrical with respect to a simultaneous permutation of space and spin coordinates. It is obvious that to satisfy the requirement we must multiply the singlet state spin function by a symmetrical space function and the triplet state spin function by an antisymmetrical space function.

The ground state space function (2.16) is symmetrical. Therefore if it is multiplied by the antisymmetrical spin function a convenient wave function is obtained. The ground state of helium can be represented by the function:

$$\Phi_{1T} = Ks(M_a)Ks(M_b)[\alpha(\omega_a)\beta(\omega_b) - \beta(\omega_a)\alpha(\omega_b)] \tag{2.39}$$

which is a singlet state. Conversely, an excited space function such as (2.20) is neither symmetrical nor antisymmetrical. Before it can be included in a wave function it must be transformed. It is obvious that the functions $Ls(M_a)K(M_b)$ is an eigenfunction of the hamiltonian H^0 just as the function $Ks(M_a)Ls(M_b)$ and corresponds to the same eigenvalue $E_2^0 = \varepsilon_1 + \varepsilon_2$. Therefore a linear combination such as:

$$\lambda Ks(M_a)Ls(M_b) + \mu Ls(M_a)Ks(M_b)$$

will also be an eigenfunction for the same eigenvalue. It is obvious that among those linear combinations:

$$\Phi_2^S = Ks(M_a)Ls(M_b) + Ls(M_a)Ks(M_b) \tag{2.40}$$

is symmetrical and can be combined with the antisymmetrical spin function to represent a singlet state and:

$$\Phi_2^A = Ks(M_a)Ls(M_b) - Ls(M_a)Ks(M_b) \tag{2.41}$$

can be combined with one of the three symmetrical spin functions to represent a triplet state. The considered singlet and triplet states are said to be produced by the same *configuration LsMs. A configuration is therefore a certain choice of atomic orbitals.*

The total wave function associated with the singlet state is:

$$\Phi_{2T}^S = N[Ks(M_a)Ls(M_b) + Ls(M_a)Ks(M_b)][\alpha(\omega_a)\beta(\omega_b) - \beta(\omega_a)\alpha(\omega_b)] \tag{2.42}$$

The total wave functions corresponding to the triplet state are:

$$\Phi_{2T}^T = N[Ks(M_a)Ls(M_b) - Ls(M_a)Ks(M_b)]\begin{cases}[\alpha(\omega_a)\beta(\omega_b) + \beta(\omega_a)\alpha(\omega_b)] \\ \alpha(\omega_a)\alpha(\omega_b) \\ \beta(\omega_a)\beta(\omega_b)\end{cases} \tag{2.43}$$

It is easy to show that if such functions are used a strong correlation is introduced between the movement of the two electrons.

Let us consider the singlet state. The probability dp_{12} of simultaneously finding electron 1 with the spin $h/4\pi$ in dv_a and electron 2 with the spin $-h/4\pi$ in dv_b is:

$$\begin{aligned} dp_{12} &= dv_a\, dv_b\, N^2[Ks(\mathrm{M}_a)Ls(\mathrm{M}_b)+Ls(\mathrm{M}_a)Ks(\mathrm{M}_b)]^2 \\ &= dv_a\, dv_b\, N^2[K^2s(\mathrm{M}_a)L^2s(\mathrm{M}_b)+2Ks(\mathrm{M}_a)Ls(\mathrm{M}_a)Ks(\mathrm{M}_b)Ls(\mathrm{M}_b) \\ &\qquad + L^2s(\mathrm{M}_a)K^2s(\mathrm{M}_b)] \end{aligned} \quad (2.44)$$

The probability dp_1 of finding electron 1 with the spin $+h/4\pi$ in dv_a is:

$$dp_1 = \int dp_{12}\, dv_b = N^2[K^2s(\mathrm{M}_a)+L^2s(\mathrm{M}_a)]\, dv_a \quad (2.45)$$

if the orthogonality of Ks and Ls is taken into account. Similarly, the probability dp_2 of finding electron 2 with the spin $-h/4\pi$ in dv_b is:

$$dp_2 = \int dp_{12}\, dv_a = N^2[K^2s(\mathrm{M}_b)+L^2s(\mathrm{M}_b)]\, dv_b \quad (2.46)$$

Therefore:

$$\begin{aligned} dp_1\, dp_2 = N^4[&K^2s(\mathrm{M}_a)K^2s(\mathrm{M}_b)+K^2s(\mathrm{M}_a)L^2s(\mathrm{M}_b) \\ &+L^2s(\mathrm{M}_a)K^2s(\mathrm{M}_b)+L^2s(\mathrm{M}_a)L^2s(\mathrm{M}_b)]\, dv_a\, dv_b \end{aligned} \quad (2.47)$$

It turns out that dp_{12} is not equal to $dp_1\, dp_2$.

The movements of the two electrons are not independent: they are correlated. The electrons of the independent electron model are not independent! Furthermore, equation (2.45) shows that the movement of electron 1 depends on both functions Ks and Ls. *When the indistinguishability of the electrons is taken into account it is not possible to associate an electron with an orbital. As a particular electron is 'guided' by all the orbitals introduced in the wave function, the shell model language becomes inconsistent.*

To underline the strong correlation effect which is contained in a function like Φ_{2T} let us consider as another example the function:

$$\Phi_{3T} = N[Ls(\mathrm{M}_a)Lp_0(\mathrm{M}_b) - Lp_0(\mathrm{M}_a)Ls(\mathrm{M}_b)]\alpha(\omega_a)\alpha(\omega_b) \quad (2.48)$$

We saw that the angular part of Lp_0 is proportional to $\cos\theta$ (see equation 1.52). To simplify the discussion let us assume that the radial parts of Ls and Lp are identical to a certain function $R(r)$. Equation (2.48) becomes:

$$\Phi_{3T} = NR(r_a)R(r_b)(\cos\theta_b - \cos\theta_a)\alpha(\omega_a)\alpha(\omega_b) \quad (2.49)$$

The wave function vanishes when:

$$\theta_a = \theta_b$$

The probability density of simultaneously finding electron 1 at a point M_a corresponding to a certain angle θ_a and electron 2 at a point M_b corresponding to the same angle is zero: the presence of an electron in the direction θ_a

forbids the presence of another electron in the same direction. On the contrary, the modulus of the function (2.49) reaches its maximum value when:

$$\theta_a = 0 \qquad \text{and} \qquad \theta_b = 180°$$

The probability density becomes high when M_a and M_b are described by such angles.

However, it is very important to remark that the symmetrization of the functions does not change the energy associated with a state. *Therefore all the energetic relationships established before that operation remain valid. The shell model language is inconsistent but leads to valid conclusions from the energetic viewpoint.*

Generalizing the independent electron model for atoms containing any number of electrons is not a difficult job. Let n be that number. By neglecting the coulombic interation an hamiltonian H^0 is obtained. It can be written as:

$$H^0 = \sum_i (T_i + F_i) \tag{2.50}$$

If Z denotes the atomic number we are led to consider the equation:

$$(T+F)\varphi_k(\mathrm{M}) = \left(-\frac{h^2}{8\pi^2 m}\Delta - \frac{Ze^2}{r}\right)\varphi_k(\mathrm{M}) = \varepsilon_k \varphi_k(\mathrm{M}) \tag{2.51}$$

The eigenfunctions $\varphi_k(\mathrm{M})$ are the classical hydrogen-like functions. A product of n functions $\varphi_k(\mathrm{M})$ like:

$$\Phi_j(\mathrm{M}_a \cdots \mathrm{M}_p) = \varphi_{k_a}(\mathrm{M}_a)\varphi_{k_b}(\mathrm{M}_b) \cdots \varphi_{k_p}(\mathrm{M}_p) \tag{2.52}$$

where j characterizes an arbitrary choice of the φ values, is an eigenfunction of the operator H^0 which corresponds to the eigenvalue:

$$E_j = \varepsilon_{k_a} + \varepsilon_{k_b} + \cdots + \varepsilon_{k_p} \tag{2.53}$$

The functions φ are called *atomic orbitals*. In order to take account of the spin, of the indistinguishability of the electrons and of the Pauli principle the following expression can be used:

$$\Phi_{jT} = \sum_P (-1)^P P \Phi_j \sigma \tag{2.54}$$

Details of the use of such an algorithm can be found in many texts.[3] If the length of the spin and its projection along the z axis are well defined σ will be an eigenvalue of both the $\mathcal{S}^2$ and $\mathcal{S}_z$ operators. The operator $\sum_P (-1)^P P$ is called an *antisymmetrizer*. It amounts to consider all the possible simultaneous permutations of the space and spin coordinates in Φ and in σ and to give the sum of all the resulting functions a positive sign when the permutation is even and a negative one when the permutation is odd. The result of the effect of such an operator on any function $\Phi\sigma$ is zero or an antisymmetrical function.

Let us go back to the helium atom. Consider a triplet state described by the spin function:

$$\sigma = \alpha(\omega_a)\alpha(\omega_b)$$

Try to produce the function:

$$\Phi_T = \sum_P (-1)^P P Ks(\mathrm{M}_a) Ks(\mathrm{M}_b)\alpha(\omega_a)\alpha(\omega_b) \tag{2.55}$$

It can readily be seen that:

$$\Phi_T = Ks(\mathrm{M}_a)Ks(\mathrm{M}_b)\alpha(\omega_a)\alpha(\omega_b) - Ks(\mathrm{M}_b)Ks(\mathrm{M}_a)\alpha(\omega_b)\alpha(\omega_a) = 0 \tag{2.56}$$

The space function $Ks(\mathrm{M}_a)Ks(\mathrm{M}_b)$ is not appropriate to represent a triplet state. No triplet state belongs to the configuration $(Ks)^2$, which is the usual symbol to denote the configuration $KsKs$. If the function:

$$\Phi_T = \sum_P (-1)^P P Ks(\mathrm{M}_a) Ls(\mathrm{M}_b)\alpha(\omega_a)\alpha(\omega_b) \tag{2.57}$$

is now considered, the equation:

$$\Phi_T = Ks(\mathrm{M}_a)Ls(\mathrm{M}_b)\alpha(\omega_a)\alpha(\omega_b) - Ks(\mathrm{M}_b)Ls(\mathrm{M}_a)\alpha(\omega_b)\alpha(\omega_a) \tag{2.58}$$

is obtained. This is exactly one of the functions (2.43).

It is possible to show that in the general case a spin function σ has the form:

$$\sum_l C_l \genfrac{}{}{0pt}{}{\alpha}{\beta}(\omega_a)\genfrac{}{}{0pt}{}{\alpha}{\beta}(\omega_b)\ldots\genfrac{}{}{0pt}{}{\alpha}{\beta}(\omega_p) \tag{2.59}$$

where the index l is associated with a given choice of the functions α or β. The calculation of the C's can be achieved by following various techniques which will not be described here. One of them is to use both projection operators[4] and branching diagrams.[5] Another one is to diagonalize some spin matrices. The use of both projection operators and spin matrices has been described by Daudel, Lefebvre and Moser.[6] It is also possible to use Young's operators.[7] Gouyet[8] proposed an algebraic procedure closely related to annihilation and creation operators. His papers contain a good bibliography about the whole problem.

If expressions (2.59) and (2.52) are introduced into equation (2.54) the function Φ_{jT} takes the form:

$$\Phi_{jT} = \sum_l C_l \sum_P (-1)^P P \varphi_{k_a}(\mathrm{M}_a)\genfrac{}{}{0pt}{}{\alpha}{\beta}(\omega_a)\cdots\varphi_{k_p}(\mathrm{M}_p)\genfrac{}{}{0pt}{}{\alpha}{\beta}(\omega_p) \tag{2.60}$$

Let us consider a specific term such as:

$$\sum_P (-1)^P P \varphi_{k_a}(\mathrm{M}_a)\alpha(\omega_a)\cdots\varphi_{k_p}(\mathrm{M}_p)\beta(\omega_p) \tag{2.61}$$

It may be written in determinantal form:

$$\begin{vmatrix} \varphi_{k_a}(\mathrm{M}_a)\alpha(\omega_a) & \cdots & \varphi_{k_p}(\mathrm{M}_a)\beta(\omega_a) \\ \varphi_{k_a}(\mathrm{M}_p)\alpha(\omega_p) & \cdots & \varphi_{k_p}(\mathrm{M}_p)\beta(\omega_p) \end{vmatrix} \tag{2.62}$$

This is a Slater determinant. Each product of an orbital φ and a spin function α or β is called a *spin orbital.* Very often a spin orbital containing an α function is simply written φ. A spin orbital containing a β function is written $\bar{\varphi}$. Thus, determinant (2.62) can be written as:

$$\det \varphi_{k_a} \cdots \bar{\varphi}_{k_p}$$

Equation (2.60) clearly shows that *some linear combinations of Slater determinants are convenient wave functions for the independent electron model.*

It is very important to note that if the same spin orbital is introduced twice in the same determinant it has two identical columns and therefore vanishes. *Therefore in a Slater determinant a given orbital cannot be used more than twice (once with the α function and once with the β function).* If each orbital is used twice with a spin function associated with the zero eigenvalues of $\mathcal{S}^2$ and $\mathcal{S}_z$ the function Φ_{jT} becomes:

$$\Phi_{jT} = \det \varphi_{k_a}\bar{\varphi}_{k_a}\varphi_{k_b}\bar{\varphi}_{k_b} \cdots \varphi_{k_p}\bar{\varphi}_{k_p} \tag{2.63}$$

The corresponding state is said to be a *closed-shell state.*

2.1.2 The Independent Electron Model for Molecules; the Born–Oppenheimer Approximation, Molecular Orbitals

Let us first discuss a simple example: the hydrogen molecule ion. A space function associated with that molecule can be written as:

$$\Psi(\mathrm{N_A}, \mathrm{N_B}, \mathrm{M})$$

when the time is not made explicit and where $\mathrm{N_A}$, $\mathrm{N_B}$ and M denote the points where proton 1, proton 2 and the electron respectively are expected.

The corresponding Schrödinger equation is:

$$\left(-\frac{h^2}{8\pi^2\mu}\Delta_\mathrm{A} - \frac{h^2}{8\pi^2\mu}\Delta_\mathrm{B} - \frac{h^2}{8\pi^2 m}\Delta + \frac{e^2}{r_\mathrm{AB}} - \frac{e^2}{r_\mathrm{A}} - \frac{e^2}{r_\mathrm{B}}\right)\Psi_i = E_i\Psi_i \tag{2.64}$$

where $\Delta_\mathrm{A}, \Delta_\mathrm{B}$ and Δ are the laplacian operators acting on the function Ψ around the points $\mathrm{N_A}$, $\mathrm{N_B}$ and M respectively, r_AB denotes the distance between $\mathrm{N_A}$ and $\mathrm{N_B}$, r_A the distance $\mathrm{MN_A}$ and r_B the distance $\mathrm{MN_B}$. Solving such an equation is a very difficult job. This is why the Born–Oppenheimer approximation is commonly used.

As the mass μ of a proton is very large in comparison with the mass m of an electron it is natural to consider first the equation:

$$\left(-\frac{h^2}{8\pi^2 m}\Delta + \frac{e^2}{r_\mathrm{AB}} - \frac{e^2}{r_\mathrm{A}} - \frac{e^2}{r_\mathrm{B}}\right)\phi_j = U_j\phi_j \tag{2.65}$$

This is called the *electronic equation* and describes the movement of an electron in the field of two fixed charges $+e$ replacing the protons. It can be written as:

$$(T_\mathrm{E} + F)\phi_j = U_j\phi_j \tag{2.66}$$

where T_E denotes the electronic kinetic energy operator and F the potential energy operator.

For a set of given positions of N_A and N_B the eigenfunction ϕ_j only depends on M. The functions ϕ_j constitute a basis in the Hilbert space associated with monoelectronic wave functions. The function Ψ_i for given positions of N_A and N_B is such a function. Therefore Ψ_i can be extended over the ϕ_j's:

$$\Psi_i = \sum_{j=1}^{j=\infty} G_j \phi_j \tag{2.67}$$

If the positions of points N_A and N_B are changed the coefficients G_j in equation (2.67) will also change. Therefore, to be more explicit, equation (2.67) can be written as:

$$\Psi_i(N_A, N_B, M) = \sum_{j=1}^{j=\infty} G_j(N_A, N_B)\phi_j(N_A, N_B, M) \tag{2.68}$$

Taking this expression into account equation (2.64) can readily be transformed into:

$$\sum (T_N + T_E + F) G_j \phi_j = E_i \sum G_j \phi_j \tag{2.69}$$

As G_j depends on M, equation (2.69) can also be written as:

$$\sum T_N G_j \phi_j + \sum G_j (T_E + F)\phi_j = E_i \sum G_j \phi_j \tag{2.70}$$

If now account is taken of equation (2.66) the foregoing expression reduces to:

$$\sum T_N G_j \phi_j + \sum G_j U_j \phi_j = E_i \sum G_j \phi_j \tag{2.71}$$

If the two members of equation (2.71) are multiplied by ϕ_k^* and integrated with respect to M it is readily seen that:

$$\sum \int \phi_k^* T_N G_j \phi_j \, dv_M + \sum G_j U_j \int \phi_k^* \phi_j \, dv_M = E_i \sum G_j \int \phi_k^* \phi_j \, dv_M \tag{2.72}$$

because G_j and U_j do not depend on M.

The functions ϕ_j are assumed to be orthonormalized; therefore:

$$\int \phi_k^* \phi_j \, dv_M = \delta_{ij} \tag{2.73}$$

and equation (2.72) reduces to:

$$\sum \int \phi_k^* T_N G_j \phi_j \, dv_M + G_k U_k = E_i G_k \tag{2.74}$$

Until now no specific approximation has been introduced. The Born–Oppenheimer approximation will now be presented. It is made of two parts. First, we assume that for any function G_j a domain D_j exists in such a way that for any point outside the domain the function vanishes and that inside the domain the derivatives of ϕ_j with respect to the coordinates of N_A and N_B are small. In other words,

$$\begin{aligned} \forall j, \exists D_j, \forall N \notin D_j, \quad & G_j = 0 \\ \forall N \in D_j, \quad & \frac{\partial \phi_j}{\partial N_l} = 0 \end{aligned} \tag{2.75}$$

N_l being one of the coordinates associated with point N. This assumption means that the G_j's (and therefore the nuclei) are localized in small volumes in which the functions ϕ_j do not vary rapidly with the positions of N_A and N_B. If this is so:

$$\frac{\partial}{\partial N_l} G_j\phi_j = \phi_j \frac{\partial G_j}{\partial N_l} + G_j \frac{\partial \phi_j}{\partial N_l}$$

reduces to:

$$\frac{\partial}{\partial N_l} G_j\phi_j = \phi_j \frac{\partial G_j}{\partial N_l} \tag{2.76}$$

as the term $G_j\ \partial\phi_j/\partial N_l$ vanishes everywhere. Therefore:

$$T_N G_j \phi_j = \phi_j T_N G_j \tag{2.77}$$

and equation (2.74) reduces to:

$$(T_N + U_k)G_{kp} = E_{kp}G_{kp} \tag{2.78}$$

Finally, it is possible to calculate the G_k's by solving the electronic equation, introducing the U_k into equation (2.78) and solving it.

The second part of the Born–Oppenheimer approximation makes the assumption that there is a leading term in equation (2.68). Then an approximate expression for Ψ_i will be:

$$\Psi_i \simeq G_{kp}\phi_k \tag{2.79}$$

if the second member is that leading term. The method of resolution of this equation will be described later on (Section 7.1.1).

As the function ϕ_k is assumed to be practically independent of N_A and N_B when G_{kp} has a non-zero value the probability density of finding a proton at a point N mainly depends on the function G. It is called the *nuclear function* and, as a consequence, equation (2.78) is called the *nuclear equation.* The localizability requirement for G is that the relative position of the protons be represented in a small volume. Interpretation of experimental results shows that such an assumption is convenient for many states of the molecules.

To summarize we can say that it is (at least in principle) possible to compute a convenient approximate total space function for a molecule by replacing it with a model in which the nuclei are replaced by fixed electric charges. The corresponding electronic equation is solved. The eigenvalues

U_k and the eigenfunctions ϕ_k depend on the positions of the charges. For a given U_k a nuclear equation can be built. It possesses eigenvalues E_{kp} and eigenfunctions G_{kp}. The product:

$$G_{kp}\phi_p$$

is an approximate total space function for the molecule.

The results which can be obtained by solving the electronic equation of various molecules have been discussed in Chapter 1. Other examples will be presented in the following chapters. The results which derive from nuclear equations, that is to say the description of the nuclear movements, will be analysed later on.

The Born–Oppenheimer approximation[9] has been improved by various authors.[10]

Let us consider a polyelectronic molecule containing n electrons and n' nuclei. A space wave function can be written as:

$$\Psi(\mathrm{N_A}, \mathrm{N_B}, \ldots, \mathrm{N_L}, \ldots, \mathrm{N_P}, \mathrm{M}_a, \mathrm{M}_b, \ldots, \mathrm{M}_l, \ldots, \mathrm{M}_q)$$

with obvious notations. The Schrödinger equation is:

$$\left(\sum_L -\frac{h^2}{8\pi^2\mu_L}\Delta_L + \sum_l -\frac{h^2}{8\pi^2 m}\Delta_l + \sum_l F_l + \sum_{l<l'} \frac{e^2}{r_{l,l'}} + \sum_{L<L'} \frac{Z_L Z_{L'} e^2}{r_{L,L'}}\right)\Psi_i = E_i \Psi_i \quad (2.80)$$

where $r_{ll'}$ represents the distance $\mathrm{M}_l\mathrm{M}_{l'}$ and $r_{L,L'}$ the distance $\mathrm{N_L N_{L'}}$, and $Z_L e$ is the charge of the nucleus expected at point $\mathrm{N_L}$. The potential operator F_l has the expression:

$$F_l = \sum_L -\frac{e^2 Z_L}{r_{l,L}} \quad (2.81)$$

where $r_{l,L}$ is the distance M_l, $\mathrm{N_L}$.

If the Born–Oppenheimer approximation is introduced we are led to consider the electronic equation:

$$\left[\sum_i \left(-\frac{h^2}{8\pi^2 m}\Delta_l + F_l\right) + \sum_{l<l'} \frac{e^2}{r_{l,l'}}\right]\phi_j = \left(U_j - \sum_{L<L'} \frac{Z_L Z_{L'}}{r_{L,L'}} e^2\right)\phi_j \quad (2.82)$$

As in the case of atoms it is not possible to solve this equation. It is useful to introduce the independent electron model which, neglecting the interaction between the electrons, considers the equation:

$$\sum_l \left(-\frac{h^2}{8\pi^2 m}\Delta_l + F_l\right)\Phi_j = \left(U_j - \sum_{L<L'} \frac{Z_L Z_{L'}}{r_{L,L'}} e^2\right)\Phi_j \quad (2.83)$$

To solve this equation we must solve the following equation:

$$\left(-\frac{h^2}{8\pi^2 m}\Delta + F\right)\varphi_k(\mathrm{M}) = \varepsilon_k \varphi_k(\mathrm{M}) \quad (2.84)$$

which describes the movement of an isolated electron in the field of charges $Z_L e$ fixed at point N_L. The functions φ_k are called *molecular orbitals.*

Everything which has been said in the case for atoms can be extended for molecules. The functions Φ_j can be obtained as a product of n orbitals. The eigenvalue:

$$U_j - \sum_{L<L'} \frac{Z_L Z_{L'}}{r_{L,L'}} e^2$$

is simply the sum of the energies of the orbitals introduced into the functions Φ_j, the electrons really being independent when that product is used. The indistinguishability of the electrons, the existence of the spin and the Pauli principle lead to the use of a linear combination of Slater determinants built on spin orbitals. *Therefore a strong correlation is introduced between the electrons of the independent electron model. Furthermore, a given orbital cannot be used more than twice.*

To exemplify the use of this model for molecules let us take the case of the hydrogen molecule. The electronic equation is:

$$\left(-\frac{h^2}{8\pi^2 m}\Delta_a - \frac{h^2}{8\pi^2 m}\Delta_b + \frac{e^2}{r_{AB}} + \frac{e^2}{r_{ab}} - \frac{e^2}{r_{aA}} - \frac{e^2}{r_{aB}} - \frac{e^2}{r_{bA}} - \frac{e^2}{r_{bB}}\right)\phi_j = U_j\phi_j \quad (2.85)$$

The electronic equation for the independent electron model reduces to:

$$\left(-\frac{h^2}{8\pi^2 m}\Delta_a - \frac{e^2}{r_{aA}} - \frac{e^2}{r_{aB}} - \frac{h^2}{8\pi^2 m}\Delta_b - \frac{e^2}{r_{bA}} - \frac{e^2}{r_{bB}}\right)\Phi_j = \left(U_j - \frac{e^2}{r_{AB}}\right)\Phi_j \quad (2.86)$$

Therefore the molecular orbitals φ are the solutions of the following equation:

$$\left(-\frac{h^2}{8\pi^2 m}\Delta_M - \frac{e^2}{r_{MA}} - \frac{e^2}{r_{MB}}\right)\varphi_k(M) = \varepsilon_k \varphi_k(M) \quad (2.87)$$

Any product of two orbitals such as:

$$\varphi_k(M_a)\varphi_{k'}(M_b) \quad (2.88)$$

is a solution of equation (2.86).

To take account of the spin and of the Pauli principle we must transform a product like (2.88) into an antisymmetrized total wave function. As the hydrogen molecule in the Born–Oppenheimer approximation framework is a two-electron problem we can use functions analogous to those obtained in the helium case (equations 2.42 and 2.43). For example, a singlet state will be described by the function:

$$\Phi_T^S = N[\varphi_k(M_a)\varphi_{k'}(M_b) + \varphi_{k'}(M_a)\varphi_k(M_b)][\alpha(\omega_a)\beta(\omega_b) - \beta(\omega_a)\alpha(\omega_b)] \quad (2.89)$$

2.1.3 Improvement of Atomic Orbitals: Variation Methods, Slater Orbitals, Self-Consistent Field Method for Atoms

We saw (equation 2.16) that the independent electron model associates the ground state of the helium atom with the function:

$$\Phi_1 = N^2 e^{-2r_a/a_0} e^{-2r_b/a_0} \tag{2.90}$$

The corresponding energy (−108 eV) is far from the experimental value (−78.6 eV). The function Φ_1 needs some improvement. To improve that function it is necessary to take account (at least partially) of the coulombic interaction between the electrons which has been completely neglected so far. Assume that at a given time the nucleus and the two electrons lie on the same straight line, one of the electrons being between the nucleus and the other electron. The coulombic repulsion between the electrons cancels a part of the attraction of the nucleus which acts on the latter electron. It is said that the former electron introduces a certain *screen effect* on the nucleus with respect to the other electron, an effect which seems to reduce the electric charge of the nucleus. This would suggest that the coulombic repulsion between the electrons could be replaced by a decrease of the nuclear charge.

Finally, we are left to try the function:

$$G_Z = N_Z^2 e^{-Zr_a/a_0} e^{-Zr_b/a_0} \tag{2.91}$$

where Z is a number smaller than 2. A question arises. What is the best value of the parameter Z? This will depend on the purpose of our calculation. If we are interested in energy, we must choose the value of Z which leads to an energy value as near as possible to the experimental value. Such is the spirit of variation methods.

To proceed further in the present case we need to demonstrate the Eckart theorem.[11] Let us consider a Schrödinger equation:

$$H\Psi_i = E_i\Psi_i \tag{2.92}$$

and an arbitrary normalized function G belonging to the same Hilbert space as the Ψ_i's. We shall show that the integral:

$$I = \int G^* HG \, dv \tag{2.93}$$

(called the variational integral) is an upper limit to the lowest eigenvalue E_1.

It is customary to assume that it is possible to build a complete set of orthonormalized eigenfunctions of any linear and hermitian operator like the hamiltonian H. Let $\Psi_1, \Psi_2, \ldots, \Psi_i, \ldots$ be that complete set.

As the function G belongs to the same Hilbert space it can be expanded in terms of the functions Ψ_j:

$$G = \sum_j C_j \Psi_j \tag{2.94}$$

Introducing this expansion into equation (2.93) it is found that:

$$I = \sum_{j} \sum_{j'} C_j^* C_{j'} \int \Psi_j^* H \Psi_{j'} \, \mathrm{d}v \tag{2.95}$$

If we now take account of equation (2.92) we find:

$$I = \sum_{j} \sum_{j'} C_j^* C_{j'} E_j \int \Psi_j^* \Psi_{j'} \, \mathrm{d}v \tag{2.96}$$

However, as the functions Ψ_i constitute an orthonormal set:

$$\int \Psi_j^* \Psi_{j'} \, \mathrm{d}v = \delta_{jj'} \tag{2.97}$$

where $\delta_{jj'}$ is the Kronecker symbol, that is to say unity when $j = j'$ and zero when $j \neq j'$. Finally, equation (2.96) reduces to:

$$I = \sum_{j} (C_j)^2 E_j \tag{2.98}$$

As G is a normalized function:

$$\sum_{j} C_j^2 = 1 \tag{2.99}$$

Therefore:

$$E_1 = \sum_{j} C_j^2 E_1 \tag{2.100}$$

Subtracting E_1 from both sides of equation (2.98) gives:

$$I - E_1 = \sum_{j} |C_j|^2 (E_j - E_1) \tag{2.101}$$

Since E_j is greater than or equal to E_1 for all values of j and the coefficients C_j^2 are of course all positive or zero, the right-hand side of equation (2.101) is positive or zero. Therefore:

$$I \geqslant E_1 \tag{2.102}$$

We have proved that I is always an upper limit to the ground state energy E_1.

This theorem is the basis of the variation method for the calculation of approximate values of the lowest energy level of a system. If we choose a number of variation functions G_a, G_b, G_c, . . . and calculate the value I_a, I_b, I_c, . . . corresponding to them, then each of these values of I is greater than the energy E_1 so that the lowest value is the nearest to E_1, that is to say the best.

Going back to the function G_Z (equation 2.91) associated with the ground state of helium, we are led to calculate the variation integral:

$$I_Z = \int G_Z^* H G_Z \, \mathrm{d}v \tag{2.103}$$

where H is the exact hamiltonian (including the coulombic repulsion between the electrons), and to search for the value of the parameter Z which corresponds to the smallest possible value of I_Z. The explicit expression of I_Z is:

$$I_Z = \int N_Z^2 e^{-Zr_a/a_0} e^{-Zr_b/a_0} \left(-\frac{h^2}{8\pi^2 m} \Delta_a - \frac{h^2}{8\pi^2 m} \Delta_b + \frac{e^2}{r_{ab}} - \frac{2e^2}{r_a} - \frac{2e^2}{r_b} \right) \times N_Z^2 e^{-Zr_a/a_0} e^{-Zr_b/a_0} \, dv_a \, dv_b \quad (2.104)$$

Simple calculation leads to:

$$I_Z = \frac{\pi^2 m e^4}{h^2} [4Z^2 + (\tfrac{5}{2} - 16)Z] \quad (2.105)$$

The minimum value of I_Z is obtained when:

$$\frac{\partial I_Z}{\partial Z} = 0 \quad (2.106)$$

that is to say:

$$Z = 2 - \tfrac{5}{16} \quad (2.107)$$

For this value of the parameter Z the variational integral I_Z takes the value -77 eV. It is a satisfactory estimation of the ground state energy of the helium atom. The difference between that value and the experimental one is only 1.6 eV. It turns out that the introduction of the parameter Z into the atomic orbital permits a very significant improvement of the wave function. It can be said that the function:

$$e^{-Zr/a_0}$$

with $Z = 2 - \frac{5}{16}$ is an *improved atomic orbital* for the description of the ground state of the helium atom.

The foregoing procedure may be extended to other atoms but the calculations become tedious. Examples of such calculations can be found in many texts.[12] Slater has described an empirical procedure of introducing the screen effect.[13] We shall not describe that procedure here, as it is now seldom used.

However, we must recall that to simplify the calculations Slater had the idea of replacing the normal hydrogen-like orbitals by some functions χ called Slater orbitals, given by the following expression:

$$\chi = r^{n-1} e^{-\xi r} \times \text{angular part in } \theta \text{ and } \varphi \quad (2.108)$$

In Slater's paper the ζ's were calculated by using empirical rules, but it is also possible to compute the best exponents by following the variation method.

If we take one exponential for each orbital we have what is called a *minimal basis set.* In principle there is no limit to the number of terms we take providing that they are of the correct symmetry. Often a linear combination of two Slater functions χ with slightly different exponents is taken to represent each atomic orbital. Such a basis set is referred to as a *double zeta basis set.* Table 2.1 gives examples of such basis sets (except for the d orbitals which are represented by only one Slater function).

Table 2.1 Typical exponents ζ for carbon, nitrogen and oxygen atoms[14]

Type	C	N	O
$1s\sigma$	5.304	6.213	7.165
$1s\sigma$	8.383	9.268	10.614
$2s\sigma$	1.269	1.468	1.601
$2s\sigma$	1.856	2.242	2.589
$2p\sigma$	1.287	1.528	1.651
$2p\sigma$	2.853	3.337	3.675
$3d\sigma$	1.895	1.935	2.103
$2p\pi$	1.287	1.529	1.651
$2p\pi$	2.836	3.337	3.675
$3d\pi$	1.175	1.429	3.019

Remember that in the case of the helium atom the best value of the variational parameter Z of equation (2.91) leads to an energy of 77 eV for the ground state, the experimental value being 78.6 eV. Therefore we must again improve the wave function to reduce the remaining gap. We might think that perhaps the exponential shape which has been chosen for the orbitals is not the best and we could then consider a function like:

$$G = \varphi_1(\mathrm{M}_a)\varphi_2(\mathrm{M}_b) \tag{2.109}$$

where the orbitals φ_1 and φ_2 are now completely unknown functions. We can therefore consider these functions to be variational ones. To calculate the best of these functions it is useful to start from the following expression:

$$H\Psi = E\Psi \Leftrightarrow \delta \int \Psi^*(H-E)\Psi \,\mathrm{d}v = 0 \tag{2.110}$$

To demonstrate this relation we must obtain another form of its second member. Let us consider a variation $\delta\Psi$ of the wave function. The integral $J = \int \Psi^*(H-E)\Psi \,\mathrm{d}v$ becomes:

$$\begin{aligned} J + \Delta J &= \int (\Psi^* + \delta\Psi^*)H - E(\Psi + \delta\Psi)\,\mathrm{d}v \\ &= J + \int \Psi^*(H-E)\,\delta\Psi\,\mathrm{d}v + \int \delta\Psi^*(H-E)\Psi\,\mathrm{d}v \\ &\quad + \int \delta\Psi^*(H-E)\,\delta\Psi\,\mathrm{d}v \end{aligned} \tag{2.111}$$

Therefore the first variation δJ is:

$$\delta J = \delta \int \Psi^*(H-E)\Psi \,\mathrm{d}v$$

$$= \int \Psi^*(H-E)\,\delta\Psi \,\mathrm{d}v + \int \delta\Psi^*(H-E)\Psi \,\mathrm{d}v \tag{2.112}$$

Since $H-E$ is an hermitian operator the last two terms in (2.112) are complex conjugates. Therefore their sum is simply twice the real part of one of them:

$$\delta J = 2R \int \delta\Psi^*(H-E)\Psi \,\mathrm{d}v \tag{2.113}$$

Now if $(H-E)\Psi = 0$ everywhere, equation (2.113) shows that δJ is zero. The direct implication $\Rightarrow$ of equation (2.110) is proved. To prove the reciprocal implication $\Leftarrow$ let us assume that δJ is zero but that $(H-E)\Psi$ does not vanish at point M and is, for example, positive. For continuity reasons there exists a small region D surrounding M in which $(H-E)\Psi$ remains positive. We can write:

$$\delta J = 2R \int_D \delta\Psi^*(H-E)\Psi \,\mathrm{d}v + 2R \int_{\mathbb{R}^{3n}D} \delta\Psi^*(H-E)\Psi \,\mathrm{d}v \tag{2.114}$$

As $\delta\Psi^*$ is an arbitrary variation of Ψ we can choose a variation such that:

$$\forall M \in D, \qquad \delta\Psi^* > 0 \qquad \text{and real}$$

$$\forall M \notin D, \qquad \delta\Psi^* = 0$$

Therefore the first term of the second member of equation (2.114) is positive and the second term is zero. It follows that δJ is positive, which is inconsistent with our hypothesis. Therefore if $\delta J = 0$, $(H-E)\Psi$ vanishes everywhere. The expression (2.110) is completely demonstrated.

Therefore for a rigorous solution of the Schrödinger equation:

$$\delta \int \Psi^*(H-E)\Psi \,\mathrm{d}v = 0 \tag{2.115}$$

It is tempting to select approximate functions in such a way that they also satisfy equation (2.112). We are led to calculate the orbitals φ_1 and φ_2 of equation (2.109) taking account of this requirement. At first sight it could be thought that we would obtain the exact solution. This is not true, because expression (2.110) is only valid when the variation $\delta\Psi$ is chosen completely arbitrarily. This is not the case here as the function must remain expressed as a product of two orbitals.

Let us see what happens when condition (2.115) is applied to function (2.109). To simplify the discussion we shall assume that the orbitals remain real functions. Equation (2.115) can be written as:

$$\int \delta\Psi(H-E)\Psi \,\mathrm{d}v = 0 \tag{2.116}$$

Consider a variation $\delta\varphi_2$ of the function φ_2; the corresponding variation of G is:

$$\delta G = \delta\varphi_2\varphi_1 \tag{2.117}$$

and equation (2.116) leads to:

$$\int \delta\varphi_2\varphi_1(H-E)\varphi_1\varphi_2 \,\mathrm{d}v = 0 \tag{2.118}$$

If we now introduce the symbol h_i to represent the operator:

$$h_i = -\frac{h^2}{8\pi^2 m}\Delta_i - \frac{2e^2}{r_i} \tag{2.119}$$

the helium hamiltonian becomes:

$$H = h_a + h_b + \frac{e^2}{r_{ab}} \tag{2.120}$$

and equation (2.119) can be written as:

$$\int \delta\varphi_2\left[\int \varphi_1\left(h_a + h_b + \frac{e^2}{r_{ab}} - E\right)\varphi_1\varphi_2 \,\mathrm{d}v_a\right]\mathrm{d}v_b = 0 \tag{2.121}$$

As $\delta\varphi_2$ is an arbitrary variation it is readily seen that the expression between the brackets must vanish. Writing:

$$e_1 = \int \varphi_1(\mathrm{M}_a)h_a\varphi_1(\mathrm{M}_a)\,\mathrm{d}v_a$$

equation (2.121) reduces to:

$$\left[h_b + \int \varphi_1^2(\mathrm{M}_a)\frac{e^2}{r_{ab}}\,\mathrm{d}v_a\right]\varphi_2(\mathrm{M}_b) = (E-e_1)\varphi_2(\mathrm{M}_b)$$
$$= \varepsilon_2\varphi_2(\mathrm{M}_b) \tag{2.122}$$

if the notation $\varepsilon_2 = E - e_1$ is introduced.

Equation (2.122) is *the Hartree equation*[15] and the operator:

$$h_b^{\mathrm{SCF}} = h_b + \int \varphi_1^2(\mathrm{M}_a)\frac{e^2}{r_{ab}}\,\mathrm{d}v_a \tag{2.123}$$

is a self-consistent field operator.

Starting from a variation $\delta\varphi_1$ it can readily be seen that:

$$h_{\mathrm{a}}^{\mathrm{SCF}}\varphi_1(\mathrm{M}_a) = \left[h_a + \int \varphi_2^2(\mathrm{M}_b)\frac{e^2}{r_{ab}}\,\mathrm{d}v_b\right]\varphi_1(\mathrm{M}_a) = \varepsilon_1\varphi_1(\mathrm{M}_a) \tag{2.124}$$

Equations (2.122) and (2.124) must be solved simultaneously by using an iterative process. Arbitrary functions φ_1^0 and φ_2^0 are chosen as zero-order approximations: hydrogen-like atomic orbitals can be such functions.

For these functions approximate self-consistent field operators $h_a^{0,\mathrm{SCF}}$ and $h_b^{0,\mathrm{SCF}}$ are calculated. Solving the equations:

$$h_a^{0,\mathrm{SCF}}\varphi_1^{(1)} = \varepsilon_1^{(1)}\varphi_1^{(1)} \qquad \text{and} \qquad h_b^{0,\mathrm{SCF}}\varphi_2^{(1)} = \varepsilon_2^{(1)}\varphi_1^{(1)}$$

permits first-order approximate functions $\varphi_1^{(1)}$ and $\varphi_2^{(1)}$ to be obtained which produce new self-consitent field operators. The procedure is continued until the difference between the $\varphi^{(n)}$'s and the $\varphi^{(n+1)}$'s becomes sufficiently small.

To represent the helium atom in its ground state it is usual to use the same orbital φ_1 twice. The function G becomes:

$$G = \varphi_1(\mathrm{M}_a)\varphi_1(\mathrm{M}_b) \tag{2.125}$$

If the foregoing iterative procedure is used the final value obtained for the variational integral:

$$I = \int G^*HG\,\mathrm{d}v \tag{2.126}$$

is 77.6 eV. The remaining gap between that value and the experimental one is only 1 eV. It is easy to show that this is the smallest possible gap for an approximate wave function expressed as a simple product of two orbitals. *The exact wave function is not such a product. The difference between the Hartree energy and the experimental one is often called the correlation energy.*

The function G given in equation (2.125) is symmetrical with respect to a permutation of the two points M_a and M_b. Therefore it is convenient to represent a physical state. The Hartree function is a good space function when the same orbital is used twice.

When two different orbitals φ_1 and φ_2 are used their product is not symmetrical nor antisymmetrical. We must multiply the product $\varphi_1\varphi_2$ by a spin function and introduce the effect of an antisymmetrizer (see Section 2.1.1).

If, for example, the spin function $\alpha(\omega_a)\alpha(\omega_b)$ is chosen the function:

$$G = \sum_P (-1)^P P\varphi_1(\mathrm{M}_a)\varphi_2(\mathrm{M}_b)\alpha(\omega_a)\alpha(\omega_b) \tag{2.127}$$

is obtained and the variational procedure must be applied to this function. When the self-consistent field approach is used with such a Slater determinant it is called the *Hartree–Fock method.*

Consider the case of a state of an atom containing an even number of electrons and represented with a Slater deerminant in which each orbital is used twice. (The function is called a *closed-shell* function.) The variational procedure enables one to reach the best Slater determinant. It is easy to show that the best *real* orbitals are solutions of the equations:

$$h^{\mathrm{SCF}}\varphi_k = \left[h + 2\sum_j (2J_j - K_j)\right]\varphi_k = \varepsilon_k\varphi_k \tag{2.128}$$

These are *the Hartree–Fock equations.* We have used coulomb and exchange operators which are defined as:

$$J_i\varphi_k(\mathrm{M}_a)=\left[\int\varphi_j^2(\mathrm{M}_b)\frac{e^2}{r_{ab}}\,\mathrm{d}v_b\right]\varphi_k(\mathrm{M}_a) \tag{2.129}$$

$$K_j\varphi_k(\mathrm{M}_a)=\left[\int\varphi_j(\mathrm{M}_b)\varphi_k(\mathrm{M}_b)\frac{e^2}{r_{ab}}\,\mathrm{d}v_b\right]\varphi_j(\mathrm{M}_a) \tag{2.130}$$

Since the operator h^{SCF} contains the answer we are seeking, the set of equations, one for each φ_j, has to be solved iteratively.

The corresponding approximate energy of the state is:

$$I=2\sum_k \varepsilon_k^N+\sum_{k<j}(2J_{kj}-K_{kj}) \tag{2.131}$$

if:

$$\varepsilon_k^N=\int\varphi_k(\mathrm{M}_a)h_a\varphi_k(\mathrm{M}_a)\,\mathrm{d}v_a \tag{2.132}$$

$$J_{kj}=\int\varphi_k^2(\mathrm{M}_b)\frac{e^2}{r_{ab}}\varphi_j^2(\mathrm{M}_a)\,\mathrm{d}v_b\,\mathrm{d}v_a \tag{2.133}$$

$$K_{kj}=\int\varphi_k(\mathrm{M}_b)\varphi_j(\mathrm{M}_b)\frac{e^2}{r_{ab}}\varphi_k(\mathrm{M}_a)\varphi_j(\mathrm{M}_a)\,\mathrm{d}v_b\,\mathrm{d}v_a \tag{2.134}$$

The name of *correlation energy* has been given to the difference between this energy and the experimental energy. *It must be pointed out that it is dangerous to use such a language* as the main part of the correlation between the position of the electrons is introduced by the antisymmetrizer in the Hartree–Fock wave functions. We have shown this fact in this section. *Therefore the correlation energy defined above is only a small part of the total correlation energy.*

To end this section let us remark that there is no need to use the same orbital twice to build an approximate wave function for the ground state of helium. We can also start from a Slater determinant built on two different orbitals φ_1 and φ_2. This is the *unrestricted Hartree–Fock* procedure. In fact this procedure is not convenient from the spin viewpoint as the wave function which is obtained is not necessarily an eigenfunction of the spin operator $\mathscr{S}^2$. It is better to project the determinant on the symmetry subspace corresponding to the state of interest. If the projection is done after the iterative procedure is achieved the technique is known as the *projected unrestricted Hartree–Fock method.* It is known as the *extended Hartree–Fock method*(16) when the projection is made before the variation. An application of the extended Hartree–Fock method to the ground state of helium has produced 90 per cent. of the 'correlation energy'.(17) Recently a simplification has been proposed in order to obtain an approximate projected unrestricted Hartree–Fock wave function.[18]

2.1.4 LCAO Approximation, Improvement of Molecular Orbitals, Self-Consistent Field Method for Molecules

We saw (Section 2.1.2) that it is easy to extend the formalism of the independent electron model to molecules. Therefore it could be anticipated that the procedure described to improve atomic orbitals can also be used to improve molecular orbitals. In principle this is so, but in practice it is not tractable. Before improving molecular orbitals it is necessary to introduce a new kind of approximation: the LCAO (linear combination of atomic orbitals) approximation.

Consider the electronic equation associated with the hydrogen molecule ion (equation 2.65):

$$\left(-\frac{h^2\Delta}{8\pi^2 m}+\frac{e^2}{r_{AB}}-\frac{e^2}{r_A}-\frac{e^2}{r_B}\right)\phi_j = U_j\phi_j \tag{2.135}$$

We know (Section 1.2.2) that equation (2.135) has nearly been solved rigorously. However, to solve such an equation is a tedious job. For larger molecules it becomes practically impossible so an approximation is necessary. Let us consider a point M in the vicinity of the nucleus A. At that point the potential $-e^2/r_A$ is much more important that the potential $-e^2/r_B$. If the latter potential is neglected the solutions of equation (2.135) are simply the wave functions Ψ_A of the hydrogen atom centred at point A. Similarly near B the functions ϕ_j must be analogous to the wave functions Ψ_B of a hydrogen atom centred at point B. Therefore, it is tempting to try an approximate function like:

$$\varphi = a\Psi_A + b\Psi_B \tag{2.136}$$

where a and b are simple coefficients.

This is the LCAO approximation. If the ground state is concerned the functions Ks will be introduced:

$$\begin{aligned}\varphi &= aKs_A + bKs_B \\ &= a'\mathrm{e}^{-r_A/a_0} + b'\mathrm{e}^{-r_B/a_0}\end{aligned} \tag{2.137}$$

This function leads to an equilibrium distance r_e of 1.32 Å (instead of 1.06) and corresponds to an electronic dissociation energy of 1.77 eV, too small by approximately 1 eV, showing that the chemical bond is not sufficiently well represented.[19]

As in the case of atoms it is useful to introduce variational exponents in the atomic orbitals. The ground state function φ becomes:

$$\varphi = a''\mathrm{e}^{-\zeta r_A} + b''\mathrm{e}^{-\zeta r_B} \tag{2.138}$$

and the exponent ζ is calculated by minimizing the variational integral:

$$I = \int \varphi^* H\varphi \, \mathrm{d}v$$

The corresponding value of r_e is 1.06 Å (the experimental value) and the electronic dissociation energy reaches 2.25 eV.[20] A further improvement is obtained if the basis of the atomic orbitals is extended by introducing, for example, L orbitals as in:

$$\varphi = a'''Ks_A + b'''Lp_{z_A} + c'''Ks_B + d'''Lp_{z_B} \tag{2.139}$$

The dissociation energy which corresponds to the best function of this kind is 2.73 eV,[21] which is very near the experimental result.

If a polyelectronic molecule is concerned one must combine the formalism of the independent electron model and LCAO approximation. The molecular wave function Φ will be approximated by a linear combination of Slater determinants built on molecular spin orbitals φ, and the molecular orbital will be extended in terms of atomic orbitals χ.

If, in this framework, a closed-shell state is considered and the self-consistent field procedure is used, *Roothan*[22] *equations* are obtained. Various rigorous derivations have been given.[23] A non-rigorous but readily comprehensible derivation is as follows. To simplify the calculations the orbitals are taken to be *real*. Such a derivation assumes that the Hartree–Fock equations (2.128) are still valid:

$$h^{SCF}\varphi_k = \varepsilon_k\varphi_k \tag{2.140}$$

If the extension of the molecular orbitals in terms of atomic orbitals are taken into account:

$$\varphi_k = \sum_l C_{kl}\chi_l \tag{2.141}$$

The following equation is readily obtained:

$$h^{SCF}\sum_l C_{kl}\chi_l = \varepsilon_k \sum_l C_{kl}\chi_l \tag{2.142}$$

If both sides of this equation are multiplied by, say, χ_m and integrated over all space:

$$\sum_l C_{kl}\int \chi_m h^{SCF}\chi_l \, dv = \varepsilon_k \sum_l C_{kl}\int \chi_m\chi_l \, dv \tag{2.143}$$

Putting:

$$h_{ml}^{SCF} = \int \chi_m h^{SCF}\chi_l \, dv \qquad \text{and} \qquad S_{ml} = \int \chi_m\chi_l \, dv \tag{2.144}$$

equation (2.143) becomes:

$$\sum_l C_{kl}(h_{ml}^{SCF} - \varepsilon_k S_{ml}) = 0 \qquad \text{for } m = 1, \ldots, n \tag{2.145}$$

if the basis contains n different atomic orbitals.

The system of equation (2.145) is called the *secular system*. As it is a system of linear and homogeneous equations with respect to the coefficient C_{kl} it has a non-trivial solution only if the determinant built on the terms

$|h_{ml}^{\mathrm{SCF}} - \varepsilon_k S_{ml}|$ vanishes. Then:

$$\det |h_{ml}^{\mathrm{SCF}} - \varepsilon_k S_{ml}| = 0 \tag{2.146}$$

which is the *secular equation*. The considered determinant contains n lines and n columns. As a consequence equation (2.146) is an algebraic equation of nth degree in ε_k. Solving equation (2.146) furnishes n roots which are possible values of ε_k.

If a given value of ε_k, obtained by solving the secular equation, is now introduced into the secular system and this system is solved, a set of coefficients C_{kl} is obtained and therefore the explicit expression of the corresponding molecular orbital:

$$\varphi_k = \sum_l C_{kl}\chi_l$$

If we look at the explicit expression for h^{SCF} (equation 2.128) we see that to solve the secular equation we need the explicit expression of the molecular orbitals, that is to say the C_{kl}'s.

We seem to be on a vicious circle. To go out from it, as in the case of atoms, we must use an iterative procedure. Let us take the simple example of the LiH molecule. It contains four electrons. Therefore to represent the ground state we can use only two molecular orbitals which produce four molecular spin orbitals. The approximate electronic wave function can be written as:

$$\Phi = \det \varphi_1(\mathrm{M}_a)\overline{\varphi_1(\mathrm{M}_b)}\varphi_2(\mathrm{M}_c)\overline{\varphi_2(\mathrm{M}_d)} \tag{2.147}$$

The complete Hartree–Fock calculation would be achieved by solving the variational equation:

$$\delta \int \Phi^* \, |H - U| \Phi \, \mathrm{d}v = 0 \tag{2.148}$$

Solution of this equation would give the *best molecular orbitals* φ. As solution of equation (2.148) is too lengthy, we must introduce the LCAO approximation. When using a minimal basis the molecular orbital can be extended as follows:

$$\varphi_k = C_{k1}Ks_{\mathrm{H}} + C_{k2}Ks_{\mathrm{Li}} + C_{k3}Ls_{\mathrm{Li}} \tag{2.149}$$

where Ks_{H} denotes an improved atomic orbital associated with the K shell of the hydrogen atom and Ks_{Li} and Ls_{Li} denote improved atomic orbitals associated respectively with the K and L shells of the lithium atom. All atomic orbitals are in this case simple functions

$$\chi_1 = N_1 \mathrm{e}^{-\zeta_1 r} \tag{2.150}$$

the N_1's being normalization factors and the ζ_1's convenient exponents (see Table 2.1).

In outline the procedure might be:

1. Guess some values of the C_{kl}'s. In so doing we have to take account of the orthonormalization condition:

$$\int \varphi_k \varphi_{k'} \, dv = \delta_{kk'} \tag{2.151}$$

The C's must satisfy the requirements:

$$\begin{aligned} \sum_l \sum_{l'} C_{kl} C_{kl'} S_{ll'} &= 1 \\ \sum_l \sum_{l'} C_{1l} C_{2l'} S_{ll'} &= 0 \end{aligned} \tag{2.152}$$

2. Compute all the various atomic integrals and hence build up the matrix elements:

$$h_{ml}^{\mathrm{SCF}} \quad \text{and} \quad S_{ml}$$

3. Solve the corresponding secular *equation* giving three possible values of ε_k.
4. Substitute the two smallest ones in the secular *system* giving new C's.
5. Go back to stage 1 and repeat *until* the values of the ε_k's converge to steady values within a certain threshold and take the values of the converged C's.
6. Calculate the value of the variational integral:

$$U_1 = \int \Phi H \Phi \, dv \tag{2.153}$$

which represents the approximate value of the total electronic energy of the molecule.

We must point out that when a self-consistent field procedure is used the total energy U is no longer the sum of the orbital energies ε_k:

$$U_1 \neq 2\varepsilon_1 + 2\varepsilon_2 \tag{2.154}$$

The foregoing relation is valid only in the framework of the independent electron model.

Obviously the obtained values of U_1 depend on the positions of the fixed charges replacing the nuclei. Repeating the whole process for any position of them, a curve can be obtained showing the value of the equilibrium internuclear distance r_e and therefore the nuclear conformation of the molecule in its ground state.

The procedure can be extended to open-shell states, and many molecules from H_2 to molecules like pyridine, naphthalene, adenine, guanine, thymine (and so on) have been treated with the self-consistent field method. We cannot describe all the results. Some excellent books have been published on the subject.[(24)]

We must add some comments. First of all, it is obvious that the iterative procedure just described is very tedious and very lengthy—the sort of problem to which computers are so well adapted. The Quantum Chemistry Program Exchange (Indiana University, Bloomington, Indiana, U.S.A.) provides quantum chemists with various molecular self-consistent field programs for electronic computers.

In principle we are not obliged to use a minimal basis set of atomic orbitals. The larger the basis, the better the results. To describe the ground state of the LiH molecule we can introduce L_H and M_{Li} orbitals, for example; an *extended basis set* is obtained. In practice the size of the basis set of atomic orbitals is limited by the price of the calculation.

In addition, it is not obligatory to use Slater orbitals. Alternative orbitals have been proposed. The gaussian functions of the form $x^l y^m z^n e^{-\alpha r^2}$ (l, m, n are integers) suggested by Boys[25] have proved to be very useful. Different self-consistent field programs computing all integrals between gaussian orbitals and solving the corresponding Roothaan equations have been written. The main disadvantage of the gaussian functions is that they do not follow the form of an atomic orbital closely. Therefore several gaussian functions must be used to replace a Slater orbital. However, as the calculation of integrals between gaussian functions is very easy it is often advantageous to use such functions.

The main problem is how to choose the exponents α. Huzinaga[26] has discussed this point. Let us consider the carbon atom as an example. Minimizing the energy of the carbon atom in the ^3P ground state Huzinaga found that, say, the $2s$ orbital can be fitted with the linear combination of gaussian functions given in Table 2.2. However, the best atomic orbitals to compute a molecular function are not necessarily the same as those which are the best for the corresponding atoms. It would be necessary rigorously to optimize the exponents and the coefficients for the molecules and for any internuclear distances. However, that would be too expensive. Clementi and Davis[27] suggested a compromise.

Table 2.2 Huzinaga exponents and coefficients for $2s$ orbital

Function	Exponent	Coefficients
1	9470	−0.0001
2	1397	−0.00076
3	307.5	−0.00418
4	85.54	−0.01701
5	26.91	−0.05399
6	9.41	−0.12134
7	3.50	−0.17554
8	1.068	0.08502
9	0.400	0.60689
10	0.135	0.43809

They considered that except for a small region near the nucleus the 2*s* function is well represented by the functions β_8, β_9 and β_{10}. They used the two functions:

$$\gamma = 0.08502\beta_8 + 0.60689\beta_9$$

and

$$\delta = \beta_{10}$$

instead of the ten β functions. An analogous procedure is used to represent the other atomic orbitals involved. When the gaussian functions are replaced by fixed linear combinations of them (like γ) it is said that a *contracted basis set* is used.

A slightly different procedure using gaussian functions is to replace the Slater orbitals by least-square fitted combinations of gaussian-type functions.[28] If an extended gaussian function basis is used the nuclear conformation calculated for a small molecule agrees very well with experimental results.

For water, Neumann and Moskowitz[29] found an HOH angle of 105° (experimental value 104°30′) and an OH interatomic distance of 1.8 a.u. (experimental value 1.81). The calculated corresponding electronic energy of the molecule is −76.0596 a.u. (experimental value −76.481). These are typical results. The gap between theoretical and experimental energy is usually less than 1 per cent., even when the minimal basis is used.

For larger molecules it becomes very tedious to calculate the nuclear conformation. When known, the experimental geometry must be used. If the geometry is unknown it is necessary to guess it.

When a wave function is obtained any observable property can be computed. It is only necessary to calculate the integral:

$$\int \Psi^* A_{op} \Psi \, dv$$

where Ψ denotes the wave function and A_{op} the operator associated with the considered property.

For water, the charge density,[30] magnetic properties,[31] polarizabilities,[32] quadrupole coupling,[31] force constants,[31] proton affinity,[33] transition electronic energies,[34] oscillator strengths[34] and ionization potential[35] have been computed. The agreement between theoretical values and experimental data is satisfactory but the precision of the calculations depends very much on the nature of the computed property.

Let us take, as an example, the case of dipole moment, which is a very sensitive test of the wave function. Table 2.3 allows comparison of the computed values and experimental results.

To enable some particular properties to be calculated simplified procedures have been proposed. The normal method used to calculate an ionization energy is to compute the energy of the molecule, the energy of the

Table 2.3 Theoretical and experimental dipole moments (D.)

Molecules	Calculated	Experimental
NH_3	1.81[36]	1.47[37]
BH_3NH_3	5.79[36]	4.9
CNH	2.11[38]	2.95[37]

corresponding ion and to take the difference. However, it can be shown (Koopmans theorem[39]) that the energies ε_k associated with the SCF orbitals of a given state of a molecule are approximations to ionization energies of that state. Table 2.4 gives a comparison of the orbital energies and ionization potentials found from photoelectron spectroscopy.

Table 2.4 Photoelectron ionization potentials and orbital energies

Molecule	Orbital and energy (eV)		Ionization potentials (eV)
H_2	$1\sigma_g$	16.18	15.88
HF	1π	17.69	16.06
CO	4σ	21.87	19.72
	5σ	15.09	14.01
N_2	$3\sigma_g$	17.28	16.96
O_2	$3\sigma_g$	20.02	20.12

Obviously the agreement is satisfactory. The differences between the experimental and theoretical results have, at least, two origins. First, the best orbitals associated with an ion are not the same as those of the molecule. Furthermore, the correlation energy for an ion is not the same as that of the molecule. A detailed discussion of the dangers of applying Koopmans theorem indiscriminately has been given by Richards and Horsley.[40]

An analogous procedure (based on the concept of virtual orbitals) has been proposed to calculate excitation energies.

2.1.5 Configuration Interaction

Consider an electronic system (molecule or atom) represented by a model in which the nuclei are replaced by fixed electric charges. With the corresponding Schrödinger equation:

$$H\Psi = E\Psi \tag{2.155}$$

the independent electron model associates the equation:

$$H^0\Phi = E\Phi \tag{2.156}$$

Products of atomic or molecular orbitals are solutions of equation (2.156) and Slater determinants or linear combinations of such determinants must be introduced to take account of the spin and of the Pauli principle. A given choice of orbitals is called a *configuration* and the various convenient linear combinations of Slater determinants produced by such a choice give approximations of wave functions associated with molecular or atomic states 'belonging to the considered configuration'.

We can improve the orbitals and have learned how to do so. However, even when using the best orbitals a small gap remains between the calculated energy and the experimental one—the so-called correlation energy (practically of the order of 1 per cent.). Although this gap is small if we are only interested in the total electronic energy, it has an important effect on the calculation of properties like atomization and isomerization energies. In some cases, the self-consistent field approximation leads to wrong results, even for qualitative predictions. For F_2, one of the earliest extended basis calculations[41] showed the Hartree–Fock energy of F_2 to lie more than 1 eV *above* the energy of two F atoms, leading to the wrong conclusion that the F_2 molecule is unstable.

More generally a gap can introduce a significant error in the calculation of all non-stationary properties because even if the approximate wave function is satisfactory from the energetic viewpoint it must be far from the exact wave function in the Hilbert space.

A new improvement is needed. Many different ways have been explored, one being to assume that the infinite set of products of orbitals which satisfy equation (2.156) forms a basis in the Hilbert space. In this way, an exact solution Ψ can be expanded in terms of such products:

$$\Psi_i = \sum_{j=1}^{j=\infty} C_{ij}\Phi_j \tag{2.157}$$

If the spin is taken into account the exact wave function including the spin will be expanded in terms of linear combinations of Slater determinants.

In practice it is not possible to introduce an infinite set of configurations. The basis is truncated and therefore only approximate wave functions can be reached. *This is the configuration interaction (CI) method.*

Let us take helium as an example and assume that the two first configurations $(1s)^2$ and $(1s)(2s)$ are introduced. If singlet states are considered the approximate wave functions are written as:

$$\begin{aligned} G_i = {} & C_{i1}\sum_P (-1)^P P 1s1s[\alpha(\omega_a)\beta(\omega_b)-\beta(\omega_a)\alpha(\omega_b)] \\ & + C_{i2}\sum_P (-1)^P P 1s2s[\alpha(\omega_a)\beta(\omega_b)-\beta(\omega_a)\alpha(\omega_b)] \end{aligned} \tag{2.158}$$

From this equation we are able to calculate the 'best values' of C_{i1} and C_{i2}. This is a variational problem.

It is useful to solve this problem in a more general framework. Let us consider a variation function G_1, which is the sum of a number n of linear independent functions $\chi_1, \chi_2, \ldots, \chi_n$ with undetermined coefficients $C_1, \ldots, C_n$:

$$G = \sum_{j=1}^{j=n} C_j \chi_j \tag{2.159}$$

The variational integral I becomes:

$$I = \frac{\int G^* H G \, \mathrm{d}v}{\int G^* G \, \mathrm{d}v} = \frac{\sum_{j,j'} C_j^* C_{j'} \int \chi_j^* H \chi_{j'} \, \mathrm{d}v}{\sum_{j,j'} C_j^* C_{j'} \int \chi_j^* \chi_{j'} \, \mathrm{d}v} \tag{2.160}$$

Putting:

$$H_{jj'} = \int \chi_j^* H \chi_{j'} \, \mathrm{d}v \qquad S_{jj'} = \int \chi_j^* \chi_{j'} \, \mathrm{d}v \tag{2.161}$$

it can readily be seen that:

$$I \sum_{j,j'} C_j C_{j'} S_{jj'} = \sum_{j,j'} C_j C_{j'} H_{jj'} \tag{2.162}$$

where to simplify the formalism we assume that the C_j's are real. To find the values of the C_j's which make I a minimum we differentiate with respect to each C_j:

$$\frac{\partial I}{\partial C_{j'}} + 2I \sum_j C_j S_{jj'} = 2 \sum_j C_j H_{jj'}, \qquad j' = 1, \ldots, n \tag{2.163}$$

The condition for a minimum is that:

$$\frac{\partial I}{\partial C_{j'}} = 0 \qquad \text{for } j' = 1, 2, \ldots, n, \ldots \tag{2.164}$$

Therefore equation (2.163) reduces to:

$$\sum_j C_j (I S_{jj'} - H_{jj'}) = 0, \qquad j' = 1, \ldots, n \tag{2.165}$$

This is a set of n simultaneous homogeneous linear equations in the n independent variables C_j. Such a set of equations is called a *secular system*. Non-trivial solutions of the system are obtained only if the determinant of the coefficients vanishes:

$$\det (I S_{jj'} - H_{jj'}) = \begin{vmatrix} IS_{11} - H_{11} & IS_{12} - H_{12} \cdots IS_{1n} - H_{1n} \\ IS_{21} - H_{21} & IS_{22} - H_{22} \cdots IS_{2n} - H_{2n} \\ \vdots & \vdots \qquad\qquad \vdots \\ IS_{n1} - H_{n1} & IS_{n2} - H_{n2} \cdots IS_{nn} - H_{nn} \end{vmatrix} = 0 \tag{2.166}$$

This is the secular equation. It is an algebraic equation which has n roots:

$$I_1, I_2, \ldots, I_n$$

Substitution of the smallest root I_1 in the secular system and solution of that system leads to a set of coefficients:

$$C_{11}, C_{12}, \ldots, C_{1n}$$

which correspond to the smallest possible value of I, and therefore (following Eckart's theorem) the value of I which gives the best approximation of the ground state energy E_1.

The function:

$$G_1 = \sum_j C_{1j}\chi_j$$

can thus be considered as an approximation of the function Ψ_I. Furthermore, McDonald[42] has demonstrated that the other roots $I_2, I_3, \ldots, I_n$ are upper limits to $E_2, E_3, \ldots, E_n$ respectively. Therefore each I_j is an approximation of an energy level of the molecule or the atom considered while the function G_i obtained by using the corresponding coefficients C_{ij} is an approximation of the wave function Ψ_i.

Obviously the quality of the approximations obtained depends on the number of configurations introduced—the greater the number the better the results. The degree of calculation precision is mainly related to the price of the computation as electronic computers are needed for such tedious calculations.

It is possible to reduce the number of configurations needed to obtain a given precision by improving the configurations (that is to say the orbitals) before variation. Another possibility is to determine simultaneously the best orbitals and the best coefficients of the configurations. This is the powerful *multiconfiguration Hartree–Fock method*,[43] but as the formalism of that method is rather involved it will be not analysed here.

Many technical procedures have been proposed to save time; Schaeffer's book[44] contains many interesting details on that problem. The concept of natural orbitals[45] provides a practical approach to the calculation of CI wave functions. The one-electron density function:

$$\rho = n \int \Psi^*(\mathrm{M}_a, \mathrm{M}_b, \ldots, \mathrm{M}_n)\Psi(\mathrm{M}_a, \mathrm{M}_b, \ldots, \mathrm{M}_n)\, \mathrm{d}v_b \ldots \mathrm{d}v_n \quad (2.167)$$

can be written as:

$$\rho = \sum_k \sum_{k'} a_{kk'}\varphi_k^*\varphi_{k'} \quad (2.168)$$

if the various Φ_j are built on a set of orbitals φ_k. The coefficients $a_{kk'}$ are the so-called first-order *density matrix*. The *natural orbitals* are those which reduce the matrix to diagonal form.

Hagstrom and Shull[46] have shown that the number of important configurations is significantly reduced when natural orbitals are used. Bender and Davidson[47], starting from that position, proposed an *iterative natural orbital method*. A set of configurations is chosen, a CI calculation is carried out and the density matrix is diagonalized to produce the natural orbitals. A new CI calculation is made on the natural orbital basis set and usually a lower energy is obtained. The iteration process is repeated until the energy reaches a minimum.

Another interesting iterative procedure has also been suggested. It is called CIPSI[48] and avoids calculation of natural orbitals. The CI matrix *restricted* to a subspace S of strongly interacting determinants is diagonalized. The eigenvector of this matrix which corresponds to the desired wave function is perturbated by determinants which do not belong to S. The most important of those determinants is added to S and the complete process is started again.

2.2 METHODS ASSUMING ELECTRON LOCALIZATION, EVENT FUNCTIONS, MATHEMATIFICATION OF THE CONCEPT OF BOND

2.2.1 Analysis of an Exact Wave Function in Terms of Loges; Leading Event, Mathematification of the Concept of Bond and Additive Properties of Molecules

We are now familiar with various procedures of calculating accurate wave functions. This makes it possible to give a precise meaning to the *concept of bond.*

Let us consider an electronic *exact* molecular wave function:

$$\Psi(\mathrm{M}_1, \omega_1, \mathrm{M}_2, \omega_2, \ldots, \mathrm{M}_n, \omega_n, t)$$

associated with a given molecule or atom containing n electrons. Consider now any *arbitrary* partition of the three-dimensional space R^3 into p volumes v_i which do not overlap. If $\mathscr{L}$ denotes the ensemble of volumes v_i we can write:

$$\mathscr{L} = \{v_i\} \qquad \text{and} \qquad R^3 = \bigcup v_i$$

$$v_i \cap v_j = \tfrac{1}{2}\, \delta_{ij}(\text{mes } v_i + \text{mes } v_j) \tag{2.169}$$

Consider also an arbitrary partition π of the integer n into p integers:

$$\alpha + \beta + \cdots + \eta + \cdots + \eta' + \cdots + \nu = n \tag{2.170}$$

Finding α electrons in v_{A} and β electrons in v_{B}, ν electrons in v_{P} is a certain physical *event* λ which can be characterized by a certain application of the ensemble $\mathscr{L}$ into the ensemble π. That application can be symbolized as follows:

$$\varepsilon_\lambda = \begin{Bmatrix} v_{\mathrm{A}}, & v_{\mathrm{B}}, & \ldots, & v_{\mathrm{H}}, & \ldots, & v_{\mathrm{K}}, & \ldots, & v_{\mathrm{P}} \\ \alpha, & \beta, & \ldots, & \eta, & \ldots, & \eta', & \ldots, & \nu \end{Bmatrix} \tag{2.171}$$

As a very simple case let us consider the helium atom:

$$n = 2$$

Consider a partition of the space into two volumes v_A and v_B. There are three possible events:

(a) The two electrons are in v_A. The application ε_1 is:

$$\varepsilon_1 = \begin{Bmatrix} v_A, & v_B \\ 2, & 0 \end{Bmatrix}$$

(b) The two electrons are in v_B:

$$\varepsilon_2 = \begin{Bmatrix} v_A, & v_B \\ 0, & 2 \end{Bmatrix}$$

(c) One electron is in v_A, the other in v_B:

$$\varepsilon_3 = \begin{Bmatrix} v_A, & v_B \\ 1, & 1 \end{Bmatrix}$$

The mean value of any property represented by the operator Ω can be written as:

$$\bar{\Omega} = \int_{v_A+v_B+\cdots+v_P} dv_1 \cdots \int_{v_A+v_B+\cdots+v_P} dv_2 \cdots \int_{v_A+v_B+\cdots+v_P} dv_n \Psi^* \Omega \Psi \tag{2.172}$$

The quantity Ω_λ defined as:

$$\bar{\Omega}_\lambda = N_\lambda \int_{v_A} dv_1 \cdots dv_\alpha \cdots \int_{v_B} dv_\xi \cdots dv_{\xi+\eta-1} \cdots \int dv_{n-\nu} \cdots dv_n \Psi^* \Omega \Psi \tag{2.173}$$

when N_λ is the number of equivalent integrals which result from the permutation of the variables and can be considered to be the contribution of the event λ to the mean value $\bar{\Omega}$. It can readily be seen that:

$$\bar{\Omega} = \sum_\lambda \Omega_\lambda \tag{2.174}$$

In particular if $\Omega = 1$ the contribution Ω_λ is the *probability of occurrence p_λ of the event λ*. In that case:

$$\bar{\Omega} = 1 = \sum_\lambda p_\lambda \tag{2.175}$$

Returning to our helium example:

$$\begin{aligned} \bar{\Omega} &= \int_{v_A+v_B} dv_1 \int_{v_A+v_B} dv_2 \Psi^* \Omega \bar{\Psi} \\ &= \int_{v_A} dv_1 \int_{v_A} dv_2 \Psi^* \Omega \bar{\Psi} \\ &\quad + \int_{v_B} dv_1 \int_{v_B} dv_2 \Psi^* \Omega \Psi + 2 \int_{v_A} dv_1 \int_{v_B} dv_2 \Psi^* \Omega \Psi \end{aligned} \tag{2.176}$$

$$\Omega_1 = \int_{v_A} dv_1 \int_{v_A} dv_2 \Psi^* \Omega \Psi \tag{2.177}$$

$$\Omega_2 = \int_{v_B} dv_1 \int_{v_B} dv_2 \Psi^* \Omega \Psi \tag{2.178}$$

$$\Omega_3 = 2 \int_{v_A} dv_1 \int_{v_B} dv_2 \Psi^* \Omega \Psi \tag{2.179}$$

Many of the foregoing formulas become obvious when it is understood that the various events correspond to a partition of the configurational space into volumes which do not overlap.

To proceed further, it is useful to also introduce a partition of the operator Ω. If it is a bielectronic operator:

$$\Omega = \sum_i \Omega_i + \sum_{i<j} \Omega_{ij} \tag{2.180}$$

we shall consider the partition:

$$\Omega = \sum_H \Omega^H + \sum_{H<H'} \Omega^{H,H'} \tag{2.181}$$

where:

$$\Omega^H = \sum_{i \in \eta} \Omega_i \tag{2.182}$$

$$\Omega^{H,H'} = \sum_{i \in \eta < j \in \eta'} \Omega_{ij} \tag{2.183}$$

It must be understood that $i \in \eta$ means that:

$$\xi \leqslant i \leqslant \xi + \eta - 1$$

(see equation 2.173). Putting:

$$\bar{\Omega}_\lambda^H = N' \int_{v_A} dv_1 \cdots dv_\alpha \cdots \int_{v_H} dv_\xi \cdots dv_{\xi+\eta-1} \cdots \times \int dv_{n-\nu} \cdots dv_n \Psi^* (\Omega^H + \Omega^{H,H}) \Psi \tag{2.184}$$

and

$$\bar{\Omega}_\lambda^{H,H\prime} = N'' \int_{v_A} dv_1 \cdots dv_\alpha \cdots \int_{v_H} dv\xi \cdots dv_{\xi+\eta-1} \cdots \times \int dv_{n-\nu} \cdots dv_n \Psi^* \Omega^{H,H'} \Psi \tag{2.185}$$

it can readily be seen that:

$$\bar{\Omega}_\lambda = \sum_H \bar{\Omega}_\lambda^H + \sum_{H<H'} \bar{\Omega}_\lambda^{H,H'} \tag{2.186}$$

Therefore, $\bar{\Omega}_\lambda^H$ can be considered to be the contribution of volume H and event λ to the mean value $\bar{\Omega}$ and $\bar{\Omega}_\lambda^{H,H'}$ the contribution of both volumes H and H' and of event λ to $\bar{\Omega}$.

Combining equations (2.186) and (2.174) we see that:

$$\bar{\Omega} = \sum_{\lambda} \left(\sum_{H} \bar{\Omega}_{\lambda}^{H} + \sum_{H<H'} \bar{\Omega}_{\lambda}^{H,H'} \right)$$
$$= \sum_{H} \left(\sum_{\lambda} \bar{\Omega}_{\lambda}^{H} \right) + \sum_{H<H'} \left(\sum_{\lambda} \bar{\Omega}_{\lambda}^{H,H'} \right) \quad (2.187)$$

This equation provides an excellent basis from which to analyse the origin of the additive properties of molecules. It reduces to

$$\bar{\Omega} = \sum_{H} \left(\sum_{\lambda} \bar{\Omega}_{\lambda}^{H} \right)$$

for monoelectronic operators.

Taking again the helium case and choosing as operator Ω the hamiltonian:

$$H = T_1 - \frac{2e^2}{r_1} + T_2 - \frac{2e^2}{r_2} + \frac{e^2}{r_{12}} \quad (2.188)$$

it is seen that, for example, the contribution of the volume v_A to the total energy associated with the third event is:

$$H_3^{A} = \int_{v_A} dv_1 \int_{v_B} dv_2 \Psi^* \left(T_1 - \frac{2e^2}{r_1} \right) \Psi \quad (2.189)$$

For the same event the contribution simultaneously due to volume A and volume B becomes:

$$H_3^{A,B} = \int_{v_A} dv_1 \int_{v_B} dv_2 \Psi^* \frac{e^2}{r_{12}} \Psi \quad (2.190)$$

If the foregoing formalism is valid for any arbitrary partition of the space, the partitions are not equivalent from the physical viewpoint. In Section 1.2 it was stated that it is interesting to select, if possible, the partition which brings the maximum amount of information about the localizability of the electrons. The missing information function I associated with a set of probabilities p_λ is given by the equation:

$$I = \sum_{\lambda} p_\lambda \log_2 p_\lambda^{-1} \quad (2.191)$$

Furthermore, we have:

$$\sum_{\lambda} p_\lambda = 1 \quad (2.192)$$

If all events have identical probability the missing information takes its largest value:

$$I_M = \log_2 q$$

where q is the number of events. Shannon[49] introduced the *relative missing information* I_r:

$$I_r = \frac{I}{I_M} \quad (2.193)$$

By minimizing the function I_r we can obtain (if it exists) the partition of the space which brings the maximum amount of information about the localizability of the electrons. *The corresponding partition is the best partition into loges of the space associated with the atom or molecule in the considered state.* Therefore we can express the *localization character* L of any partition of the space as:

$$L = 1 - I_r \tag{2.194}$$

The value of L *associated with the best partition into loges is the localizability character of the considered state.*

In general, it is not easy to obtain the best partition into loges as it is not possible to demonstrate its existence. In practice, however, the symmetry of the electronic system suggests a kind of convenient partition, as shown (Section 1.2.3) by solution of the problem for the ground state of helium.

The property of the function I_r is such that when it reaches a minimum, one event has a probability much higher than the probability of any other event. It is the leading event.

We have shown that the various volumes $v_A, v_B, \ldots, v_H, \ldots, v_P$ *corresponding to the best partition into loges correspond to atomic cores, localized or delocalized bonds, lone pairs,* etc. *The best partition into loges produces a mathematification of chemical intuitions.* Equation (2.187) makes it possible to separate any property of a molecule into increments associated with atomic cores, bonds, lone pairs, etc., and (if the property is bielectronic) into increments representing interacting terms associated with the various pairs of foregoing volumes. In order to show the power of that method we shall use it to establish a general relationship between the isomerization energies of saturated hydrocarbons. Let us consider a normal molecule and its isoderivative:

```
       H  H                     H
       |  |                     |
 ···C—C—C—C···            ···C—C—C···
       |  |                     |
       H  H                  H—C—H
                                |
                                H
```

Such molecules contain carbon core loges and CC and CH localized bond loges. If the energy is to be calculated we have to introduce quantities like the contribution of a carbon core:

$$\gamma_C = \sum_\lambda H_\lambda^C \tag{2.195}$$

the contribution of the various bonds:

$$\gamma_{CC} = \sum_\lambda H_\lambda^{CC} \quad \text{and} \quad \gamma_{CH} = \sum_\lambda H_\lambda^{CH} \tag{2.196}$$

and interacting terms like:

$$\gamma_{CC,CH} = \sum_{\lambda} H_{\lambda}^{CC,CH} \tag{2.197}$$

We shall assume that analogous terms are equal. For example, all the γ_C will be assumed to be equal. Furthermore, we shall take into account only the interacting terms between bond loges starting from the same cores, as the interactions decrease with distance. With such approximation the part of the normal molecule which is represented below corresponds to the energy:

$$\begin{aligned} E_n = 4\gamma_C + 3\gamma_{CC} + 4\gamma_{CH} + 4\gamma_{C,CH} + 4\gamma_{C,CC} \\ + 8\gamma_{CC,CH} + 2\gamma_{CH,CH} + 2\gamma_{CC,CC} \end{aligned} \tag{2.198}$$

For the isomolecule the analogous energy is:

$$\begin{aligned} E_i = 4\gamma_C + 3\gamma_{CC} + 4\gamma_{CH} + 4\gamma_{C,CH} + 4\gamma_{C,CC} \\ + 6\gamma_{CC,CH} + 3\gamma_{CH,CH} + 3\gamma_{CC,CC} \end{aligned} \tag{2.199}$$

The isomerization energy is:

$$\Delta E = E_i - E_n = \gamma_{CH,CH} + \gamma_{CC,CC} - 2\gamma_{CC,CH}$$

It must not depend on the length of the chain. The experiment supports that result, as the isomerization energy (normal → iso) for all paraffins is approximately 1.7 kcal mol^{-1}. In the same way it can readily be found that the isomerization energy $\Delta E'$ between a normal molecule and its 2,2-dimethyl derivative:

```
          H
          |
       H—C—H
          |
   ···C—C—C···
          |
       H—C—H
          |
          H
```

is:

$$\Delta E' = 3(\gamma_{CC,CC} + \gamma_{CH,CH} - 2\gamma_{CC,CH}) \tag{2.200}$$

Therefore:

$$\Delta E' = 3\,\Delta E \tag{2.201}$$

Once more experimental findings support that conclusion as experiments give:

$$\Delta E' \approx 4.7 \text{ kcal mol}^{-1}$$

These results were previously obtained by Brown[(50)] but the method used was a rough LCBO procedure. Our discussion shows that they stay valid in the framework of a much more elaborate analysis.

To end this section the notion of *event function* will be introduced. The function Ψ_λ which is equal to the exact function Ψ when α points are in v_A, β points in $v_B, \ldots, \nu$ points in v_P is called the event function associated with the event λ. It is obvious that:

$$\Psi = \sum_\lambda \Psi_\lambda \tag{2.202}$$

The exact wave function is simply the sum of all possible event functions.

We can also use the normalized event function defined as:

$$\varphi_\lambda = N_\lambda \Psi_\lambda \tag{2.203}$$

with:

$$N_\lambda = \frac{\varepsilon_\lambda}{\sqrt{(P_\lambda)}} \quad \text{and} \quad |\varepsilon_\lambda|^2 = 1$$

Equation (2.202) becomes:

$$\Psi = \sum_\lambda \varepsilon_\lambda \sqrt{(P_\lambda)}\varphi_\lambda \tag{2.204}$$

The event functions are not continuous at the frontier of the loges. It is possible to express the various foregoing increments in terms of the Ψ_λ or of the φ_λ but we have to integrate over *open* volumes to avoid the discontinuity problem.

Let us remark that it would be easy to take account of the spin in the definition of an event. An event would correspond to a certain distribution of the electrons in the loges with a certain organization of their spin.

2.2.2 Analysis of Approximate Wave Functions in Terms of Loges; Most Localized Orbitals, 'Size' of an Electron; Covalent and Dative Bonds

The formalism described in Section 2.2.1 is general and applies to any wave function. It is valid not only for an exact wave function but also for any approximate wave function. In principle we have nothing to add. In practice it is different because some kinds of approximate wave function make it possible to obtain information about a good division of a molecule into loges by following a very simple procedure. This is the case of wave functions expressed in the simple form of a unique Slater determinant.

Before proceeding further it is necessary to introduce a new mathematical concept. A set of normalized functions f_i defined on a given space is said to be localized with a precision ε in an ensemble of volume V_i whose intersection has zero measure if:

$$\int_{V_i} f_i^* f_i \, \mathrm{d}v \geqslant 1 - \varepsilon \tag{2.205}$$

Consider a two-electron system represented by the function:

$$\Phi = \frac{1}{\sqrt{2}} \det \varphi_1(\mathrm{M})\varphi_2(\mathrm{M}') \tag{2.206}$$

φ_1 and φ_2 being respectively localized in V_1 and V_2 with precision ε. The probability P_{12} of finding electron 1 in V_1 and electron 2 in V_2 is:

$$\begin{aligned} P_{12} &= \int_{V_1} \mathrm{d}v_\mathrm{M} \int_{V_2} |\Phi|^2 \, \mathrm{d}v_{\mathrm{M}'} \\ &= \tfrac{1}{2} \int_{V_1} \varphi_1^*(\mathrm{M})\varphi_1(\mathrm{M}) \, \mathrm{d}v_\mathrm{M} \int_{V_2} \varphi_2^*(\mathrm{M}')\varphi_2(\mathrm{M}') \, \mathrm{d}v_{\mathrm{M}'} \\ &\quad + \tfrac{1}{2} \int_{V_1} \varphi_2^*(\mathrm{M})\varphi_2(\mathrm{M}) \, \mathrm{d}v_\mathrm{M} \int_{V_2} \varphi_2^*(\mathrm{M}')\varphi_2(\mathrm{M}') \, \mathrm{d}v'_\mathrm{M} \\ &\quad - \int_{V_1} \varphi_1^*(\mathrm{M})\varphi_2(\mathrm{M}) \, \mathrm{d}v_\mathrm{M} \int_{V_2} \varphi_1^*(\mathrm{M}')\varphi_2(\mathrm{M}') \, \mathrm{d}v_{\mathrm{M}'} \end{aligned} \tag{2.207}$$

Since:

$$\int \varphi_1^*\varphi_2 \, \mathrm{d}v = 0 \qquad \text{and} \qquad \int_{V_1} \varphi_1^*\varphi_2 \, \mathrm{d}v = -\int_{V_2} \varphi_1^*\varphi_2 \, \mathrm{d}v \tag{2.208}$$

then:

$$\begin{aligned} P_{12} &= \tfrac{1}{2} \int_{V_1} \varphi_1^*\varphi_1 \, \mathrm{d}v \int_{V_2} \varphi_2^*\varphi_2 \, \mathrm{d}v \\ &\quad + \tfrac{1}{2} \int_{V_1} \varphi_2^*\varphi_2 \, \mathrm{d}v \int_{V_2} \varphi_1^*\varphi_1 \, \mathrm{d}v \\ &\quad + \left| \int_{V_1} \varphi_1^*\varphi_2 \right| \mathrm{d}v^2 \end{aligned} \tag{2.209}$$

The last two terms in equation (2.209) are positive; therefore:

$$\begin{aligned} P_{12} &\geqslant \tfrac{1}{2} \int_{V_1} \varphi_1^*\varphi_1 \, \mathrm{d}v \int_{V_2} \varphi_2^*\varphi_2 \, \mathrm{d}v \\ &\geqslant \tfrac{1}{2}(1-\varepsilon)^2 \end{aligned} \tag{2.210}$$

Finally, the probability P of simultaneously finding one electron in V_1 and the other in V_2 is such that:

$$P = P_{12} + P_{21} \geqslant (1-\varepsilon)^2 \tag{2.211}$$

If the orbitals φ are well localized ε is small and P is high, there is a leading event, the missing information function will have a small value, and V_1 and V_2 correspond to a good division into loges. *When a wave function is expressed in terms of well-localized functions the volumes in which they are localized correspond to a good division into loges.* We shall return to this point later.

The molecular orbitals which are obtained by solving the Roothaan equations are usually not well localized. However, it is easy to transform them into more localized orbitals. It is well known that a determinant such as:

$$\Phi = \det \varphi_1 \varphi_2 \tag{2.212}$$

does not change if the φ's are replaced by functions f derived from the φ's with the help of a unitary transform T such as:

$$f_i = \sum_j T_{ij} \varphi_j \tag{2.213}$$

Therefore:

$$\det f_1 f_2 = \det \varphi_1 \varphi_2 = \Phi \tag{2.214}$$

We need to find the transform T which gives the *most localized orbitals f*. From the loge viewpoint the most appropriate procedure to obtain the most localized orbitals would be to search for the transform T which minimizes the product of the various $\int |f_i|\,|f_j|\,\mathrm{d}v$ because it is possible to show[51] that if all $\int |f_i|\,|f_j|\,\mathrm{d}v$ are small:

$$\exists\{V_i\}, \qquad V_i \cap V_j = \varnothing, \qquad i \neq j,$$

$$\forall i, \qquad \int f_i^* f_i \,\mathrm{d}v \approx 1$$

Unfortunately, this criterion has not been used. Brion and Daudel[52] suggested minimization of the exchange energy, a criterion commonly used.[53] Foster and Boys[54] proposed another criterion which maximizes the distances between the charge centres of the various orbitals.

We saw that one advantage of the loge theory is to provide an idea of the space needed by an electron in the various parts of a molecule. To obtain that information we divide the volume of a given loge by the mean number of electrons which are found in that loge. However that procedure is not valid for superificial loges as they are going to infinity. Following Robb, Haines and Csizmadia,[55] the size of an electron in a given loge can be represented by the volume of a sphere of radius equal to the root mean square of the distance of an electron in the loge from the gravity centre of that loge. If V denotes the loge, the size v of an electron in that loge is:

$$v = \tfrac{4}{3}\pi \left| \int_V \mathrm{d}v_{\mathrm{M}_1} \int_{R^{3(n-1)}} \Psi^* r_{\mathrm{M}_1}^2 \Psi \,\mathrm{d}v_2\,\mathrm{d}v_3 \cdots \mathrm{d}v_n \right|^{3/2} \tag{2.215}$$

where r_{M_1} denotes the distance between the point M_1 and the gravity centre of the loge V.

If the function Ψ is approximated with a single Slater determinant expressed in terms of well-localized orbitals the expression (2.215) reduces approximately to:

$$v = \frac{4}{3}\pi\left(\int_V \varphi^* r_M^2 \varphi \, dv_M\right)^{3/2} \quad (2.216)$$

where φ denotes the orbital localized in the loge V.

Robb, Haines and Csizmadia calculated many electronic sizes. They observed, for example, that the size of an electron (represented simply by the expectation value of r_M^2) in a two-electron bond loge of ammonia is 2.052 a.u. For the lone pair loge of the same molecule the size reaches 2.505 a.u. This result is in very good agreement with the qualitative ideas set forth in Gillespie's book.[56]

The loge theory has recently been used to propose a criterion to distinguish between covalent and dative bonds. The classical criterion that permits such a distinction is related to a possible mechanism of the formation of the bond. It is said, for example, that a two-electron bond is dative when it results from a lone pair shared by two atoms. Such a mechanism is classically assumed in the case of borazane:

$$H_3B + NH_3 \rightarrow H_3B \leftarrow NH_3$$

In many cases several mechanisms of formation of the same bond may be admitted, which leads to different conclusions concerning the nature of the bond. Furthermore, the relation between the nature of a bond and the mechanism of its formation is not necessarily direct, because a certain reorganization of the electron density can follow the establishment of the bond. It is not obvious that a bond must have a good remembrance of its history.

For all these reasons it is better to search for a criterion based directly on the actual structure of the molecules. Daudel and Veillard proposed such a criterion.[57] *A two-electron bond loge is said to correspond to a covalent bond if it is established between two groups of loges containing the same total charge* $+e$ *during the leading event. Following the same criterion the loge corresponds to a dative bond if the total charges of the two groups are* 0 *and* $+2e$ *respectively for the leading event* (*for a non-cyclic molecule*).

Aslangul *et al.*[58] have determined good divisions into loges for borazane $H_3B \leftarrow NH_3$ and aminoborane H_2B—NH_2 by calculating SCF wave functions and searching for the best localized orbitals. They found that the gravity centre of the BN localized orbital always lies between the middle of the bond and the nitrogen nucleus, but at distances of 0.59 a.u. and 0.45 a.u. from that middle in borazane and aminoborane respectively. As expected, the BN dative bond is more polar than the covalent bond, but the direction of the dipole moment $B \mapsto N$ is *opposite* to that of the classical arrow $H_3B \leftarrow NH_3$. It turns out that this arrow must be interpreted as an indication of an electron transfer during the formation of the bond following the classical mechanism and not the actual dipole moment of the bond.

2.2.3 A Division into Loges as a Starting Point to Calculate Elaborate Wave Functions

The concept of loges has been introduced to analyse a known wave function. However, it can be used to build elaborate wave functions.

Let us assume that we already have an idea about the topology of a good division into loges. Information about the topology could be a convenient starting point. Such information may result from a knowledge of an approximate wave function already computed and could be anticipated for a large molecule by analogy with previous results concerning smaller molecules of the same family.

Furthermore, many experimental measurements can be interpreted in terms of loges. For example, some interatomic distances can be associated with the presence of a single localized bond. We saw (Section 1.2.6) that such a bond corresponds to a two-electron localized loge.

Let us assume that (for example) Figure 1.18 represents the topology of a good division into loges of a sixteen-electron system containing the four nuclei A, B, C and D. Equation (2.204) suggests a variation function like:

$$\Psi = \sum_{\lambda} C_{\lambda} \Phi_{\lambda} \tag{2.217}$$

to represent the molecule. If, for example, during the leading event there are two electrons in each core loge, two electrons in the unshared loge adjacent to the core A, two electrons in the bond loge localized between core A and core B and four electrons in the delocalized bond loge extended over the three cores B, C and D, the generating space function associated with that event can be written as:

$$\Upsilon_1 = A(1, 2)B(3, 4)C(5, 6)D(7, 8)P(9, 10)L_{\mathrm{AB}}(11, 12)L_{\mathrm{BCD}}(13, 14, 15, 16) \tag{2.218}$$

The various functions $A, B, C, \ldots, L$ are called loge functions.[59] To take account of the spin and of the Pauli principle the corresponding event function must be written as:

$$\Phi_1 = \sum_{P} (-1)^P \Upsilon_1 \sigma \tag{2.219}$$

where σ is a convenient spin function. Following such a procedure we can associate an event function to each event and obtain a more explicit expression of equation (2.217).

The coefficients C_{λ} of equation (2.217) and the various loge functions can in principle be calculated by solving the usual variational equation:

$$\delta \int \Psi^*(H - U)\Psi \, \mathrm{d}v = 0 \tag{2.220}$$

This is a generalization of the Hartree–Fock procedure.

In principle the loge functions must be localized in the corresponding loges. However, as equation (2.217) can only be understood as a starting point to built a variational family of functions we can only calculate the loge functions in order to satisfy orthogonality conditions. In that case it is not necessary to know the frontiers of the loges.

From the practical viewpoint, it is useful to expand the various loge functions in terms of monoelectronic functions (as atomic orbitals or gaussian orbitals). We shall write, for example:

$$L_{\mathrm{BCD}} = \sum_i \sum_j \sum_k \sum_l s_{ijkl} \chi_i(13) \chi_j(14) \chi_k(15) \chi_l(16) \tag{2.221}$$

The variational procedure will furnish the coefficients s_{ijkl} and therefore the function L_{BCD}.

Many classical procedures of calculating wave functions only take into consideration the leading event and neglect the others. The variational function reduces to:

$$\Psi = \Phi_1 = \sum_P (-1)^P P Y_1 \sigma \tag{2.222}$$

We shall now analyse specific methods derived from equation (2.222).

2.2.3.1 The atomic case

Consider the beryllium atom in its ground state. The best division into loges is made of a sphere (the two-electron K loge) and of the remaining part of the space (a two-electron L loge). The corresponding wave function is therefore:

$$\Psi = \sum_P (-1)^P P K(1,2) L(3,4)$$

if all events except the leading one are neglected. Ludeña and Amzel[60] calculated completely localized loge functions and found an energy of -14.58 a.u. (slightly better than the Hartree–Fock one which corresponds to -14.56 a.u.). The experimental value is -14.667 a.u. Various authors used non-completely localized two-electron loge functions, usually called *geminals*. The geminals can be calculated in order to satisfy the *soft orthogonality condition*:

$$\int K(1,2) L(1,2)\, \mathrm{d}v_1\, \mathrm{d}v_2 = 0 \tag{2.223}$$

or the *strong orthogonality condition*

$$\int K(1,2) L(1,3)\, \mathrm{d}v_1 = 0 \tag{2.224}$$

If no orthogonality condition is required the geminals are said to be *free geminals*.[61] Table 2.5 shows some results. It can be seen that all methods give satisfactory results. The quality of the wave function depends mainly on

Table 2.5 Ground state energy of beryllium

Method	Energy (a.u.)	References
Orthogonalized geminals	−14.611	62
	−14.657	63
Free geminals	−14.621	62
	−14.650	64
Completely localized loge functions	−14.58	60
Experimental value	−14.667	

the number of monoelectronic functions introduced in the expansion of the two-electron loge functions.

It is easy to see that if the expansion is reduced to one term only:

$$K(1,2) = 1s(1)1s(2)$$

$$L(1,2) = 2s(1)2s(2)$$

and if the $1s$ and $2s$ are calculated by following a variational procedure, the wave function reduces to the Hartree–Fock one.

Let us remark that when the loge functions are not completely localized *the wave function allows some electronic configurations which correspond to non-leading events.* This is due to overlapping which appears between the loge functions, an overlapping introducing (for example) a non-zero probability of finding four electrons in the K loge.

2.2.3.2 Representation of cores and lone pairs

We saw (Section 1.2.5) that it is easy to obtain a good idea of the topology of the best division into loges of atoms. Therefore there is no problem in writing the loge function associated with atomic cores in a molecule. A K loge is usually represented by a geminal. In some rough function we can replace the geminal by the first term in the expansion in terms of monoelectronic functions. The K loge function reduces to:

$$K(1,2) = 1s(1)1s(2)$$

A lone pair is also usually represented by a geminal. If we want to reduce its expression to one term we must select that term in such a way as to fill the space of the loge. The usual atomic orbitals s, p_0, $p_{\pm1}$ are not convenient: the first one has a spherical symmetry, the second one is localized near the z axis and the others are mainly located near a plane (Section 1.2.1). However, unitary transforms make it possible to produce directional orbitals.

It is usual to give the expression of these directional orbitals in terms of the functions s, p_x, p_y, p_z which form a basis equivalent to s, p_0, $p_{\pm1}$ but containing only real functions. Let us recall that:

$$p_x = \frac{1}{\sqrt{2}}(p_{+1} + p_{-1}) = R(r)\frac{\sqrt{3}}{2\sqrt{\pi}}\sin\theta\cos\varphi$$
$$p_y = \frac{1}{i\sqrt{2}}(p_{+1} - p_{-1}) = R(r)\frac{\sqrt{3}}{2\sqrt{\pi}}\sin\theta\sin\varphi \qquad (2.225)$$

From that basis it is possible to build the three following sets of *hybrid orbitals* which possess interesting directional properties:

$$\text{Digonal orbitals:}\begin{cases} d_1 = \frac{1}{\sqrt{2}}(s + p_x) \\ d_2 = \frac{1}{\sqrt{2}}(s - p_x) \end{cases} \qquad (2.226)$$

$$\text{Trigonal orbitals:}\begin{cases} t_1 = \frac{1}{\sqrt{3}}s + \frac{\sqrt{2}}{\sqrt{3}}p_x \\ t_2 = \frac{1}{\sqrt{3}}s - \frac{1}{\sqrt{6}}p_x + \frac{1}{\sqrt{2}}p_y \\ t_3 = \frac{1}{\sqrt{3}}s - \frac{1}{\sqrt{2}}p_x - \frac{1}{\sqrt{6}}p_y \end{cases} \qquad (2.227)$$

$$\text{Tetrahedral orbitals:}\begin{cases} te_1 = \tfrac{1}{2}(s + p_x + p_y + p_z) \\ te_2 = \tfrac{1}{2}(s + p_x - p_y - p_z) \\ te_3 = \tfrac{1}{2}(s - p_x + p_y - p_z) \\ te_4 = \tfrac{1}{2}(s - p_x - p_y + p_z) \end{cases} \qquad (2.228)$$

The two hybrids d_1 and d_2 are localized near the x axis, the first one mainly along the positive part of the axis and the second one mainly along the negative one.

The three hybrids t_1, t_2 and t_3 have their maximum values in the xy plane, the first points along the x axis (positive direction) and the other two along the directions forming an angle of 120° with the x axis.

The tetrahedral hybrids te_1, te_2, te_3, te_4 are pointing towards the vertices of a regular tetrahedron centred at the origin of the coordinates, the first hybrid pointing along the axis of the triad of x, y, z.

Let us assume that we want to represent the lone pair of NH_3. As there are four two-electron loges surrounding the K core in that molecule (Section 1.2.6) we need the tetrahedral hybrids built on the basis $2s$, $2p_x$, $2p_y$ and $2p_z$. If the axis of the lone pair loge coincides with the axis of the triad of x, y, z we can represent that loge with the simple function:

$$P(1, 2) = te_1(1)te_1(2)$$

2.2.3.3 Representation of localized bonds

A two-electron localized bond loge is usually represented by a geminal. Therefore, molecules for which a good division into loges is made of K loges, two-electron localized bond loges and lone pair loges are conveniently represented by a wave function built on orthogonalized geminals. Such a procedure was used by Ahlrichs and Kutzelnigg[65] to study LiH, BeH_2, BH_3 and CH_4. In the case of LiH the results are highly accurate. In other cases the results are not so spectacular, but the method does appear to offer a significant improvement over the independent electron model, at little extra cost in item and effort, both as an end in itself and as a starting point for more accurate calculations.

We can also use hybrids to represent roughly localized bonds. Going back to our molecule NH_3, it is convenient to represent one of the bond loges by the linear combination:

$$\lambda te_2 + \mu h_2$$

where h_2 denotes the orbital $1s$ of the hydrogen atom contributing to the NH bond located along the axis of the function te_2 and λ and μ are variational coefficients. As a consequence a very simple wave function to represent the NH_3 molecule could be:

$$\begin{aligned}\Psi = \sum_P (-1)^P P 1s(1)1s(2)te_1(3)te_1(4) \\ \times[\lambda te_2(5)+\mu h_2(5)][\lambda te_2(6)+\mu h_2(6)][\lambda te_3(7)+\mu h_3(7)] \\ \times[\lambda te_3(8)+\mu h_3(8)][\lambda te_4(9)+\mu h_4(9)] \\ \times[\lambda te_4(10)+\mu h_4(10)] \qquad (2.229)\end{aligned}$$

With such a simple wave function we have only to calculate the ratio λ/μ. The variational calculation is therefore very short.

Obviously more elaborate combinations of hybrids can be used. Klessinger and McWeeny[66] used the following loge functions:

$$L_{CH} = \{ah(1)h(2) + bte(1)te(2) + C[h(1)te(2) + te(1)h(2)]\} \qquad (2.230)$$

The variational calculation leads to an energy of -53.48 eV, slightly better than the Roothaan equations which, however, are more difficult to solve.

It turns out that it is advantageous to take account of the localizability of electrons. By doing so, it is possible to obtain very quickly simple but satisfactory functions or very elaborate ones at a reduced cost.

Following an idea of Frost,[67] Barthelat and Durand[68] have calculated a wave function for the methane molecule in which the CH bond loges are represented by a simple gaussian function:

$$\Psi = \sum_P (-1)^P PK(1,2)\chi_1(3)\chi_1(4)\chi_2(5)\chi_2(6)\chi_3(7)\chi_3(8)\chi_4(9)\chi_4(10) \qquad (2.231)$$

Table 2.6 Energy of methane as a function of the CH bond length d_{CH}

d_{CH}(Å)	α_{opt}	$\frac{Cg_{opt}}{CH}$	U_{min}(a.u.)	Virial $\left(-\frac{E}{T}\right)$
1.00	0.39	0.604	−38.182	0.98
1.05	0.37	0.601	−38.222	1
1.093	0.356	0.597	−38.237	1.008
1.15	0.337	0.593	−38.235	1.02
1.20	0.320	0.587	−38.216	1.03

The energy U is minimized with respect to the exponent α in the gaussian functions and the positions of the gravity centres g of these functions which are kept on the various CH nuclear lines. Table 2.6 shows how the energy depends on the CH internuclear distance. It can be seen that the energy reaches a minimum when:

$$d_{CH} = 1.093 \text{ Å}$$

This is exactly the same as the experimental value. That simple function is therefore able to give the right nuclear conformation of the molecule.

2.2.3.4 Representation of delocalized bonds

Let us take, as an example, the case of the diborane molecule B_2H_6 discussed in Section 1.2.6. It contains two-electron loges delocalized over one hydrogen and the two boron cores. One of the two loge functions can be expanded as:

$$L_{B_1HB_2} = \sum_{i,j} s_{ij}\chi_i(1)\chi_j(2) \tag{2.232}$$

the χ's being a basis of monoelectronic functions.

To simplify the calculations we can select some important terms in that expansion by taking a simple product of molecular orbitals:

$$L_{B_1HB_2} = \varphi(1)\varphi(2) \tag{2.233}$$

the molecular orbital φ being taken as a linear combination of atomic orbitals:

$$\varphi = a\chi_{B_1} + b\chi_H + c\chi_{B_2} \tag{2.234}$$

With such an approximation $L_{B_1HB_2}$ reduces to:

$$\begin{aligned} L_{B_1HB_2} = {} & a^2\chi_{B_1}(1)\chi_{B_1}(2) + b^2\chi_H(1)\chi_H(2) \\ & + c^2\chi_{B_2}(1)\chi_{B_2}(2) \\ & + ab[\chi_{B_1}(1)\chi_H(2) + \chi_H(1)\chi_{B_1}(2)] \\ & + ac[\chi_{B_1}(1)\chi_{B_2}(2) + \chi_{B_2}(1)\chi_{B_1}(2)] \\ & + bc[\chi_H(1)\chi_{B_2}(2) + \chi_{B_2}(1)\chi_H(2)] \end{aligned} \tag{2.235}$$

It is obvious that the second member of equation (2.235) is a paticular case of the expansion (2.232), the coefficients s_{ij} being linked by certain particular relationships.

As previously, convenient hybrids can be introduced into equation (2.235). In outline the procedure might be:

1. Guess the topology of a good division into loges and the nature of the leading event. Knowledge of the frontiers of the loges is not necessary.
2. Associate with each loge a loge function, the number of points in that function being the number of electrons which is found in the loge during the leading event.
3. Expand each loge function in terms of a basis of monoelectric functions (atomic orbitals, gaussian functions). If elaborate wave functions are not necessary it is possible to reduce the expansion to a small number of terms selected in order to fill the part of the space roughly associated with the loge. Hybrids are useful to achieve this end.
4. Derive the total wave function from the product of loge functions by taking account of the spin function and of the Pauli principle.
5. Calculate by following a variational procedure various parameters: coefficients of the expansions and eventually exponents of the bases.

The McWeeny group function technique shows a nice particular way to make the effective computation.[69] *An interesting feature lies in the fact that when a loge function is finally obtained for a given molecule it can be used as a starting point for the iteration procedure necessary to calculate an analogous loge function in another molecule.*

2.2.3.5 Taking account of all events

The foregoing procedure only takes account of the leading event, at least as a starting point. In fact, it has been stated that, if loge functions are not completely localized, other events are introduced. It is also possible to introduce systematically all events or at least some which seem to be important.

To describe the first triplet state of helium (for example) it would be possible to try a variational wave function like:

$$\Psi = \sum_P (-1)^P P[C_1 K(1)L(2) + C_2 K(1,2) + C_3 L(1,2)]\alpha(1)\alpha(2) \tag{2.236}$$

including all possible events. The perturbative configuration interaction localized orbitals (PCILO) method introduced by Claverie *et al.*[70] demonstrates something of this kind.

A function associated with the leading event is used as a zero-order term. Other events are introduced following the Rayleigh–Schrödinger perturbation expansion. A more detailed analysis of the PCILO method in the framework of loge theory is given in a review.[71]

Another very powerful procedure of calculating molecular properties is the $X\alpha$ method proposed in an excellent review by Slater[72] as well as by others. This introduces a statistical exchange in the self-consistent field and bears some relationship to loge theory. However as this method amounts to a change in the expression for the hamiltonian it will not be analysed here.

2.2.4 Semiempirical Calculations: LCBO, LCVO and Del Ré Methods

In Section 1.2.6 it was stated that in a saturated hydrocarbon C_nH_{2n+2} the number N of electrons which is found outside the atomic core loges during the leading event is twice the number of pairs of adjacent cores. This is why a good partition of the molecular space into loges contains only two-electron localized bond loges: CC loges and CH loges.

If only the leading event is taken into account it is therefore tempting to introduce the following generating function:

$$\begin{aligned} \Upsilon = K_1(1,2)K_2(3,4)\cdots L_{C_1H_a}(p,p+1) \\ \times L_{C_1H_b}(p+2,p+3)\cdots L_{C_1C_2}(q,q+1) \\ \times L_{C_2C_3}(q+2,q+3)\cdots \end{aligned} \tag{2.237}$$

where the K are the various carbon core functions, the L_{CH} the loge functions associated with the CH bonds and the L_{CC} the loge functions associated with the CC bonds. Following the method described in Section 2.2.3, the various two-electron functions can be expanded on a basis set of monoelectronic functions:

$$L = \sum_i \sum_j s_{ij}\chi_i\chi_j$$

If the expansion is limited to the first term the generating function becomes:

$$\begin{aligned} \Upsilon = 1s_1(1)1s_1(2)1s_2(3)1s_2(4)\cdots \\ \chi_{C_1H_a}(p)\chi_{C_1H_a}(p+1)\cdots\chi_{C_1C_2}(q)\chi_{C_1C_2}(q+1) \end{aligned} \tag{2.238}$$

For the ground state of a paraffin (a singlet state) this generating function leads to the following wave function:

$$\Psi = \det 1s_1\overline{1s_1}1s_2\overline{1s_2}\cdots\chi_{C_1H_a}\bar{\chi}_{C_1H_a}\chi_{C_1C_2}\bar{\chi}_{C_1C_2}\cdots \tag{2.239}$$

The various χ's are called *bond orbitals*. Let T be a unitary transform and ϕ_j the functions which derive from the χ_i under the effect of this unitary transform:

$$\phi_j = \sum_i C_{ij}\chi_i \tag{2.240}$$

The determinant:

$$\det \phi_1\bar{\phi}_1\phi_2\bar{\phi}_2\cdots\phi_j\bar{\phi}_j\cdots \tag{2.241}$$

is identical to the second member of equation (2.239). Therefore we can write:

$$\Psi = \det \phi_1\bar{\phi}_1\phi_2\bar{\phi}_2 \cdots \phi_j\bar{\phi}_j \cdots \tag{2.242}$$

If the variational equation:

$$\delta\langle\Psi|\, H - E\, |\Psi\rangle = 0$$

is introduced, it can readily be seen that the coefficients C_{ij} are solutions of Roothaan's equations. As each ϕ is a linear combination of bond orbitals the method under consideration is called the *LCBO method.*

Hall[73] proposed a semiempirical procedure based on that formalism. If we assume that the self-consistent field operator h^{SCF} is known and fixed it is possible to replace Roothaan's equations by the usual secular system:

$$\sum_i C_{ij}(\langle\chi_i|\, h^{\text{SCF}}\, |\chi_j\rangle - \varepsilon\langle\chi_i \mid \chi_j\rangle) = 0, \qquad \text{for } i = 1, 2, \ldots, n \tag{2.243}$$

Snce we saw that the paraffin bonds are well individualized they correspond to good loges. The bond orbitals are therefore rather well localized and it is reasonable to neglect the overlap integrals:

$$\langle\chi_i \mid \chi_j\rangle, \qquad i \neq j$$

If the methane molecule is taken as an example the symmetry properties of this molecule lead to simplifications. The quantity $\langle\chi_i|\, h^{\text{SCF}}\, |\chi_j\rangle$ $(i \neq j)$ does not depend on the indices. This is also true for $\langle\chi_i|\, h^{\text{SCF}}\, |\chi_i\rangle$. Hence we can set:

$$\alpha_{\text{CH}} = \langle\chi_i|\, h^{\text{SCF}}\, |\chi_i\rangle$$

and

$$\beta_{\text{CH}} = \langle\chi_i|\, h^{\text{SCF}}\, |\chi_j\rangle, \qquad i \neq j$$

Putting:

$$\alpha_{\text{CH}} = \alpha \qquad \text{and} \qquad \beta_{\text{CH}} = \beta$$

and assuming that the lowest LCBO reduces to the 1s orbital, the secular equation becomes:

$$\begin{vmatrix} \alpha-\varepsilon & \beta & \beta & \beta \\ \beta & \alpha-\varepsilon & \beta & \beta \\ \beta & \beta & \alpha-\varepsilon & \beta \\ \beta & \beta & \beta & \alpha-\varepsilon \end{vmatrix} = 0 \tag{2.244}$$

The roots of this equation are:

$$\varepsilon_1 = \alpha + 2\beta$$
$$\varepsilon_2 = \alpha - \beta$$

the last one being a triplet root. Following Koopmans' theorem, these roots give approximations to the ionization energies. If we want to reproduce experimental values (13 and 20 eV) we must put:

$$\alpha = -14.75 \text{ eV}$$

and

$$\beta = -1.75 \text{ eV}$$

This gives a first example of a *semiempirical procedure* in which some theoretical parameters are chosen to reproduce experimental values.

At first sight this procedure could be considered to be uninteresting. In fact, it is very useful because the parameters obtained from experimental results for a given molecule can be used to calculate unmeasured properties for other molecules.

Let us consider the family of normal saturated hydrocarbons. To evaluate ionization energies of such compounds we can assume that the parameters α and β remain approximately constant in that family. We also need the values of three other parameters:

$$c = \langle \chi_{CC'} | \, h^{SCF} \, | \chi_{CC'} \rangle$$
$$d = \langle \chi_{CC'} | \, h^{SCF} \, | \chi_{CH} \rangle$$
$$e = \langle \chi_{CC'} | \, h^{SCF} \, | \chi_{C'C''} \rangle$$

if interactions between non-adjacent loges are neglected. Hall has determined the values of the five parameters α, β, c, d and e which give the best fit over eight molecules. Table 2.7 shows how excellent is that fit, which gives confidence in the whole theory.

Table 2.7 Ionization energies of n-alkanes

	First ionization energy (eV)	
Hydrocarbon	Calculated	Measured
Propane	11.214	11.21
Butane	10.795	10.80
Pentane	10.554	10.55
Hexane	10.412	10.43
Heptane	10.323	10.35
Octane	10.265	10.24
Nonane	10.224	10.21
Decane	10.194	10.19

Lennard Jones and Hall[74] have used that LCBO method to study the distribution of electric charge in positive paraffin ions. If the wave function is given by equation (2.242), it can readily be seen (equation 2.46) that the probability of finding one electron in a volume dv is:

$$\begin{aligned} dp &= 2(n_1 |\phi_1|^2 + n_2 |\phi_2|^2 + \cdots + n_j |\phi_j|^2 + \cdots)\, dv \\ &= 2 \sum_j n_j |\phi_j|^2 \, dv \end{aligned} \tag{2.245}$$

where n_j represents the number of orbitals ϕ_i introduced in the determinant. If equation (2.240) is taken into account equation (2.245) becomes:

$$dp = \sum_j n_j \left| \sum_i C_{ij}\chi_i \right|^2 dv \tag{2.246}$$

where n_j again represents the number of orbitals ϕ_j introduced in the determinant. If, now, it is assumed that each χ_i is completely localized in the corresponding bond loge the products $\chi_i\chi_j (i \neq j)$ vanish and equation (2.246) reduces to:

$$dp = 2 \sum_j n_j \sum_i |C_{ij}|^2 |\chi_i|^2 \, dv \tag{2.247}$$

Integrating dp in the volume v_i of the loge associated with the bond loge i leads to:

$$q_i = \int_{v_i} dp = \sum_j n_j |C_{ij}|^2 \tag{2.248}$$

where q_i is the amount of electronic charge contained by the loge i.

Let us consider a paraffin in its ground state. It has been stated that equations (2.239) and (2.242) are equivalent. We can apply equation (2.248) using the expression of the wave function given by equation (2.239) to give:

$$q_i = 2 \tag{2.249}$$

Each loge contains an electronic charge of two electrons. This result is consistent with the fact that we are taking only the leading event into account.

For a positive paraffin ion in its ground state equation (2.239) is no longer valid, because the number of 'bonding electrons' is no longer twice the number of pairs of adjacent atomic cores. However, the LCBO treatment of positive ions assumes that an expression similar to equation (2.242) is still valid: each orbital ϕ_j is used twice except for the last one which is introduced only one time in the wave function, the number of electrons being odd. If the ion contains $2n-1$ electrons the wave function is written as:

$$\Psi = \det \phi_1\bar{\phi}_1\phi_2\bar{\phi}_2 \cdots \phi_j\bar{\phi}_j \cdots \phi_n \tag{2.250}$$

Therefore the loss of electronic charge which appears in each loge by comparison to the neutral paraffin is simply:

$$q_{in} = |C_{in}|^2 \tag{2.251}$$

Lennard Jones and Hall[74] assumed that the parameters of the ion are not significantly different from the parameters of the neutral molecules. Therefore they calculated the values of the C_{in} by solving the secular equations with the values of the parametes α, β, c, d, e obtained for the neutral molecule. If the CC bond loges of normal octane are numbered by starting from one end of the molecule, the corresponding positive charges are:

0.035 in bond 1

0.115 in bond 2

0.200 in bond 3

0.234 in bond 4

This example shows how the semiempirical approach permits calculation of unmeasured properties.

The same kind of method has been used by Brown[75] to determine the atomization energies of saturated hydrocarbons. Overlap integrals are no longer neglected. The following notations were used:

$$S=\langle\chi_{\mathrm{CH}}\chi_{\mathrm{CH'}}\rangle \qquad \alpha=\langle\chi_{\mathrm{CH}}h^{\mathrm{SCF}}\chi_{\mathrm{CH}}\rangle$$
$$\alpha+h\gamma=\langle\chi_{\mathrm{CC'}}h^{\mathrm{SCF}}\chi_{\mathrm{CC'}}\rangle \qquad \beta=\langle\chi_{\mathrm{CH}}h^{\mathrm{SCF}}\chi_{\mathrm{CH'}}\rangle$$
$$\gamma=\beta-S\alpha \qquad \theta\beta=\langle\chi_{\mathrm{CH}}h^{\mathrm{SCF}}\chi_{\mathrm{CC'}}\rangle$$
$$\eta\beta=\langle\chi_{\mathrm{CC'}}h^{\mathrm{SCF}}\chi_{\mathrm{C'C''}}\rangle \qquad \eta\mathbf{S}=\langle\chi_{\mathrm{CC'}}\chi_{\mathrm{C'C''}}\rangle$$

Thus it is assumed that the interaction between two given loges possessing the same relative geometrical situation (a CH bond and a CC′ bond starting from the same carbon core, for example) does not depend on the position of the loges in the molecule. Furthermore, interaction between non-adjacent loges are neglected. Finally, the total energy is calculated by finding the sum of the energies associated with the various LCBO orbitals introduced in the wave function. This is a drastic approximation and is only valid in the framework of the independent electron model.

Table 2.8 gives the expression of atomization energies in terms of the various parameters. It can be seen that if the overlap integral S is neglected the atomization energy is simply:

$$E=2\alpha N_{\mathrm{CC}}+(2\alpha+2h\gamma)N_{\mathrm{CH}} \tag{2.252}$$

where N_{CC} and N_{CH} denote the number of CC bond loges and the number

Table 2.8

Hydrocarbon	Atomization energies
Methane	$8\alpha-24\gamma S$
Propane	$20\alpha+4h\gamma-(28+40\theta^2+4\eta^2)\gamma S$
Butane	$26\alpha+6h\gamma-(32+56\theta^2+8\eta^2)\gamma S$
Isobutane	$26\alpha+6h\gamma-(36+48\theta^2+12\eta^2)\gamma S$

of CH bond loges respectively. With such an approximation all isomers will possess the same atomization energies. Therefore, the term related to S yields the isomerization energies.

This isomerization energy appears to be:

$$\Delta E = (4 - 8\theta^2 + 4\eta^2)\gamma S$$

when the following pairs of isomers are compared: butane and isobutane, pentane and isopentane, and hexane and isohexane. The experimental values are: 1.7, 1.9 and 1.7 kcal mol^{-1} respectively. The fact that ΔE is about the same for all pairs is in good agreement with the theoretical expression and encourages the use of the same parameter values for all saturated hydrocarbons.

If the two pairs—normal pentane and 2,2-dimethylpropane; hexane and 2,2-dimethylbutane—are now considered the same theory leads to the following expression of the isomerization energies:

$$\begin{aligned}\Delta E' &= (12 - 24\theta^2 + 12\eta^2)\gamma S \\ &= 3\Delta E\end{aligned}$$

The experiment confirms this result as $\Delta E'$ is found to be 4.7 kcal mol^{-1}.

We saw (equation 2.201) that the same expression can be derived from the exact wave function by introducing much lower approximations than in the present treatment. As in the LCBO method only one event is taken into account, the electronic charge for a saturated molecule in its ground state in each loge is (see equation 2.249):

$$q = 2$$

This result shows a limitation of the method. For instance, it will not be possible to use it to study the effect on the charge of a CC bond of the substitution of a fluorine atom in a saturated hydrocarbon.

To avoid this difficulty Sandorfy and Daudel[76] proposed a new procedure, sometimes called the LCVO (linear combination of valence orbitals) method. In this method the bond orbitals are expressed as a linear combination of atomic orbitals. For example, a CH bond orbital χ_{CH} is expressed as:

$$\chi_{CH} = s_1 h + s_2 te$$

where h refers to the 1s orbital of the hydrogen and te the tetrahedral hybrid of the carbon atom pointing in the direction of the hydrogen nucleus (equation 2.228).

The general expression for a molecular orbital becomes:

$$\phi_j = \sum_i \sum_P C_{ijP} \psi_{iP} \tag{2.253}$$

where ψ_{iP} is one of the atomic hybrids introduced in the bond orbital i.

A very drastic approximation is usually made, the overlap between the hybrids being neglected. We shall see later that such an approximation is very often made in the framework of semiempirical methods and some analysis of this nature will be presented. It is easy to see that with such an approximation the quantity $|C_{ijP}|^2$ represents the contribution of the molecular orbital j and the atomic orbital P in the electronic charge of the loge i. The calculation of the charge of a loge can be made by summing the various contributions:

$$q_i = \sum_j \sum_P |C_{ijP}|^2 \tag{2.254}$$

The calculation of the C_{ijP} requires knowledge of the parameters:

$$\alpha_{iP} = \langle \psi_{iP} h^{\text{SCF}} \psi_{iP} \rangle$$

and

$$\beta_{lP,mQ} = \langle \psi_{lP} h^{\text{SCF}} \psi_{mQ} \rangle$$

which can be fitted to experimental data like dipole moments.

Figure 2.1 shows the bond charge calculated for a typical substituted propane in which the atom X is more electronegative than carbon, i.e. possesses a more negative α value. Clearly, the effect of the substitutent X is to reduce the population of the CC bond. *This effect is not oscillatory and dies away rapidly.*

$$\mathrm{X} \overset{2.016}{\text{———}} \mathrm{C} \overset{1.985}{\text{———}} \mathrm{C} \overset{1.999}{\text{———}} \mathrm{C}$$

Figure 2.1 Charge distribution in a substituted propane molecule

More details about this method and many other problems related to semiempirical procedures have been given by Daudel and Sandorfy.[77]

The Del Ré method[78] applies the LCVO method to each bond as if it was on its own. A bond orbital χ_i is written as:

$$\chi_i = C_{ip}\psi_P + C_{iQ}\psi_Q$$

and the coefficients are calculated by using the McDonald theorem. This leads to the secular system:

$$\begin{aligned} C_{iP}(\alpha_{iP} - \varepsilon) + C_{iQ}\beta_{iP,iQ} &= 0 \\ C_{iP}\beta_{iP,iQ} + C_{iQ}(\alpha_{iQ} - \varepsilon) &= 0 \end{aligned} \tag{2.255}$$

To obtain non-trivial solutions of this determinant we must solve the secular equation:

$$(\alpha_{iP} - \varepsilon)(\alpha_{iQ} - \varepsilon) - \beta^2_{iP,iQ} = 0 \tag{2.256}$$

Therefore, we obtain expressions of the energies ε as functions of the parameters α and β. By putting the obtained values in the secular system, sets of coefficients C_{iP} are found. The energy of the molecule is calculated as the sum of the various ε associated with the bonds of the molecule. The main problem is the choice of the parameters α and β. Del Ré proposed a process to calculate and then to reproduce the bond electric dipole moment.

2.2.5 Semiempirical Calculations: CNDO and Extended Hückel Approximations

Let us return to the Roothaan's equations for a closed-shell system (Section 2.1.4). The molecular orbitals φ_k are expanded on a basis set of atomic orbitals χ_l:

$$\varphi_k = \sum_l C_{kl}\chi_l \tag{2.257}$$

They must satisfy the equations:

$$\sum_l C_{kl}(h_{ml}^{\mathrm{SCF}} - \varepsilon_k S_{ml}) = 0, \qquad \text{for } m = 1, \ldots, n \tag{2.258}$$

The self-consistent field operator is:

$$h^{\mathrm{SCF}} = h + 2\sum_j J_j - \sum_j K_j \tag{2.259}$$

where:

$$\begin{aligned} J_j\varphi_k(\mathrm{M}_a) &= \left[\int \varphi_j^2(\mathrm{M}_b)\frac{e^2}{r_{ab}}\,\mathrm{d}v_b\right]\varphi_k(\mathrm{M}_a) \\ K_j\varphi_k(\mathrm{M}_a) &= \left[\int \varphi_j(\mathrm{M}_b)\varphi_k(\mathrm{M}_b)\frac{e^2}{r_{ab}}\,\mathrm{d}v_b\right]\varphi_j(\mathrm{M}_a) \end{aligned} \tag{2.260}$$

More explicitly:

$$h_{ml}^{\mathrm{SCF}} = \langle\chi_m h\chi_l\rangle + 2\sum_j \langle\chi_m J_j\chi_l\rangle - \sum_j \langle\chi_m K_j\chi_l\rangle \tag{2.261}$$

An integral like $\langle\chi_m J_j\chi_n\rangle$ can be written as:

$$\langle\chi_m J_j\chi_l\rangle = \int \chi_m(\mathrm{M}_a)\int \varphi_j^2(\mathrm{M}_b)\frac{e^2}{r_{ab}}\,\mathrm{d}v_b\chi_l(\mathrm{M}_a)\,\mathrm{d}v_a \tag{2.262}$$

Taking account of equation (2.257):

$$\begin{aligned} \langle\chi_m J_j\chi_l\rangle &= \sum_{p,q} C_{jp}C_{jq}\int \chi_m(\mathrm{M}_a)\chi_p(\mathrm{M}_b)\chi_q(\mathrm{M}_b)\frac{e^2}{r_{ab}}\,\chi_l(\mathrm{M}_a)\,\mathrm{d}v_a\,\mathrm{d}v_b \\ &= \sum_{p,q} C_{jp}C_{jq}\left\langle \chi_m(\mathrm{M}_a)\chi_l(\mathrm{M}_a)\frac{e^2}{r_{ab}}\,\chi_p(\mathrm{M}_b)\chi_q(\mathrm{M}_b)\right\rangle \end{aligned} \tag{2.263}$$

and putting:

$$(ml/pq) = \left\langle \chi_m(\mathrm{M}_a)\chi_l(\mathrm{M}_a)\frac{e^2}{r_{ab}}\chi_p(\mathrm{M}_b)\chi_q(\mathrm{M}_b)\right\rangle$$
$$P_{pq} = 2\sum_j C_{jp}C_{jq} \tag{2.264}$$

it can readily be seen that:

$$2\sum_j \langle \chi_m J_j \chi_l\rangle = \sum_{p,q} P_{pq}(ml/pq) \tag{2.265}$$

Analogously, it can be seen that:

$$\sum_j \langle \chi_m K_j \chi_l\rangle = \tfrac{1}{2}\sum_{pq} P_{pq}(mp/lq) \tag{2.266}$$

Finally,

$$h_{ml}^{\mathrm{SCF}} = h_{ml} + \sum_{p,q} P_{pq}[(ml/pq) - \tfrac{1}{2}(mp/lq)] \tag{2.267}$$

For a large molecule the number of molecular integrals (ml/pq) which must be computed is extremely large. The CNDO (complete neglect of differential overlap) approximation has been introduced to reduce the computation time. It is based on two main approximations: a core approximation and the zero-differential overlap (ZDO) approximation.

The core approximation can easily be understood from loge theory. Let us consider a good partition of the molecule into loges. Various atomic core loges will appear. During the leading event they contain, say, r electrons. We can replace the molecule by a model in which each core is replaced by an electrostatic distribution creating approximately the same electric field. If the molecule contains n electrons the model will contain $n-r$ electrons.

The Roothaan's equations will keep the same shape but now h_{ml} will have the expression:

$$h_{ml} = \langle \chi_m h^{\mathrm{core}} \chi_l\rangle \tag{2.268}$$

where h^{core} represents the potential due to the whole electrostatic distribution replacing the atomic core loges.

Furthermore, as repulsion integrals (ml/pq) involving an overlap distribution $\chi_m\chi_l$ $(m \neq l)$ have values near zero it has become customary to neglect them, putting:

$$(ml/pq) = (mm/pp)\,\delta_{ml}\,\delta_{pq} \tag{2.269}$$

This is the zero-differential overlap (ZDO) approximation.[79] In addition, the corresponding overlap integrals $S_{ml}(m \neq l)$ are neglected. The Roothaan's equations reduce to:

$$\sum_l C_{kl} h_{ml}^{\mathrm{SCF}} = C_{km}\varepsilon_k \tag{2.270}$$

with:

$$
\begin{aligned}
h_{ll}^{\mathrm{SCF}} &= h_{ll} + \sum_{p} P_{pp}(ll/pp) - \tfrac{1}{2}P_{ll}(ll/ll) \\
h_{ml}^{\mathrm{SCF}} &= h_{ml} - \tfrac{1}{2}P_{ml}(mm/ll), \qquad m \neq l
\end{aligned}
\tag{2.271}
$$

Other less severe approximations are made using the CNDO method. A complete discussion of these has been given by Pople and Beveridge.[80]

In summary the CNDO approximations are:

(a) Replacing the overlap matrix by the unit matrix in the Roothaan equations and neglecting the overlap integrals S_{lm} in normalizing the molecular orbitals.

(b) Neglecting differential overlap in all two-electron integrals so that:

$$(ml/pq) = (mm/pp)\,\delta_{ml}\,\delta_{pq}$$

(c) Reducing the remaining set of Coulomb-type integrals to one value per atom pair:

$$(ll/mm) = \gamma_{\mathrm{AB}}$$

if χ_l is related to atom A and χ_m to atom B.

(d) Neglecting monatomic differential overlap in the interaction integrals involving the cores of other atoms. If V_{B} denotes the potential replacing the atomic core loge B:

$$\langle \chi_l |\, V_{\mathrm{B}}\, | \chi_m \rangle = 0, \qquad \text{if } m \neq l \text{ for } \chi_l, \chi_m \text{ on A}$$

Furthermore, a unique value:

$$v_{\mathrm{AB}} = \langle \chi_l |\, V_{\mathrm{B}}\, | \chi_l \rangle$$

will be introduced for all χ_l associated with a given atom A.

(e) Taking diatomic off-diagonal core matrix elements to be proportional to the corresponding overlap integrals:

$$h_{ml} = \beta_{\mathrm{AB}}^{0} S_{ml}, \qquad \text{for } \chi_m \text{ on A, } \chi_l \text{ on B}$$

An effective CNDO calculation requires values for the overlap integrals S_{ml}, the core hamiltonain elements, the repulsion integrals γ_{AB} and the bonding parameters β_{AB}. Various kinds of parameterization have been proposed: CNDO/1,[81] CNDO/2,[82] Jaffé parameterization[83] and others.[84] Here we shall only describe the CNDO/2 parameterization which has been found to be very successful. The others have been described by Pople and Beveridge[80] and by Daudel and Sandorfy.[77]

The atomic orbitals are Slater-type orbitals (STO). The exponents are calculated by following Slater's rules except that for hydrogen 1.2 is used. The overlap integrals S_{lm} are explicitly calculated. The repulsion integrals γ_{AB} are approximated by using the expression:

$$\gamma_{\mathrm{AB}} = \int s_{\mathrm{A}}^{2}(\mathrm{M}_a)\,\frac{e^2}{r_{ab}}\,s_{\mathrm{B}}^{2}(\mathrm{M}_b)\,\mathrm{d}v_a\,\mathrm{d}v_b \tag{2.272}$$

where the function s are s valence orbitals. The integrals v_{AB} are estimated from the approximate equation:

$$v_{AB} = Z_B \gamma_{AB} \tag{2.273}$$

where Z_B denotes the charge of the core B.

The expression of h_{ll} needs to be made explicit:

$$h_{ll} = \langle \chi_l | -\frac{h^2}{8\pi^2 m}\Delta - V_A - \sum_{B \neq A} V_B | \chi_l \rangle, \qquad \text{for } \chi_l \text{ on A}$$

Using equation (2.273):

$$-\langle \chi_l | \sum_{B \neq A} V_B | \chi_l \rangle = -\sum_{B \neq A} Z_B \gamma_{AB}$$

The remaining parameter:

$$U_{ll} = \langle \chi_l | -\frac{h^2}{8\pi^2 m}\Delta - V_A | \chi_l \rangle, \qquad \text{for } \chi_l \text{ on A}$$

is related to the Mulliken electronegativities by the approximate equation:

$$U_{ll} = -\tfrac{1}{2}(I_l + A_l) - (Z_A - \tfrac{1}{2})\gamma_{AA} \tag{2.274}$$

where I_l is the ionization energy associated with the orbital χ_l (Koopmans theorem) and A_l is the electron affinity. Table 2.9 shows the values adopted for $I_l + A_l$.

Table 2.9 Values of $\frac{1}{2}(I + A)$ (in electronvolts)

	H	Li	Be	B	C	N	O	F
$\frac{1}{2}(I_s + A_s)$	7.176	3.106	5.946	9.594	14.051	19.316	25.390	32.272
$\frac{1}{2}(I_p + A_p)$		1.258	2.563	4.001	5.572	7.275	9.111	11.080

The parameters β^0_{AB} are completely empirical. They obey the equation:

$$\beta^0_{AB} = \tfrac{1}{2}(\beta^0_A + \beta^0_B)$$

The parameters β^0_P are given in Table 2.10. The CNDO/2 parameterization is due to Pople and Segal[81] and has been extended to third-row elements by Santry and Segal.[85]

Table 2.10 Values of β^0_P (in electronvolts)

	H	Li	Be	B	C	N	O	F
$-\beta^0_P$	9	9	13	17	21	25	31	39

It can readily be seen that, taking account of all these approximations, equation (2.271) becomes:

$$h_{ll}^{SCF} = -\tfrac{1}{2}(I_l + A_l) + \sum_{B \neq A} (P_{BB} - Z_B)\gamma_{AB} + [(P_{AA} - Z_A) - \tfrac{1}{2}(P_{ll} - 1)]\gamma_{AA} \quad (2.275)$$

and

$$h_{lm}^{SCF} = \beta_{AB}^{0} S_{lm} - \tfrac{1}{2} P_{lm} \gamma_{AB}$$

It is understood that the orbital χ_l belongs to atom A and the orbital χ_m to atom B. Furthermore, the following notations have been introduced:

$$P_{BB} = \sum_{l \text{ on } B} P_{ll} \quad (2.276)$$

Less severe approximations have been also suggested: intermediate neglect of differential overlap (INDO)[86] and neglect of diatomic differential overlap (NDDO).[87] They will not be discussed here. Programs have been written for the different computers used to calculate CNDO and INDO molecular orbitals. Some have been presented by Pople and Beveridge.[80]

To show the usefulness of the various approximations described in this section it would be best to present various applications. We know that by minimizing the total energy with respect to internuclear distances the most stable nuclear conformation can be obtained. Table 2.11 makes it possible to compare calculated and experimental results. There is no doubt that the agreement between experiment and theory is quite satisfactory.

Table 2.11 Nuclear conformations

	Equilibrium bond length (Å)			Angle	
Molecule	CNDO	INDO	Exp.	CNDO	Exp.
H_2	0.746	0.746	0.742		
Li_2	2.179	2.134	2.672		
O_2^+	1.095	1.100	1.123		
CN	1.169	1.174	1.172		
H_2O				107	104.45
CO_2				180	180
O_3				114	116.8
OF_2				99.2	103.8
				CNDO	INDO
NH_3				106.7	109.7
BH_3				120	120

We saw in Section 2.2.4 that when overlap is neglected the quantity $|C_{jp}|^2$ represents the amount of electronic charge which the orbital φ_j brings to the atom region p. As this section deals with ZDO approximations we can use this result and say that P_{BB} represents the electronic charge which all

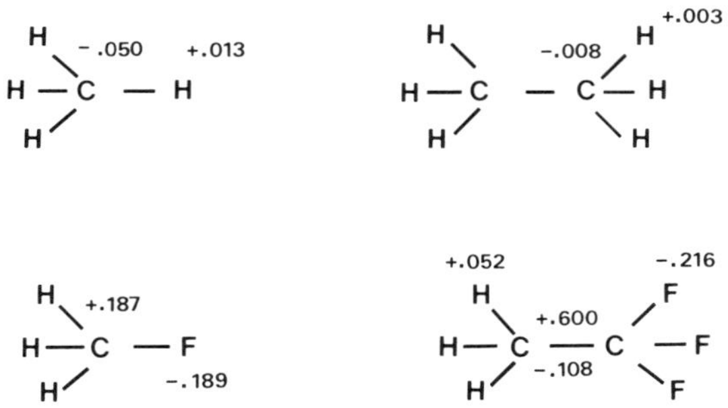

Figure 2.2 Charge distribution in saturated molecules

valence orbitals bring to the atom B. Therefore the net charge Δq_A of each atom A appears to be:

$$\Delta q_A = Z_A - P_{AA} \tag{2.277}$$

where Z_A is, as always, the charge of the core loge A. Figure 2.2 shows the distribution of some atomic charges. It can be seen that in the saturated hydrocarbon the hydrogen atoms are slightly positive and that in the fluorine compound there is a very marked inductive effect. We can have confidence in these distributions of charges as the CNDO and INDO wave functions correspond to dipole moments, in agreement with experiment (Table 2.12).

Table 2.12 Theoretical and experimental dipole moments

	Dipole moment (debyes)		
Molecule	CNDO	INDO	Exp.
LiH	−6.16	−6.20	−5.88
CH	1.87	1.69	1.46
LiF	7.90	7.86	6.6
NH_3	1.97		1.468
NCN	2.48		2.98
Nitrobenzene	5.33		4.28
Formyl fluoride	38.2		41

For a non-polar molecule the various net charges which appear in equations (2.275) are small. If they are neglected:

$$h_{ll}^{SCF} = -\tfrac{1}{2}(I_l + A_l) \tag{2.278}$$

This equation explains the success of the Hoffmann method,[88] usually called the extended Hückel (EH) method. In the EH method the h_{ll}^{SCF}

matrix elements are simply taken as the approximate valence state ionization potentials. Furthermore, the h_{lm}^{SCF} elements are written as:

$$h_{lm}^{\text{SCF}} = 0.5K(h_{ll}^{\text{SCF}} + h_{mm}^{\text{SCF}})S_{lm} \tag{2.279}$$

where K is usually chosen to be 1.75 in order to give the best agreement with the greatest possible number of experimental values. The Roothaan's equations become:

$$c_{km}(I_l - \varepsilon_k) + \sum_{l \neq m} C_{kl}[0.87(I_l + I_m) - \varepsilon_k S_{lm}] = 0, \qquad \text{for } m = 1, \ldots, n \tag{2.280}$$

It is no longer necessary to use an iterative procedure to solve this very simple secular system as the coefficients of the C_{kl}'s do no more depend on the unknowns.

A very rapid computer program allows application of the method to a wealth of molecules with all valence electrons taken into account. The total energy is usually calculated by summing the various orbital energies involved, as in the simple independent electron model. As Hoffmann has said: 'A Hückel calculation of this type, carried out as a function of inter-nuclear distance, gives rise to a potential curve having a minimum not far from the correct experimentally determined geometry of the molecule'. Binding energies are in less satisfactory agreement with experimental values and the moduli of net atomic electron charges are usually too high. An iterative extended Hückel method has been proposed. For details see Daudel and Sandorfy's book.[77]

The kind of parameterization discussed in this section can also be introduced in the framework of the configuration interaction method. A CNDO CI procedure has been proposed. The PCILO method, discussed at the end of Section 2.2.3, can also be used with this kind of parameterization. The CNDO PCILO method has been found to be very successful in predicting nuclear conformation of molecules. Pullman[89] used it with success to calculate the nuclear conformation of amino acids, nucleosides and nucleotides, polypeptides, polysaccharides and other biomolecules.

2.2.6 Semiempirical Calculations: π,σ Separation, Piology, Pariser–Parr Approximation

Let us consider the ethylene molecule. The ground state nuclear conformation of this molecule is planar (Figure 2.3). A minimal basis of atomic orbitals is made of h_1, h_2, h_3, h_4, $1s_1$, $2s_1$, $2p_{1x}$, $2p_{1y}$, $2p_{1z}$, $1s_2$, $2s_2$, $2p_{2x}$, $2p_{2y}$, $2p_{2z}$ if the h's denote the $1s$ orbitals associated with the hydrogen nuclei, other notations being obvious. The coordinate axes are evoked in the figure. It can readily be seen that all atomic orbitals, the $2p_z$ orbitals excepted, are *symmetrical* with respect to the nuclear plane. On the other

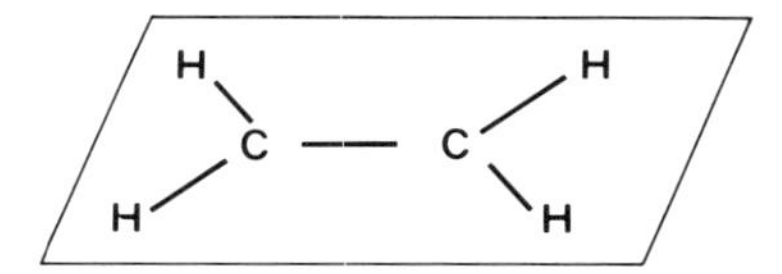

Figure 2.3 Ethylene geometry

hand, the $2p_z$ orbitals are antisymmetrical with respect to this plane as $\cos\theta$ is positive above the plane and negative below. Therefore we can consider two classes of LCAO molecular orbitals: the linear combinations of symmetrical atomic orbitals which are called *σ orbitals* and the linear combinations of antisymmetrical atomic orbitals which are called *π orbitals*. The σ and π orbitals belong to two different irreducible linear representations of the symmetry group of the nuclear conformation (D_{2h}).

It has been shown by Roothaan[90] that when solving the Roothaan equations *if we start from π and σ orbitals* for a given iteration *we obtain π and σ orbitals* at the end of the iteration. In other words, the iteration procedure never mixes the π and σ orbitals. This is the *π, σ separation*, the main basis of the piology.

If now the Roothaan's equations are solved, starting from π and σ orbitals, it is seen that the ground state can be represented by seven σ orbitals and one π orbital, each of them being used twice:

$$\Psi = \det \varphi_{\sigma 1}\bar{\varphi}_{\sigma 1}\varphi_{\sigma 2}\bar{\varphi}_{\sigma 2}\varphi_{\sigma 3}\bar{\varphi}_{\sigma 3}\,\varphi_{\sigma 4}\bar{\varphi}_{\sigma 4}\varphi_{\sigma 5}\bar{\varphi}_{\sigma 5}\varphi_{\sigma 6}\bar{\varphi}_{\sigma 6}\varphi_{\sigma 7}\bar{\varphi}_{\sigma 7}\varphi_{\pi}\bar{\varphi}_{\pi} \tag{2.281}$$

The π orbital corresponds to the highest orbital energy ε. It is useful to understand qualitatively why this is so. Let t_c, t_1, t_2 be the trigonal hybrids associated with C_1 pointing towards C_2, H_1 and H_2 respectively and t'_c, t_3, t_4 hybrids associated with C_2 pointing towards C_1, H_3 and H_4.

We introduce the following orbitals (called *equivalent orbitals* because during a symmetry operation they are exchanged between them):

$$1s_1,\ 1s_2$$

$$\begin{aligned}
\chi_1 &= \lambda h_1 + \mu t_1 & \chi'_1 &= \lambda' h_1 - \mu' t_1 \\
\chi_2 &= \lambda h_2 + \mu t_2 & \chi'_2 &= \lambda' h_2 - \mu' t_2 \\
\chi_3 &= \lambda h_3 + \mu t_3 & \chi'_3 &= \lambda' h_3 - \mu' t_3 \\
\chi_4 &= \lambda h_4 + \mu t_4 & \chi'_4 &= \lambda' h_4 - \mu' t_4 \\
l &= N_\sigma(t_c + t'_c) & l^* &= N_{\sigma^*}(t_c - t'_c) \\
\pi &= N_\pi(2p_{1z} + 2p_{2z}) & \pi^* &= N_{\pi^*}(2p_{1z} - 2p_{2z})
\end{aligned}$$

and from them derive the symmetry orbitals (that is to say orbitals belonging to the various linear representations of the symmetry group D_{2h}):

$$\varphi_1 = K(1s_1 + 1s_2)$$
$$\varphi_2 = K'(1s_1 - 1s_2)$$
$$\varphi_3 = a(\chi_1 + \chi_2 + \chi_3 + \chi_4)$$
$$\varphi_4 = b(\chi_1 + \chi_2 - \chi_3 - \chi_4)$$
$$\varphi_5 = c(\chi_1 - \chi_2 - \chi_3 + \chi_4)$$
$$\varphi_6 = d(\chi_1 - \chi_2 + \chi_3 - \chi_4)$$
$$\varphi_{11} = a'(\chi_1' + \chi_2' + \chi_3' + \chi_4')$$
$$\varphi_{12} = b'(\chi_1' + \chi_2' - \chi_3' - \chi_4')$$
$$\varphi_{13} = c'(\chi_1' - \chi_2' - \chi_3' + \chi_4')$$
$$\varphi_{14} = d'(\chi_1' - \chi_2' + \chi_3' - \chi_4')$$
$$\varphi_7 = l$$
$$\varphi_{10} = l^*$$
$$\varphi_8 = \pi$$
$$\varphi_9 = \pi^*$$

All normalizing factors introduced in the orbital are assumed to be positive.

Let us characterize by S_x, S_y, S_p, A_x, A_y, A_p the orbitals which are symmetrical or antisymmetrical with respect to axis Ox, Oy or the plane of the molecule respectively. Six kinds of symmetry orbital will appear:

$\varphi_1, \varphi_3, \varphi_7, \varphi_{11}$	are	S_x, S_y, S_p
$\varphi_2, \varphi_4, \varphi_{10}, \varphi_{12}$	are	S_x, A_y, S_p
φ_5, φ_{13}	are	A_x, S_y, S_p
φ_6, φ_{14}	are	A_x, A_y, S_p
φ_8	is	S_x, S_y, A_p
φ_9	is	S_x, A_y, A_p

Each class of orbitals forms a basis for a linear irreducible representation of the point group D_{2h}. By generalizing what has been said for the π and σ orbitals it may be expected that it is possible to manage the calculation of the self-consistent field solutions in such a way as to avoid mixing the different classes of orbital. This is true.

To go farther, it is necessary to use two rules. The first one is based on the differentiation between *bonding* and *antibonding orbitals.* An orbital is said to be bonding when it favours the presence of electrons between adjacent cores, antibonding in the opposite case. Consider an equivalent orbital like π, the $2p_{1z}$ and $2p_{2z}$ orbitals having their positive parts above the plane and their negative parts below. The overlapping between them creates an increase in the modulus of the orbital π between the carbon cores; *π is a* *bonding orbital.* The reverse situation appears with π^* as the overlap occurs between $2p_{1z}$ and $-2p_{2z}$; π^* *is an antibonding orbital.* It can readily be seen that in our list of equivalent orbitals the first column corresponds to bonding orbitals, the second column to antibonding orbitals. As the region between two adjacent cores corresponds to a relatively low value of the potential energy *the presence of a bonding orbital in a molecular orbital is a stabilizing factor,* i.e. it decreases the value of the energy ε associated with the orbital. This is the first rule. As a consequence of this rule we may anticipate that φ_3, φ_4, φ_5, φ_6 are more 'stable' than φ_{11}, φ_{12}, φ_{13}, φ_{14}, that φ_7 is more stable than φ_{10} and that φ_8 is more stable than φ_9.

The second rule is that orbitals of very different energies do not mix much. φ_1 and φ_2 are rich in $1s$ orbitals and therefore have very low energy in comparison to all other symmetry orbitals. φ_2 is theoretically antibonding but as the overlapping between the two $1s$ orbitals is negligible it will have about the same energy as φ_1. Therefore, φ_1 will not mix very much with other orbitals of the first class. Furthermore, φ_{11} will also not mix very much in this class for the opposite reason. The first class will produce the following molecular orbitals:

$$\varphi_{\sigma 1} \approx \varphi_1$$

$$\varphi_{\sigma 3} = l\varphi_3 + m\varphi_7$$

$$\varphi_{\sigma 7} = l'\varphi_3 + m'\varphi_7$$

$$\varphi_{\sigma 11} \approx \varphi_{11}$$

The same kind of discussion shows that:

$$\varphi_{\sigma 2} \approx \varphi_2$$

$$\varphi_{\sigma 4} \approx \varphi_4$$

$$\varphi_{\sigma 10} = n\varphi_{10} + \sigma\varphi_{12}$$

$$\varphi_{\sigma 12} = n'\varphi_{10} + \sigma'\varphi_{12}$$

$$\varphi_{\sigma 5} \approx \varphi_5$$

$$\varphi_{\sigma 13} \approx \varphi_{13}$$

$$\varphi_{\sigma 6} \approx \varphi_6$$

$$\varphi_{\sigma 14} \approx \varphi_{14}$$

$$\varphi_{\pi 8} = \varphi_8 = \pi$$

$$\varphi_{\pi 9} = \varphi_9 = \pi^*$$

From all these orbitals $\varphi_{\sigma 1}$ and $\varphi_{\sigma 2}$ will be associated with the lowest energy. The group $\varphi_{\sigma 3}$, $\varphi_{\sigma 4}$, $\varphi_{\sigma 5}$, $\varphi_{\sigma 6}$, $\varphi_{\sigma 7}$ and $\varphi_{\sigma 8}$ will have higher energies but we cannot predict the energy order in this group without further analysis. All other molecular orbitals are mainly antibonding and will close the list of orbitals classified according to the increasing energy order.

The calculation shows that the overlap between $2p_{1z}$ and $2p_{2z}$ is smaller than between t_c and t_c' or between h_i and t_i. This explains why, finally, the order of the molecular orbitals is:

$$(\varphi_{\sigma 1}, \varphi_{\sigma 2}), (\varphi_{\sigma 3}, \varphi_{\sigma 4}, \varphi_{\sigma 5}, \varphi_{\sigma 6}, \varphi_{\sigma 7}), \varphi_{\pi 8}, \varphi_{\pi 9}, (\varphi_{\sigma 10}, \varphi_{\sigma 11}, \varphi_{\sigma 12}, \varphi_{\sigma 13}, \varphi_{\sigma 14})$$

We can now understand quite well why the ground state of ethylene is conveniently represented by the function (2.281). We can anticipate that the configuration $(\varphi_{\sigma 1})^2(\varphi_{\sigma 2})^2(\varphi_{\sigma 3})^2 \cdots (\varphi_{\sigma 7})^2\varphi_{\pi 8}$ will represent the ground state of the ethylene positive ion and that the configurations:

$$(\varphi_{\sigma 1})^2 \cdots (\varphi_{\sigma 7})^2 \varphi_{\pi 8}\varphi_{\pi 9}$$

$$(\varphi_{\sigma 1})^2 \cdots (\varphi_{\sigma 7})^2 (\varphi_{\pi 9})^2$$

will represent the first excited states of the molecule.

It appears that the reorganization of the electronic structure of ethylene under the effect of ionization or light excitation is mainly represented by a change in the orbital part of the wave function insofar as small energy changes are concerned. This is the basis of the piology, in which there is an approximation assuming that for such a change the σ part of the wave function remains unchanged. The importance of the piology is due to the fact that what we have said for ethylene can be extended to many planar or quasi-planar molecules or to the planar region of molecules, the number of π orbitals involved depending on the size of the electronic system concerned.

To continue the discussion for the ethylene example, we can state that in the self-consistent field operator h^{SCF} the part:

$$h + \sum_{j=1}^{j=7} (2J_j - K_j)$$

remains the same for the ethylene ground state, the ethylene positive ion ground state and the first ethylene excited states. As a consequence this part of the operator is called the *core operator.* We set:

$$h^{\mathrm{c}} = h + \sum_{j=1}^{j=7} (2J_j - K_j) \cdots \tag{2.282}$$

Goeppert-Mayer and Sklar[91] proposed an interesting approximation for this operator. The effect of the hydrogen atoms is neglected as they are far from the π region. The rest of the core is replaced by the coulomb field produced by a charge distribution equal to the difference between that of the two carbons in a ^5S state (spherical symmetry) and that of the orbitals $2p_{1z}$ and $2p_{2z}$. The core operator then reduces to:

$$h^{\mathrm{c}} = -\frac{h^2\Delta}{8\pi^2 m} + u_1^+ + u_2^+ \tag{2.283}$$

where:

$$u_i^+ = u_i - \int [2p_{iz}(\mathrm{M}_b)]^2 \frac{e^2}{r_{ab}} \, \mathrm{d}v_b \tag{2.284}$$

u_i being the coulomb potential of a carbon at i in the ^5S state. Furthermore, it is reasonable to assume that as a first approximation we have:

$$\langle 2p_{iz}| -\frac{h^2\Delta}{8\pi^2 m} + u_i^+ |2p_{iz}\rangle = W_{2p} \quad (2.285)$$

where W_{2p} represents the ionization energy of a carbon atom in a ^5S state. With such approximation a matrix element like:

$$\langle 2p_{1z}| h^c |2p_{1z}\rangle$$

reduces to:

$$\begin{aligned}\langle 2p_{1z}| h^c |2p_{1z}\rangle &= W_{2p} + \langle 2p_{1z}| u_2^+ |2p_{1z}\rangle \\ &= W_{2p} + \langle 2p_{1z}| u_2 |2p_{1z}\rangle - \int 2p_{2z}^2(\mathrm{M}_b)\frac{e^2}{r_{ab}} 2p_{1z}^2(\mathrm{M}_a)\, \mathrm{d}v_{ab} \quad (2.286)\end{aligned}$$

To avoid the calculation of various polycentric integrals Pariser and Parr[92] introduced at that level a ZDO approximation which consists of neglecting both the overlap between the orbitals $2p_{1z}$ and $2p_{2z}$ and all the bielectronic integrals containing a product of the form $2p_{2z}(i)\,2p_{1z}(i)$ at least once.

Finally the matrix element $\langle 2p_{1z}| h^c |2p_{2z}\rangle$ is considered to be an empirical parameter. The value chosen can be that which gives the best possible duplication with non-empirical or with experimental results.

The Pariser and Parr approximation has been found to be useful for the treatment of conjugated molecules.

REFERENCES

1. T. Kato, *Trans. Amer. Math. Soc.*, **70,** No. 2, 212 (1951).
2. L. Pauling and E. B. Wilson, *Introduction to Quantum Mechanics*, McGraw-Hill, 1935.
3. R. Daudel, *Adv. Quantum Chemistry*, **1,** 115 (1964).
4. P. O. Lowdin, *Phys. Rev.*, **97,** 1509 (1955).
5. E. M. Corson, *Perturbation Method in the Quantum Mechanics of n-electron Systems*, Hafner, New York, 1951.
6. R. Daudel, R. Lefebvre and C. Moser, *Quantum Chemistry*, Wiley-Interscience, 1959.
7. F. A. Matsen, *J. Phys. Chem.*, **68,** 3282 (1964); J. I. Musher, *Journal de Physique*, **31,** C4–51 (1970).
8. J. F. Gouyet, *Phys. Rev. A.*, **2,** 139, 1286 (1970).
9. M. Born and J. R. Oppenheimer, *Ann. Physik*, **84,** 457 (1927); M. Born, *Gött. Nachr. Math Phys.*, **K1,** 1 (1951).
10. R. Daudel and S. Bratoz, *Cahier de Physique*, **75, 76,** 39 (1956); W. D. Hobey and A. D. McLachlan, *J. Chem. Phys.*, **33,** 1965 (1960); R. Lefebvre and M. Garcia-Sucre, *Int. J. Quantum Chem.*, **1s,** 339 (1967).
11. C. Eckart, *Phys. Rev.*, **36,** 878 (1930).

12. W. E. Duncanson and C. A. Coulson, *Proc. Roy. Soc., Edinburgh*, **62,** 37 (1944), *Nature*, **164,** 1003 (1949).
13. J. C. Slater, *Phys. Rev.*, **36,** 57 (1930).
14. R. K. Nesbet, *J. Chem. Phys.*, **40,** 3619 (1964).
15. D. R. Hartree, *Proc. Cambridge Phil. Soc.*, **24,** 111, 426 (1928), **25,** 225, 310 (1929).
16. P. O. Lowdin, *Phys. Rev.*, **97,** 1509 (1955).
17. R. Lefebvre and Y. G. Smeyers, *Int. J. Quantum Chemistry*, **1,** 403 (1967).
18. Y. G. Smeyers, *Anales de Fisica*, **67,** 17 (1971).
19. L. Pauling, *Chem. Rev.*, **5,** 173 (1928).
20. B. N. Finkelstein and G. E. Horowitz, *Z. f. Phys.*, **48,** 118 (1928).
21. B. N. Dickenson, *J. Chem. Phys.*, **1,** 317 (1933).
22. C. C. J. Roothaan, *Rev. Mod. Phys.*, **23,** 69 (1951).
23. R. Daudel, R. Lefebvre and C. Moser, *Quantum Chemistry*, Wiley-Interscience, 1959, p. 552.
24. W. G. Richards and J. A. Horsley, *Ab Initio Molecular Orbital Calculations for Chemists*, Clarendon Press, Oxford, 1970; W. G. Richards, T. E. H. Walker and R. K. H. Hinkley, *A Bibliography of Ab Initio Molecular Wave Functions*, Clarendon Press, Oxford, 1971.
25. S. F. Boys, *Proc. Roy. Soc.*, **A200,** 542 (1950).
26. S. Huzinaga, *J. Chem. Phys.*, **42,** 1293 (1965); A. Veillard, *Theor. Chim. Acta*, **12,** 405 (1968); H. Baschoc, J. Hornback and J. W. Moskowitz, *J. Chem. Phys.*, **51,** 1511 (1969).
27. E. Clementi and D. R. Davis, *J. Chem. Phys.*, **45,** 2593 (1966).
28. J. A. Pople, in *Aspects de la Chimie Quantique Contemporaine*, CNRS, 1971, p. 17.
29. D. Neumann and J. W. Moskowitz, *J. Chem. Phys.*, **49,** 2056 (1968).
30. E. Switkes, R. M. Stevens and W. N. Lipscomb, *J. Chem. Phys.*, **51,** 5229 (1969).
31. C. W. Kern and R. L. Matcha, *J. Chem. Phys.*, **49,** 2081 (1968).
32. G. P. Arrighini, M. Maestro and R. Moccia, *Symp. Faraday Soc.*, **2,** 45 (1968).
33. A. C. Hopkinson, N. K. Holbrook, K. Yates and I. G. Czimadia, *J. Chem. Phys.*, **49,** 3596 (1968).
34. K. J. Miller, S. R. Miekzarrek and M. Krauss, *J. Chem. Phys.*, **51,** 26 (1969).
35. J. Andriesson, *Chem. Phys. Letters*, **3,** 257 (1969).
36. A. Veillard and R. Daudel, *La Nature et les Propriétés des Liaisons de Coordination*, CNRS, 1970, p. 23.
37. A. L. McClellan, *Tables of Experimental Dipole Moments*, Freeman and Co., 1963.
38. W. E. Palke and W. N. Lipscomb, *J. Amer. Chem. Soc.*, **89,** 2384 (1966).
39. T. Koopmans, *Physica*, **1,** 104 (1933).
40. W. G. Richards and J. A. Horsley, *Ab Initio Molecular Orbital Calculations for Chemists*, Clarendon Press, Oxford, 1970, Chap. 10.
41. A. C. Wahl, *J. Chem. Phys.*, **41,** 2600 (1964).
42. J. K. L. McDonald, *Phys. Rev.*, **43,** 830 (1933).
43. G. Das and A. C. Wahl, *Phys. Rev. Letters*, **24,** 440 and also references quoted (1970); B. Lévy, *Chem. Phys. Letters*, **18,** 59 (1973).
44. H. F. Schaeffer III, *The Electronic Structure of Atoms and Molecules*, Addison-Wesley, 1972.
45. P. O. Lowdin, *Phys. Rev.*, **97,** 1474 (1955).
46. S. Hagstrom and H. Shull, *Rev. Mod. Phys.*, **35,** 624 (1963).
47. C. F. Bender and E. R. Davidson, *J. Phys. Chem.*, **70,** 2675 (1966).
48. B. Huron, J. P. Malrieu and P. Rancurel, *J. Chem. Phys.*, **58,** 5745 (1973).

49. C. E. Shannon, Prediction and entropy of printed English, *Bell Sytem Technical Journal*, **30,** 50 (1951).
50. R. D. Brown, *J. Chem. Soc.*, **1953,** 2615 (1953).
51. R. Daudel, *Les Fondements de la Chimie Théorique*, Gauthier Villars, 1956 (English ed., *The Fundamentals of Theoretical Chemistry*, Pergamon, Oxford).
52. H. Brion and R. Daudel, *Compt. Rend. Acad. Sci.*, **237,** 457 (1953).
53. C. E. Edminston and K. Ruedenberg, *Rev. Mod. Phys.*, **35,** 457 (1963).
54. J. M. Foster and S. F. Boys, *Rev. Mod. Phys.*, **32,** 300 (1960).
55. M. A. Robb, N. J. Haines and I. G. Csizamadia, *J. Amer. Chem. Soc.*, **95,** 42 (1973).
56. R. J. Gillespie, *Molecular Geometry*, Van Nostrand Reinhold, 1973.
57. R. Daudel and A. Veillard, *Nature et Propriétés des Liaisons de Coordination*, CNRS, 1970, p. 21.
58. C. Aslangul, A. Veillard, R. Daudel and F. Gallais, *Theor. Chim. Acta*, **23,** 211 (1971).
59. R. Daudel, *Les Fondements de la Chimie Théorique*, Gauthier Villars, Paris, 1956 (English ed., *The Fundamentals of Theoretical Chemistry*, Pergamon, Oxford).
60. E. Ludeña and V. Amzel, *J. Chem. Phys.*, **52,** 5923 (1970).
61. F. Bopp, *Z. Phys.*, **156,** 348 (1959).
62. R. McWeeny and B. T. Sutcliffe, *Proc. Roy. Soc.*, **A273,** 103 (1963).
63. K. Miller and K. Ruedenberg, Communication at the Sanibel Meeting, 1965.
64. J. M. Leclercq, Thèse de 3ème Cycle, Sorbonne, 1966.
65. R. Ahlrichs and W. Kutzelnigg, *Theor. Chim. Acta.*, **10,** 337 (1968), *Chem. Phys. Lett.*, **1,** 651 (1969).
66. M. Klessinger and R. McWeeny, *J. Chem. Phys.*, **42,** 3343 (1965).
67. A. A. Frost, *J. Chem. Phys.*, **47,** 3707 (1967), *J. Chem. Phys.*, **47,** 3714 (1967), *J. Phys. Chem.*, **72,** 1289 (1968).
68. J. C. Barthelat and Ph. Durand, *Theoretica Chim. Acta*, **27,** 109 (1972).
69. R. McWeeny, *Proc. Roy. Soc.*, **A253,** 242 (1959), *Rev. Mod. Phys.*, **32,** 335 (1960).
70. P. Claverie, S. Diner and J. P. Malrieu, *Int. J. Quantum Chem.*, **1,** 751 (1967); S. Diner, J. P. Malrieu and P. Claverie, *Theoret. Chim. Acta*, **13,** 1 (1969); S. Claverie, J. P. Malrieu, J. P. Jordan and M. Gilbert, *Theoret. Chim. Acta*, **15,** 100 (1969); F. Jordan, M. Gilbert, J. P. Malrieu and U. Pincelli, *Theoret. Chim. Acta.*, **15,** 211 (1969).
71. C. Aslangul *et al.*, *Adv. Quantum Chem.*, **6,** 93 (1972).
72. J. C. Slater, *Adv. Quantum Chemistry*, **6,** 1 (1972).
73. G. G. Hall, *Proc. Phys. Soc.*, **A205,** 541 (1951).
74. J. Lennard Jones and G. G. Hall, *Trans. Far. Soc.*, **48,** 581 (1952).
75. R. D. Brown, *J. Chem. Soc.*, **1953,** 2615 (1953).
76. C. Sandorfy and R. Daudel, *Compt. Rend. Acad. Sci.*, **238,** 93 (1954); C. Sandorfy, *Canadian J. Chem.*, **33,** 1337 (1955).
77. R. Daudel and C. Sandorfy, *Semiempirical Wave-Mechanical Calculations on Polyatomic Molecules*, Yale University Press, 1971.
78. G. Del Ré, *J. Chem. Soc.*, **1958,** 4031 (1958), *Electronic Aspects of Biochemistry*, Academic Press, New York, 1964, p. 221.
79. R. G. Parr, *J. Chem. Phys.*, **20,** 239 (1952).
80. J. Pople and D. Beveridge, *Approximate Molecular Orbital Theory*, McGraw-Hill, 1970.
81. J. A. Pople and G. A. Segal, *J. Chem. Phys.*, **43,** 8136 (1965).
82. J. A. Pople and G. A. Segal, *J. Chem. Phys.*, **44,** 3289 (1966).
83. J. Del Bene and H. H. Jaffe, *J. Chem. Phys.*, **48,** 1807 (1968).

84. H. Fischer and H. Kollmar, *Theoret. Chimica Acta*, **13,** 213 (1969).
85. D. P. Santry and G. A. Segal, *J. Chem. Phys.*, **47,** 158 (1967).
86. J. A. Pople, D. L. Beveridge and P. A. Dobosh, *J. Chem. Phys.*, **47,** 2026 (1967).
87. J. A. Pople, D. P. Santry and G. A. Segal, *J. Chem. Phys.*, **43,** 8125 (1965).
88. R. Hoffmann, *J. Chem. Phys.*, **39,** 1397 (1963).
89. B. Pullman, in *Aspect de la Chimie Quantique Contemporaine*, CNRS, 1971, p. 261.
90. C. C. J. Roothaan, *Rev. Mod. Phys.*, **23,** 80 (1951).
91. H. Goeppert-Mayer and A. L. Sklar, *J. Chem. Phys.*, **6,** 645 (1938).
92. R. Pariser and R. Parr, *J. Chem. Phys.*, **21,** 466, 767 (1953).

CHAPTER 3

A First Insight into Electronic Structure of Molecules

3.1. ELECTRONIC DENSITY AND BADER PARTITIONING

3.1.1 Behaviour of Electronic Density in Atoms and Molecules

The electronic density at point M in an atom or molecule is the product of the electronic charge e and the density of the probability of finding an electron at point M. The density $\rho(\mathrm{M})$ is obviously the sum of the density $\rho^+(\mathrm{M})$ of the probability of finding an electron at point M with a spin of $\frac{1}{2}h/2\pi$ and of the density $\rho^-(\mathrm{M})$ of the probability of finding an electron at point M with a spin of $-\frac{1}{2}h/2\pi$.

If $\Psi(\mathrm{M}_a, \omega_a, \mathrm{M}_b, \omega_b, \ldots, \mathrm{M}_n, \omega_n)$ denotes the electronic wave function associated with an atom or molecule it is readily seen that:

$$\rho^+(\mathrm{M}) = \sum_i \int \mathrm{d}v_a\, \mathrm{d}\omega_a \cdots \int \mathrm{d}v_{i-1}\, \mathrm{d}\omega_{i-1} \int \mathrm{d}v_{i+1}\, \mathrm{d}\omega_{i+1} \cdots \int \mathrm{d}v_n\, \mathrm{d}\omega_n$$
$$\times \Psi^*(\mathrm{M}_a, \omega_a, \ldots, \mathrm{M}_{i-1}, \omega_{i-1}, \mathrm{M}, +\tfrac{1}{2}h/2\pi, \mathrm{M}_{i+1}, \omega_{i+1}, \ldots, \mathrm{M}_n, \omega_n)$$
$$\times \Psi(\mathrm{M}_a, \omega_a, \ldots, \mathrm{M}_{i-1}, \omega_{i-1}, \mathrm{M}, +\tfrac{1}{2}h/2\pi, \mathrm{M}_{i+1}, \omega_{i+1}, \ldots, \mathrm{M}_n, \omega_n)$$

and

$$\rho^-(\mathrm{M}) = \sum_i \int \mathrm{d}v_a\, \mathrm{d}\omega_a \cdots \int \mathrm{d}v_{i-1}\, \mathrm{d}\omega_{i-1} \int \mathrm{d}v_{i+1}\, \mathrm{d}\omega_{i+1} \cdots \int \mathrm{d}v_n\, \mathrm{d}\omega_n$$
$$\times \Psi^*(\mathrm{M}_a, \omega_a, \ldots, \mathrm{M}_{i-1}, \omega_{i-1}, \mathrm{M}, -\tfrac{1}{2}h/2\pi, \mathrm{M}_{i+1}, \omega_{i+1}, \ldots, \mathrm{M}_n, \omega_n)$$
$$\times \Psi(\mathrm{M}_a, \omega_a, \ldots, \mathrm{M}_{i-1}, \omega_{i-1}, \mathrm{M}, -\tfrac{1}{2}h/2\pi, \mathrm{M}_{i+1}, \omega_{i+1}, \ldots, \mathrm{M}_n, \omega_n)$$

As the wave function Ψ is antisymmetric with respect to a permutation of the space and spin coordinates associated with two electrons, each electron gives the same contribution to a density at a given point. Therefore $\sum_i$ can be replaced by n (the number of electrons), the various integrals being identical.

Figure 3.1 shows some contour maps of electronic density distributions for homonuclear diatomic molecules.[1,2] These maps have been calculated from a SCF wave function. The contours increase in value from the outermost one to the innermost one. The value of the density is given in Table 3.1.

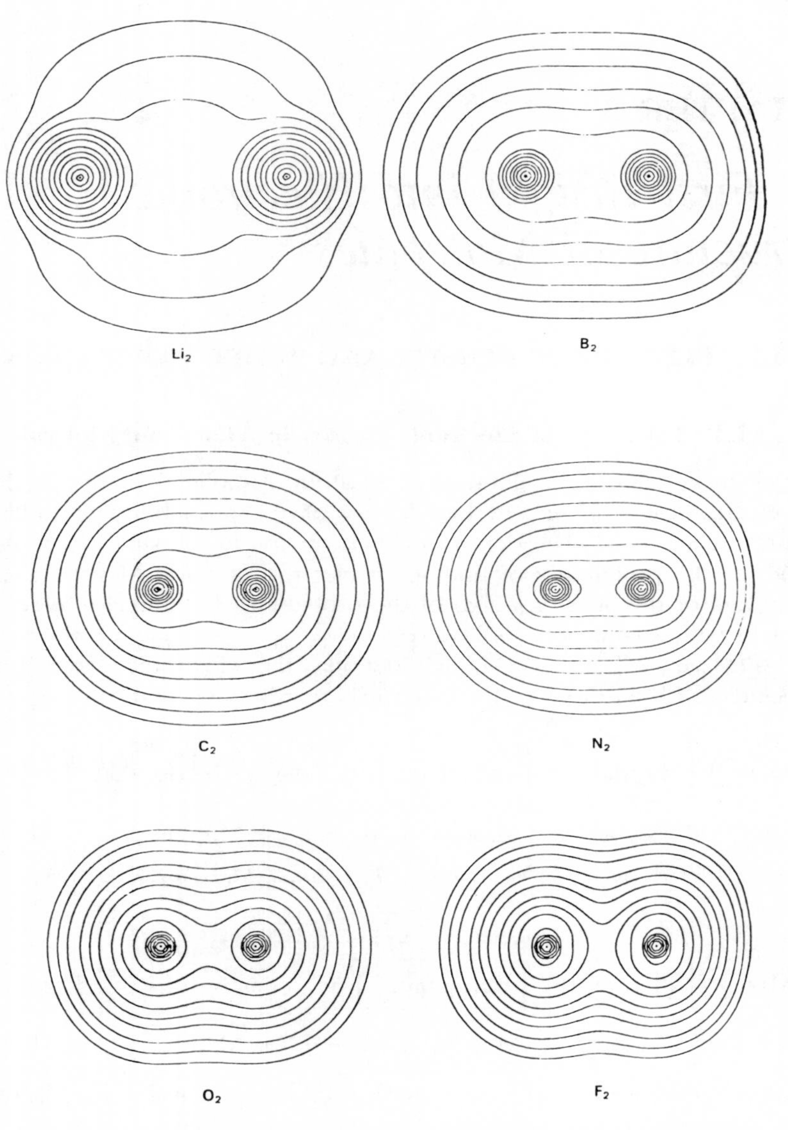

Figure 3.1 Electronic density in various small molecules (after Bader[1]. Reproduced by permission of Clarke, Irwin & Co Limited

Figure 3.2 shows analogous maps for heteronuclear molecules. In all cases a saddle point appears in the bonding region between the nuclei. The density has a maximum value at each nucleus.

Some calculations have been made to show the effect of configuration interaction on the calculation of electronic density in a molecule[3].

Table 3.1 Electronic density of the various contours

Contour number beginning with outermost one	Value of the density (a.u.)	Contour number beginning with outermost one	Value of the density (a.u.)
1	0.002	8	0.4
2	0.004	9	0.8
3	0.008	10	2
4	0.02	11	4
5	0.04	12	8
6	0.08	13	20
7	0.2		

The effect of the configuration interaction is a charge transfer of 0.06 electron from the oxygen to the carbon. This is not a negligible effect. It is sufficient to produce a change of sign of the electric dipole moment. The electric dipole moment of CO calculated from a Hartree–Fock function is $+0.15\,D$ (C^+O^-). It becomes $-0.17\,D$ (C^-O^+) when a configuration interaction wave function is built from this Hartree–Fock function (including 138 doubly excited configurations and 62 singly excited ones). The experimental value deduced from a microwave experiment is $-0.12\,D$.

3.1.2 Molecular Difference Density Function δ(M)

The difference density function has been introduced by Daudel and others[4]. The difference density $\delta(M)$ at point M in a molecule is simply the difference between the actual density $\rho(M)$ and the virtual density $\rho^f(M)$ which would result from the addition of the densities in the free atoms:

$$\delta(M) = \rho(M) - \rho^f(M)$$

Therefore, *in a point where $\delta(M)$ is positive the bonding leads to an increase of the electronic density. In a point where $\delta(M)$ is negative the bonding leads to a decrease of the electronic density.* For that reason the δ function is also called the *bond density function*; it shows the effect of the formation of the bonds on the electronic distribution.

Many authors have calculated this function for various molecules.[5] Figure 3.3 shows typical results of such calculations. For an homonuclear molecule like N_2 the effect of binding is:

(a) An increase of the electronic density in a central region of the molecule (it corresponds to an increase of the order of 0.1 electron)
(b) An increase of the electronic density in the 'lone pair' regions.
(c) A decrease near each nucleus.

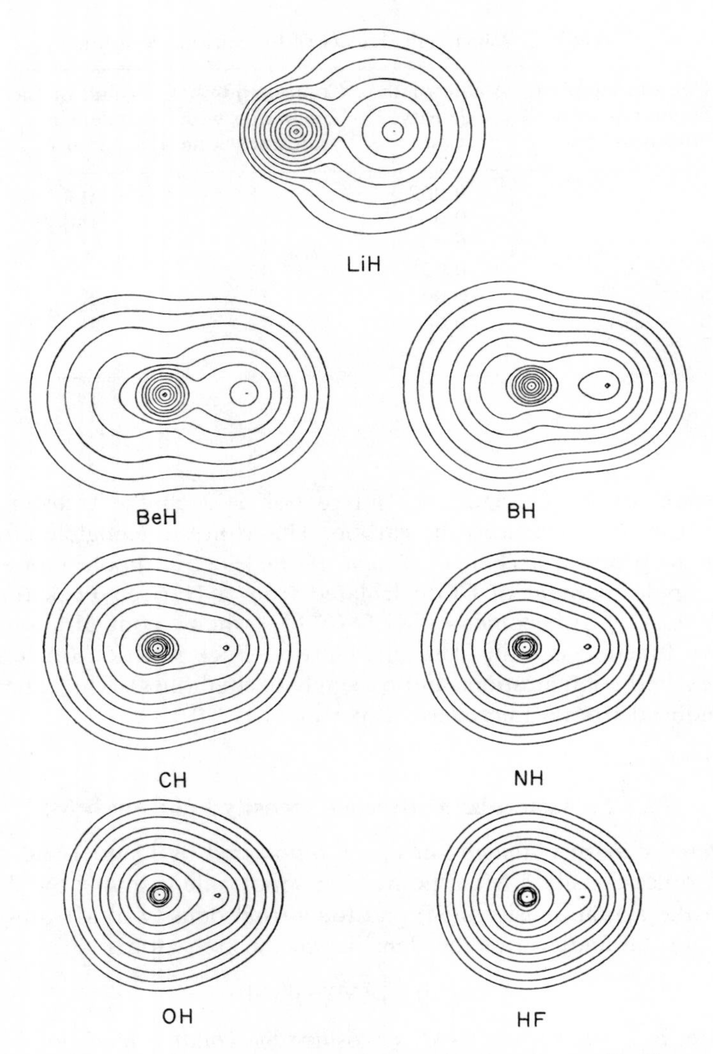

Figure 3.2 Contour maps of the molecular charge distributions of the diatomic hydride molecules LiH to HF. The proton is the nucleus on the right-hand side (from Bader[2]). Reproduced by permission of Clarke, Irwin & Co Limited

For an heteronuclear molecule like LiF we see:

(a) An increase of the electronic density in the whole fluorine region except for a small volume near the nucleus.
(b) A decrease of the electronic density in the lithium region except for a small volume not far from the nucleus. The overall effect amounts to a transfer of charge from the lithium atom region to the fluorine one.

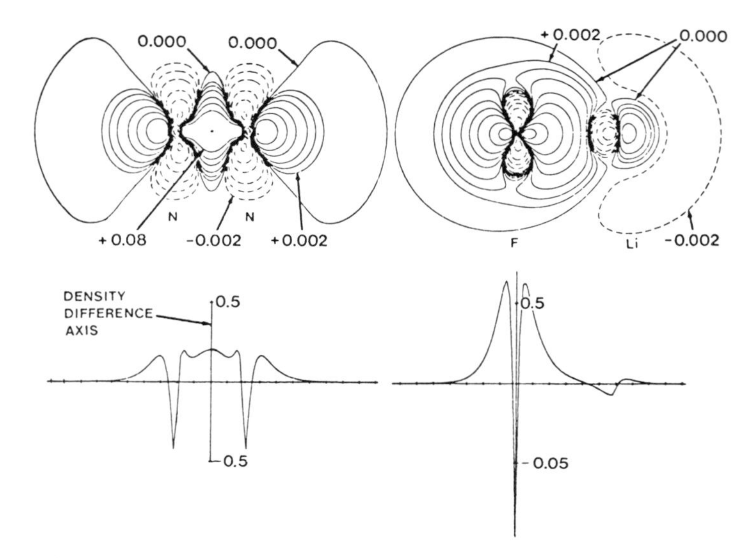

Figure 3.3 Density difference function in N_2 and FLi (from Bader[2]. Reproduced by permission of Clark, Irwin & Co Limited

The effect of bonding on the electronic density has been proved experimentally by measuring the radioactive decay of various compounds. The effect of chemical bond on the rate of decay of some radioactive nuclei has been predicted by Daudel[6] in the case of electron capture and in the case of isomeric transitions. This prediction was based upon the fact that these radioactive decays depend on the electronic distribution near the radioactive nuclei. The prediction for electron capture has also been made by Segré.[7] Measurements carried out in France and at the Brookhaven National Laboratory confirmed that prediction: the decay rate of ^{7}Be was found to be larger in metallic Be than in BeO and BeF_2.

The relation between the rate of decay of a radioactive nucleus and the electronic structure of a molecule has been found to be so sensitive that the measurement of decay is now a powerful procedure of chemical analysis of very small amounts of radioactive compounds.[8]

The measurement of the δ function outside the nuclear region is difficult. In molecular crystals it can be achieved by X-ray diffraction. However, it is also necessary to obtain precise information about the movement of the nuclei in the crystal. Therefore neutron diffraction experiments are also needed.[9]

The comparison between experiment and calculation needs very elaborate wave functions because the difference density function is very sensitive to the quality of the wave function.

Cade[5] has shown that the minimal basis set and even saturated sp basis sets are inadequate for the calculation of $\delta(M)$ in the bond regions. Addition

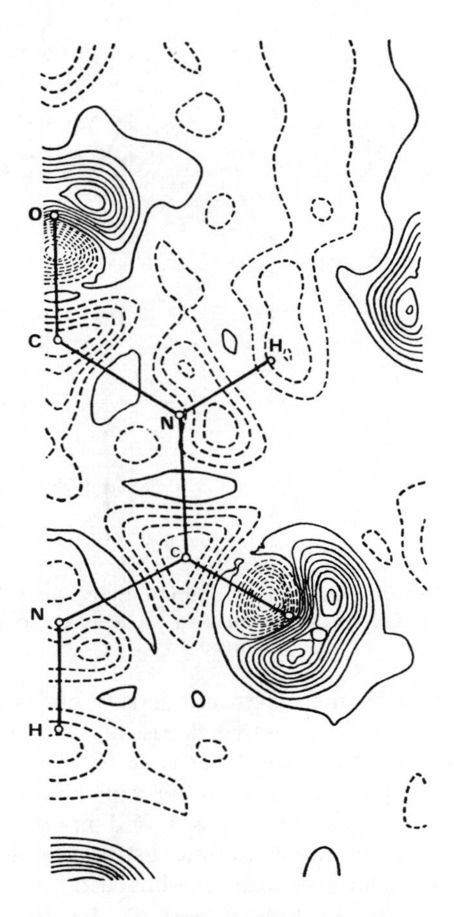

Figure 3.4 Density difference in cyanuric acid: calculation (from Coppens[9]). Reproduced by permission of Butterworth and Co

of polarized functions like *d* orbitals in the case of first-row atoms is needed. Furthermore, it has been stated that the effect of CI is far from being negligible.

Figure 3.4 corresponds to the theoretical difference density in the plane of the cyanuric acid molecule according to a minimal basis set calculation. Figure 3.5 shows the X–N map, that is to say the experimental results.[9] It can clearly be seen that the minimal basis set underestimates the increase of electronic density in the bond regions and overestimates the increase in the lone pair regions.

Figure 3.6 represents the isodensity difference curves for a formamide dimer.[10] It analyses the effect of the hydrogen bond on the distribution of the electronic density. The main effect is a flying-away of the electrons from

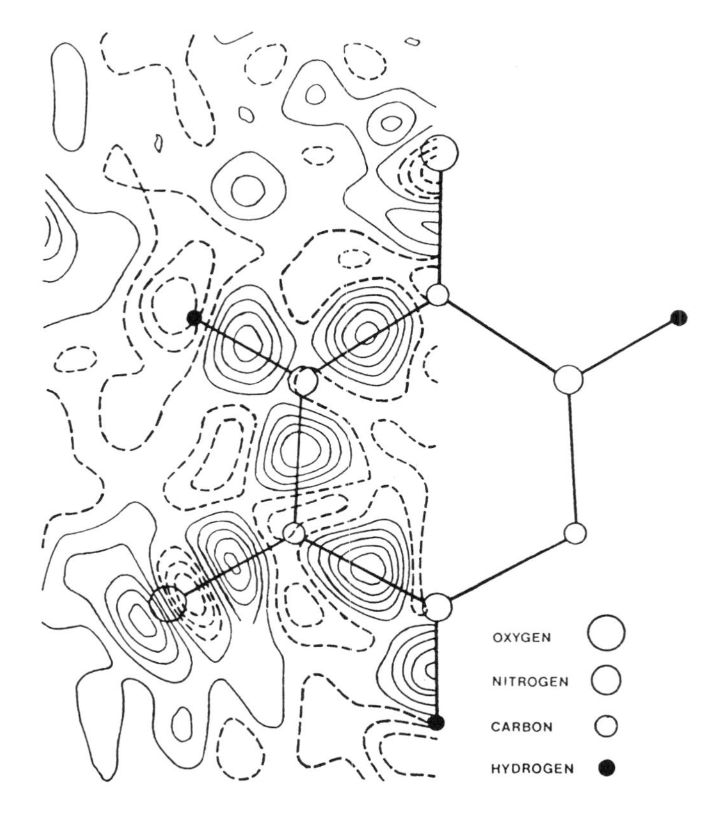

Figure 3.5 Density difference in cyanuric acid: measurement (from Coppens[9]). Reproduced by permission of Butterworth and Co

the intermolecular region. As a result the proton is denuded and the NH bond becomes more polar. There are also delocalized effects due to rearrangements over the whole molecular periphery.

A difference density function can also be introduced to represent the effect of ionization or excitation on the electronic density in a molecule. It is simply the difference between the density $\rho_f(M)$ in the ionized or excited state and the density $\rho_i(M)$ in the initial state:

$$\delta(M) = \rho_f(M) - \rho_i(M)$$

3.1.3 Bader Partitioning

Bader has introduced a procedure of partitioning a molecule in fragments in which the virial theorem is locally satisfied. Such virial or Bader partitioning is based on topographical features of the electronic density distribution.

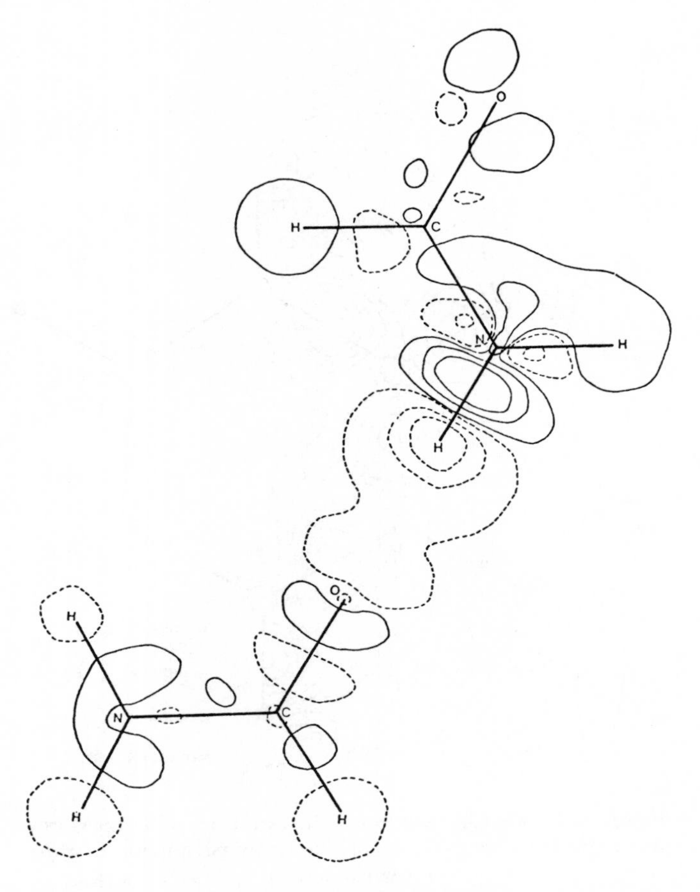

Figure 3.6 Density difference function in the formamide dimer (from Pullman[10]). Reproduced by permission of C.N.R.S.

A partitioning surface is defined as that collection of all gradient paths which is traced out by the vectors $\nabla\rho(M)$ which originate and terminate at stationary points in the density distribution,[11] a stationary point being a point at which $\nabla\rho(M)=0$. We have said that there is generally a saddle point in $\rho(M)$ between each pair of bonded nuclei. This corresponds to a stationary point, and the collection of partitioning surfaces through such points divides a system into fragments. The mathematical requirement for a partitioning surface S is given by the zero-flux requirement:

$$\nabla\rho(M)\cdot\mathbf{n}=0 \qquad \forall M\in S$$

where **n** is the vector normal to S.

It has been demonstrated that a fragment bounded by a surface of zero-flux has a well-defined kinetic energy and that this kinetic energy

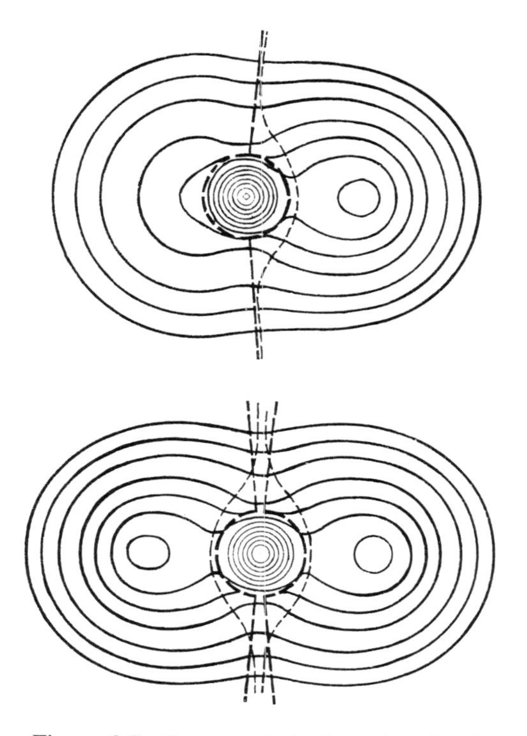

Figure 3.7 Loge and virial partitioning in BeH and HBeH

satisfies the virial relationship. It can be assumed that such a fragment possesses a certain autonomy. This is true. The virial fragments are approximately transferable from one molecule to another.

The short dashed lines on Figure 3.7 correspond to the partitioning surfaces in BeH and in HBeH. It can be seen that the H fragments in HBeH are very similar to the H fragment in BeH. The total electronic charge of fragment H in BeH is 1.868. In HBeH the corresponding charge is 1.861.

3.2 LOGE PARTITIONING, TRANSFERABILITY OF BONDS, STEREOCHEMISTRY AND GILLESPIE THEORY

Another procedure of partitioning a molecule into fragments is the loge theory, which has already been analysed. We shall now compare the loge partitioning and the virial partitioning.[12] This will give us an opportunity to present some new aspects of loge theory.

Let us consider the BH molecule. It is a six-electron problem, already discussed by Daudel *et al.*[13] Figures 3.8 and 3.9 illustrate the results of a search for the best spherical loge centered at the boron nucleus in a

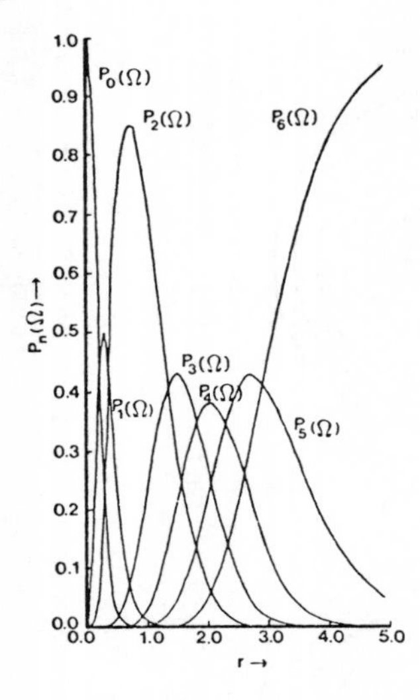

Figure 3.8 Probability of occurrence of the various electronic events (from Daudel *et al.*[13])

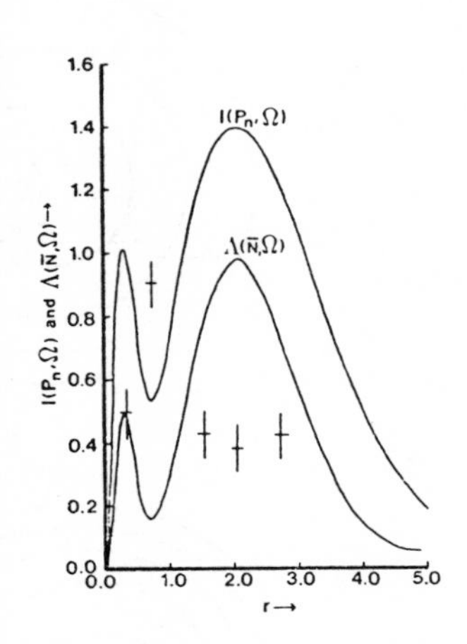

Figure 3.9 Three-loge partition (from Daudel *et al.*[13])

two-loge partitioning of this system. For such a partitioning there are seven electronic events.

Figure 3.8 shows the variation with the radius r of the spherical loge of the probability P_n of the event that n electrons will be found in that loge, the other ones being in the remaining part of the space. It is seen that only P_2 reaches a large value corresponding to a leading event. The other probabilities never reach a value above 0.5.

Figure 3.9 shows the variation of the missing information function I. It is seen that the function I has its unique minimum value for $r = 0.7$ precisely when P_2 reaches its maximum (0.85). The corresponding spherical loge is the *core loge.* Also plotted on Figure 3.9 is the fluctuation of the number N of electrons in the central loge. That fluctuation Λ is the difference between the average value $\overline{N^2}$ of the square of that number and the square $(\bar{N})^2$ of the average value of N. It can be seen that Λ and I reach their minimum for the same value of r. *The best partitioning into loges corresponds to a minimum of the fluctuation of the number of electrons in the loges.*

To go farther we can search for a three-loge partitioning, including a partitioning of the valence charge distribution. The central spherical loge of radius r will be considered again. The remaining part of the space will be partitioned by a cone of angle α centred on the B nucleus and possessing the BH line as an axis. Figure 3.10 makes it possible to compare Bader partitioning and the best three-loge partitioning.

The core loge of the three-loge partitioning has about the same radius as the two-loge partitioning. The angle α is 73°. The leading event corresponds

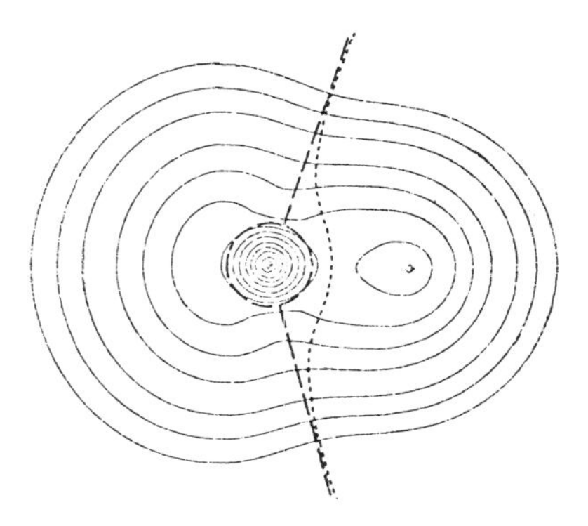

Figure 3.10 Loge and virial partitioning in BH (from Daudel *et al.*[13])

to two electrons in each loge. One can say that the valence electronic distribution is partitioned into a two-electron bond loge BH and a lone pair loge. It is seen that the lone pair loge is much more bulky than the bond loge. This is in accord with one of the basic postulates of Gillespie's theory of molecular geometry[(14)] (see also Section 2.2.2).

Furthermore, it appears that the BH virial fragment is very similar to the BH bond loge. When such agreement arises one can say that an aristic (from the Greek αριοτος meaning the best) partitioning is obtained because for such a fragment:

(a) The virial theorem is satisfied.
(b) The missing information is a minimum.
(c) The fluctuation is a minimum.
(d) It is transferable from one molecule to another.
(e) The electronic correlation is a maximum in the fragment and a minimum between two different fragments.

These five criteria give a real physical meaning to the concept of chemical bond. The last one has been proved by Bader.[(12)]

It was stated in Section 2.1.1 that the correlation occurs if the density of the probability $P(\mathrm{M}_a, \mathrm{M}_b)$ of finding one electron at point M_a *and* one electron at point M_b differs from the product of the density $p(\mathrm{M}_a)$ of finding one electron at point M_a and of the density $p(\mathrm{M}_b)$ of finding one electron at point M_b. Therefore if it is written:

$$P(\mathrm{M}_a, \mathrm{M}_b) = p(\mathrm{M}_a)p(\mathrm{M}_b)[1 + f(\mathrm{M}_a, \mathrm{M}_b)]$$

the function $f(\mathrm{M}_a, \mathrm{M}_b)$ directly measures the correlation effect. Very often the function f has negative values as the correlation reduces the probability of finding two electrons in the same small region of the space.

Starting from this expression Bader has been able to show that the fluctuation Λ of the number of the electrons in a loge Ω is given by the following expression:

$$\Lambda = \bar{N} + \int_{\Omega} p(M_a)p(M_b)f(M_a, M_b)\, dv_a\, dv_b$$

The integral usually has a negative value, its measure being the 'Fermi hole' of correlation. Therefore the minimum value of the fluctuation Λ corresponds to the maximum of the modulus of the correlation integral. As good loges correspond to small values of Λ they correspond to a strong correlation inside the loges. As a consequence the correlation between two loges is minimized.

We can now put the Gillespie theory of molecular geometry on a sound basis. Let us consider a molecule like NH_3. From what has been found for BH it may be anticipated that a two-electron core loge will be found surrounding the nitrogen nucleus. For the leading event eight electrons remain outside this loge: four with a spin of $+\frac{1}{2}h/2\pi$ and four with a spin of $-\frac{1}{2}h/2\pi$.

From what has been said about the most probable set of electronic positions in an atom (Section 1.2.5) it can be anticipated that the electrons possessing the same spin will tend to form tetrahedral angles (109°28′) between them and the nitrogen nucleus. However, in the most probable set of electronic positions there is usually a pair of electrons with opposite spins at a nucleus. Therefore near each hydrogen nucleus we can predict the presence of such a pair. The situation is described in Figure 3.11, where the three hydrogen nuclei tend to form angles of 109°28′ with the nitrogen nuclei. We can predict for the NH_3 molecule the presence of a lone pair loge and of three NH two-electron loges surrounding the nitrogen core loge. Furthermore, as the lone pair is more bulky than the bond loges we can predict that in fact the HNH angle will be slightly smaller than 109°28′. The experimental value is 107°.

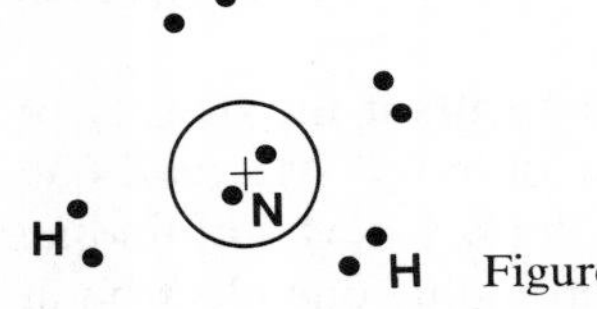

Figure 3.11 Leading electronic event for the molecule NH_3

The analysis suggests the possibility of predicting the geometry of molecules by using the following postulates which have been widely used by Gillespie.[14] If for a good partition into loges of a molecule there are only two electron bond loges and lone pair loges surrounding a core, the axes of the loges tend to form the largest possible angles. Furthermore, as the lone pair loges are bulky they tend to surround the core to the maximum possible extent. Therefore the lone pair loges will occupy those positions in which there is most room for them. From the first rule it may be anticipated that the arrangement of the loges surrounding the core will mainly depend on their number:

(a) If there are two the arrangement is linear.
(b) If there are three it is planar triangular.
(c) If there are four it is tetrahedral.
(d) If there are five it is trigonal bipyramidal.
(e) If there are six it corresponds to a regular octahedra

Table 3.2 gives examples of bond angles experimentally found for atoms having four two-electron loges surrounding the core.

Table 3.2 Some typical bond angles

OH_2	$104.5 \pm 0.1°$
NH_3	107.3 ± 0.2
OF_2	103.2 ± 1
OCl_2	110.8 ± 1
$O(CH_3)_2$	$111 \quad \pm 3$

The bond angles are not far from the tetrahedral angle.

To show the efficiency of the second rule let us consider the case of five two-electron loges surrounding a central core (Figure 3.12). The loges will follow a bipyramidal arrangement. It is easy to see that for such an arrangement there is more room in the equatorial positions than in the axial

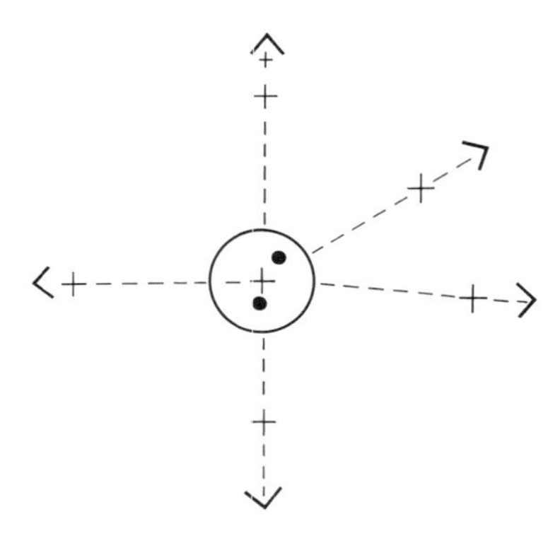

Figure 3.12 Trigonal bipyramid organization

positions: an electron pair in an equatorial position has only two neighbouring pairs at 90°; in an axial position it has three neighbouring pairs at 90°. This is why for all compounds shown on Figure 3.12 the lone pairs always occupy the equatorial positions. Many other interesting examples are described in Gillespie's book.

REFERENCES

1. R. F. W. Bader (private communication).
2. R. F. W. Bader, *An Introduction to the Electronic Structure of Atoms and Molecules*, Clarke, Irwin and Co., 1970.
3. F. Grimaldi, A. Lecourt and C. Moser, *Internat. J. Quantum Chem.*, **1,** 153 (1967).
4. R. Daudel, *Compt. Rend. Acad. Sci.* **235,** 886 (1952); M. Roux and R. Daudel, *Compt Rend. Acad. Sci.*, **240,** 90 (1955); M. Roux, S. Besnainou and R. Daudel, *J. Chimie Phys.*, **54,** 218 (1956).
5. M. Roux, M. Cornille and L. Burnelle, *J. Chem. Phys.*, **37,** 1009 (1962); R. F. W. Bader, W. H. Henneker and P. E. Cade, *J. Chem. Phys.*, **46,** 3341 (1967); R. F. W. Bader, I. Keaveny and P. E. Cade, *J. Chem. Phys.*, **46,** 3381 (1967); B. J. Ransil and J. J. Sinai, *J. Chem. Phys.*, **46,** 4050 (1967); M. J. Hazelrigg and P. Politzer, *J. Amer. Chem. Soc.*, **73,** 1009 (1969); P. E. Cade, *Trans. Am. Crystallogr. Ass.* **8,** 1 (1972).
6. R. Daudel, *Rev. Scientifique*, **87,** 162 (1947).
7. E. Segré, *Phys. Rev.*, **71,** 274 (1947).
8. J. I. Vargas, *M.T.P. Int. Rev. Sci. Série* 1, **8,** 75 (1971).
9. P. Coppens, *M.T.P. Informational Review of Science*, Physical Chemistry Series II, Butterworths, London, 1975.
10. A. Pullman, in *Aspect de la Chimie Quantique Contemporaire* (Eds. A. Pullman and R. Daudel), CNRS, Paris, 1971.
11. R. F. W. Bader and P. M. Beddall, *J. Chem. Phys.*, **56,** 3320 (1972); R. F. W. Bader, P. M. Beddall and J. Jr. Peslak, *J. Chem. Phys.*, **58,** 557 (1973); R. F. W. Bader, A. J. Duke and R. R. Messer, *J. Amer. Chem. Soc.*, **95,** 7715 (1973); G. R. Runtz, Ph.D. Thesis, McMaster University, 1973.
12. R. Bader, *Localization and Delocalization in Quantum Chemistry*, Reidel, 1975.
13. R. Daudel, R. F. W. Bader, M. E. Stephens and D. S. Borrett, *Can. J. Chem.*, **52,** 1310 (1974).
14. R. J. Gillespie, *Molecular Geometry*, Van Nostrand Reinhold, London, 1972.

PART II

METHODS AND APPLICATIONS OF QUANTUM CHEMISTRY

CHAPTER 4

A Survey of the ab initio Methods of Quantum Chemistry

4.1 THE CALCULATION OF APPROXIMATE ELECTRONIC WAVE FUNCTIONS

It would be a good idea at this point to briefly review the main topics described in the first part. Therefore before introducing details of calculations and explicit applications, we shall look again at key formulas and the underlying approximations.

One of the aims of quantum molecular physics is to resolve the time-independent Schrödinger equation:

$$H\Gamma_i = W_i\Gamma_i \tag{4.1}$$

In the case of a molecule containing N nuclei and n electrons, equation (4.1) may be written more explicitly using classical notations:

$$\left(-\sum_{J=1}^{N}\frac{\hbar^2}{2M_J}\nabla_J^2-\frac{\hbar^2}{2m}\sum_{j=1}^{n}\nabla_j^2-\sum_{j=1}^{n}\sum_{J=1}^{N}\frac{Z_Je^2}{r_{jJ}}+\frac{1}{2}\sum_{\substack{j=1\\ j\neq j'}}^{n}\sum_{j'=1}^{n}\frac{e^2}{r_{jj'}}+\frac{1}{2}\sum_{\substack{J=1\\ J\neq J'}}^{N}\sum_{J'=1}^{N}\frac{Z_JZ_{J'}e^2}{r_{JJ'}}\right)\Gamma_i = W_i\Gamma_i \tag{4.2}$$

The total wave functions Γ_i depend on the $3N+3n$ nuclear and electronic coordinates. The set of W_i's is the spectrum of the hamiltonian operator H associated with the total energy of the molecule. This hamiltonian can also be written as:

$$H = T_{\text{N}} + T_{\text{e}} + V_{\text{Ne}} + V_{\text{ee}} + V_{\text{NN}} \tag{4.3}$$

where T_{N} and T_{e} are the kinetic operators related to nuclei and electrons and V_{NN}, V_{ee}, V_{Ne} are the potential operators of nuclear, electronic or mixed origin.

As we have already shown (Section 2.1.2), the Born–Oppenheimer approximation allows the separation of electronic and nuclear motions. This means that the total wave function may be separated into electronic and nuclear parts such as:

$$\Gamma_i(1, 2, \ldots, J, \ldots, N; 1, 2, \ldots, j, \ldots, n) = \phi_i(1, 2, \ldots, J, \ldots, N)\psi_i(1, 2, \ldots, j, \ldots, n) \tag{4.4}$$

Electronic wave functions are obtained by solving the corresponding electronic equation:

$$(T_e + V_{Ne} + V_{ee} + V_{NN})\psi_i = u_i\psi_i \tag{4.5}$$

for any nuclear configuration.

In this way a set of eigenfunctions ψ_i and a set of eigenvalues u_i can be obtained for any choice of nuclear coordinates. Therefore each u_i will correspond to a curve or a hypersurface depending on the number of nuclei of the molecule (two, three or more than three). Nuclear wave functions are obtained by solving the so-called nuclear equation:

$$(T_N + u_i)\phi_i = W_i\phi_i \tag{4.6}$$

where the potential term u_i is nothing more than the total electronic energy of a given state (i) of the system. This is why u_i is commonly called, in the general case, the potential energy hypersurface of the molecule. Once u_i is known numerically or analytically the nuclear equation may be solved. It must be recalled here that ϕ_i depends on $3N$ variables from which 3 may be associated with the translation of the centre of mass and 3 (or 2 for linear molecules) with the rotation of the whole system. So ϕ_i $(i = 1, 2, \ldots, N)$ may be further simplified since it may be written in terms of a translation function and a rotation–vibration function. The former are well known as they are simply the solutions of the Schrödinger equation of a particle in a box.

The solutions for the rotation–vibration equation depend on the form of the potential energy and usually contain rotational, vibrational and coupling terms. This will be analysed in more detail later (Section 4.2.5). Let us return to the electronic equation. For a given nuclear configuration, the potential term V_{NN} is constant and equation (4.5) may be written in the simplified form:

$$(T_e + V_{Ne} + V_{ee})\psi_i = E_i\psi_i \tag{4.7}$$

or:

$$H\psi_i = E_i\psi_i \tag{4.8}$$

where:

$$E_i = u_i - V_{NN} \tag{4.9}$$

More explicitly, the electronic equation is written in the atomic units system as:

$$\left(-\frac{1}{2}\sum_j \nabla_j^2 - \sum_j\sum_J \frac{Z_J}{r_{jJ}} + \sum\sum_{j<j'} \frac{1}{r_{jj'}}\right)\psi_i = E_i\psi_i \tag{4.10}$$

This equation finds an exact solution only when $n = 1$. Nevertheless, exact solutions can always be expanded in terms of Slater determinants which combine into configurations (Section 2.1.5):

$$\psi_i = \sum_{k=1}^{\infty} C_{ik}\Phi_k \tag{4.11}$$

where Φ_k is an antisymmetrized product of n spin orbitals. In practice, the Φ_k basis is truncated and approximate wave functions can be obtained using the configuration interaction method:

$$\psi_{i_{\text{app}}} = \sum_{k=1}^{p} C_{ik}\Phi_k \tag{4.12}$$

Expansion coefficients C_{ik} and therefore approximate solutions of equation (4.10) are reached by the classical variational procedure which leads to the resolution of a secular system and a secular equation.

The configuration interaction method needs the evaluation of integrals between Slater determinants estimated by the well-known Slater rules.[1] Instead of using the expression (4.12) one may retain only one configuration, i.e. a unique determinant in the case of closed-shell systems, such as, applying the variation theorem, the orbitals introduced in this determinant are the best possible in the energetic sense. This leads to the restricted Hartree–Fock (RHF) method for closed-shell systems ($2n$ electrons):

$$\psi_{i_{\text{app}}} = \det \phi'_1\bar{\phi}'_1 \cdots \phi'_n\bar{\phi}'_n \tag{4.13}$$

where the orbitals ϕ'_i are obtained by solving the Hartree–Fock equations:

$$h^{\text{HF}}\phi'_i = \sum_j \varepsilon_{ij}\phi'_j \tag{4.14}$$

where the Hartree–Fock operator is given by:

$$h^{\text{HF}}_{(\mu)} = -\frac{1}{2}\nabla^2_\mu - \sum_{J=1}^{N}\frac{Z_J}{r_{\mu J}} + \sum_j [2J'_j(\mu) - K'_j(\mu)] \tag{4.15}$$

or

$$h^{\text{HF}}_{(\mu)} = h^{\text{N}}(\mu) + \sum_j [2J'_j(\mu) - K'_j(\mu)] \tag{4.16}$$

As this hamiltonian depends (through the bielectronic operators) on the solution ϕ'_i the equations can only be solved by an iterative procedure until self-consistency is obtained. Furthermore, for technical reasons, Hartree–Fock equations are solved in an equivalent form:

$$h^{\text{HF}}_{(\mu)}\phi_i = \varepsilon_{ii}\phi_i \tag{4.17}$$

This is possible, even for an orthornormalized set of monoelectronic functions, because the basis set is not uniquely defined in Hartree-Fock space. Therefore equations (4.14) may lead to many different types of orbitals. The pseudoeigenvalue equation (4.17) will lead to the so-called canonical orbitals delocalized on the whole molecule.

Until now, the explicit form of the ϕ_i functions has not been specified. This is due to the fact that equation (4.17) is generally unsolvable in molecular systems. Therefore we shall expand once more the unknown functions (ϕ_i) in a certain basis set. We have shown (Section 2.1.4) that, in the LCAO approximation, molecular orbitals are developed in terms of functions located on atoms, the atomic orbitals (χ). Thus:

$$\phi_i = \sum_{p=1}^{m} C_{ip}\chi_p \tag{4.18}$$

The χ_p functions being analytically known, the unknowns will be the coefficients. We have therefore displaced the problem, and in the framework of the Hartree–Fock method we will talk about the best coefficients rather than the best molecular orbitals. Furthermore, the Hartree–Fock equations will be replaced by Roothaan's equations (Section 2.1.4):

$$\sum_{p=1}^{m} C_{ip} \int \chi_q^* h^{\mathrm{HF}} \chi_p \, dv = \sum_{p=1}^{m} C_{ip}\varepsilon_{ii} \int \chi_q^* \chi_p \, dv \tag{4.19}$$

or

$$\sum_{p} C_{ip}(h_{pq}^{\mathrm{HF}} - \varepsilon_i S_{pq}) = 0 \tag{4.20}$$

and, simpler, in matricial notation:

$$\mathbf{HC} = \mathbf{\Delta C}E \tag{4.21}$$

The matrix elements of the Hartree–Fock operator in the chosen atomic basis set may be explicitly written as (for real χ's):

$$\begin{aligned} h_{pq}^{\mathrm{HF}} = & \int \chi_p(1)\left(-\frac{\nabla_1^2}{2}\right)\chi_q(1)\, dv_1 + \sum_J \int \chi_p(1)\left(-\frac{Z_J}{r_{1J}}\right)\chi_q(1)\, dv_1 \\ & + \sum_j \sum_r \sum_s C_{jr}C_{js}\left[2 \int \chi_p(1)\chi_q(1)\frac{1}{r_{12}}\chi_r(2)\chi_s(2)\, dv_1\, dv_2 \right. \\ & \left. - \int \chi_p(1)\chi_r(1)\frac{1}{r_{12}}\chi_q(2)\chi_s(2)\, dv_1\, dv_2\right] \end{aligned} \tag{4.22}$$

or, more concisely:

$$h_{pq}^{\mathrm{HF}} = h_{pq}^{\mathrm{N}} + \sum_r \sum_s D_{rs}(2\langle pq \mid rs\rangle - \langle pr \mid qs\rangle) \tag{4.23}$$

with:

$$D_{rs} = \sum_j C_{jr}C_{js}$$

expressing the dependence of the Hartree–Fock operator with respect to the solutions.

The pseudosecular system (4.20) must be solved iteratively until self-consistency is reached, which is why Roothaan's method is known as the SCF–LCAO–MO procedure.

To summarize, the electronic equation is usually solved for molecules, first at the monodeterminantal level using the LCAO approximation in a basis set chosen as well as possible to approach the Hartree–Fock limit. Afterwards, the SCF molecular orbitals so obtained allow generation of many other Slater determinants which might be used at a more or less CI sophisticated level.

Until now, the approximate Hartree–Fock wave functions have been discussed for closed-shell systems where all molecular orbitals are doubly occupied. The best molecular orbitals are obtained in the framework of the RHF procedure by solving equations (4.14). It must be stressed here once more that the explicit form of the monoelectronic operator depends on the solutions and thus also on the way the orbitals are occupied. Therefore for open-shell systems, the Hartree–Fock operator (4.16) does not apply, as the bielectronic operators will be of a different nature depending on the chosen configuration. Two methods have been proposed to study the open-shell systems: the open-shell (spin) restricted Hartree–Fock method of Roothaan[2] and the open-shell (spin) unrestricted Hartree–Fock method of Pople and Nesbet.[3] We shall briefly describe these two approaches of open-shell systems.

In order to preserve eigenfunctions for the spin operators, the RHF method partitions the wave function in a closed-shell part described by doubly occupied molecular orbitals and an open-shell part by singly occupied molecular orbitals, the wave function obtained being the best one in the Hartree–Fock sense. So for a doublet state:

$$\psi^{\text{RHF}} = |\phi_1\bar{\phi}_1 \cdots \phi_n\bar{\phi}_n\phi_{n+1}| \tag{4.24}$$

The energy of such a determinant contains a closed-shell contribution, an open-shell contribution and a coupling term. For a wave function such as (4.24) the energy will be:

$$E = 2\sum_{i=1}^{n} h_i^{\text{N}} + \sum_{i=1}^{n}\sum_{j=1}^{n} (2J_{ij} - K_{ij}) + h_{n+1}^{\text{N}} + \sum_{j=1}^{n} (2J_{jn+1} - K_{jn+1}) \tag{4.25}$$

Minimization of the energy towards variation of the orbitals leads to the resolution of two equations, one for the closed-shell orbitals and one for the open-shell orbitals. Nevertheless, after mathematical transformations, these two equations may be reduced to an eigenvalue problem of the same kind as for closed shells, although the Hartree–Fock operator is slightly more complicated. The interested reader should consult the original paper of Roothaan which presents the problem quite nicely.

The unrestricted Hartree–Fock procedure adopts a quite different point of view. As the number of electrons of one spin is different from the number of electrons of the other spin, and as the mutual repulsions between electrons is different following their spin state, there is no reason why in the closed-shell part the orbitals associated with different spin functions should be identical. A spin unrestricted wave function such as:

$$\psi^{\mathrm{UHF}} = |a_1(1) \cdots a_{n+1}(n+1)b_1(n+2) \cdots b_n(2n+1)| \tag{4.26}$$

will then be proposed, where a_i and b_i are respectively the spin orbitals of α and β spin functions. The energy associated is simply:

$$E = \sum_i^{\alpha+\beta} h_i^{\mathrm{N}} + \frac{1}{2} \sum_i^{\alpha+\beta} \sum_j^{\alpha+\beta} J_{ij} - \frac{1}{2}\left(\sum_i^{\alpha} \sum_j^{\alpha} + \sum_i^{\beta} \sum_j^{\beta}\right) K_{ij} \tag{4.27}$$

The best functions a_i and b_i being obtained from resolution of two eigenvalues problems:

$$h^{\alpha} a_i = \varepsilon_i^{\alpha} a_i \qquad \text{and} \qquad h^{\beta} b_i = \varepsilon_i^{\beta} b_i \tag{4.28}$$

where:

$$h^{\alpha} = n^{\mathrm{N}} + \sum_j^{\alpha} J_j + \sum_j^{\beta} J_j - \sum_j^{\alpha} K_j \tag{4.29}$$

and

$$h^{\beta} = h^{\mathrm{N}} + \sum_j^{\beta} J_j + \sum_j^{\alpha} J_j - \sum_j^{\beta} K_j \tag{4.30}$$

Unfortunately the wave function obtained is not an eigenfunction of the S^2 spin operator and therefore does not describe a pure spin state. It must be said that nowadays open-shell calculations at the Hartree–Fock level are still cumbersome as the methodology is not uniquely defined and the orbitals obtained cannot be, even in canonical form, interpreted as easily as in closed-shell calculations.

Let us now discuss briefly the question of the choice of the method to be used for solving practical problems. As the quantum chemist has usually a well-defined problem to solve (configurational and conformational analysis, spectra interpretation, search of reaction pathways, etc.) he already knows the nature of the molecules (or atoms) he will study as well as the different states which will be of interest. This will suggest the method to be employed (CI, Hartree–Fock open or closed shell, etc.) and the portion of the hypersurface(s) to explore (if any). A sequence of operations and choices will then be performed. These are summarized below in order to familiarize the reader with the applications that will be developed later on:

(a) As presumably the LCAO approximation will be used, the choice of the basis set will be performed. This is determined by the precision requested, by the size of the problem (i.e. the number of electrons) and by the kind of information to be gathered.
(b) The basis set being known, the Hartree–Fock–Roothaan (and CI) equations can be solved for one or a set of nuclear configurations in the chosen electronic state(s). This will need the computation of many atomic (and molecular) mono- and bielectronic integrals, followed by a SCF (and CI) calculation. This step can be realized by using computer programs put at the disposal of the scientific community by 'software libraries' such as the Quantum Chemistry Program Exchange[4] and could be quite time-consuming. This will generate numerically the desired portion of the potential energy hypersurface.
(c) Approximate wave function(s) and hypersurface(s) being known, a more

or less sophisticated analysis can be performed. For example, an attempt to solve the nuclear equation may be done; molecular properties may be obtained from the wave function; external perturbation may be introduced and the response of the wave function analysed; a reaction mechanism may be deduced from a reaction pathway.

To conclude this section we will make some comments on the semiempirical approach to the problems. It is clear that all that has been said until now is very dependent on the number of electrons in the system. This number will indeed determine the number of molecular orbitals involved, the number of configurations possibly built in CI and the size of the equations to be solved. In some chemical systems the number of electrons may even be so large that the solutions become unreachable by actual technology. Quantum chemists will try to overcome this difficulty by introducing further simplifications.

A possible way to do this is to introduce core approximations (see Sections 2.2.4, 2.2.5 and 2.2.6) in which certain electrons are frozen in atomic or molecular cores. The remaining electrons moving in the potential field of the positively charged core(s) are considered to be responsible for the interesting properties of the system. All the equations written until now may remain valid except that we have to replace h^{N} everywhere by h^{c} (the core hamiltonian). The monoelectronic hamiltonian is then written as:

$$h = h^{\mathrm{c}} + \sum_{i} \sum_{j} (2J_{ij} - K_{ij}) \tag{4.31}$$

where h^{c} depends implicitly on the unknown orbitals of the frozen electrons and $J_{ij}(K_{ij})$ contains only the contributions of the remaining electrons. These may be valence or π electrons leading to different levels of approximation.

Obviously some assumptions must be made for calculating the core integrals. Two possibilities arise: comparison with explicit calculations gives an idea of the behaviour of the matrix elements of h^{c} and comparison with experimental properties shows how these matrix elements should behave. The first viewpoint leads to simulation of the Hartree–Fock results, the second to what is called a pure semiempirical approach. The latter has been extensively used and leads to the well-known semiempirical methods such as CNDO, INDO, NDDO, MINDO, extended Hückel at the valence electrons level or as Hückel, Pariser–Parr–Pople, etc., at the π level.

Further parameterization with experiments may even be introduced. Reliable information has been reached by these semiempirical methods, depending on the quality of the parameterization and on the kind of property under investigation.

4.2 THE CALCULATION OF MOLECULAR PROPERTIES

We shall see in this section how to estimate some molecular properties, starting from a known wave function describing a definite state of the system.

At this point, a molecular property will be described as an observable value, measurable by some experiment and related by some operator as shown in Section 1.1.1. The expectation value for the observable will be:

$$\langle G\rangle = \int \psi^* G_{op} \psi \, d\tau \tag{4.32}$$

or

$$\langle G\rangle = \langle \psi | \, G_{op} \, | \psi \rangle \tag{4.33}$$

with

$$\langle \psi \mid \psi \rangle = 1 \tag{4.34}$$

the wave function being normalized.

4.2.1 Density Matrices

Before proceeding further, let us recall some features of the wave function. The normalization of the function:

$$\int \psi^*(x_1 \cdots x_n)\psi(x_1 \cdots x_n) \, dx_1 \cdots dx_n = 1 \tag{4.35}$$

where the symbol x_i includes all variables of electron i, simply means that there exists in space a set of n electrons. If we associate with every variable a definite value (position in space and spin), the expression:

$$\psi^*(x_1 \cdots x_n)\psi(x_1 \cdots x_n) \, dx_1 \cdots dx_n \tag{4.36}$$

gives the probability of finding these n particles in the chosen configuration. If we integrate over all possible situations for particle n, we will reach the probability that there are $(n-1)$ particles in that configuration, the last one being anywhere. The successive integrations over the other particles will decrease the probability function by one order at a time, preserving nevertheless the same meaning to the function. Finally, although we shall arrive at a normalization integral, two very interesting functions will appear, i.e. the first- and second-order density matrices. Multiplied by dx_1, the function:

$$\int \psi^*(x_1 \cdots x_n)\psi(x_1 \cdots x_n) \, dx_2 \cdots dx_n \tag{4.37}$$

will give the probability of finding the first particle with configuration x_1 (position and spin), the others being anywhere.

This development could of course also apply to any other particle i, all particles being indistinguishable. With this in mind we multiply the function (4.27) by the number of particles and define the function:

$$\gamma(x_1 x_1') = n \int \psi(x_1 \cdots x_n)\psi^*(x_1' \cdots x_n) \, dx_2 \cdots dx_n, \qquad \text{for } x_1 = x_1' \tag{4.38}$$

known as the first-order density matrix. This function no longer defines a probability as it does not integrate to unity but to the number of particles (electrons) in the system. The presence of the two indices, the second marked by a prime, allows us to distinguish the two functions under the integration operand—this will be useful later on. As this first function reduces the information contained in the wave function to a new function which is monoelectronic, it will be of great interest for all properties depending intrinsically on the variables of only one electron, whatever that is.

As some properties depend also on the behaviour of a couple of particles, e.g. through their interdistance, it is important to introduce in the same way as above a function depending on the variables of two electrons. Thus:

$$\int \psi(x_1x_2\cdots x_n)\psi^*(x_1x_2\cdots x_n)\,\mathrm{d}x_3\cdots\mathrm{d}x_n \tag{4.39}$$

will give some information on the probability of finding the couple (1, 2) in a given configuration. Furthermore, a second-order density matrix may be defined by the relation:

$$\Gamma(x_1x_2x_1'x_2') = n(n-1)\int \psi(x_1x_2\cdots x_n)\psi^*(x_1'x_2'\cdots x_n)\,\mathrm{d}x_3\cdots\mathrm{d}x_n \tag{4.40}$$

for $x_1' = x_1$ and $x_2' = x_2$. Higher order density matrices may be defined, but as no properties depend intrinsically on a triplet of electrons these would not be of any use in this section.

Both density matrices $\gamma(x_1x_1')$ and $\Gamma(x_1x_1'x_2x_2')$ may be further reduced as the variable x stays for spatial and spin variables. Integration over spin leads to reduced density matrices. Indeed, if $x_1 = r_1 \times s_1$ and $x_2 = r_2 \times s_2$, one will find:

$$\gamma(r_1r_1') = n\int \psi(1,\ldots,n)\psi^*(1',\ldots,n)\,\mathrm{d}s_1\,\mathrm{d}x_2\cdots\mathrm{d}x_n \tag{4.41}$$

$$\Gamma(r_1r_2r_1'r_2') = n(n-1)\int \psi(1,2,\ldots,n)\psi^*(1',2',\ldots,n)\,\mathrm{d}s_1\,\mathrm{d}s_2\,\mathrm{d}x_3\cdots\mathrm{d}x_n \tag{4.42}$$

We shall now go back to relation (4.32) and comment on the use of density matrices. Taking a general operator G_{op}, this relation will always be a sum of simpler operators depending on none, one or two electrons at most:

$$G_{\mathrm{op}} = G_0 + \sum_i G_1(i) + \frac{1}{2}\sum_{i,j} G_2(i,j) \tag{4.43}$$

Introducing this relation into (4.32) and referring to the definition of first-

and second-order density matrices, gives:

$$\langle G\rangle = G_0 + \int_{x_1'=x_1} G_1(1)\gamma(x_1x_1')\,dx_1 + \frac{1}{2}\int_{\substack{x_1'=x_1\\x_2'=x_2}} G_2(1,2)\Gamma(x_1x_2x_1'x_2')\,dx_1\,dx_2 \quad (4.44)$$

The meaning of the prime is now clear as relation (4.32) explicitly shows that the operator only operates on $\psi(x_1 \cdots x_n)$ and may even transform this function through derivation, for example. The prime allows the function which the operator will transform to be labelled, and as soon as this is done the integration(s) may take place, equating x_1' and x_1.

Let us consider the hamiltonian operator H and the corresponding energy E:

$$H = H_0 + H_1 + H_2 \quad (4.45)$$

where:

$$H_0 = \sum_{J<J'}^{N}\sum^{N} \frac{Z_J Z_{J'}}{r_{JJ'}} \quad (4.46)$$

$$H_1 = \sum_{i=1}^{n} h(i) = \sum_{i=1}^{n}\left[-\frac{1}{2}\nabla^2(i) - \sum_{J=1}^{N}\frac{Z_J}{r_{iJ}}\right] \quad (4.47)$$

$$H_2 = \sum_{i=1}^{n}\sum_{j=1}^{n} h(i,j) = \frac{1}{2}\sum_{\substack{i=1\\i\neq j}}^{n}\sum_{j=1}^{n}\frac{1}{r_{ij}} \quad (4.48)$$

Then:

$$E = H_0 + \int_{x_1'=x_1}\left[-\frac{1}{2}\nabla^2(1) - \sum_{J=1}^{N}\frac{Z_J}{r_{1J}}\right]\gamma(x_1x_1')\,dx_1 + \frac{1}{2}\int_{\substack{x_1'=x_1\\x_2'=x_2}}\frac{1}{r_{12}}\Gamma(x_1x_2, x_1'x_2')\,dx_1\,dx_2 \quad (4.49)$$

It should be noticed that the development above did not introduce any approximation on the wave function. Therefore, at this point it is completely general. What would happen if one were to use some simpler expressions for the wave function? In the case of the Hartree–Fock wave function, written, for example, as:

$$\psi^{\mathrm{HF}}(1, 2, \ldots, n) = |\phi_1(x_1)\phi_2(x_2)\cdots\phi_n(x_n)| \quad (4.50)$$

application of formula (4.38) leads, after some determinantal algebra, to the relation:

$$\gamma^{\mathrm{HF}}(x_1x_1') = \sum_i \phi_i(x_1)\phi_i^*(x_1') \quad (4.51)$$

the spin orbitals being assumed orthonormal. The reader may convince himself in many ways. A good exercise would be to treat a function of three particles explicitly with and or without the antisymmetrizer. An easy demonstration can be done as follows:

$$\psi(x_1 x_2 \cdots x_n) = |\phi_1(x_1)\phi_2(x_2) \cdots \phi_n(x_n)| \tag{4.52}$$

or, more explicitly:

$$\psi(x_1 x_2 \cdots x_n) = \frac{1}{\sqrt{n!}} \begin{vmatrix} \phi_1(x_1)\phi_2(x_1) \cdots \phi_n(x_1) \\ \phi_1(x_2)\phi_2(x_2) \cdots \phi_n(x_2) \\ \\ \phi_1(x_n)\phi_2(x_n) \cdots \phi_n(x_n) \end{vmatrix}$$

$$= \frac{1}{\sqrt{n}} \left\{ \phi_1(x_1) \left[\frac{1}{\sqrt{(n-1)!}} \min \phi_1(x_1) \right] - \phi_2(x_1) \times \left[\frac{1}{\sqrt{(n-1)!}} \min \phi_2(x_1) \right] \cdots \pm \phi_n(x_1) \left[\frac{1}{\sqrt{(n-1)!}} \min \phi_n(x_1) \right] \right\} \tag{4.53}$$

It can be seen that:

$$\frac{1}{\sqrt{(n-1)!}} \min \phi_1(x_1) \tag{4.54}$$

is nothing other than a determinantal function of $(n-1)$ particles built on the same orbitals except ϕ_1. The same is true for all the other functions between brackets. We shall note these functions for commodity $\Phi^i(x_2 \cdots x_n)$ where i stands for the unused orbital. Therefore $\gamma^{\mathrm{HF}}(x_1 x_1')$ may be written as:

$$\gamma^{\mathrm{HF}}(x_1 x_1') = n \times \frac{1}{n} \int [\phi_1(x_1)\Phi^1(x_2 \cdots x_n) - \phi_2(x_1)\Phi^2(x_2 \cdots x_n) \cdots \pm \phi_n(x_1)\Phi^n(x_2 \cdots x_n)]^2 \, \mathrm{d}x_2 \cdots \mathrm{d}x_n \tag{4.55}$$

or

$$\begin{aligned} \gamma^{\mathrm{HF}}(x_1 x_1') &= \phi_1^*(x_1')\phi_1(x_1) \int \Phi^{1*}(x_2 \cdots x_n)\Phi^1(x_2 \cdots x_n) \, \mathrm{d}x_2 \cdots \mathrm{d}x_n \\ &+ \phi_2^*(x_1')\phi_2(x_1) \int \Phi^{2*}(x_2 \cdots x_n)\Phi^2(x_2 \cdots x_n) \, \mathrm{d}x_2 \cdots \mathrm{d}x_n \\ &\cdots + \phi_n^*(x_1')\phi_n(x_1) \int \phi^{n*}(x_2 \cdots x_n)\Phi^n(x_2 \cdots x_n) \end{aligned} \tag{4.56}$$

$\pm$ double products. Of course:

$$\int \Phi^{i*}(x_2 \cdots x_n)\Phi^i(x_2 \cdots x_n) \, \mathrm{d}x_2 \cdots \mathrm{d}x_n = 1 \tag{4.57}$$

and

$$\int \Phi^{i*}(x_2 \cdots x_n)\Phi^j(x_2 \cdots x_n)\, dx_2 \cdots dx_n = 0 \tag{4.58}$$

as the two determinants differ only in one column by orthogonal functions which will integrate to zero. Thus:

$$\gamma^{HF}(x_1x_1') = \sum_i \phi_i(x_1)\phi_i^*(x_1') \tag{4.59}$$

This demonstration is really nothing else than an application of Slater rules.

Let us now introduce the LCAO approximation (Roothaan's method). In this case:

$$\phi_i(x) = \sum_{p=1}^{m} C_{ip}\chi_p(x) \tag{4.60}$$

where the atomic basis set $\chi_p(x)$ is not orthogonal, so that:

$$S_{pq} = \int \chi_p^*(x)\chi_q(x)\, dx \tag{4.61}$$

Relation (4.51) obviously becomes:

$$\gamma^{HFR}(x_1x_1') = \sum_i \sum_p \sum_q C_{ip}C_{iq}\chi_p(x_1)\chi_q^*(x_1') \tag{4.62}$$

or

$$\gamma^{HFR}(x_1x_1') = \sum_p \sum_q D_{pq}\chi_p(x_1)\chi_q^*(x_1') \tag{4.63}$$

where D_{pq} is the density matrix expressed in the discrete atomic basis set, commonly known as the 'density matrix' of SCF calculations. Putting this in CI formalism:

$$\psi^{CI}(x_1 \cdots x_n) = \sum_i C_i\Phi_i(x_1 \cdots x_n) \tag{4.64}$$

where Φ_i are Slater determinants or combinations of these describing the different configurations. The obtention of the first-order density matrix is a bit more tedious, although it is simply an application of the already mentioned Slater rules. For many determinantal functions, one finds:

$$\gamma^{CI}(x_1x_1') = \sum_i \sum_j a_{ij}\phi_i(x_1)\phi_j^*(x_1') \tag{4.65}$$

where a_{ij} are the coefficients of the molecular orbital products obtained by an appropriate combination of the CI expansion coefficients. We introduce these because the ϕ_i basis set is usually transformed by diagonalization of the a_{ij} matrix in the natural basis set θ_k.[5]

The first-order density matrix then takes a more useful expression very close to the Hartree–Fock one. Indeed:

$$\gamma^{CI}(x_1x_1') = \sum_k \eta_k \theta_k(x_1)\theta_k^*(x_1') \tag{4.66}$$

with $0 \leqslant \eta_k \leqslant 1$, where η_k is the occupation number of the orbital θ_k. This formula becomes very useful in computing the one-electron properties from sophisticated CI calculations. As the natural orbitals are expanded over the same atomic basis set as ϕ_i, it becomes:

$$\theta_k = \sum_{p=1}^{m} C_{kp}\chi_p \tag{4.67}$$

Expression (4.66) becomes:

$$\gamma^{CI}(x_1x_1') = \sum_p \sum_q \sum_k \eta_k C_{kp} C_{kq} \chi_p(x_1)\chi_q^*(x_1') \tag{4.68}$$

and finally:

$$\gamma^{CI}(x_1x_1') = \sum_p \sum_q D_{pq}^{CI} \chi_p(x_1)\chi_q^*(x_1') \tag{4.69}$$

which is formally equivalent to (4.63). Thus, for a chosen atomic basis set, the computation of the properties is independent of the method used and the same computational programs will hold.

The expression of the second-order density matrix in these different approximations is more difficult to obtain and so the reader is directed to more specialized literature.[6] Although the principle is equivalent, the deduction is much longer. We shall only give the expression of the second-order density matrix in the Hartree–Fock model, which is:

$$\Gamma^{HF}(x_1x_2x_1'x_2') = (1 - P_{12})\gamma^{HF}(x_1x_1')\gamma^{HF}(x_2x_2') \tag{4.70}$$

where P_{12} is the permutation responsible for the exchange terms in the energy expression. The very simple form of this second-order matrix is simply the product of two first-order matrices.

4.2.2 Some One-electron Properties

We shall now study some internal properties of the electronic distribution. By internal properties we mean properties of isolated systems without any response to an external perturbation such as the electric field.

4.2.2.1 Electron density

From the definition of the first-order density matrices it is clear that the electron density is nothing more than the reduced first-order density matrix

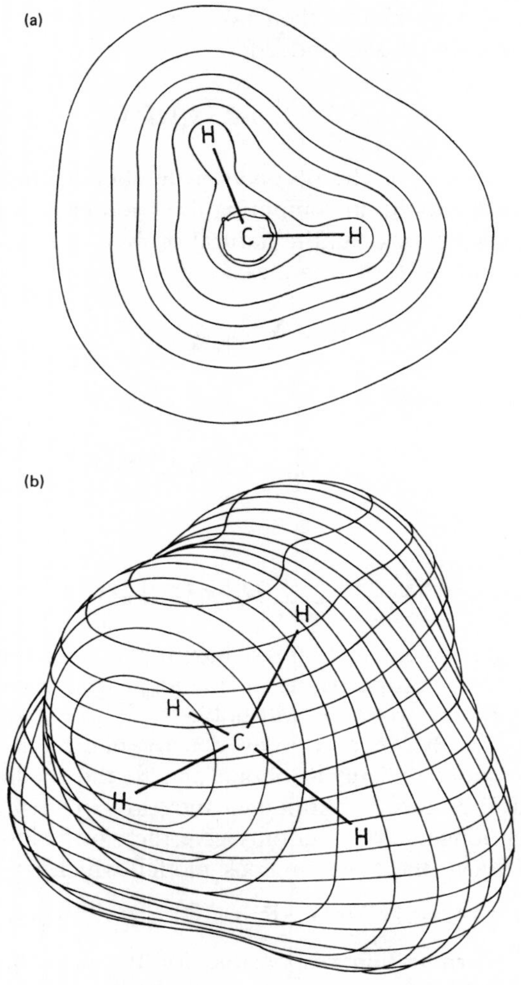

Figure 4.1 Electron density of the methane molecule (a) in the HCH plane and (b) in three dimensions

(4.41) where $r_1 = r'_1$. Thus:

$$\rho(\mathrm{M}) = \gamma(\mathrm{MM}'), \qquad \text{for } \mathrm{M} = \mathrm{M}' \tag{4.71}$$

where M is a point of space R^3. For approximate wave functions we find:

$$\rho^{\mathrm{HF}}(\mathrm{M}) = \sum_i \eta_i \phi_i^2(\mathrm{M}) \qquad (\eta_i = \text{occupation number 1 or 2}) \tag{4.72}$$

$$\rho^{\mathrm{HFR}}(\mathrm{M}) = \sum_p \sum_q D_{pq}^{\mathrm{HF}} \chi_p^*(\mathrm{M}) \chi_q(\mathrm{M}) \tag{4.73}$$

$$\rho^{\mathrm{CI}}(\mathrm{M}) = \sum_k \eta_k \theta_k^2(M) = \sum_p \sum_q D_{pq}^{\mathrm{CI}} \chi_p^*(M) \chi_q(\mathrm{M}) \qquad (4.74)$$

These electron densities are commonly shown in the literature by maps presenting isovalue contours for a selected plane of space or in three dimensions by a selected isovalue volume as shown in Figure 4.1 for the methane molecule.

4.2.2.2 Mulliken population analysis

If the electron density $\rho(\mathrm{M})$ is integrated over certain volumes of space, a mean number of electrons contained in these volumes will be found and a partition of space into loges will be defined, shown extensively in the first part. Of course, if integration takes place over all space the total number of electrons of the system will be found.

It is then tempting to try to find how the electrons are distributed in the molecular space. This is done using the Mulliken population analysis in the LCAO approach. Although this is not a property as no experiment allows the population to be analysed precisely, nevertheless, it will be discussed here as it provides a good insight into molecular spaces and correlates easily to well-known experimental models such as bond properties, polarization, inductive and mesomeric effects, and so on. Thus:

$$n = \int \rho(\mathrm{M})\, \mathrm{d}v_{\mathrm{M}} \qquad (4.75)$$

or

$$n = \sum_p \sum_q D_{pq}^{\mathrm{HF}} \int \chi_p^*(\mathrm{M}) \chi_q(\mathrm{M})\, \mathrm{d}v_{\mathrm{M}} \qquad (4.76)$$

and finally:

$$n = \sum_p \sum_q D_{pq}^{\mathrm{HF}} S_{pq} \qquad (4.77)$$

If the atomic orbitals are distinguished by their atomic centres A, B, ..., relation (4.77) becomes:

$$n = \sum_{\mathrm{A}} \sum_{\mathrm{B}} \left(\sum_{p \in \mathrm{A}} \sum_{q \in B} D_{pq}^{\mathrm{HF}} S_{pq} \right) \qquad (4.78)$$

Defining a population by:

$$P_{\mathrm{AB}} = \sum_{p \in \mathrm{A}} \sum_{q \in \mathrm{B}} D_{pq}^{\mathrm{HF}} S_{pq} \qquad (4.79)$$

gives a partitioning of the total number of electrons in one- and two-centre contributions, P_{AA} being known as the atomic population and P_{AB} as the overlap population between A and B:

$$n = \sum_{A} P_{AA} + \sum_{A} \sum_{B \neq A} P_{AB} \tag{4.80}$$

Further, a gross atomic population can be defined by:

$$P_A = \sum_{B} P_{AB} \tag{4.81}$$

so the n electrons will be spread through the Mulliken population analysis over the different atoms:

$$n = \sum_{A} P_A \tag{4.82}$$

The charge of the atom will appear by comparison between P_A and the atomic number Z_A, the net atomic charge being equal to:

$$\delta_A = P_A - Z_A \qquad (\times e = 1 \text{ in a.u.}) \tag{4.83}$$

Even though this approach is attractive, one should note that there are some drawbacks to it. It is very dependent on the chosen atomic basis set and results from one basis to another may seem contradictory. The analysis assumes that the atomic orbitals are well-localized functions on atoms and do not contribute to a description of neighbours. This is certainly not the case for diffuse functions which are commonly used in extended basis sets. It must be remembered that the maximum of a p or d function may lie very far from the corresponding atom.

Other population analysis models[7] try to correct these drawbacks but their description will not be given here.

4.2.2.3 Difference density function δ(M)

Another way to look at the electronic reorganization in a molecule is to substract the sum of atomic densities built in the same nuclear configuration from the molecular electron density (see Section 3.1) so that the difference density function δ(M) is defined as:

$$\delta(\text{M}) = \rho(\text{M}) - \sum_{A} \rho_A(\text{M}) \tag{4.84}$$

Figure 4.2 shows an example of δ(M) for the methane molecule. This information will be reliable only if the used wave function is accurate enough.

4.2.2.4 Spin properties

In open-shell systems the number of electrons of one spin is different from the number of electrons of the other spin. Therefore, some spin polarization

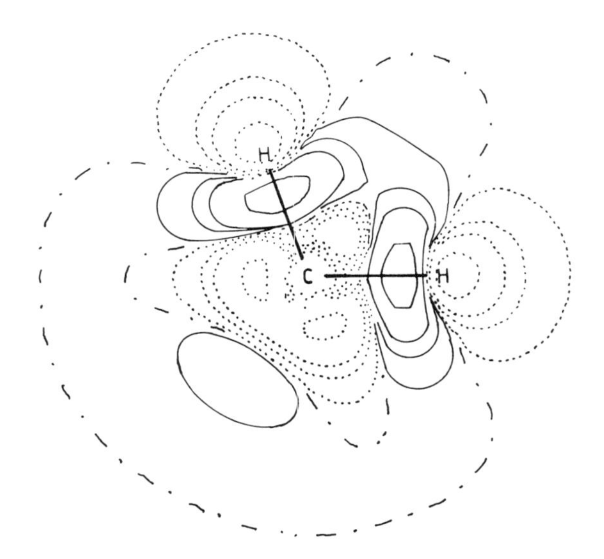

Figure 4.2 Difference density function of methane in the HCH plane (STO-3G)

will appear in the electron distribution. This property is quite interesting as the recent development of magnetic resonance techniques offers many new experiments. The S_Z operator is surely the most interesting one for this property as it distinguishes between both spins. Let us introduce it in relation (4.32) and use the first-order density matrix:

$$\langle S_Z \rangle = \int_{\substack{r_1'=r_1 \\ s_1'=s_1}} S_Z(1)\gamma(r_1 s_1, r_1' s_1')\, dr_1\, ds_1 \tag{4.85}$$

As spin functions can only be $\alpha(s_1)$ or $\beta(s_1)$ for pure spin states:

$$\gamma(r_1 s_1, r_1' s_1') = \gamma^{\alpha}(r_1 r_1')\alpha(s_1)\alpha^*(s_1') + \gamma^{\beta}(r_1 r_1')\beta(s_1)\beta^*(s') \tag{4.86}$$

Integration over spin leads to:

$$\langle S_Z \rangle = \int_{r_1'=r_1} \frac{1}{2}[\gamma^{\alpha}(r_1 r_1') - \gamma^{\beta}(r_1 r_1')]\, dr_1 \tag{4.87}$$

We shall call spin density the function:

$$\rho^{S}(M) = \gamma^{\alpha}_{M=M'}(M, M') - \gamma^{\beta}_{M=M'}(M, M') \tag{4.88}$$

which integrates to the number of unpaired electrons and describes the spin organization in space. Here also the approximate functions are largely used and lead to a spin population analysis. In the LCAO approximation it appears that:

$$D^{HF}_{pq} = D^{\alpha}_{pq} + D^{\beta}_{pq}$$

Thus:

$$\langle S_Z\rangle = \frac{1}{2}\int_{r'=r} \sum_p \sum_q (D^{\alpha}_{pq} - D^{\beta}_{pq})\chi_p(r)\chi^*_q(r')\,\mathrm{d}r \tag{4.89}$$

and

$$\rho^{S}(M) = \sum_p \sum_q (D^{\alpha}_{pq} - D^{\beta}_{pq})\chi^*_p(M)\chi_q(M) \tag{4.90}$$

After integration over r, relation (4.89) gives:

$$\langle S_Z\rangle = \frac{1}{2}\sum_p \sum_q (D^{\alpha}_{pq}S_{pq} - D^{\beta}_{pq}S_{pq}) \tag{4.91}$$

where $D^{\alpha,\beta}_{pq}$ will lead to an alpha or beta population analysis comparable to the Mulliken analysis.

The atomic spin density will simply be:

$$P^{S}_{A} = P^{\alpha}_{A} - P^{\beta}_{B} \tag{4.92}$$

and

$$\langle S_Z\rangle = \frac{1}{2}\sum_A P^{S}_{A} \tag{4.93}$$

It is also interesting to define an orbital spin density as:

$$D^{S}_{p} = P^{\alpha}_{p} - P^{\beta}_{p} \tag{4.94}$$

where:

$$P^{\alpha,\beta}_{p} = \sum_q D^{\alpha,\beta}_{pq}S_{pq} \tag{4.95}$$

Figure 4.3 and Table 4.1 provide an example of results obtained for CH_3 by the unrestricted Hartree–Fock method. It is indeed important to notice that there is a great difference between the results obtained in restricted and unrestricted open-shell calculations as the spin density reduces in RHF to the density of the singly occupied orbitals while in UHF all the orbitals contribute to the spin properties.

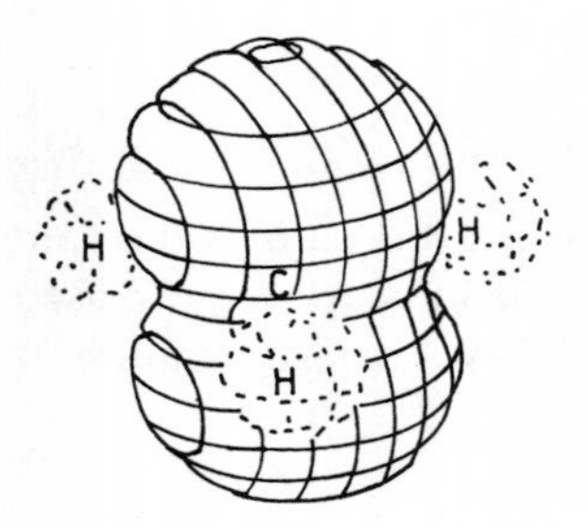

Figure 4.3 The spin density of methyl radical in three dimensions (STO-3G)

Table 4.1 Spin properties of methyl radical (4-31G)

Atom(A)	Atomic orbital (p)	P_p^α	P_p^β	P_p^S	P_A^S	$\rho^S(A)$
H	$1s$	0.3695	0.4775	−0.1080	−0.1080	−0.034
C	$1s$	0.9980	0.9977	0.0003	—	—
	$2s$	0.7560	0.5880	0.1680	—	—
	$2p_x(\pi)$	1.0000	0.0000	1.0000	—	—
	$2p_y$	0.5688	0.4910	0.0780	—	—
	$2p_z$	0.5688	0.4910	0.0780	1.3243	0.2293

4.2.2.5 Dipole moment

Permanent electric dipole moments are often present in neutral molecules. They appear only because the centre of charge of the electronic distribution is not at the same point as that of the nuclei.

The dipole moment is classically given by the relation:

$$\boldsymbol{\mu} = Q(\langle \mathbf{r}_N \rangle - \langle \mathbf{r}_e \rangle) \tag{4.96}$$

where Q is the total charge of electrons (or nuclei) and $\langle \mathbf{r}_N \rangle$ and $\langle \mathbf{r}_e \rangle$ are the centres of charge of nuclei and electrons respectively. The operators we shall use are clearly the position operators and the expectation value for the dipole moment will be:

$$\langle \boldsymbol{\mu} \rangle = Q\langle \mathbf{r}_N \rangle - \int_{r=r'} \mathbf{r}\gamma(rr')\,\mathrm{d}r \tag{4.97}$$

where:

$$\langle \mathbf{r}_N \rangle = \frac{\sum_{J=1}^{N} Z_J \mathbf{r}_J}{\sum_{J=1}^{N} Z_J} \tag{4.98}$$

The reduced density matrix may be used as no spin dependency appears.

This becomes in the Hartree–Fock–Roothaan approximation:

$$\langle \boldsymbol{\mu} \rangle = Q\langle \mathbf{r}_N \rangle - \sum_{i=1}^{n} \eta_i \int \phi_i^*(r)\mathbf{r}\phi_i(r)\,\mathrm{d}r \tag{4.99}$$

or

$$\langle \boldsymbol{\mu} \rangle = Q\langle \mathbf{r}_N \rangle - \sum_p \sum_q D_{pq}^{HF} \int \chi_p^*(r)\mathbf{r}\chi_q(r)\,\mathrm{d}r \tag{4.100}$$

The dipole moment is of particular interest if one is looking at the change in energy due to a small perturbation (such as, for example, an infinitesimal test charge) of the hamiltonian. This is usually examined through perturbation theory.[8] Nevertheless, some information can be obtained from the first-order density matrix, assuming that this static perturbation acts only on the electron distribution and has no effect on the mutual repulsion between electrons (i.e. on the second-order density matrix). If the potential is developed in series such as:

$$V = V_0 + \left(\frac{\partial V}{\partial r}\right)_0 \bar{r} + \bar{r}' \frac{1}{2}\left(\frac{\partial^2 V}{\partial r^2}\right)_0 \bar{r} + \cdots \tag{4.101}$$

then

$$\Delta E = \left(\frac{\partial V}{\partial r}\right)_0 \left[\int_{r'=r} \bar{r}\gamma(rr')\,\mathrm{d}r + \sum_{J=1}^{N} Z_J \bar{r}_J\right]$$
$$+ \frac{1}{2}\left(\frac{\partial^2 V}{\partial r^2}\right)\left[\int_{r'=r} \bar{r}^2\gamma(r_1 r')\,\mathrm{d}r + \sum_{J=1}^{N} Z_J \bar{r}_J^2\right] \tag{4.102}$$

The first term on the right contains the dipole moment contribution; the second term contains the quadrupole moment contribution. Higher order contributions are neglected here although they are easily computed.

Expression (4.102) is simply the one obtained from an expansion in multipoles of the particle distribution.

4.2.2.6 Molecular electrostatic potential

We will now consider another property not experimentally observable but of interest for chemists—the electrostatic potential.

Let us consider that a perturbation on a molecule is due to a unitary point charge located at position M in space. The interaction energy between this point charge and the molecular charge distribution is given by:

$$V(\mathrm{M}) = \sum_J \frac{Z_J}{r_{J\mathrm{M}}} + \int_{r'=r} \frac{1}{r_{\mathrm{M}}}\gamma(r, r')\,\mathrm{d}r \tag{4.103}$$

This value is positive or negative following the repulsion or attraction appearing between the charge and the molecule. It will simulate the interaction with a proton or some electrophilic agent, revealing the nucleophilic and electrophilic domains in the molecular environment at a first-order approximation.

The interaction energy $V(\mathrm{M})$, commonly called the molecular electrostatic potential, may be directly computed (without considering a multipole expansion) for a complete grid of points pertinently chosen by the user. This is not very difficult as formula (4.103) may be expressed in a Hartree–Fock or Roothaan approximation as:

$$V(\mathrm{M}) = \sum_J \frac{Z_J}{r_{J\mathrm{M}}} + \sum_i n_i \left\langle \phi_i \right| -\frac{1}{r_{\mathrm{M}}} \left| \phi_i \right\rangle \tag{4.104}$$

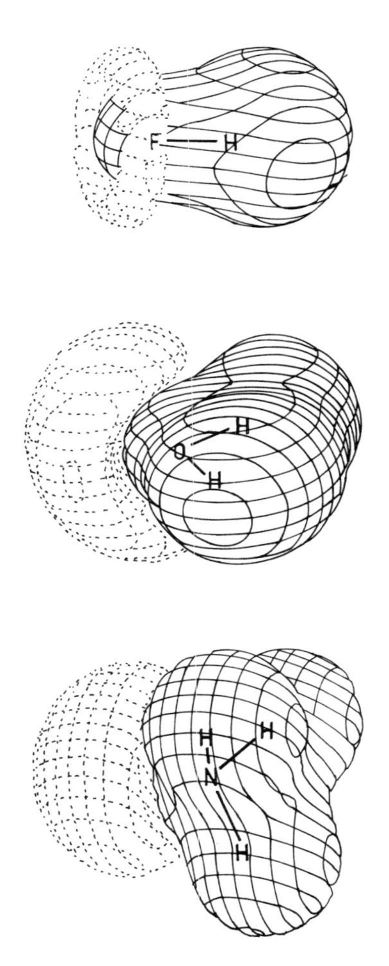

Figure 4.4 Electrostatic potential of HF, H_2O and NH_3 molecules in three dimensions

or

$$V(\mathrm{M}) = \sum_J \frac{Z_J}{r_{JM}} + \sum_p \sum_q D_{pq}^{\mathrm{HF}} \left\langle \chi_p \right| -\frac{1}{r_{\mathrm{M}}} \left| \chi_q \right\rangle \qquad (4.105)$$

The integrals appearing in (4.105) are strictly identical to those needed to compute the nuclear attraction energy. Three-dimension representations of some molecular electrostatic potentials (for HF, H_2O and NH_3) are shown in Figure 4.4. More details on this concept will be found in the reference paper of Scrocco and Tomasi.[9]

Many other first-order properties may be computed but it would be useless to examine them all as the interested user would find much more about them in specialized articles.

4.2.3 Two Electron Properties

Many properties are dependent on electronic interdistances and thus need the second-order density matrix to be estimated explicitly. Energy is the only one which will be briefly analysed in this book.

Formula (4.49) gave an explicit expression of the energy using the density matrices formalism. This can be rewritten in the Hartree–Fock approximation as:

$$E = H_0 + \int_{x_1'=x_1} h^{\mathrm{N}}(1)\gamma^{\mathrm{HF}}(x_1x_1')\,\mathrm{d}x_1 + \frac{1}{2}\int_{\substack{x_1'=x_1\\x_2'=x_2}} \left(\frac{1-P_{12}}{r_{12}}\right)\gamma^{\mathrm{HF}}(x_1x_1')\gamma^{\mathrm{HF}}(x_2x_2')\,\mathrm{d}x_1\,\mathrm{d}x_2 \quad (4.106)$$

where:

$$h^{\mathrm{N}}(1) = -\frac{1}{2}\nabla^2(1) - \sum_J \frac{Z_J}{r_{1J}} \quad (4.107)$$

The terms:

$$\frac{1}{2}\frac{1}{r_{12}} \quad \text{and} \quad -\frac{1}{2}\frac{P_{12}}{r_{12}} \quad (4.108)$$

are responsible for Coulomb and exchange bielectronic contributions. Equation (4.106) may be further rearranged:

$$E = H_0 + \int_{x_1'=x_1} \left[h^{\mathrm{N}}(1) + \frac{1}{2}\int_{x_2'=x_2} \left(\frac{1-P_{12}}{r_{12}}\right)\gamma^{\mathrm{HF}}(x_2x_2')\,\mathrm{d}x_2\right]\gamma^{\mathrm{HF}}(x_1x_1')\,\mathrm{d}x_1 \quad (4.109)$$

The expression between brackets is nothing more than the Hartree–Fock operator.

Expressing $\gamma^{\mathrm{HF}}(x_1x_1')$ as in (4.86) and integrating over spin, it can be found for a closed-shell problem, i.e. $\gamma^{\alpha}(rr') = \gamma^{\beta}(rr') = \gamma(rr')$, that:

$$E = H_0 + \int_{r_1'=r_1} \left[2h^{\mathrm{N}}(1) + \int_{r_2'=r_2} \left(\frac{2-P_{12}}{r_{12}}\right)\gamma(r_2r_2')\,\mathrm{d}r_2\right]\gamma(r_1r_1')\,\mathrm{d}r_1 \quad (4.110)$$

As now:

$$\gamma(r_1r_1') = \sum_i \phi_i(r_1)\phi_i^*(r_1') \quad (4.111)$$

then:

$$E = H_0 + 2\sum_i \langle\phi_i|\, h^{\rm N}\, |\phi_i\rangle + 2\sum_i\sum_j \langle\phi_i|\, J_j\, |\phi_i\rangle - \sum_i\sum_j \langle\phi_i|\, K_j\, |\phi_i\rangle \quad (4.112)$$

where:

$$J_j = \int \phi_j^*(r_2)\frac{1}{r_{12}}\phi_j(r_2)\,\mathrm{d}r_2 \quad (4.113)$$

and

$$K_j = \int \phi_j^*(r_2)\frac{P_{12}}{r_{12}}\phi_j(r_2)\,\mathrm{d}r_2 \quad (4.114)$$

or

$$E = H_0 + 2\sum_i \langle\phi_i|\, h^{\rm N}\, |\phi_i\rangle + \sum_i\sum_j 2\left\langle\phi_i\phi_j\right|\frac{1}{r_{12}}\left|\phi_j\phi_i\right\rangle - \sum_i\sum_j \left\langle\phi_i\phi_j\right|\frac{1}{r_{12}}\left|\phi_i\phi_j\right\rangle \quad (4.115)$$

This is more commonly written as:

$$E = H_0 + 2\sum_i \varepsilon_i^{\rm N} + \sum_i\sum_j (2J_{ij} - K_{ij}) \quad (4.116)$$

Noticing that the eigenvalues of the Fock operator are:

$$\varepsilon_i = \varepsilon_i^{\rm N} + \sum_j (2J_{ij} - K_{ij}) \quad (4.117)$$

the energy may also be written as:

$$E = H_0 + 2\sum_i \varepsilon_i - \sum_i\sum_j (2J_{ij} - K_{ij}) \quad (4.118)$$

and finally:

$$E = H_0 + \sum_i (\varepsilon_i^{\rm N} + \varepsilon_i) \quad (4.119)$$

All these formulas are of interest, depending on the information needed or given.

In the LCAO approximation, as:

$$\gamma(r_1 r_1') = \sum_p\sum_q\sum_i C_{ip}C_{iq}\chi_p(r_1)\chi_q^*(r_1') = \sum_p\sum_q D_{pq}\chi_p(r_1)\chi_q^*(r_1') \quad (4.120)$$

then:

$$E = H_0 + \sum_p \sum_q 2D_{pq} \langle \chi_p | h^N | \chi_q \rangle + 2 \sum_p \sum_q \sum_r \sum_s D_{pq} D_{rs}$$
$$\times \left\langle \chi_p \chi_r \left| \frac{1}{r_{12}} \right| \chi_s \chi_q \right\rangle - \sum_p \sum_q \sum_r \sum_s D_{pq} D_{rs} \left\langle \chi_p \chi_r \left| \frac{1}{r_{12}} \right| \chi_q \chi_s \right\rangle \quad (4.121)$$

or

$$E = H_0 + \sum_p \sum_q 2D_{pq} h^N_{pq} + \sum_p \sum_q \sum_r \sum_s D_{pq} D_{rs} (2\langle pq \mid rs \rangle - \langle ps \mid qr \rangle) \quad (4.122)$$

In comparison to (4.118) and (4.119) this expression can also be written as:

$$E = H_0 + \sum_p \sum_q 2D_{pq} h_{pq} - \sum_p \sum_q \sum_r \sum_s D_{pq} D_{rs} (2\langle pq \mid rs \rangle - \langle ps \mid qr \rangle) \quad (4.123)$$

or

$$E = H_0 + \sum_p \sum_q D_{pq} (h^N_{pq} + h_{pq}) \quad (4.124)$$

The fact must be remembered that D_{pq} as defined does not include the occupation number of the molecular orbital which is the coefficient 2 appearing in some formulas. One has the evident relation:

$$D^{HF}_{pq} = 2D_{pq} \quad (4.125)$$

Now that the energy expression for a given electronic state is known, we may turn our interest to some other energetic properties such as transition energies, ionization potentials and electron affinities. Following the level of approximation of the wave function, here also the results will be different. We shall restrict ourselves to the Hartree–Fock model starting from a closed-shell configuration. Other interesting energetic properties such as reaction energies, isomerization barriers, activation barriers, etc., will be discussed later on.

4.2.4 Energy of Excited and Ionized States

Starting from a wave function expressed in terms of molecular orbitals, we may build approximate, non-variational, excited wave functions by changing the occupations of the orbitals and adapting spin functions to the desired spin state.

In this way the following wave functions for singlet and triplet excited states and for positive and negative ions may be obtained:

$$\Psi^S = \frac{1}{\sqrt{2}} [|\phi_1 \bar{\phi}_1 \cdots \phi_k \bar{\phi}_l \cdots \phi_n(2n-1) \bar{\phi}_n(2n)|$$
$$- |\phi_1 \bar{\phi}_1 \cdots \bar{\phi}_k \phi_l \cdots \phi_n(2n-1) \bar{\phi}_n(2n)|] \quad (4.126)$$

$$\Psi^{\mathrm{T}}(S_Z=0)=\frac{1}{\sqrt{2}}[|\phi_1\bar{\phi}_1\cdots\phi_k\bar{\phi}_l\cdots\bar{\phi}_n(2n-1)\bar{\phi}_n(2n)|$$

$$+|\phi_1\bar{\phi}_1\cdots\bar{\phi}_k\phi_l\cdots\phi_n(2n-1)\bar{\phi}_n(2n)|] \quad (4.127)$$

$$\Psi^{+}=|\phi_1\bar{\phi}_1\cdots\phi_n(2n-1)| \quad (4.128)$$

$$\Psi^{-}=|\phi_1\phi_1\cdots\phi_n(2n+1)\bar{\phi}(2n)\phi_{n+1}(2n+1)| \quad (4.129)$$

The corresponding configurations may be represented by the following energetic diagrams:

	E_0	E^{S}	E^{T}	E^{+}	E^{-}
ϕ_l		↓	↑		
ϕ_{n+1}					↑
ϕ_n	↑↓	↑↓	↑↓	↑	↑↓
ϕ_k	↑↓	↑	↑	↑↓	↑↓
ϕ_2	↑↓	↑↓	↑↓	↑↓	↑↓
ϕ_1	↑↓	↑↓	↑↓	↑↓	↑↓

The energy associated with each of these configurations can be computed using (4.109) with a proper spin integration. It can be shown that:

(a) The excitation energies for singlet and triplet are respectively:

$$\Delta E^{\mathrm{S}}=E^{\mathrm{S}}-E_0=\varepsilon_l-\varepsilon_k-J_{kl}+2K_{kl} \quad (4.130)$$

$$\Delta E^{\mathrm{T}}=E^{\mathrm{T}}-E_0=\varepsilon_l-\varepsilon_k-J_{kl} \quad (4.131)$$

Since K_{kl} is a positive value it follows from the Hund rule that the triplet is of lower energy than the corresponding excited singlet.

(b) The first ionization potential is given by:

$$IP=E^{+}-E^{0}=-\varepsilon_n \quad (4.132)$$

This relation is very easy to demonstrate as:

$$E^{+}=2\sum_{i=1}^{n-1}\varepsilon_i^{\mathrm{N}}+\sum_{i=1}^{n-1}\sum_{j=1}^{n-1}(2J_{ij}-K_{ij}) \qquad \text{(closed part)}$$

$$+\varepsilon_n^{\mathrm{N}}+\sum_{j=1}^{n-1}(2J_{nj}-K_{nj}) \qquad \text{(open part)} \quad (4.133)$$

$$E^{+}-E_0=-\varepsilon_n^{\mathrm{N}}-\sum_{j-1}^{n-1}(2J_{nj}-K_{nj})-(2J_{nn}-K_{nn})=-\varepsilon_n \quad (4.134)$$

Thus the first ionization potential assuming that the molecular orbitals for the ion and the molecule are identical is minus the energy of the highest occupied molecular orbital (HOMO). This is known as the Koopmans theorem and is always valid whatever may be the orbital from which the electron is extracted.

(c) The electron affinity of the molecule is simply given by:

$$\mathrm{EA} = E^- - E_0 = \varepsilon_{n+1} \tag{4.135}$$

which is the counterpart of the preceding formula and can easily be demonstrated.

Thus, at first approximation, one may have an idea of some energetical properties but it should be emphasized that the molecular orbitals for excited and ionized species are not related and the correlation effects due to the Hartree–Fock method would not be negligible. To obtain a better insight into these properties much more sophisticated calculations (e.g. large CI) should be undertaken.

4.2.5 The Rotational and Vibrational Energy Levels

Let us now return to equation (4.6) and consider the nuclear functions and the corresponding energy levels for the simplest case, the diatomic molecule. The general problem of nuclear motions in a polyatomic molecule and supermolecule will be discussed more rigorously in the last part of this book.

For a diatomic molecule, the nuclear wave equation which describes the translational, rotational and vibrational motions of the nuclei may be written as:

$$\left[-\frac{\hbar^2}{2M_\mathrm{A}}\nabla_\mathrm{A}^2 - \frac{\hbar^2}{2M_\mathrm{B}}\nabla_\mathrm{B}^2 + u(\mathrm{A},\mathrm{B})\right]\phi(\mathrm{A},\mathrm{B}) = W\phi(\mathrm{A},\mathrm{B}) \tag{4.136}$$

It is formally identical to the Schrödinger equation for the hydrogen atom and will be solved in a similar way (see Section 1.2.1). Here, again, the gravity centre theorem applies and equation (4.136) can be separated into two equations, one describing the translational motion of the system and the other its internal motion. The former is simply the Schrödinger equation of 'the particle in the box' problem:

$$-\frac{\hbar^2}{2(M_\mathrm{A}+M_\mathrm{B})}\nabla_{xyz}^2 F(xyz) = \varepsilon_\mathrm{T} F(xyz) \tag{4.137}$$

where xyz are the cartesian coordinates of the gravity centre of the molecule.

For a cubic box of edge a, the translational energy levels depend on three quantum numbers (n_1, n_2, n_3) according to the simple relation:

$$\varepsilon_\mathrm{T}(n_1 n_2 n_3) = \frac{h^2}{8(M_\mathrm{A}+M_\mathrm{B})a^2}(n_1^2+n_2^2+n_3^2) \tag{4.138}$$

The Schrödinger equation describing the internal motion of the molecule has the form:

$$-\frac{\hbar^2}{2\mu}\nabla_{xyz}^2\chi(xyz) + u(xyz)\chi(xyz) = \varepsilon\chi(xyz) \tag{4.139}$$

where xyz are the relative coordinates and μ is the reduced mass of the system. Moreover:

$$W = \varepsilon_T + \varepsilon \tag{4.140}$$

Equation (4.139) can also be written:

$$-\frac{\hbar^2}{2\mu}\left(\frac{\partial^2}{\partial r^2} + \frac{2}{r}\frac{\partial}{\partial r} + \frac{1}{r^2}L\right)\chi(r\theta\phi) + u\chi(r\theta\phi) = \varepsilon\chi(r\theta\phi) \tag{4.141}$$

where r, θ, ϕ are polar coordinates of one nucleus relative to the other as the origin and L is the Laplace operator:

$$L = \frac{1}{\sin\theta}\frac{\partial}{\partial\theta}\left(\sin\theta\frac{\partial}{\partial\theta}\right) + \frac{1}{\sin^2\theta}\frac{\partial^2}{\partial\phi^2} \tag{4.142}$$

The equation of internal motion (4.141) can only be solved if the potential energy $u = u(r)$ has been previously obtained.

4.2.5.1 The separation of rotation and vibration

Approximate solutions of equation (4.141) may be reached by neglecting the coupling between rotation and vibration. In order to study the rotational motion alone, one can put (for example):

$r = r_e$ (equilibrium distance between A and B)

On the other hand, pure vibrational energy levels can be obtained by taking fixed values for the angles θ and ϕ.

By putting $r = r_e$, equation (4.141) becomes:

$$-\frac{\hbar^2}{2\mu r_e^2}LY(\theta, \phi) = [\varepsilon_R - u(r_e)]Y(\theta, \phi) \tag{4.143}$$

which is simply the Schrödinger equation of a rigid rotator. Then it follows that:

$$\varepsilon_R - u(r_e) = K(K+1)\frac{\hbar^2}{2\mu r_e^2} \tag{4.144}$$

or

$$\varepsilon_R(K) = K(K+1)\frac{\hbar^2}{2I_e} + u(r_e) \tag{4.145}$$

where K is the rotational quantum number ($K = 0, 1, 2, \ldots$) and $I_e = \mu r_e^2$ is the equilibrium moment of inertia of the molecule.

Taking θ and ϕ constant and assuming a Hooke law potential energy function:

$$u(r) = \frac{k}{2}(r - r_e)^2 \tag{4.146}$$

one may reduce equation (4.141) to the Schrödinger equation of a harmonic oscillator whose well-known energy levels are:

$$\varepsilon_V(v) = \left(v + \frac{1}{2}\right)h\nu \tag{4.147}$$

One obtains a very simple expression for the energy levels of the diatomic molecule:

$$W = \varepsilon_T + \varepsilon_R + \varepsilon_V + u(r_e) \tag{4.148}$$

and, more explicitly, for the fundamental electronic state:

$$W_0(n_1 n_2 n_3 hv) = \frac{\hbar^2}{8(M_A + M_B)a^2}(n_1^2 + n_2^2 + n_3^2) + K(K+1)\frac{\hbar^2}{2I_e} + \left(v + \frac{1}{2}\right)h\nu + u_0(r_e) \tag{4.149}$$

4.2.5.2 Correct resolution with model potential

The wave equation for the rotation and vibration of a diatomic molecule (4.141) can be separated into two equations, one containing only the angular variables and the other depending only on the radial variable. The former can be written as:

$$LY = -K(K+1)Y \tag{4.150}$$

where:

$$Y = Y_{KM}(\theta, \phi) \tag{4.151}$$

are the well-known spherical harmonics and K and M play the same role as the azimuthal and magnetic quantum numbers in the hydrogen atom. Their allowed values are:

$$K = 0, 1, 2, \ldots \quad \text{and} \quad M = -K, -(K+1), \ldots, K-1, K \tag{4.152}$$

The radial equation has the form:

$$\frac{d^2R}{dr^2} + \frac{2}{r}\frac{dR}{dr} + \left[\frac{2\mu}{\hbar^2}(\varepsilon - u) - \frac{K(K+1)}{r^2}\right]R = 0 \tag{4.153}$$

and

$$\chi(R\theta\phi) = R(r)Y_{KM}(\theta, \phi) \tag{4.154}$$

Different potential energy functions may be used to solve the radial equation (4.146). The function $u(r)$ can be obtained numerically by solving the electronic equation (4.5) for different nuclear configurations, and its analytical representation is generally easy to find. Unfortunately the radial equation can only be solved analytically by using very simple model potentials such as the harmonic and the Morse potentials. For small displacements of the nuclei from their equilibrium positions one can assume a Hooke law potential energy function (4.146) where k is the so-called force constant which is nothing else than the second derivative of the potential function at the equilibrium distance $r = r_e$:

$$k = \left(\frac{\partial^2 u}{\partial r^2}\right)_{r=r_e} \tag{4.155}$$

The value of k is easy to obtain by using the potential energy curve deduced from the solutions of the electronic equation. Introducing the harmonic potential in the radial equation one obtains the following rotational–vibrational energy levels for the ground state of the diatomic molecule:

$$\varepsilon_{K,V} = K(K+1)\frac{\hbar^2}{2I_e} + \left(v + \frac{1}{2}\right)h\nu_e - \frac{K^2(K+1)^2\hbar^4}{8\pi^2\nu_e^2 I_e^3} + u_0 \tag{4.156}$$

where:

$$\nu_e = \frac{1}{2\pi}\sqrt{\left(\frac{k}{\mu}\right)} \tag{4.157}$$

is the vibrational frequency of the molecule. The first term is obviously the rotation energy of a rigid rotator, the second one is the vibrational energy of an harmonic oscillator and the third one is a coupling term taking account of the stretching of the molecule due to the rotation.

The harmonic potential is not very good for large deformations of the molecule. The Morse curve represents the calculated potential energy function much better, taking account of the anharmonicity of the vibration. The Morse function:

$$u(r) = D[1 - e^{-a(r-r_e)}]^2 \tag{4.158}$$

is plotted in Figure 4.5.

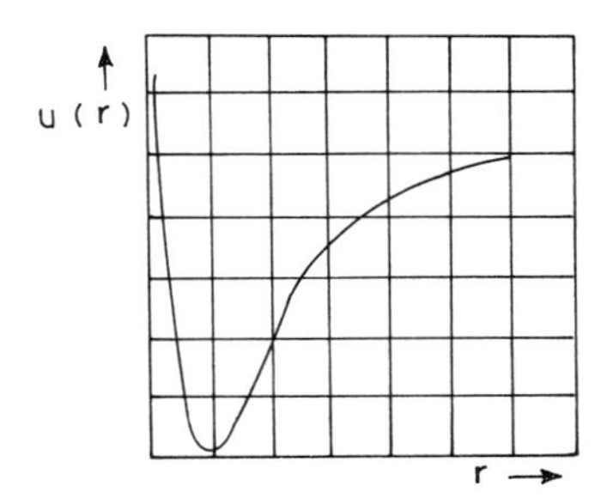

Figure 4.5 The Morse function

The constants D and a can be calculated from theoretical or experimental data. D is simply the electronic dissociation energy and it is easy to show that a has the explicit form:

$$a = \sqrt{\left(\frac{k}{2D}\right)} \tag{4.159}$$

The solutions of the radial equation, using the Morse potential, may be written with the notations employed in molecular spectroscopy:

$$\varepsilon_{k,V} = hc\left[\tilde{\nu}_e\left(\nu+\frac{1}{2}\right) - x_e\tilde{\nu}_e\left(v+\frac{1}{2}\right)^2 + K(K+1)B_e + D_eK^2(K+1)^2 - \alpha_e\left(\nu+\frac{1}{2}\right)K(K+1)\right] \tag{4.160}$$

in which:

$$\tilde{\nu}_e = \frac{\nu_e}{c} = \frac{1}{2\pi c}\sqrt{\left(\frac{k}{\mu}\right)} = \frac{a}{2\pi c}\sqrt{\left(\frac{2D}{\mu}\right)}$$

$$x_e = \frac{hc\nu_e}{4D}$$

$$B_e = \frac{h}{8\pi^2 I_e c} \tag{4.161}$$

$$D_e = -\frac{h^3}{128\pi^6\mu^3\tilde{\nu}_e^2c^3r_e^6}$$

$$\alpha_e = \frac{3h^2\tilde{\nu}_e}{16\pi^2\mu r_e^2 D}\left(\frac{1}{ar_e} - \frac{1}{a^2r_e^2}\right)$$

The successive terms in equation (4.160) are respectively:

(a) The vibrational energy of the harmonic oscillator,
(b) An anharmonic term,
(c) The rotational energy of the rigid rotator,
(d) A term taking account of the stretching of the molecule due to the rotation,
(e) A true rotation–vibration coupling term.

The energy levels corresponding to equation (4.160) are represented in Figure 4.6. Some quantitative illustrations of this formalism will be given in a later section.

Further details on the calculation of rotational–vibrational energy levels of diatomic molecules can be found in many textbooks.[10]

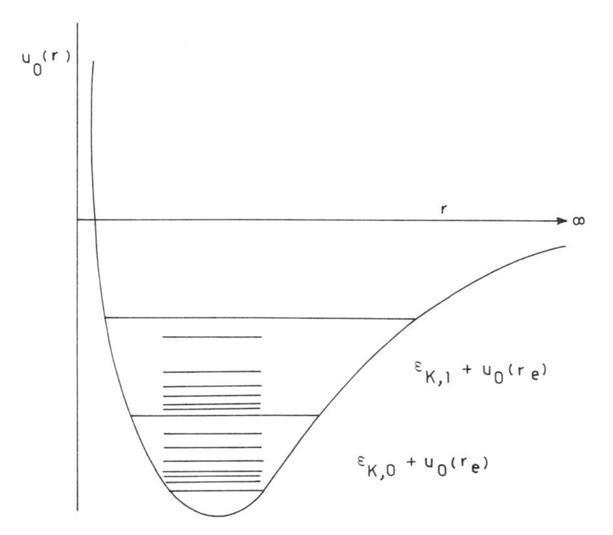

Figure 4.6 Schematic representation of the rotation–vibration energy levels of a diatomic molecule in its fundamental electronic state

4.2.6 Thermodynamical Properties

Up to now we have been concerned with the properties of hypothetical isolated molecules. In order to compare theoretical results and experimental data, it is often necessary to consider a collection of molecules in a dilute gas phase (in order to avoid complications due to molecular interactions) at a certain temperature. We shall be interested here in the calculation of some state functions such as internal energy and enthalpy of one mole of perfect gas and also in the calculation of the equilibrium constant.

4.2.6.1 Internal energy and enthalpy

The internal energy of a thermodynamical system at temperature T can be separated into two contributions such as:

$$U(T) = U(0) + [U(T) - U(0)] \tag{4.162}$$

The first term represents the energy of the molecules at 0 K and the second one defines the so-called thermal corrections. The explicit expression for $U(T)$ can be deduced in statistical thermodynamics. In the case of diatomic molecules it can be written as:

$$U(T) = U_{\mathrm{T}} + U_{\mathrm{R}} + U_{\mathrm{V}} + U_{\mathrm{e}} \tag{4.163}$$

if the reference state of zero energy is taken as the nuclei and electrons are completely separated. The first three terms are the translational, rotational and vibrational contributions which may be written respectively as:

$$U_T = \frac{3}{2} RT$$

$$U_R = \frac{2}{2} RT \tag{4.164}$$

$$U_V = N\frac{h\nu}{2} + N\frac{h\nu}{e^{h\nu/kT} - 1}$$

where:

$$k = \frac{R}{N} \tag{4.165}$$

is the Boltzmann constant and ν is the vibrational frequency of the diatomic molecule. The term:

$$N\frac{h\nu}{2} \tag{4.166}$$

is commonly called the zero point energy (ZPE). We shall note that this is simply the difference between the electronic dissociation energy and the experimental one at 0 K.

We now have:

$$U(T) = \frac{3}{2} RT + \frac{2}{2} RT + N\frac{h\nu}{2} + N\frac{h\nu}{e^{h\nu/kT} - 1} + U_e(T) \tag{4.167}$$

If the temperature is not too high the molecules will be in their fundamental electronic state

$$U_e(T) = U_e(0) = Nu_0(r_e) \tag{4.168}$$

and

$$U(0) = N\frac{h\nu}{2} + Nu_0(r_e) \tag{4.169}$$

The thermal corrections are then easy to calculate using the following relation:

$$U(T) - U(0) = \frac{3}{2} RT + \frac{2}{2} RT + N\frac{h\nu}{e^{h\nu/kT} - 1} \tag{4.170}$$

Finally, the internal energy of one mole of perfect diatomic gas is related to the theoretical electronic energy $u_0(r_e)$ according to the formula:

$$U(T) = Nu_0(r_e) + N\frac{h\nu}{2} + [U(T) - U(0)] \tag{4.171}$$

The corresponding enthalpy is simply:

$$H(T) = U(T) + RT \tag{4.172}$$

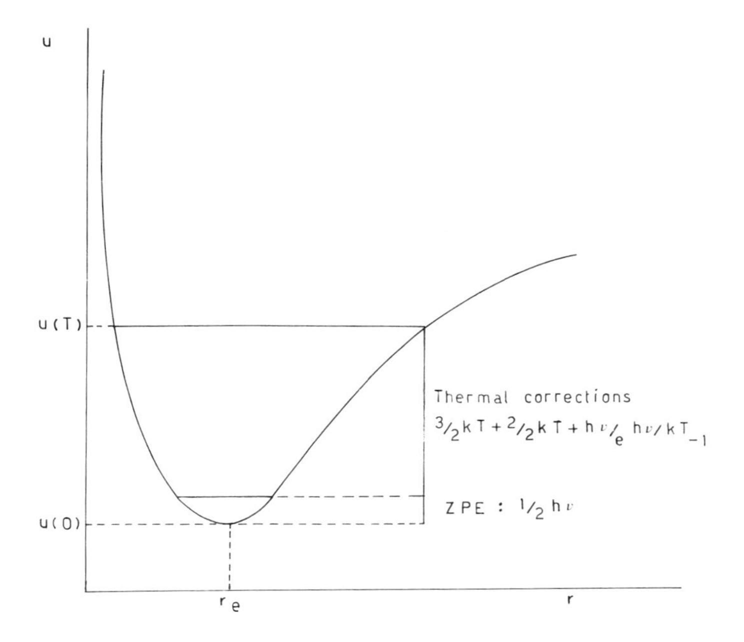

Figure 4.7 The different contributions to $U(T)$

or

$$H(T) = Nu_0(r_e) + N\frac{h\nu}{2} + [H(T) - H(0)] \tag{4.173}$$

where:

$$H(T) - H(0) = \frac{7}{2}RT + N\frac{h}{e^{h\nu/kT} - 1} \tag{4.174}$$

Only energy or enthalpy differences such as heats of reaction can be obtained experimentally. An illustration of equation (4.171) is given in Figure 4.7.

4.2.6.2 Equilibrium constants

The equilibrium constant of a gas phase reaction in terms of partial pressures is defined by the well-known expression:

$$K_p = \exp\left(-\frac{\Delta G^\circ}{RT}\right) \tag{4.175}$$

where ΔG° is the standard Gibbs free energy change of the chemical reaction which can be calculated from the standard molar Gibbs free energy G_i° of any component of the gaseous mixture. This quantity is related to the corresponding partition function at standard pressure by:

$$G^\circ = U^\circ(0) - RT \ln \frac{Z^\circ}{N} \tag{4.176}$$

where:

$$Z^\circ = \sum_j \omega_j e^{-\Delta\varepsilon_j/kT} \tag{4.177}$$

ω_j being the degeneracy of level j whose relative energy is $\Delta\varepsilon_j = \varepsilon_j - \varepsilon_0$. Thus:

$$\Delta G^\circ(T) = \sum_i k_i N_i \left[U_i^\circ(0) - RT \ln \frac{Z_i^\circ}{N} \right] \tag{4.178}$$

where N_i is the stoichiometric coefficient of component i and $k_i = +1$ for any product and -1 for any reactant. Hence:

$$\Delta G^\circ(T) = \Delta U^\circ(0) - RT \ln \prod_i \left(\frac{Z_i^\circ}{N} \right)^{k_i N_i} \tag{4.179}$$

and

$$K_p(T) = \prod_i \left(\frac{Z_i^\circ}{N} \right)^{k_i N_i} \exp\left(-\frac{\Delta U^\circ(0)}{RT} \right) \tag{4.180}$$

The heat of reaction at 0 K, $\Delta U^\circ(0)$, is easy to obtain from theoretical results. For example, for isotopic exchange equilibria:

$$A_2 + B_2 \rightleftarrows 2\ AB \tag{4.181}$$

the heat of reaction is

$$\Delta U^\circ(0) = \frac{Nh}{2}(2\nu_{AB} - \nu_{A_2} - \nu_{B_2}) + N(2u_{^\circ AB} - u_{^\circ A_2} - u_{^\circ B_2}) \tag{4.182}$$

In this particular case the second term vanishes since A_2, B_2 and AB have the same electronic energy, $U_0(r_e)$.

The partition functions of each component can also be calculated from molecular properties theoretically estimated, such as I_e and ν. Numerical examples will be given in Section 5.4.

REFERENCES

1. R. Daudel, R. Lefebvre and C. Moser, *Quantum Chemistry, Methods and Applications*, Interscience, 1959, p. 470.
2. C. C. J. Roothaan, *Rev. Mod. Phys.*, **32,** 179 (1960).
3. J. A. Pople and R. K. Nesbet, *J. Chem. Phys.*, **22,** 571 (1954).
4. Quantum Chemistry Program Exchange (QCPE), Chemistry Department Room 204, Indiana University, Bloomington, Indiana 47401, U.S.A.
5. P. O. Löwdin, *Phys. Rev.*, **97,** 1474 (1955).

6. J. E. Harriman, *J. Chem. Phys.*, **40,** 2827 (1964); A. Hardisson and J. E. Harriman, *J. Chem. Phys.*, **46,** 2639 (1967); R. McWeeny and B. T. Sutcliffe, *Methods of Molecular Quantum Mechanics*, Academic Press, London, New York, 1976; E. R. Davidson, *Reduced Density Matrices in Quantum Chemistry*, Academic Press, London, New York, 1976.
7. I. Cohen, *J. Chem. Phys.*, **57,** 5076 (1972); K. R. Roby, *Mol. Struct.*, **27,** 81 (1974).
8. P. Swanstrom and F. Hegelund, in *Computational Techniques in Quantum Chemistry and Molecular Physics* (Eds. G. H. F. Dierksen, B. T. Sutcliffe and A. Veillard), Reidel, 1975, p. 299. R. McWeeny and B. T. Sutcliffe, *Methods of Molecular Quantum Mechanics*, Academic Press, London, New York, 1976, pp. 200–230.
9. E. Scrocco and J. Tomasi, *Topics in Current Chemistry*, **42,** 95 (1973).
10. L. Pauling and E. B. Wilson, *Introduction to Quantum Mechanics*, McGraw-Hill, 1935; G. Herzberg, *Molecular Spectra and Molecular Structure*, Vol. 1, Van Nostrand, 1950.

CHAPTER 5

The Practice of Quantum Chemistry Calculations

5.1 ATOMIC AND MOLECULAR BASIS SETS

Until now, we have presented wave functions built over orbitals and orbitals further developed over basis functions, but the kind of basis functions one would introduce in an expansion have not been introduced. The main goal of this section is to familiarize the reader with these functions and to give some simple rules to follow when choosing suitable basis functions adapted to the nature of the problem to be solved.

We shall nevertheless restrict ourselves to monoelectronic functions, discarding correlated functions, group functions, and so on. We shall consider functions usually used in *ab initio* calculations at the Hartree–Fock level, followed or not by some configuration interaction.

5.1.1 Some Mathematical Considerations

Our main problem is to fit the molecular orbitals ϕ with a combination of a given set of functions χ:

$$\phi(\bar{x}) = \sum_{p=1}^{m=\infty} C_p \chi_p(\bar{x}) \tag{5.1}$$

where m is the dimension of the complete basis set, which in our case is infinite. One may stress that the eigenfunctions of the hydrogen atom are known and form a complete set, so they would be a convenient choice, but the completeness of this set needs inclusion of the continuum, which is not at all convenient to use. So one has to truncate the set to a limited number of functions. This introduces the possible occurrence of false minimum and convergence problems.

If we limit the basis functions set, we shall have to find those functions which will best fit into the function $\phi(\bar{x})$, i.e. the basis set must be as complete as possible although the number of functions should remain as small as possible to avoid useless calculations. Thus:

$$\phi(\bar{x}) \simeq \phi_m(\bar{x}) = \sum_{p=1}^{m} C_p \chi_p(\bar{x}) \tag{5.2}$$

The most natural way to find the coefficients C_p is to minimize the mean square deviation:

$$D = \int_a^b |\phi(\bar{x}) - \phi_m(\bar{x})|^2 \, d\bar{x} \tag{5.3}$$

The first derivative of D with respect to each of the coefficient C_p leads to a set of equations whose solution gives the best coefficients (assuming a check has been made on the second derivative to ensure the minimum).

The number of functions needed may be decided by fixing *a priori* a threshold on D. This method presents nevertheless some important drawbacks: indeed, the function $\phi(\bar{x})$ must be known and this is rarely the case in molecular computations; on the other hand, such a fitting of a function holds only 'in the mean'. This means not only that punctual values of the fitting function could be quite a long way from the actual value but also that it could be worthwhile to refine the fitting in certain regions of more interest for the treated problem, discarding other regions (e.g. near a nucleus or long-range behaviour of the function, etc.).

Some more care should be taken to solve another problem which may arise. Sometimes, to improve the agreement some extra function χ is added. This must be linearly independent of the others, otherwise the determinant associated to the metric matrix:

$$S_{pq} = \int \chi_p^*(\bar{x}) \chi_q(\bar{x}) \, dx \tag{5.4}$$

will be zero, or nearly zero in the case of approximate linear dependency. An easy way to get rid of it, as proposed by Löwdin,[1] is by 'canonical orthogonalization', which consists in diagonalizing the metric matrix and discarding the eigenvalues below an arbitrary small value. This allows the basis set to be refined and excludes undesirable functions. This point will be examined later on with other basis set transformations.

5.1.2 Atomic Orbitals and Basis Functions

As molecules contain atoms and as atoms are well known, theoretically and experimentally, it is quite natural to start our study with atoms and then try to transpose the information gained to molecules. Literature about basis sets is quite abundant, so there is no need for us to describe it all. Nevertheless some important features have to be recalled.

First, the eigenfunctions for the hydrogen atom are well known. Furthermore, the Hartree–Fock orbitals for atoms are also known, but only numerically. Many types of functions have been used and described[2] in the development of (5.2) but we shall restrict ourselves to the commonly used Slater-type orbitals (STO) and gaussian-type orbitals (GTO), discussing their advantages and drawbacks. In the atomic problem, we are faced with a spherical symmetry problem, so the functions will contain a radial and an angular function, the latter being simply the spherical harmonics $Y_{lm}(\theta, \phi)$. We shall thus write an atomic orbital:

$$\chi_{nlr}(r\theta\phi) = R_{nl}(r)Y_{lm}(\theta\phi) \tag{5.5}$$

and the different types of orbitals will differ by the form of the radial function R. The atomic orbitals also present some peculiarities at the nucleus (i.e. a cusp) and the basis functions used will try to satisfy those conditions as well as possible.

5.1.2.1 Slater-type orbitals (STO)

The first type of radial dependency is given by the so-called Slater orbitals:

$$R_n(r\alpha) = \frac{(2\xi)^{n+1/2}}{(2n!)^{1/2}} * r^{n-1}e^{-\xi r} \tag{5.6}$$

These functions arise from a potential:

$$V(r) = -\frac{\xi n}{r} + \frac{n(n-1) - l(l+1)}{2r^2} \tag{5.7}$$

and depend, besides r, on a non-linear parameter ξ. This will be quite important for optimizing the function, as it may be added as a variational parameter to the determination of the best coefficients. It must be noted that when $l = n - 1$, the potential (5.7) is proportional to r^{-1} and the functions are identical to hydrogen-like functions. Furthermore, these functions present some cusps at their origin and are thus adequate for a good description of the wave function near the nucleus. Unfortunately the s functions are nodeless and consequently not orthogonal; nevertheless they may be orthogonalized by a simple procedure. The $1s$, $2s$, $2p$ functions are then very near to the Hartree–Fock solutions; for higher quantum numbers some discrepancies arise.

The STOs are of great interest for fitting atomic orbitals but their transposition to molecules is not easy as the evaluation of integrals over polycentric functions is quite difficult. Further information on this has been given by Shavitt and Karplus.[3] Figure 5.1 gives the shape of some real Slater functions for various values of n.

5.1.2.2 Gaussian-type orbitals (GTO)

These very commonly used functions have a radial dependency given by:

$$R_n(r, \alpha) = N_\alpha r^{n-1}e^{-\alpha r^2} \tag{5.8}$$

with:

$$N_\alpha = 2^{n+1}[(2n-1)!!^{-1/2}(2\pi)^{-1/4}\alpha^{(2n+1)/4}]$$

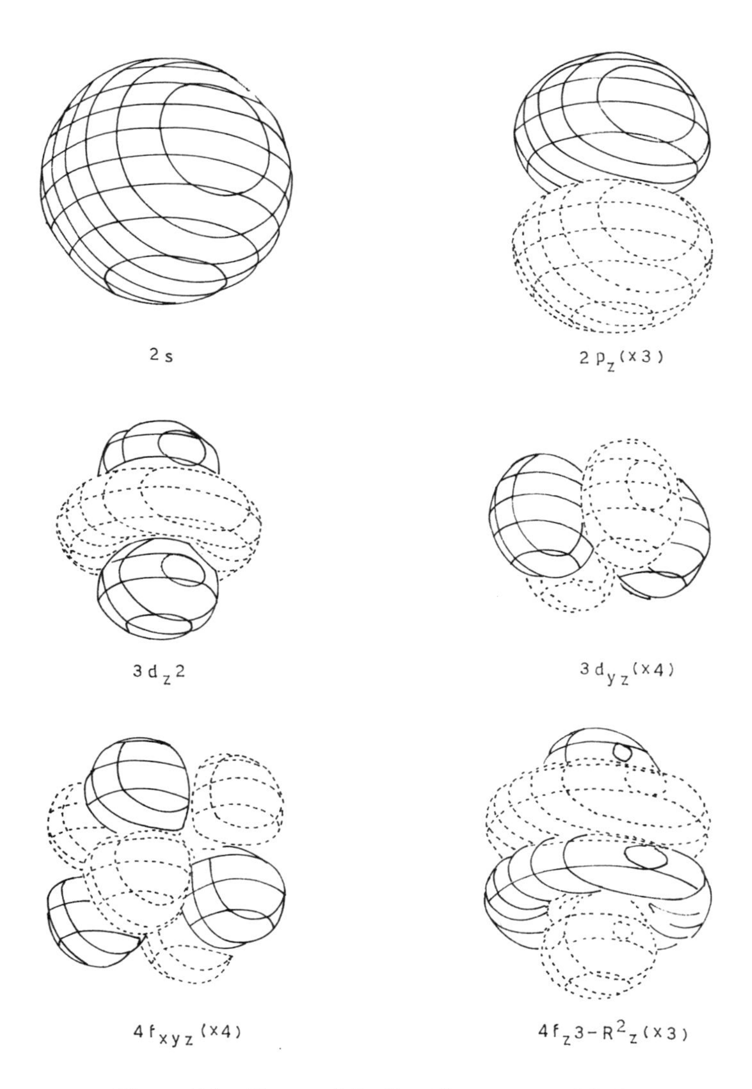

Figure 5.1 Shape of the first Slater-type orbitals

arising from a potential:

$$V(r) = 2\alpha^2 r^2 + \frac{n(n-1) - l(l+1)}{2r^2} \tag{5.9}$$

Such functions have the great advantage of giving easily integrable polycentric functions, which is the reason why they are so commonly used in molecular computations. Nevertheless, they present some important disadvantages. As these functions are steeper than the STOs they do not adequately describe outer regions of atoms, and so more GTOs will be needed to fit the atomic orbitals. The second important disadvantage is the absence of cusps, affecting their ability to describe properties relying on the wave function close to the nucleus. Some authors[4] tried to overcome this difficulty by adding a polynomial ensuring a cusp to the GTO.

Nevertheless the advantages easily overcome the disadvantages in molecular calculations and totally justify their use. It must also be noticed that although the relation (5.5) is convenient, the gaussian functions are usually not given in this form, but rather in a cartesian form where the spherical harmonics are replaced by cubic harmonics.[5] Then:

$$\chi_{nlm}(xyz\alpha) = N_{(\alpha lmn)} x^l y^m z^n e^{-\alpha r^2} \tag{5.10}$$

5.1.2.3 Lobe functions

To avoid the use of complicated spherical harmonics, some authors proposed to choose spherical functions and simulate the angular dependency simply by the superposition of a number of symmetrically distributed gaussian[6] or Slater functions[7] around the nucleus. In addition to the α parameter, the position of the function becomes a new variational parameter which may be optimized by obtention of the best atomic orbitals. As the integrals over polycentric *s* functions are much easier to obtain, this method may overcome the difficulties inherent to the STOs.

Furthermore, the use of a unique type of function is very adequate for programming the solutions. The only remaining problem arises from the transposition to molecules. Indeed, the position of the lobe functions, adequate only for atomic orbitals, should be reoptimized. In molecules, these lobe functions often become bond functions as one tries to position them on the bond axes, allowing them to float (FSGO—Floating Spherical Gaussian Orbitals).[8]

5.1.3 The Contraction of Basis Functions

Even though a single STO already gives a fair representation of the atomic orbitals, the use of two STOs will improve the accuracy, and a very high accuracy will be obtained by four or five STOs. To distinguish these levels of accuracy, one can talk about a minimal basis set, a double zeta basis set, triple zeta, and so on, depending on the number of functions used.

If we now consider the GTOs, the situation will be quite different as these orbitals are not the natural solution of the central field problem. Therefore many more functions will be needed to obtain the same degree of accuracy as the STOs. Unfortunately, the larger the basis set, the more time-consuming will be the obtention of the orbitals. It appears, however, that results of nearly the same accuracy can be obtained by contracting the gaussian functions into new functions called contracted gaussian-type orbitals (CGTO). This is symbolized by:

$$\chi = \sum_{p=1}^{n} C_p G_p = \sum_{q=1}^{n' \ll n} C'_q \left(\sum_{j=1}^{n_q} a_j^{(q)} G_j^{(q)} \right) \qquad (5.11)$$

The set of $G_j^{(q)}$ functions is of course identical to the set of G_p functions, but the number of variational coefficients C' is drastically reduced, the a_j coefficients being fixed once and for all. The obtention of the variational coefficients is now easier from a computational point of view. It seems that there is only a little loss of variational freedom in comparing uncontracted GTO sets to the corresponding CGTO set. The gain in computer time makes the latter highly desirable for molecular calculations.

Two schemes of contraction are mainly used. The first minimizes energy for the atoms. The second fits, by least squares, the CGTOs onto a highly accurate function. Recently, this fit has been made directly onto STOs.[9] Contraction may be realized at any level of desired accuracy (i.e. minimal, double zeta, etc., basis set); nevertheless none of these contractions will overcome the poor behaviour of gaussian functions near the nucleus.

The introduction of different kinds of basis sets and contractions needs some new conventions.[10] Commonly in the literature, the size of the primitive uncontracted basis set is given in parentheses and those specifying the contracted basis in square brackets, e.g. a 8*s*, 4*p*, 1*d* basis may be represented by (8, 4, 1) and further contracted to a [4, 2, 1] basis set of a double zeta type augmented by a polarization function.

If hydrogens and heavy atoms are present, two different kinds of information will be given: a first group between the brackets describes the heavy centres, the second group describes the hydrogen. For example, for the methane molecule we may have the following types of basis sets:

[4, 2, 1/2, 1]	Basis set or double zeta with polarization
[4, 2/2]	Basis set or double zeta
[3, 2/2]	Basis set or split-valence basis set (i.e. a set for which the atomic cores are represented at a minimal level, while valence orbitals are at a double zeta level)
[2, 1/1]	Basis set or minimal

Although these conventions are widely used, another terminology has been introduced by Pople and coworkers[9] which has become quite familiar to quantum chemists. Their basis sets, at minimal level, are labeled STO-*N*G, meaning any Slater type orbital is fit by N gaussian functions as for example in the very popular STO-3G basis set. They are extended at double zeta level through the use of STO *l*-*mn*G where *l*, *m*, *n* indicate respectively the number of gaussian functions fitting the core orbitals, the inner valence orbitals and the outer valence orbital.

For example a STO 4-31G basis set for a carbon atom contains 4 gaussian functions describing the 1*s* orbital, 3 gaussian functions describing a set of 2*s* and 2*p* orbitals while the second set of 2*s* and 2*p* orbitals contains a single gaussian function. As an hydrogen atom contains no core orbitals, the 31G part of the basis set suffices to describe it.

Further polarization of the basis set is shown by the use of a single asterisk (*) when a set of d functions is added to the heavy centres, or by the use of a double asterisk (**) when a set of d functions is added to the heavy centres and a set of p functions to the hydrogen atoms.

These polarization functions consist commonly of a single gaussian function. This is the case in the STO 6-31G* and STO 6-31G** basis sets.

Before going any further, it must be pointed out that what has been described until now rests on the fact that the wave function minimizes the energy; however, the basis set obtained from such a criterion may have a poor behaviour for short-and long-range properties of the orbitals. If we are interested in those behaviours, our basis would not be of any help, and we should use some other basis set leading perhaps to a poorer energy but to a better result in the desired region. Such orbitals are known as property optimized[11] and may be obtained by a least square fit over an expansion weighted in the desired region, i.e. formula (5.3) is replaced by the relation:

$$D = \int_0^{\infty} \left(\chi - \sum_{p=1}^{n} C_p G_p\right)^2 * \omega(r) r^2 \, dr \qquad (5.12)$$

where $\omega(r)$ is r^n, n being any positive or negative integer. This remark will be of importance later on when, for the first time, we abandon energy optimizations for more specific goals.

5.1.4 Transposition to Molecules

It is well known that the molecular problem is quite different from the atomic one. Indeed, as more than one centre of chemical interest appears, each of these must be well described, the number of functions in the expansion being much more important. On the other hand, a molecule is more than simply juxtaposed atoms, so more flexibility must be introduced in order to account for molecular stability. Thus, at first sight, the expansion (5.1) must have the following properties:

(a) The basis set must be complete enough to ensure convergence.
(b) Basis functions must be important where the molecular system needs them (adequate location and extension of the functions).
(c) The basis set must be well balanced, i.e. we must find a balanced representation of all the centres which avoids favouring one with respect to the other.
(d) As the functions will be spread everywhere, approximate linear dependencies must be avoided.
(e) The needed integrals must be easy to evaluate.

Keeping these properties in mind, we should now look at some molecular basis sets. Many have been published in the literature.[10,12] They go from simple minimal basis sets to highly sophisticated ones; from light atoms to heavy transition metals. With such a multitude, a good choice is hardly possible, so must be made after a good definition of the 'objectives'. One would not use the same basis set for a small molecule of 10 electrons such as methane and for a system of 68 electrons such as naphthalene. Neither would we if we were looking for a general insight into the molecule or wanted high precision on some specific property. In addition, as we are depending on our technology, we will be restricted by the amount of computing time, by the program and by the computer we have at our disposal.

Two directions could now be taken to introduce enough flexibility into the basis set. We could use a basis set of atomic origin and introduce enough functions (double zeta, triple zeta with polarization functions) to ensure completeness of the chosen basis or we could use a relatively small basis set and reoptimize the non-linear parameters $\bar{\alpha}$ to adapt them on the studied molecule by varying the spatial extension of the functions.

The first approach leads to results near the Hartree–Fock limit, but some rules suggested by Roos and Sieghbahn[13] for first-row atoms need to be followed:

1. Start with a contracted gaussian set. At least five s-type functions are needed to represent the $1s$ orbital and three s- and p-type functions to represent the $2s$ and $2p$ orbitals respectively.
2. Add one polarization function to each hydrogen.
3. Add one polarization function to each heavy centre.

Further extensions could be made by:

4. Increase the s, p basis set.
5. Add further polarization functions to all centres.

By polarization functions, experience shows that the following must be meant:

(a) Polarization is obtained by d functions on heavy atoms, p functions on hydrogen atoms and s functions in bond regions.
(b) A single d function is satisfactory.
(c) The placement of bond functions is fairly insensitive to exact location along the bond axis, as long as the atoms are well represented by the atomic functions.

Experience further shows that relaxing the contraction of gaussian functions generally does not help much. Further suggestions for second-row atoms may be found in the paper of Roos and Sieghbahn.[13] For heavier atoms, the review by Veillard and Demuynck[14] should be consulted.

The aim of the approach described above was mainly to obtain the best wave function from the energetical point of view, but many other studies have been devoted to a systematic analysis of the quality of other properties. This is essential for chemists who would like to have an idea of the reliability of all interesting properties. A recent study resting on information theory tried to introduce a new insight into the problem[15] and to assess how well experimentally known values are reproduced by approximate wave functions; in this way an evaluation was made of the limitations of the theory. We do not intend here to present information theory but only to draw on the main conclusions of this study.

If we select some basis sets, each will give an energy and an approximate wave function. For any peculiar property P_i, each wave function will generate a value and thus contain some information $J(P_i)$ which is zero for the worst result and infinity for the correct result (by correct result, we mean the experimental value). The quality of the result in terms of information $J(P_i)$ is thus a real positive number related to the ability of a basis set to reproduce the experimental value. Considering that property P_i, the different basis sets may be ordered according to their quality. The same considerations hold for a couple of properties P_i, P_j, and so on, for any set of properties $\{P\}$. The performance of each basis set may now be evaluated.

Table 5.1 gives some examples for the HF molecule. Energy, equilibrium distance and dipole moment are the considered properties. The computed properties themselves lead to a somewhat confused image. Energies are always sufficiently well determined but other properties show variations that cannot easily be explained. For instance, bond lengths become shorter as energy information increases. In order to follow this change in quality, one has to rely on distance information, or better still on the performance of the basis set for the coupled properties, i.e. $J(E_e, r_e)$.

The overall quality of a basis set will thus be given by the pair $J(E_e)$, $J(\{P\})$ in our example $J(E_e)$, $J(E_e, r_e, \mu_e)$. Following this point of view some definitions may be given:

(a) A relatively well-balanced basis set is a set for which $J(E_e) \simeq J(\{P\})$.
(b) At the SCF level, there exists an upper limit for information which allows us to define a well-balanced basis set as a set for which $J(\{P\})$ lies between $J(E_e)$ and the upper limit, and is thus greater than $J(E_e)$.

The same kind of analysis may be extended beyond the SCF level to calculations including electron correlation (CI, perturbation, etc.). In the preliminary study it was found that the most sophisticated basis sets, although they reach a very good energy, do not always give good properties for the studied molecule. Furthermore, an analysis in terms of information does not only give the quality of the basis set chosen but also gives the methodology used to go beyond the Hartree–Fock limit. Information theory thus presents a powerful technique to criticize the theoretical framework and to provide the kind of methodology which should be used to attain some desired objective.

Table 5.1 Information quantities of HF in several basis sets

Primitive set	Contracted set	E_e(a.u.)	r_e(a.u.)	μ_e(a.u.)	$J(E_e)'$†	$J(E_e)$	$J(R_e)$	$J(\mu_e)$	$J(E_e r_e)$	$J(E_e r_e \boldsymbol{\mu}_e)$
$6s3p/3s$ [9]	$2s1p/1s$	−98.572844	1.8055	0.49258	0.0000	0.0000	0.0000	0.1482	0.0000	0.0000
$12s6p/6s$ [9]	$2s1p/1s$	−99.501718	1.8028	0.51000	1.3963	0.9285	0.0536	0.2279	0.3658	0.2708
$8s4p/4s$ [16]	$3s2p/2s$	−99.887286	1.7410	0.89971	3.0292	1.6065	3.1587	0.4288	2.0272	1.0328
$10s4p/4s$ [16]	$3s2p/2s$	−99.983425	1.7386	0.90487	4.1001	1.8403	3.6781	0.3888	2.2859	1.0381
$9s5p/4s$ [17]	$3s2p/3s$	−100.018895	1.7467	0.95544	4.8518	1.9371	2.3951	0.0468	2.1303	0.7171
$9s5p/4s$ [17]	$3s2p/2s$	−100.020169	1.7475	0.96334	4.8876	1.9407	2.3182	0.0000	2.1050	0.6724
$9s5p/5s$ [17]	$3s2p/3s$	−100.020665	1.7376	0.96256	4.9019	1.9421	3.9411	0.0046	2.3983	0.7005
$9s5p/4s$ [17]	$4s3p/2s$	−100.022946	1.7390	0.93645	4.9691	1.9485	3.5850	0.1658	2.3775	0.8479
$11s6p/5s$ [18]	$4s2p/3s$	−100.026364	1.7422	0.91244	5.0761	1.9583	2.9703	0.3321	2.2997	0.9899
$9s5p/4s2p$ [17]	$3s2p/2s1p$	−100.034266	1.7257	0.87851	5.3588	1.9811	3.3370	0.6056	2.3786	1.2380
$10s6p/5s$ [19]	$5s3p/3s$	−100.036872	1.7380	0.93757	5.4656	1.9887	3.8301	0.1588	2.4346	0.8462
$10s6p/5s$ [19]	$5s4p/3s$	−100.037008	1.7371	0.93656	5.4715	1.9891	4.1255	0.1653	2.4527	0.8536
$9s5p/4s2p$ [17]	$4s3p/2s1p$	−100.040470	1.7046	0.83604	5.6275	1.9993	1.3561	1.0422	1.6083	1.3209
$11s6p/5s2p$ [18]	$4s2p/3s1p$	−100.044050	1.7168	0.84243	5.8089	2.0099	2.1756	0.9675	2.0880	1.4594
$10s6p/5s2p$ [19]	$5s4p/3s1p$	−100.044751	1.7206	0.81251	5.8473	2.0120	2.5628	1.3560	2.2360	1.7677
$9s5p2d/4s2p$ [17]	$3s2p1d/2s1p$	−100.049112	1.7053	0.74383	6.1119	2.0250	1.3951	3.1467	1.6434	1.8463
$9s5p2d/4s2p$ [17]	$4s3p1d/2s1p$	−100.049799	1.7046	0.74154	6.1579	2.0270	1.3561	3.2708	1.6162	1.8271
$11s6p2d/5s2p$ [18]	$4s2p1d/3s1p$	−100.057755	1.7036	0.69515	6.8466	2.0511	1.3088	3.5601	1.5884	1.8117
$10s6p1d/5s2p$ [19]	$5s4p1d/3s1p$	−100.059724	1.7078	0.74436	7.0833	2.0572	1.5297	3.1196	1.7463	1.9408
$10s6p2d/5s2p$ [19]	$5s3p1d/3s1p$	−100.062343	1.7027	0.74871	7.4737	2.0652	1.2630	2.9147	1.5579	1.7513

† $J(E_e)'$ refers to the Hartree–Fock limit.

5.1.5 Orthogonalization of Basis Functions

The basis functions we have previously examined will now be useful in the computation of approximate wave functions and of many properties. They will appear in developments, integrals, and so on. The nature of these functions will be of some importance as, sometimes, easy computation of a property will follow. As mentioned earlier, the wave functions and molecular orbitals present, for a commodity, the property of being orthonormal. When we presented the STOs we mentioned that they are nodeless, and thus not orthogonal, but that they could easily be orthonormalized. We shall now describe briefly the commonly used orthonormalization procedures. If $\{\chi\}$ is a set of basis functions, we shall associate a 'metric matrix' to it. This matrix is better known to chemists as the 'overlap' matrix $\mathbf{S}$:

$$S_{pq} = \int \chi_p^*(\bar{x})\chi_q(\bar{x})\,\mathrm{d}\bar{x} \tag{5.13}$$

or, in matricial notation:

$$\mathbf{S} = \langle \boldsymbol{\chi} \mid \boldsymbol{\chi} \rangle \tag{5.14}$$

We shall suppose that, in the general case, the basis functions are linearly independent and that the inverse matrix $\mathbf{S}^{-1}$ exists.

Let us now consider a non-singular linear transformation $\mathbf{A}$ of the basis set $\boldsymbol{\chi}$ leading to a new basis set $\boldsymbol{\chi}^{\perp}$ which is orthogonal:

$$\langle \boldsymbol{\chi}^{\perp} \mid \boldsymbol{\chi}^{\perp} \rangle = \mathbf{E} = \begin{pmatrix} 1 & & 0 \\ & \ddots & \\ 0 & & 1 \end{pmatrix} \tag{5.15}$$

Thus, as:

$$\boldsymbol{\chi}^{\perp} = \boldsymbol{\chi}\mathbf{A} \tag{5.16}$$

equation (5.15) becomes:

$$\langle \boldsymbol{\chi}\mathbf{A} \mid \boldsymbol{\chi}\mathbf{A} \rangle = \mathbf{A}^{+}\langle \boldsymbol{\chi} \mid \boldsymbol{\chi} \rangle \mathbf{A} = \mathbf{A}^{+}\mathbf{S}\mathbf{A} = \mathbf{E} \tag{5.17}$$

Our problem reduces now to the obtention of the transformation matrix $\mathbf{A}$. This may be done in several ways:

5.1.5.1 Schmidt procedure on 'successive orthonormalization'

This very simple procedure consists in orthonormalizing each member of the set against all the previous functions and subsequently normalizing. This orthogonalization extracts from the function the orthogonal component against the previous function subset by subtracting the undesirable contributions. Assuming that the basis functions χ are normalized, the first function will be the starting point:

$$\chi_1 = \chi_1^{\perp}$$

and

$$\langle \chi_1 | \chi_1 \rangle = 1 = \langle \chi_1^\perp | \chi_1^\perp \rangle \tag{5.18}$$

The second function χ_2 becomes:

$$\chi_2^\perp = \chi_2 - a\chi_1^\perp \tag{5.19}$$

As:

$$\langle \chi_1^\perp | \chi_2^\perp \rangle = 0$$

then:

$$\langle \chi_1^\perp | \chi_2 \rangle - a\langle \chi_1^\perp | X_1^\perp \rangle = S_{12} - a = 0 \tag{5.20}$$

and

$$a = S_{12} \tag{5.21}$$

Furthermore:

$$\begin{aligned} \langle \chi_2^\perp | \chi_2^\perp \rangle &= 1 - 2S_{12}\langle \chi_1 | \chi_2 \rangle + S_{12}^2 \\ &= 1 - S_{12}^2 \end{aligned} \tag{5.22}$$

Normalization on $\chi_2^\perp$ leads to:

$$\chi_2^\perp = \frac{1}{(1 - S_{12}^2)^{1/2}} (\chi_2 - S_{12}\chi_1^\perp) \tag{5.23}$$

The third function becomes:

$$\chi_3^\perp = N_3(\chi_3 - a\chi_2^\perp - b\chi_1^\perp) \tag{5.24}$$

As:

$$\langle \chi_1^\perp | \chi_3^\perp \rangle = 0 \qquad \text{and} \qquad \langle \chi_2^\perp | \chi_3^\perp \rangle = 0$$

one easily finds that:

$$a = S_{23} \qquad \text{and} \qquad b = S_{13} \tag{5.25}$$

The normalization implies:

$$N_3 = (1 - S_{13}^2 - S_{23}^2)^{-1/2} \tag{5.26}$$

Thus for any function χ_{n+1} one will find:

$$\chi_{n+1}^\perp = N_{n+1}\left(\chi_{n+1} - \sum_{i=1} S_{i,n+1}\chi_i^\perp\right) \tag{5.27}$$

and

$$N_{n+1} = \left(1 - \sum_{i=1} S_{i,n+1}\right)^{-1/2}$$

The calculations based on this formula are rather cumbersome and lead to a triangular form of the transformation matrix. We will not go any further with this procedure but will now turn to more current approaches.

5.1.5.2 Lowdin's approach

Let us return to relations (5.16) and (5.17):

$$\chi^{\perp} = \chi \mathbf{A}$$
$$\mathbf{A}^{+}\mathbf{S}\mathbf{A} = \mathbf{E}$$

We may try a solution of the form:

$$\mathbf{A} = \mathbf{S}^{-1/2}\mathbf{B} \tag{5.28}$$

Substitution of **A** leads to:

$$\mathbf{E} = \mathbf{B}^{+}(\mathbf{S}^{-1/2})^{+}\mathbf{S}\mathbf{S}^{-1/2}\mathbf{B} = \mathbf{B}^{+}\mathbf{B} \tag{5.29}$$

A general solution of the problem thus has the form:

$$\chi^{\perp} = \chi \mathbf{S}^{-1/2}\mathbf{B} \tag{5.30}$$

where **B** could be any arbitrarily chosen unitary matrix. The obtention of $\mathbf{S}^{-1/2}$ itself is not very difficult. Since **S** is a self-adjoint matrix which is positive definite, it may be brought to the diagonal form $\mathbf{S}_0$ by a unitary transformation **T**:

$$\mathbf{T}^{+}\mathbf{S}\mathbf{T} = \mathbf{S}_0 \tag{5.31}$$

or

$$\mathbf{S} = \mathbf{T}\mathbf{S}_0\mathbf{T}^{+} \tag{5.32}$$

An arbitrary function f of **S** is then defined through the relation:

$$f(\mathbf{S}) = \mathbf{T}f(\mathbf{S}_0)\mathbf{T}^{+} \tag{5.33}$$

or

$$\mathbf{S}^{-1/2} = \mathbf{T}\mathbf{S}_0^{-1/2}\mathbf{T}^{+} \tag{5.34}$$

The special choice $\mathbf{B} = \mathbf{E}$ in (5.30) leads to symmetric orthonormalization[20] which gives orthogonal functions resembling the initial functions as closely as possible. The other choice $\mathbf{B} = \mathbf{T}$ leads to canonical orthonormalization[1] which has already been presented in Section 5.1.1 to discuss the linear dependency problem. In this case:

$$\chi^{\perp} = \chi \mathbf{S}^{-1/2}\mathbf{T} = \chi \mathbf{T}\mathbf{S}_0^{-1/2} \tag{5.35}$$

or

$$\chi_p^{\perp} = \frac{1}{\sqrt{S_{op}}} \sum_i \chi_i T_{ip} \tag{5.36}$$

The sum of the absolute squares of the coefficients is nothing more than $(S_{op})^{-1}$ and the orthogonalized function corresponding to a vanishing eigenvalue of **S** blows up. One may then arrange the functions after decreasing values of $\mathbf{S}_0$, exclude the lowest eigenvalues and delete those functions in order to reduce the basis functions to a linearly independent set.

5.1.6 Further Transformations of Atomic Orbitals—Hybridization

It is a matter of fact that when chemists discuss molecules they like to refer to a geometrical configuration. Unfortunately the atomic functions used are not adapted to such a description. A spherical s **function** and three p functions which are perpendicular to each other do not justify a tetrahedral arrangement of atoms. This is the reason why a further transformation was introduced.

As hybridization is not used very much, we shall describe it only briefly here. Many procedures have been developed, one being the 'maximum overlap principle',[21] which states that the strength of a chemical bond increases with the overlap between the orbitals involved and allows in that sense the generation of hybridized atomic orbitals, and another being the 'maximum projection technique',[22] which searches for the hybrid orbitals having the maximum projection in the subset of the functions describing the atom to which it is bonded. The interested reader should consult the given references and many others, but we shall restrict our investigations to some 'standard hybrids'.

The problem of hybridization reduces, then, to the transformation of valence atomic functions in a minimal basis set by the relation:

$$\mathbf{h} = \boldsymbol{\chi}\mathbf{V} \tag{5.37}$$

where $\boldsymbol{\chi}$ is the conventional orthonormal valence basis set centered on the concerned atom. The hybrids $\mathbf{h}$ should satisfy the same conditions as $\boldsymbol{\chi}$, i.e.:

$$\langle \boldsymbol{\chi} \mid \boldsymbol{\chi} \rangle = \mathbf{E} \Rightarrow \langle \mathbf{h} \mid \mathbf{h} \rangle = \mathbf{E} \tag{5.38}$$

The system of equations (5.37) implies the knowledge of 16 coefficients which will be fixed by some symmetry conditions restricted by the constraint (5.38). In the standard hybridization approach the constancy of valence angles is assumed (i.e. 109°28′, 120° and 180° for tetrahedral, trigonal and digonal arrangements respectively).

Considering the methane molecule in a given orientation (Figure 5.2), we

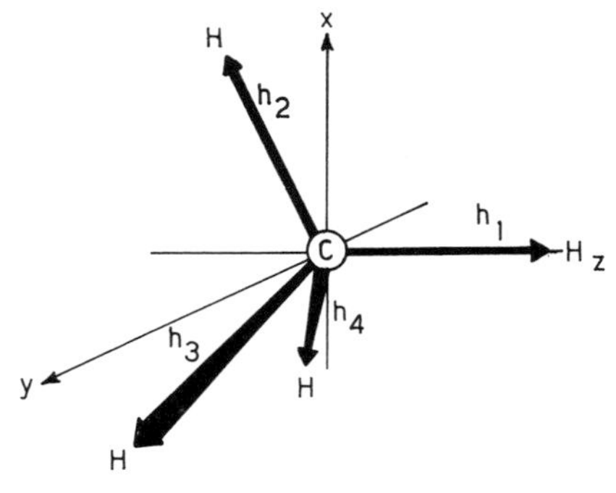

Figure 5.2 Hybrid orbitals of methane

can write the system of equations:

$$h_1 = V_{11}s + V_{21}p_x + V_{31}p_y + V_{41}p_z$$
$$h_2 = V_{12}s + V_{22}p_x + V_{32}p_y + V_{42}p_z$$
$$h_3 = V_{13}s + V_{23}p_x + V_{33}p_y + V_{43}p_z$$
$$h_4 = V_{14}s + V_{24}p_x + V_{34}p_y + V_{44}p_z$$

with the symmetry constraints:

$$V_{11} = V_{12} = V_{13} = V_{14} = a$$
$$V_{21} = V_{31} = V_{32} = 0$$
$$V_{41} = b$$
$$V_{22} = c$$
$$V_{42} = V_{43} = V_{44} = d$$
$$V_{33} = -V_{34} = e$$
$$V_{23} = V_{24} = f$$

Relation (5.38) leads immediately to:

$$a = \frac{1}{2} \qquad b = \frac{\sqrt{3}}{2} \qquad c = \frac{\sqrt{2}}{\sqrt{3}}$$

$$d = \frac{-1}{2\sqrt{3}} \qquad e = \frac{1}{\sqrt{2}} \qquad f = \frac{-1}{\sqrt{6}}$$

Thus:

$$h_1 = \frac{1}{2}s + 0p_x + 0p_y + \frac{\sqrt{3}}{2}p_z$$

$$h_2 = \frac{1}{2}s + \frac{\sqrt{2}}{\sqrt{3}}p_x + 0p_y - \frac{1}{2\sqrt{3}}p_z$$

$$h_3 = \frac{1}{2}s - \frac{1}{\sqrt{6}}p_x + \frac{1}{\sqrt{2}}p_y - \frac{1}{2\sqrt{3}}p_z$$

$$h_4 = \frac{1}{2}s - \frac{1}{\sqrt{6}}p_x - \frac{1}{\sqrt{2}}p_y - \frac{1}{2\sqrt{3}}p_z$$

It must be mentioned that any change of the orientation of the molecule will yield another transformation matrix **V**. For example, in Figure 5.3 the orientation of the molecule will lead to the hybrids:

$$h_1 = \tfrac{1}{2}s + \tfrac{1}{2}p_x + \tfrac{1}{2}p_y + \tfrac{1}{2}p_z$$
$$h_2 = \tfrac{1}{2}s - \tfrac{1}{2}p_x - \tfrac{1}{2}p_y + \tfrac{1}{2}p_z$$
$$h_3 = \tfrac{1}{2}s + \tfrac{1}{2}p_x - \tfrac{1}{2}p_y - \tfrac{1}{2}p_z$$
$$h_4 = \tfrac{1}{2}s - \tfrac{1}{2}p_x + \tfrac{1}{2}p_y - \tfrac{1}{2}p_z$$

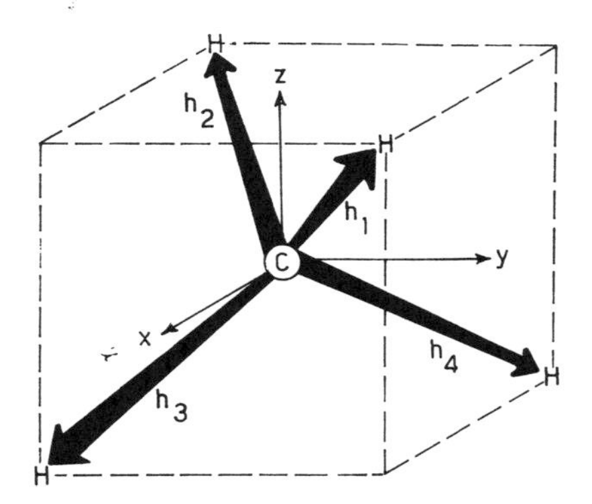

Figure 5.3 Reoriented hybrids of methane

The coefficients of s functions are of course unaffected, but the others rearrange following the rotation of the axes. This simply means that the geometrical dependency is rejected in the transformation matrix, the hybrids themselves becoming independent of the axes. This leads to an important use of hybrid orbitals—their transferability from one compound to another either on the function itself or on its properties (i.e. matrix elements of an operator in the hybrid basis set). The same reasoning holds for digonal or trigonal hybrids. For example, the hybrid atomic orbitals for acetylene, referring to C_1 in Figure 5.4, will be:

$$h_1 = \frac{1}{\sqrt{2}} s + \frac{1}{\sqrt{2}} p_x + 0p_y + 0p_z$$

$$h_2 = \frac{1}{\sqrt{2}} s - \frac{1}{\sqrt{2}} p_x + 0p_y + 0p_z$$

$$h_3 = 0s + 0p_x + 1p_y + 0p_z$$

$$h_4 = 0s + 0p_x + 0p_y + 1p_z$$

This procedure may be refined by the introduction of experimental values for the bond angles rather than the tetrahedral angle. The transformation matrix will then be obtained by constraint (5.38) and by trigonometrical relations introducing the empirical parameter. In the case of ethane, for

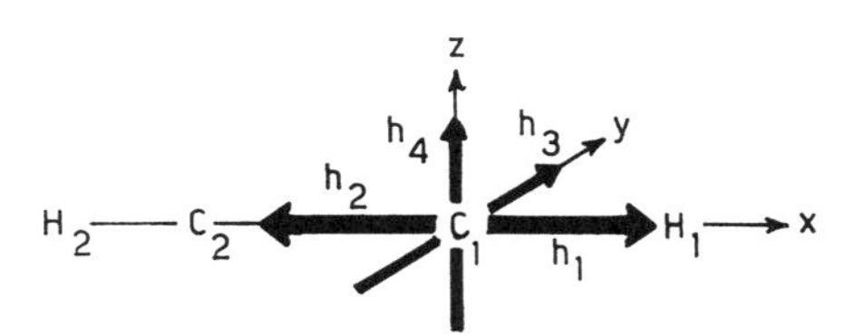

Figure 5.4 Hybrid orbitals in acetylene

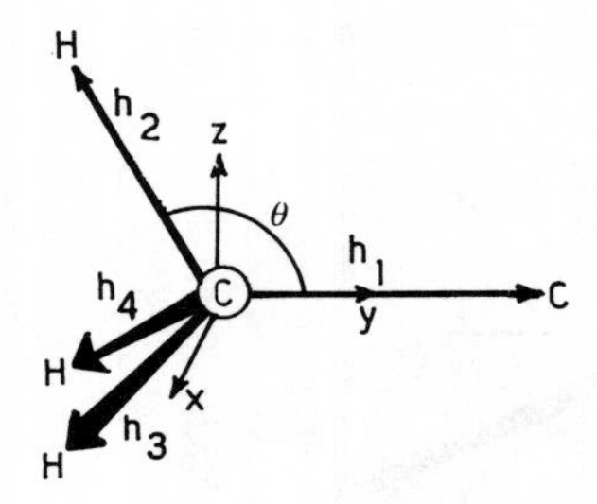

Figure 5.5 Adapted hybrids in ethane

example, we can distinguish three equivalent hybrids pointing towards the hydrogens and a fourth hybrid pointing towards the other carbon atom. The angle θ between two hybrids is expressed by the well-known cosine relation:

$$\cos \theta = \frac{\mathbf{a} \cdot \mathbf{b}}{|\mathbf{a}|\,|\mathbf{b}|} = \frac{V_{21}V_{22} + V_{31}V_{32} + V_{41}V_{42}}{(V_{21} + V_{31} + V_{41})^{1/2}(V_{22} + V_{32} + V_{42})^{1/2}} \tag{5.39}$$

where **a** and **b** are the vectors obtained through the components on the coordinate axes. Referring to Figure 5.5, one may build a symbolic transformation matrix depending on a geometrical parameter a. The hybrids are then obtained by:

$$(h_1, h_2, h_3, h_4) = (s, p_x, p_y, p_z)$$

$$\begin{bmatrix} a & \sqrt{\left(\frac{1-a^2}{3}\right)} & \sqrt{\left(\frac{1-a^2}{3}\right)} & \sqrt{\left(\frac{1-a^2}{3}\right)} \\ 0 & 0 & \frac{\sqrt{2}}{2} & \frac{\sqrt{2}}{2} \\ \sqrt{(1-a^2)} & \frac{a}{\sqrt{3}} & \frac{a}{\sqrt{3}} & \frac{a}{\sqrt{3}} \\ 0 & \sqrt{\left(\frac{2}{3}\right)} & \frac{1}{\sqrt{6}} & \frac{1}{\sqrt{6}} \end{bmatrix}$$

It follows that:

$$\cos \theta = \frac{a}{(a^2+2)^{1/2}} \Leftrightarrow a = \operatorname{cotg} \theta * \sqrt{2} \tag{5.40}$$

As θ is known, the transformation matrix is uniquely defined and adapted to the chosen nuclear configuration.

This procedure nevertheless contains some drawbacks as it assumes that hybrids lie along the internuclear axes. This is not always true, namely for strained compounds such as microcycles. In that case, one must rely on the procedures previously mentioned.

5.2 THE GAUSSIAN-TYPE INTEGRALS

5.2.1 Integrals in Atomic Basic Sets

Let us define an atomic basis set ($\{\chi\}$) where each basis function is a linear combination of gaussian-type functions (G_ν):

$$\chi_p = N_p \sum_\nu C_\nu^{(p)} G_\nu^{(p)} \tag{5.41}$$

where N_p is a normalization factor such that $\int \chi_p^* \chi_p \, dr = 1$ and $\{C_\nu^{(p)}\}$ is a given set of expansion coefficient. The explicit form of any $G_\nu^{(p)}$ may be written as:[5]

$$G_\nu^{(p)}(r) = N_A x_A^l y_A^m z_A^n \exp(-\alpha r_A^2) \tag{5.42}$$

where N_A is a normalization factor such that $\int G_\nu^{(p)*} G_\nu^{(p)} \, dr = 1$ and α is the gaussian exponent.

The vectors $\mathbf{A}$ and $\mathbf{r}$ correspond to the position vectors respectively of the gaussian centre and of any point in the molecular fixed cartesian space. Their components are (see also Figure 5.6):

$$\mathbf{A} = \{A_x; A_y; A_z\}$$
$$\mathbf{r} = \{x; y; z\}$$

If we define $\mathbf{r}_A$ as $\mathbf{r} - \mathbf{A}$, then:

$$\mathbf{r}_A = \{x_A; y_A; z_A\} = \{x - A_x; y - A_y; z - A_z\}$$

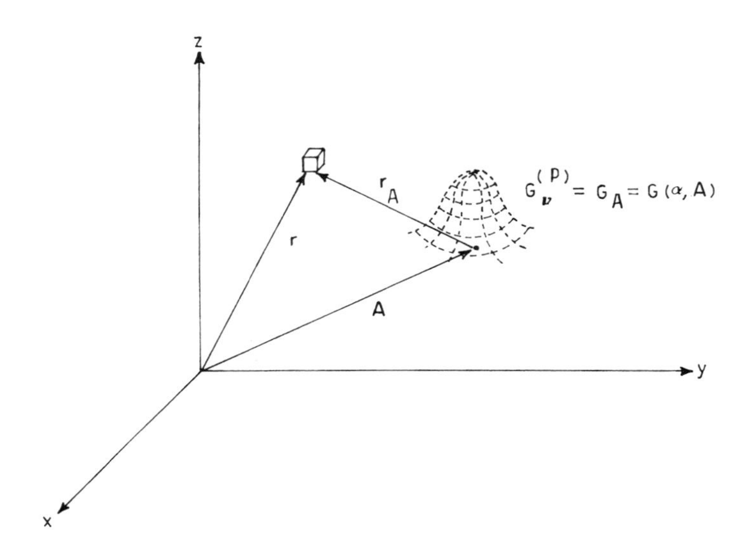

Figure 5.6 Vector definition for any gaussian function

Using this kind of functional basis the integrals to be solved may be written as:

$$I = \int \chi_p^* O_1(r)\chi_q \, dr = N_p^* N_q \sum_\mu \sum_\nu C_\mu^{(p)*} C_\nu^{(q)} \int G_\mu^{(p)*} O_1(r) G_\nu^{(q)} \, dr \quad (5.43)$$

for the one-electron integrals. For the two-electron integrals:

$$\begin{aligned} I &= \int \chi_p^*(r_1)\chi_q(r_1)O_2(r_1r_2)\chi_q^*(r_2)\chi_s(r_2)\, dr_1\, dr_2 \\ &= N_p^* N_q N_r^* N_s \sum_\mu \sum_\nu \sum_o \sum_\pi C_N^{(p)*} C_\nu^{(q)} C_o^{(r)*} C_\pi^{(s)} \\ &\quad \times \int G_\mu^{(p)*}(r_1) G_\nu^{(q)}(r_1) O_2(r_1r_2) G_o^{(r)*}(r_2) G_\pi^{(s)}(r_2)\, dr_1\, dr_2 \end{aligned} \quad (5.44)$$

In the next section we consider only integrals between the gaussian-type functions. Relations (5.43) and (5.44) enable us to go through the integrals between the atomic orbitals.

5.2.2 Integrals between Gaussian Functions[23]

5.2.2.1 The product theorem

One advantage of using the gaussian-type function is given by the product theorem. It tells us that 'the gaussian product remains a gaussian'. If we write:

$$\exp(-\alpha r_A^2)\exp(-\beta r_B^2) = K \exp(-\gamma r_P^2) \quad (5.45)$$

it is easy to prove that this expression becomes valid for:

$$K = \exp\left[-\frac{\alpha\beta}{\alpha+\beta}(\mathbf{A}-\mathbf{B})^2\right] = \exp\left(-\frac{\alpha\beta}{\alpha+\beta}\overline{AB}^2\right)$$

$$\gamma = \alpha + \beta$$

and

$$r_P = \mathbf{r} - \mathbf{P}$$

where:

$$\mathbf{P} = \frac{\alpha\mathbf{A} + \beta\mathbf{B}}{\alpha+\beta}$$

Moreover, we can generalize the former gaussian product theorem by inserting the angular part expressed as cubic harmonical functions: $(x_A^l y_A^m z_A^n)(x_B^{l'} y_B^{m'} z_B^{n'})$. To reach this goal let us express x_A as:

$$\begin{aligned} x_A &= x - A_x \\ &= (x - P_x) + (P_x - A_x) \\ &= x_P + \overline{PA}_x \end{aligned}$$

Then x_A^l becomes:

$$x_A^l = (x_P + \overline{PA}_x)^l$$
$$= \sum_{i=0}^{l} \overline{PA}_x^{l-i} \binom{l}{i} x_P^i \tag{5.46}$$

where:

$$\binom{l}{i} \text{ stands for } \frac{l!}{i!\,(l-i)!}$$

For $x_A^l x_B^{l'}$ we have:

$$x_A^l x_B^{l'} = \sum_{i=0}^{l} \sum_{j=0}^{l'} \overline{PA}_x^{l-i} \overline{PB}_x^{l'-j} \binom{l}{i} \binom{l'}{j} x_P^{i+j}$$
$$= \sum_{k=0}^{l+l'} \left[\sum_{\substack{i=0 \\ i+\gamma=k}}^{l} \sum_{j=0}^{l'} \overline{PA}_x^{l-i} \overline{PB}_x^{l'-j} \binom{l}{i} \binom{l'}{j} \right] x_P^k$$
$$= \sum_{k=0}^{l+l'} f_k(l, l', \overline{PA}_x, \overline{PB}_x) x_P^k \tag{5.47}$$

The product theorem finds its general expression:

$$G_A G_B = N_A N_B x_A^l y_A^m z_A^n \exp(-\alpha r_A^2) x_B^{l'} y_B^{m'} z_B^{n'} \exp(-\beta r_B^2)$$
$$= N_A N_B K \sum_{k_x=0}^{l+l'} \sum_{k_y=0}^{m+m'} \sum_{k_z=0}^{n+n'} f_{k_x} f_{k_y} f_{k_z} x_P^{k_x} y_P^{k_y} z_P^{k_z} \exp(-\gamma r_P^2) \tag{5.48}$$

If we split this product into its cartesian components we have:

$$G_A G_B = N_A N_B K G_P^x G_P^y G_P^z$$

where:

$$G_P^h = \sum_{k_h} f_{k_h} h_P^{k_h} \exp(-\gamma h_P^2) \tag{5.49}$$

if h stands for x, y or z.

5.2.2.2 The complete gamma functions and related integrals

According to the product theorem, it appears that we expect to solve integrals of the following form:

$$F_k(\gamma) = \int_{-\infty}^{+\infty} \exp(-\gamma t^2) t^k \, dt \tag{5.50}$$

where t stands for $h_P = h - P_h$ and dt for dh. This integral is non-vanishing as long as k is an even integer. In such a case we can write:

$$F_k(\gamma) = 2\int_0^{\infty} \exp(-\gamma t^2) t^k \, dt$$
$$= \gamma^{-(k+1)/2} \int_0^{\infty} e^{-u} u^{(k-1)/2} \, du \qquad (\text{if } u = \gamma t^2)$$
$$= \gamma^{-(k+1)/2} \Gamma\left(\frac{k+1}{2}\right) \tag{5.51}$$

This gamma function has a recurrence formula:

$$\Gamma(n) = (n-1)\Gamma(n-1)$$

which is valid as long as n is an integer strictly greater than 1. We can derive this as follows:

$$\Gamma(n) = \int_0^\infty u^{n-1} \exp(-u)\, du$$

Putting:

$$r = u^{n-1} \qquad \text{and} \qquad dv = \exp(-u)\, du$$

we find:

$$\Gamma(n) = [u^{n-1} \exp(-u)]_0^\infty + (n-1)\int_0^\infty \exp(-u)u^{n-2}\, du = (n-1)\Gamma(n-1)$$

Thus:

$$F_k(\gamma) = \gamma^{-(k+1)/2}\left(\frac{k-1}{2}\right)!!\, \Gamma(\tfrac{1}{2})$$

where:

$$\left(\frac{k-1}{2}\right)!! \text{ stands for } \left(\frac{k-1}{2}\right)\left(\frac{k-3}{2}\right)\left(\frac{k-5}{2}\right)\cdots\left(\frac{3}{2}\right)\left(\frac{1}{2}\right)$$

In the particular case where $k = 0$ we find:

$$F_0(\gamma) = \int_{-\infty}^{\infty} \exp(-\gamma t^2)\, dt = \sqrt{\left(\frac{\pi}{\gamma}\right)}$$
$$= \gamma^{-1/2}\Gamma(\tfrac{1}{2})$$

From this we conclude that:

$$\Gamma(\tfrac{1}{2}) = \sqrt{\pi} \tag{5.52}$$

and for $F_k(\gamma)$ the final expression now becomes:

$$F_k(\gamma) = \gamma^{-(k+1)/2}\frac{(k-1)!!}{2^{k/2}}\sqrt{\pi}, \qquad \text{for } k = \text{even integer}$$

$$F_k(\gamma) = 0 \qquad \text{for } k = \text{odd integer} \tag{5.53}$$

5.2.2.3 The overlap integrals

The overlap integral between two gaussian functions G_A and G_B respectively centered at point A of position vector **A** and point B of position vector **B** may be expressed as:

$$
\begin{aligned}
S_{AB} &= \int_{-\infty}^{+\infty} G_A G_B \, dr \\
&= K N_A N_B \int_{-\infty}^{+\infty} G_P(x)\,dx \int_{-\infty}^{+\infty} G_P(y)\,dy \int_{-\infty}^{+\infty} G_P(z)\,dz \\
&= K N_A N_B \sum_{k_x=0}^{(l+l')/2} f_{2k_x} \frac{(2k_x-1)!!}{2^{k_x}} \sum_{k_y=0}^{(m+m')/2} f_{2k_y} \frac{(2k_y-1)!!}{2^{k_y}} \\
&\quad \times \sum_{k_z=0}^{(n+n')/2} f_{2k_z} \frac{(2k_z-1)!!}{2^{k_z}} \gamma^{-(2k_x+2k_y+2k_z+3)/2} (\pi)^{3/2}
\end{aligned}
\tag{5.54}
$$

For normalization reasons we write, for $G_A = G_B = G = N_g x^l y^m z^n \exp(-\alpha r_A^2)$:

$$
\int G^* G \, dr = 1
$$

In such a case, the former integral can be reduced to:

$$
1 = N_g^2 \gamma^{-(l+m+n+3/2)} \frac{(l-1)!!\,(m-1)!!\,(n-1)!!}{2^{(l+m+n)/2}} (\pi)^{3/2} \tag{5.55}
$$

where:

$$
\gamma = 2\alpha
$$

This expression enables us to select the appropriate normalization factor N for any gaussian-type functions. Finally, the normalization factor previously introduced (N_p) for the atomic orbital can also be deduced from a similar relationship:

$$
\begin{aligned}
I &= \int_{-\infty}^{+\infty} \chi_p^* \chi_p \, dr \\
&= N_p^2 \sum_{\mu} \sum_{\nu} C_\mu^{(p)} C_\nu^{(p)} \int_{-\infty}^{+\infty} G_\mu^{(p)*} G_\nu^{(p)} \, dr \\
&= N_p^2 \sum_{\mu \neq \nu} C_\mu^{(p)} C_\nu^{(p)} \int_{-\infty}^{+\infty} G_\mu^{(p)} G_\nu^{(p)} \, dr + \sum_{\mu} (C_\mu^{(p)})^2
\end{aligned}
\tag{5.56}
$$

Before closing this section we give in Table 5.2 the explicit form of some overlap integrals between s and p gaussian-type functions.

5.2.2.4 The kinetic integrals

The kinetic energy integrals are of the form:

$$
\begin{aligned}
T &= -\tfrac{1}{2} \int G_A^* \left(\frac{\partial^2}{\partial x^2} + \frac{\partial^2}{\partial y^2} + \frac{\partial^2}{\partial z^2} \right) G_B \, dr \\
&= -\tfrac{1}{2} \left(\langle G_A | \frac{\partial^2}{\partial x^2} | G_B \rangle + \langle G_A | \frac{\partial^2}{\partial y^2} | G_B \rangle + \langle G_A | \frac{\partial^2}{\partial z^2} | G_B \rangle \right)
\end{aligned}
\tag{5.57}
$$

Table 5.2 Some simple overlap integrals

$S=\langle G_A \mid G_B\rangle$	α; **A** l m n	β; **B** l' m' n'	$\int_{-\infty}^{+\infty} G_A^* G_B\,\mathrm{d}r$
$\langle s \mid s\rangle$	0 0 0	0 0 0	$KN_AN_B\left(\frac{\pi}{\gamma}\right)^{3/2}$
$\langle x \mid s\rangle$	1 0 0	0 0 0	$-\frac{\beta}{\gamma}\overline{AB}_x\langle s \mid s\rangle$
$\langle x \mid x\rangle$	1 0 0	1 0 0	$\left(\frac{1}{2\gamma}-\frac{\alpha\beta}{\gamma^2}\overline{AB}_x^2\right)\langle s \mid s\rangle$
$\langle x \mid y\rangle$	1 0 0	0 1 0	$-\frac{\alpha\beta}{\gamma^2}\overline{AB}_x\overline{AB}_y\langle s \mid s\rangle$

Any part of this expression may be written in the x, y and z components of the gaussian-type function. If:

$$G_A = N_A G_A^x G_A^y G_A^z$$

and

$$G_B = N_B G_B^x G_B^y G_B^z$$

then:

$$t^x = \langle G_A^x | \frac{\partial^2}{\partial x^2} | G_B^x \rangle\langle G_A^y \mid G_B^y\rangle\langle G_A^z \mid G_B^z\rangle N_A N_B$$

$$t^y = \langle G_A^x \mid G_B^x\rangle\langle G_A^y | \frac{\partial^2}{\partial y^2} | G_B^y\rangle\langle G_A^z G_B^z\rangle N_A N_B$$

$$t^z = \langle G_A^x \mid G_B^x\rangle\langle G_A^y \mid G_B^y\rangle\langle G_A^z | \frac{\partial^2}{\partial z^2} | G_B^z\rangle N_A N_B$$

and

$$t = -\tfrac{1}{2}(t^x + t^y + t^z) \tag{5.58}$$

For symmetry convenience let us introduce the new variables u and v such that:

$$u = G_A^h \quad \text{and} \quad \mathrm{d}v = \frac{\partial^2}{\partial h^2} G_B^h$$

$$\mathrm{d}u = \frac{\partial G_A^h}{\partial h}\mathrm{d}h \quad \text{and} \quad v = \frac{\partial G_B^h}{\partial h}$$

Then we have:

$$\int u\,\mathrm{d}v = [uv] - \int v\,\mathrm{d}u$$

and we can write:

$$\langle G_A^h | \frac{\partial^2}{\partial h^2} | G_B^h \rangle = -\left[G_A^h \frac{\partial G_B^h}{\partial h} \right]_{-\infty}^{+\infty} - \left\langle \frac{\partial}{\partial h} G_A^h \,\middle|\, \frac{\partial}{\partial h} G_B^h \right\rangle$$

The first term in the right member always vanishes and the second term may be further developed. For example, as:

$$G_A^x = x_A^l \exp(-\alpha x_A^2)$$

then:

$$\frac{\partial G_A^x}{\partial x} = \frac{\partial G_A^x}{\partial x_A} \frac{\partial x_A}{\partial x} = \frac{\partial G_A^x}{\partial x_A}$$

$$= l x_A^{l-1} \exp(-\alpha x_A^2) - 2\alpha x_A^{l+1} \exp(-\alpha x_A^2) \qquad \text{for } l \geq 1$$

$$= -2\alpha x_A^{l+1} \exp(-\alpha x_A^2) \qquad \text{for } l = 0$$

For:

$$\left\langle \frac{\partial G_A^x}{\partial x} \right|$$

we adopt a more condensed expression:

$$\left\langle \frac{\partial G_A^x}{\partial x} \right| = \left\langle \frac{\partial}{\partial x} l \right|$$

$$= l\langle l-1| - 2\alpha \langle l+1|$$

where:

$$\langle l-1| = 0 \qquad \text{if } l = 0$$

The former integral:

$$\langle G_A^x | \frac{\partial^2}{\partial x^2} | G_B^x \rangle$$

becomes:

$$\langle G_A^x | \frac{\partial^2}{\partial x^2} | G_B^x \rangle = -\left\langle \frac{\partial G_A^x}{\partial x} \,\middle|\, \frac{\partial G_B^x}{\partial x} \right\rangle = -ll'\langle l-1 \mid l'-1 \rangle$$

$$| + 2\beta l \langle l-1 \mid l'+1 \rangle + 2\alpha l' \langle l+1 \mid l'-1 \rangle - 4\alpha\beta \langle l+1 \mid l'+1 \rangle$$

Finally, t^x may be written as:

$$t^x = N_A N_B(-ll'\langle l-1 \mid l'-1 \rangle + 2\beta l \langle l-1 \mid l'+1 \rangle + 2\alpha l' \langle l+1 \mid l'-1 \rangle$$
$$| - 4\alpha\beta \langle l+1 \mid l'+1 \rangle) \langle m \mid m' \rangle \langle n \mid n' \rangle \qquad (5.59)$$

Similar expressions may be found for t^y and t^z.

The kinetic energy integral can therefore be expanded as a linear combination of overlap integrals. As an example we give in Table 5.3 the resulting expressions for some kinetic integrals.

Table 5.3 Some simple kinetic integrals

T	α; **A** $l\ m\ n$	β; **B** $l'\ m'\ n'$	Integral value
$-\frac{1}{2}\langle s\vert \Delta \vert s\rangle$	0 0 0	0 0 0	$\frac{\alpha\beta}{\gamma}\left(3-\frac{2\alpha\beta}{\gamma}\overline{AB}^2\right)\langle s \vert s\rangle$
$-\frac{1}{2}\langle x\vert \Delta \vert s\rangle$	1 0 0	0 0 0	$\frac{\alpha\beta}{\gamma}\left(3-2\frac{\alpha\beta}{\gamma}\overline{AB}^2\right)\langle x \vert s\rangle-\left(\frac{2\alpha\beta^2\overline{AB}_x}{\gamma^2}\right)\langle s \vert s\rangle$
$-\frac{1}{2}\langle x\vert \Delta \vert x\rangle$	1 0 0	1 0 0	$\frac{\alpha\beta}{\gamma^2}\langle s \vert s\rangle-\left(2\frac{\alpha\beta^2}{\gamma^2}\overline{AB}_x\right)\langle s \vert x\rangle+\left(2\frac{\alpha^2\beta}{\gamma^2}\overline{AB}_x\right)\langle x \vert s\rangle$ $+\frac{\alpha\beta}{\gamma}\left(3-2\frac{\alpha\beta}{\gamma}\overline{AB}^2\right)\langle x \vert x\rangle$
$-\frac{1}{2}\langle x\vert \Delta \vert y\rangle$	1 0 0	0 1 0	$\left(2\frac{\alpha^2\beta}{\gamma^2}\overline{AB}_y\right)\langle x \vert s\rangle-\left(2\frac{\alpha\beta^2}{\gamma^2}\overline{AB}_x\right)\langle s \vert y\rangle$ $+\frac{\alpha\beta}{\gamma}\left(3-2\frac{\alpha\beta}{\gamma}\overline{AB}^2\right)\langle x \vert y\rangle$

5.2.2.5 The moment integrals

The calculation of some molecular properties (such as the dipole moment, the quadrupole moment, etc.) requires a knowledge of the moment integrals. The most general expression of those integrals is:

$$M(\lambda)=M(l''m''n'')=\langle G_A\vert\, x^{l''}y^{m''}z^{n''}\,\vert G_B\rangle \tag{5.60}$$

where λ stands for $l''+m''+n''$. We can transform any product $x^{l''}y^{m''}z^{n''}$ (as previously shown; see 5.46) into a polynomial expansion:

$$x^{l''}=\sum_{ii=0}^{l''}P_x^{l''-ii}\binom{l''}{ii}x_P^{ii}$$

Thus we find:

$$x_A^l x_B^{l'} x^{l''}=\sum_{i=0}^{l}\sum_{j=0}^{l'}\sum_{ii=0}^{l''}\overline{PA}_x^{l-i}\,\overline{PB}_x^{l'-i}\,P_x^{l''-ii}\binom{l}{i}\binom{l'}{j}\binom{l''}{ii}x_P^{i+j+ii}$$

$$=\sum_{k=0}^{l+l'+l''}f_{k_x}(ll'l''\overline{PA}_x\overline{PB}_xP_x)x_P^k$$

where:

$$f_{k_x}=\sum_{\substack{i=0\\ i+j+ii=k_x}}^{l}\sum_{j=0}^{l'}\sum_{ii=0}^{l''}\overline{PA}_x^{l-i}\,\overline{PB}_x^{l'-i}\,P_x^{l''-ii}\binom{l}{i}\binom{l'}{j}\binom{l''}{ii}$$

The moment integral becomes:

$$M(\lambda)=KN_AN_B\sum_{k_x}\sum_{k_y}\sum_{k_z}f_{k_x}f_{k_y}f_{k_z}\int x_P^{k_x}y_P^{k_y}z_P^{k_z}\exp(-\gamma\bar{r}_P^2)\,dx_P\,dy_P\,dz_P$$

Table 5.4 Some simple first-order ($\lambda = 1$) moment integrals

$M(\lambda = 1)$	α; **A** l	m	n	l''	m'	n''	β; **B** l'	m'	n'	Integral value
$\langle s \vert x \vert s \rangle$	0	0	0	1	0	0	0	0	0	$P_x \langle s \mid s \rangle$
$\langle x \vert x \vert s \rangle$	1	0	0	1	0	0	0	0	0	$\left(\overline{PA}_x P_x + \frac{1}{2\gamma}\right) \langle s \mid s \rangle$
$\langle y \vert x \vert s \rangle$	0	1	0	1	0	0	0	0	0	$\overline{PA}_y P_x \langle s \mid s \rangle$
$\langle x \vert x \vert x \rangle$	1	0	0	1	0	0	1	0	0	$P_x \left(\frac{1}{\gamma} + \overline{PA}_x \overline{PB}_x\right) \langle s \mid s \rangle$
$\langle x \vert x \vert y \rangle$	1	0	0	1	0	0	0	1	0	$\overline{PB}_y \left(\frac{1}{2\gamma} + \overline{PA}_x P_x\right) \langle s \mid s \rangle$
$\langle y \vert x \vert y \rangle$	0	1	0	1	0	0	0	1	0	$P_x \left(\frac{1}{2\gamma} + \overline{PA}_y \overline{PB}_y\right) \langle s \mid s \rangle$

This shows that the moment integrals may be reduced to a linear combination of overlap integrals. Here also only the even k values give non-vanishing integrals and have further to be considered:

$$M(\lambda) = KN_A N_B \sum_{k_x=0}^{(l+l'+l'')/2} \sum_{k_y=0}^{(m+m'+m'')/2} \sum_{k_z=0}^{(n+n'+n'')/2} f_{2k_x} f_{2k_y} f_{2k_z} \times \frac{(2k_x - 1)!!\,(2k_y - 1)!!\,(2k_z - 1)!!}{\gamma^{(k_x+k_y+k_z+3/2)} \times 2^{k_x+k_y+k_z}} \pi^{3/2} \tag{5.61}$$

In Table 5.4 we report some simple moment integrals useful for the dipole moment calculation.

5.2.2.6 The Laplace transform

The remaining one- and two-electron integrals we have to solve depend on $1/r$ terms. The Laplace transform[24] enables us to replace those kinds of terms by a gaussian function. To do that we introduce the new integration variable ρ. For a more general purpose we use $(1/r)^\lambda$ instead of $(1/r)^1$. The explicit form of Laplace's transform is:

$$\mathscr{L}(t) = \int_0^\infty \exp(-t\rho) g(\rho)\, d\rho \tag{5.62}$$

Let us, for our actual purpose, replace t by r^2 and $g(\rho)$ by $\rho^{\lambda/2-1}$; we find:

$$\mathscr{L}(r^2) = \int_0^\infty \exp(-r^2\rho)\rho^{\lambda/2-1}\, d\rho$$

or if $u = \rho r^2$:

$$\mathscr{L}(r^2) = \left(\frac{1}{r}\right)^{\lambda} \int_0^{\infty} \exp(-u) u^{\lambda/2-1}\, \mathrm{d}u$$

From expression (5.51) which defines the gamma function we find that:

$$\mathscr{L}(r^2) = \left(\frac{1}{r}\right)^{\lambda} \Gamma\left(\frac{\lambda}{2}\right)$$

Then we write:

$$\left(\frac{1}{r}\right)^{\lambda} = \frac{1}{\Gamma(\lambda/2)} \int_0^{\infty} \exp(-r^2\rho)\rho^{\lambda/2-1}\, \mathrm{d}\rho \tag{5.63}$$

Another convenient form of the last expression can be obtained by replacing ρ by u^2:

$$\left(\frac{1}{r}\right)^{\lambda} = \frac{1}{\Gamma(\lambda/2)} \int_0^{\infty} \exp(-r^2u^2)u^{\lambda-1}\, \mathrm{d}u \tag{5.64}$$

5.2.2.7 The potential integrals

The potential integrals are two- or three-centre integrals of the form:

$$V = -\sum_{\mathrm{c}}^{\text{all nuclei}} Z_{\mathrm{c}} V_{\mathrm{c}} = -\sum_{\mathrm{c}} Z_{\mathrm{c}} \langle G_{\mathrm{A}} | \frac{1}{r_{\mathrm{c}}} | G_{\mathrm{B}} \rangle \tag{5.65}$$

where:

$$r_{\mathrm{c}}^2 = (x - C_x)^2 + (y - C_y)^2 + (z - C_z)^2$$

and **C** stands for the position vector of any nucleus c as shown in Figure 5.7.

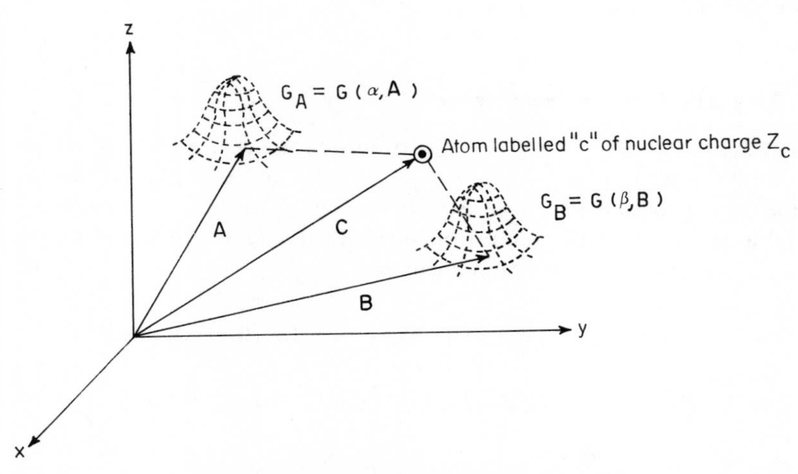

Figure 5.7 The potential integral calculation

By using the Laplace transform (in the particular case of $\lambda = 1$) defined by (5.63) and the value of $\Gamma(1/2)$ given in (5.52) the V_c integral becomes:

$$V_c = \frac{1}{\sqrt{\pi}} \int_0^{\infty} d\rho\, \rho^{-1/2} \int_{-\infty}^{+\infty} dx\, dy\, dz \exp(-\rho r_c^2) G_A G_B$$

$$= \frac{N_A N_B}{\sqrt{\pi}} \int_0^{\infty} d\rho\, \rho^{-1/2} \int_{-\infty}^{+\infty} dx\, x_A^l x_B^{l'} \exp(-\rho x_C^2) \exp(-\alpha x_A^2) \exp(-\beta x_B^2)$$

$$\times \int_{-\infty}^{+\infty} dy \cdots \int_{-\infty}^{+\infty} dz \cdots$$

Having collected together all the terms which depend on x, y and z separately, we find:

$$V_c = \frac{N_A N_B}{\sqrt{\pi}} \int_0^{\infty} d\rho\, \rho^{-1/2} U_x^{(V)} U_y^{(V)} U_z^{(V)}$$

where, if h stands for x, y or z and if $i(i')$ replaces the l, m or n (l', m' or n') angular momenta, we write for $U_h^{(V)}$:

$$U_h^{(V)} = \int_{-\infty}^{+\infty} dh \exp(-\rho h_C^2 - \alpha h_A^2 - \beta h_B^2) h_A^i h_B^{i'}$$

Let us now apply the product theorem; according to (5.45) we introduce the following definitions:

$$\gamma = \alpha + \beta \qquad \theta = \gamma + \rho$$

$$K_h = \exp\left(-\frac{\alpha\beta}{\gamma} \overline{AB}_h^2\right) \qquad \Lambda_h = \exp\left(-\frac{\gamma\rho}{\theta} \overline{PC}_h^2\right)$$

$$P_h = (\alpha A_h + \beta B_h)/\gamma \qquad Q_h = (\gamma P_h + \rho C_h)/\theta$$

$$h_P = h - P_h \qquad h_Q = h - Q_h$$

$$K = K_x K_y K_z \qquad \Lambda = \Lambda_x \Lambda_y \Lambda_z$$

The integrals $U_h^{(V)}$ become:

$$U_h^{(V)} = K_h \Lambda_h \int_{-\infty}^{+\infty} dh\, h_A^i h_B^{i'} \exp(-\theta h_Q^2) = K_h \Lambda_h J_h^{(V)}$$

For V_c we now have:

$$V_c = \frac{N_A N_B K}{\sqrt{\pi}} \int_{-\infty}^{+\infty} d\rho \Lambda(\rho) \rho^{-1/2} J_x J_y J_z$$

Let us now define the new variable t as:

$$\rho = \frac{\gamma t^2}{1 - t^2}$$

or

$$t^2 = \frac{\rho}{\theta} \tag{5.66}$$

Then:

$$dt = \frac{\gamma\, d\rho}{2\theta^{3/2}\rho^{1/2}}$$

and

$$\rho^{-1/2}\, d\rho = 2\gamma^{1/2}(1-t^2)^{-3/2}\, dt$$

It follows that:

$$V_c = \frac{2N_A N_B K}{\gamma\sqrt{\pi}} \int_0^1 dt\, \Lambda(t^2)\left(\frac{\gamma}{1-t^2}\right)^{3/2} J_x^{(V)} J_y^{(V)} J_z^{(V)} \tag{5.67}$$

At this point two attempts at solution have been carried out. In the first (the most classical one), we solve analytically the J_x, J_y and J_z integrals. As for equation (5.47) we expand the product $h_A^i h_B^{i'}$ as:

$$h_A^i h_B^i = \sum_{k_h=0}^{i+i'} f_{k_h}(ii'\overline{QA}_h\overline{QB}_h) h^{k_h}$$

where:

$$f_{k_h}(ii'\overline{QA}_h\overline{QB}_h) = \sum_{\substack{j=0\\ j+j'=k_h}}^{i} \sum_{j'=0}^{i'} \overline{QA}_h^{i-j}\overline{QB}^{i'-j'}\binom{i}{j}\binom{i'}{j'}$$

Now we introduce this expression in $J_h^{(V)}$:

$$J_h^{(V)} = \sum_{k_h=0}^{i+i'} f_{k_h} \int_{-\infty}^{+\infty} dh_Q h_Q^{k_h} \exp(-\theta h_Q^2)$$

As only the even integrals are non-vanishing we write from (5.53):

$$J_h^{(V)} = \sum_{k_h=0}^{(i+i')/2} f_{2k_h}\theta^{-(2k_h+1)/2} \frac{(2k_h-1)!!}{2^{k_h}}$$

From (5.66) we can easily prove that:

$$\overline{QA}_h = \overline{PA}_h - t^2\overline{PC}_h$$
$$\overline{QB}_h = \overline{PB}_h - t^2\overline{PC}_h$$
$$\theta = \frac{\rho}{t^2} = \frac{\gamma}{1-t^2}$$

We now have that:

$$\left(\frac{\gamma}{1-t^2}\right)^{1/2} J_h^{(v)} = \sum_{k_h=0}^{(i+i')/2} \left(\frac{\gamma}{1-t^2}\right)^{-k_h} \frac{(2k_h-1)!!}{2^{k_h}}$$
$$\times \sum_{\substack{j=0\\ j+j'=2k_h}}^{i} \sum_{j'=0}^{i'} \binom{i}{j}\binom{i'}{j'}(\overline{PA}_h - t^2\overline{PC}_h)^{i-j}(\overline{PB}_h - t^2\overline{PC}_h)^{i'-j'}$$

And for V_c we find:

$$V_c = \frac{2N_A N_B K}{\gamma\sqrt{\pi}} \sum_{k_x=0}^{(l+l')/2} \sum_{k_y=0}^{(m+m')/2} \sum_{k_z=0}^{(n+n')/2} \int_0^1 dt\, \Lambda(t)$$
$$\times \left(\frac{1-t^2}{\gamma}\right)^{k_x+k_y+k_z} \frac{(2k_x-1)!!}{2^{k_x}} \frac{(2k_y-1)!!}{2^{k_y}} \frac{(2k_z-1)!!}{2^{k_z}}$$
$$\times \left[\sum_{\substack{j_x \; j'_x \\ j_x+j'_x=2k_x}}^{l} \sum^{l'} \binom{l}{j_x}\binom{l'}{j'_x} (\overline{PA}_x - t^2\overline{PC}_x)^{l-j_x} (\overline{PB}_x - t^2\overline{PC}_x)^{l'-j'_x} \right.$$
$$\left. + \sum_{j_y} \sum_{j'_y} + \cdots + \sum_{j_z} \sum_{j'_z} + \cdots \right] \tag{5.68}$$

For convenience we summarize this expression in powers of t as:

$$V_c = \frac{2N_A N_B K}{\gamma\sqrt{\pi}} \sum_{j=0}^{\lambda} C_j F_j(\gamma\overline{PC}^2) \tag{5.69}$$

where:

$$F_j(\gamma\overline{PC}^2) = \int_0^1 \exp(-\gamma\overline{PC}^2 t^2) t^{2j}\, dt$$
$$\lambda = l + l' + m + m' + n + n'$$

and C_j must be determined according to the previous development. Examples are given in Table 5.5.

Table 5.5 Some simple potential integrals

$V_c = \langle G_A \vert \frac{1}{r_c} \vert G_B \rangle$	α; **A** l	m	n	β; **B** l'	m'	n'	Integral value
$\langle s \vert \frac{1}{r_c} \vert s \rangle$	0	0	0	0	0	0	$KN_A N_B \pi \frac{2}{\gamma} F_0(\gamma\overline{PC}^2)$
$\langle x \vert \frac{1}{r_c} \vert s \rangle$	1	0	0	0	0	0	$-KN_A N_B \pi \frac{2}{\gamma}\left[\frac{\beta}{\gamma}\overline{AB}_x F_0(\gamma\overline{PC}^2) + \overline{PC}_x F_1(\gamma\overline{PC}^2)\right]$
$\langle x \vert \frac{1}{r_c} \vert x \rangle$	1	0	0	1	0	0	$KN_A N_B \pi \frac{2}{\gamma}\left[\left(\frac{1}{2\gamma} - \frac{\alpha\beta}{\gamma^2}\overline{AB}_x^2\right) F_0(\gamma\overline{PC}^2) - \left(\frac{1}{2\gamma} + \frac{\alpha-\beta}{\gamma}\overline{AB}_x\overline{PC}_x\right) F_1(\gamma\overline{PC}^2) + \overline{PC}_x^2 F_2(\gamma\overline{PC}^2)\right]$
$\langle x \vert \frac{1}{r_c} \vert y \rangle$	1	0	0	0	1	0	$KN_A N_B \pi \frac{2}{\gamma}\left[-\frac{\alpha\beta}{\gamma^2}\overline{AB}_x\overline{AB}_y F_0(\gamma\overline{PC}^2) + \frac{\beta\overline{PC}_y\overline{AB}_x - \alpha\overline{PC}_x\overline{AB}_y}{\gamma} F_1(\gamma\overline{PC}^2) + \overline{PC}_x\overline{PC}_y F_2(\gamma\overline{PC}^2)\right]$

The $F_j(\gamma\overline{PC}^2)$ *may be computed as follows. If*

$$F_m(W) = \int_0^1 \exp(-Wt^2)t^{2m}\,\mathrm{d}t \qquad (5.70)$$

by successive substitution of the type $u = \exp(-Wt^2)$ *and* $\mathrm{d}v = t^{2m}\,\mathrm{d}t$ *we find a general formulation:*

$$F_m(W) = \frac{\exp(-W)}{2}\Gamma(m+1/2)\sum_{i=0}^{\infty}\frac{W^i}{\Gamma(m+i+3/2)} \qquad (5.71)$$

with:

$$F_\infty(W) = 0$$

$$F_m(0) = \frac{1}{2m+1}$$

$$\frac{\mathrm{d}F_m(W)}{\mathrm{d}W} = -F_{m+1}(W)$$

Particular expressions are available for small and large values of W:

(a) *For small* W *we can develop* $F_m(W)$ *in Taylor series:*

$$F_m(W_0+\Delta W) = F_m(W_0) + \left(\frac{\mathrm{d}F_m(W)}{\mathrm{d}W}\right)_{W_0}(W_0+\Delta W) + \cdots$$

which becomes for $W_0 = 0$:

$$F_m(W) = \sum_{i=0}^{\infty}\frac{(-W)^i}{i!\,(2m+2i+1)} \qquad (5.72)$$

(b) *For large* W *we write:*

$$F_m(W) = \int_0^\infty \exp(-Wt^2)t^{2m}\,\mathrm{d}t - \int_1^\infty \exp(-Wt^2)t^{2m}\,\mathrm{d}t$$

$$= \frac{\Gamma(m+1/2)}{2W^{m+1/2}} - \frac{1}{2W^{m+1/2}}\int_W^\infty \exp(-u)u^{m-1/2}\,\mathrm{d}u$$

$$= \left(\frac{1}{2W^{m+1/2}}\right)[\Gamma(m+1/2) - f_{\mathrm{err}}(m+1/2, W)] \qquad (5.73)$$

where $f_{\mathrm{err}}(a, W)$ *is the well-known error function which can be evaluated as a continuous fraction:*[(25)]

$$f_{\mathrm{err}}(a, W) = \int_W^\infty \exp(-u)u^{a-1}\,\mathrm{d}u$$

$$= W^a \exp(-W)\left(\frac{1}{W+}\,\frac{1-a}{1+}\,\frac{1}{W+}\,\frac{2-a}{1+}\,\frac{2}{W+}\cdots\right)$$

(c) *For very large value of* W, *as:*

$$\lim_{W\to\infty} f_{\mathrm{err}}(a, W) = 0$$

Table 5.6 Example of $F_m(W)$ function evaluation†

m	W	Computational method for $F_m(W)$‡ Small W (5.72)	Large W (5.73)	Very large W§ (5.74)
0	1	0.74682(10)	0.74549(25)	0.88623(1)
	2	0.59814(13)	0.59711(14)	0.62666(1)
	3	0.50434(16)	0.50419(11)	0.51166(1)
	4	0.44104(19)	0.44101(9)	0.44311(1)
	5	0.39571(22)	0.39571(8)	0.39633(1)
	6	0.36161(25)	0.36161(7)	0.36180(1)
	15	0.22883(49)	0.22883(5)	0.22883(1)
1	4	0.05284(19)	0.05289(8)	0.05539(1)
	15	0.00763(49)	0.00763(4)	0.00763(1)
2	4	0.01753(19)	0.01774(8)	0.02077(1)
	15	0.00076(49)	0.00076(5)	0.00076(1)

† Difference between small W and large W formula results comes from numerical accuracy.
‡ In parentheses we report the extension of the sum over i which stabilizes the result.
§ No accurate results under $W = 8$.

we write:

$$F_m(W) = \frac{\Gamma(m+1/2)}{2W^{m+1/2}} = \frac{(2m-1)!!}{2^{m+1}} \frac{\sqrt{\pi}}{W^{m+1/2}} \tag{5.74}$$

In practice, a table of the $W^{m+1/2}F_m(W)$ function is first generated for some values of m and W; interpolation methods furnish the appropriate value of F_m for any particular W. In Table 5.6 we give some examples of the $F_m(W)$ function evaluation.

More material concerning the evaluation of $F_m(W)$ is available.[26]

In order to avoid this long expansion (which becomes hard working for high quantum numbers) as well as the evaluation of the $F_m(W)$ functions, another approach of the integration problem is available from the quadrature method. Let us go back to the previous expression for V_c (5.67) where we replace $\sqrt{[\gamma/(1-t^2)]}\, J_h^{(V)}$ by $I_h^{(V)}$:

$$V_c = \frac{2N_A N_B K}{\gamma\sqrt{\pi}} \int_0^1 dt\, \Lambda(t^2) I_x^{(V)} I_y^{(V)} I_z^{(V)}$$

The product I_x, I_y, I_z is a polynomial of degree λ (λ is equal to the sum of the angular momenta of the gaussian functions: $\lambda = l + l' + m + m' + n + n'$) in the variable t^2. Then:

$$I_x^{(V)} I_y^{(V)} I_z^{(V)} = P_\lambda(t^2)$$

The integral over ρ in the V_c expression may be evaluated by the Rys quadrature method.[27] Let t_a and w_a be the zero and the associated weight factors of the kth Rys polynomial (with k greater than $\lambda/2$); then it appears that:

$$\int_0^1 P_\lambda(t^2)\,\Lambda(t^2)\,dt = \sum_{a=1}^{k} I_x^{(V)}(t_a) I_y^{(V)}(t_a) I_z^{(V)}(t_a) w_a \tag{5.75}$$

Finally, the one-dimensional integrals $I_h^{(V)}(t_a)$ are calculated by the Gauss–Hermite quadrature:

$$I_h^{(V)}(t_a) = \sum_{b=1}^{q} \left\{\sqrt{\left[\frac{\sigma_b(1-t_a^2)}{\gamma}\right]} + \overline{QA}_h\right\}^i \left\{\sqrt{\left[\frac{\sigma_b(1-t_a^2)}{\gamma}\right]} + \overline{QB}_h\right\}^{i'} \omega_b \tag{5.76}$$

where σ_b and ω_b are the zero and the associated weight of the qth Hermite polynomial (if $2q$ is greater than $i+i'$). This technique seems to require less computer time than the former one using the error function evaluation specially for functions having large angular momenta such as the d gaussian-type functions.

5.2.2.8 The electron repulsion integrals

The general form of the two-electron integrals between gaussian-type functions is given by:

$$\mathcal{R} = \int_{-\infty}^{+\infty} G_A^*(\mathbf{r}_1) G_B^*(\mathbf{r}_1) \frac{1}{r_{12}} G_C(\mathbf{r}_2) G_D(\mathbf{r}_2)\, d\mathbf{r}_1\, d\mathbf{r}_2 \tag{5.77}$$

The two variables $\mathbf{r}_1$, $\mathbf{r}_2$ and the four gaussian centres **A**, **B**, **C**, **D** are reported in Figure 5.8 as well as the r_{12} term which is the norm of the $\mathbf{r}_{12}$

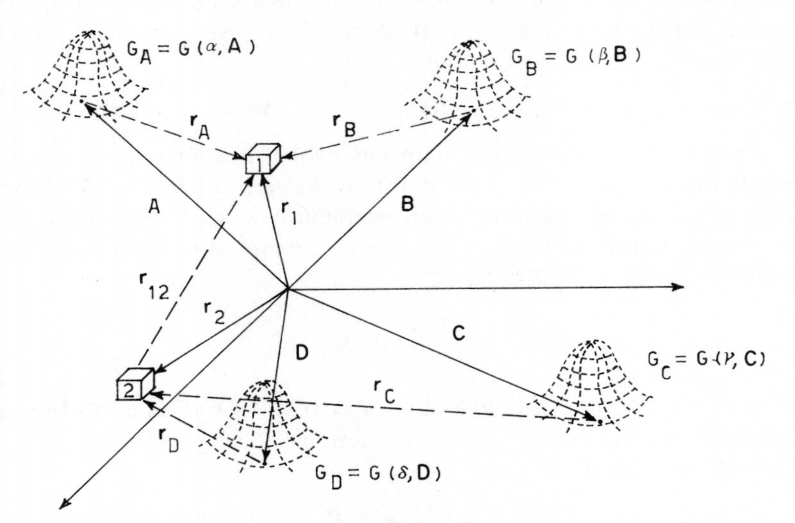

Figure 5.8 Vector definition for integral evaluation of the two electrons

vector. The cartesian components of those vectors are:

$$\mathbf{A}=\{A_x;\,A_y;\,A_z\}$$
$$\mathbf{B}=\{B_x;\,B_y;\,B_z\}$$
$$\mathbf{C}=\{C_x;\,C_y;\,C_z\}$$
$$\mathbf{D}=\{D_x;\,D_y;\,D_z\}$$
$$\mathbf{r}_1=\{x_1;\,y_1;\,z_1\}$$
$$\mathbf{r}_2=\{x_2;\,y_2;\,z_2\}$$

For further use let us introduce the following conventions:

$$\mathbf{r}_A=\mathbf{r}_1-\mathbf{A}=\{x_A;\,y_A;\,z_A\}$$
$$\mathbf{r}_B=\mathbf{r}_1-\mathbf{B}=\{x_B;\,y_B;\,z_B\}$$
$$\mathbf{r}_C=\mathbf{r}_2-\mathbf{C}=\{x_C;\,y_C;\,z_C\}$$
$$\mathbf{r}_D=\mathbf{r}_2-\mathbf{D}=\{x_D;\,y_D;\,z_D\}$$
$$\mathbf{r}_{12}=\mathbf{r}_1-\mathbf{r}_2=\{x_{12};\,y_{12};\,z_{12}\}$$

According to the gaussian product theorem (5.45), the four gaussian functions may be replaced by two new ones:

$$\exp(-\alpha r_A^2)\exp(-\beta r_B^2)=K_1\exp(-\varepsilon_1 r_P^2)$$
$$\exp(-\gamma r_C^2)\exp(-\delta r_D^2)=K_2\exp(-\varepsilon_2 r_Q^2)$$

where:

$$\varepsilon_1=\alpha+\beta \qquad \varepsilon_2=\gamma+\delta$$
$$K_1=\exp\left(-\frac{\alpha\beta}{\alpha+\beta}\,\overline{AB}^2\right) \qquad K_2=\exp\left(-\frac{\gamma\delta}{\gamma+\delta}\,\overline{CD}^2\right)$$
$$\mathbf{r}_P=\mathbf{r}_1-\mathbf{P}=\{x_P;\,y_P;\,z_P\} \qquad \mathbf{r}_Q=\mathbf{r}_2-\mathbf{Q}=\{x_Q;\,y_Q;\,z_Q\}$$
$$\mathbf{P}=(\alpha\mathbf{A}+\beta\mathbf{B})/(\alpha+\beta) \qquad \mathbf{Q}=(\gamma\mathbf{C}+\delta\mathbf{D})/(\gamma+\delta)$$

According to (5.63) we also replace $1/r_{12}$ by:

$$\frac{1}{r_{12}}=\frac{1}{\sqrt{\pi}}\int_0^\infty \exp(-r_{12}^2\rho)\rho^{-1/2}\,\mathrm{d}\rho$$

The electron repulsion integral becomes:

$$\mathcal{R}=\frac{N_A N_B N_C N_D K_1 K_2}{\sqrt{\pi}}\int_0^\infty \mathrm{d}\rho\,\rho^{-1/2}U_x^{(R)}U_y^{(R)}U_z^{(R)} \tag{5.78}$$

where $U_x^{(R)}$, $U_y^{(R)}$ and $U_z^{(R)}$ are of the form:

$$U_h^{(R)}=\int \mathrm{d}h_1\int \mathrm{d}h_2\,h_A^{k_h}h_B^{l_h}h_C^{m_h}h_D^{n_h}\exp[-\rho(h_2-h_1)^2]$$
$$\times\exp[-\varepsilon_1(h_1-P_h)^2]\exp[-\varepsilon_2(h_2-Q_h)^2] \tag{5.79}$$

This expression is useful for classical derivation of the integral value. If we remember that:

$$h_2 - h_1 = (h_Q + Q_h) - (h_P + P_h)$$
$$= h_Q - h_P - \overline{PQ}_h$$

then:

$$U_h^{(R)} = \exp(-\rho\overline{PQ}_h^2)\int dh_1 h_A^{k_h} h_B^{l_h} \exp[-(\varepsilon_1+\rho)h_P^2 - 2\rho\overline{PQ}_h h_P]$$
$$\times \int dh_2 h_C^{m_h} h_D^{n_h} \exp[-(\varepsilon_2+\rho)h_Q^2 + 2\rho\overline{PQ}_h h_Q + 2\rho h_P h_Q] \qquad (5.80)$$

This double integral can be solved in a similar way to the one used for the nuclear attraction potential integrals. For example, for s type functions ($k_h = l_h = m_h = n_h = 0$ where $h = x$, y or z) we find:

$$U_h^{(R)}(ssss) = \exp(-\rho\overline{PQ}_h^2)\int dh_1 \exp\left[-(\varepsilon_1+\rho)h_P^2 - 2\rho\overline{PQ}_h h_P + \frac{\rho^2(h_P+\overline{PQ}_h)^2}{\varepsilon_2+\rho}\right]$$
$$\times \int dZ \exp[-(\varepsilon_2+\rho)Z^2]$$

where Z stands for $h_Q - \rho(h_P - \overline{PQ}_h)/(\varepsilon_2+\rho)$. The integral over Z has as solution: $\sqrt{[\pi/(\varepsilon_2+\rho)]}$. New substitution similar to the last one leads to the final result:

$$U_h^{(R)}(ssss) = \frac{\pi}{[\varepsilon_1\varepsilon_2 + (\varepsilon_1+\varepsilon_2)\rho]^{1/2}} \exp\left[-\frac{\varepsilon_1\varepsilon_2\overline{PQ}_h^2\rho}{\varepsilon_1\varepsilon_2+(\varepsilon_1+\varepsilon_2)\rho}\right]$$

Collecting together the $U_x^{(R)}$, $U_y^{(R)}$ and $U_z^{(R)}$ results and replacing $1+[\rho(\varepsilon_1+\varepsilon_2)]/\varepsilon_1\varepsilon_2$ by $(1-t^2)^{-1}$, we find, according to (5.70):

$$\langle ss|\frac{1}{r}|ss\rangle = \frac{2N_A N_B N_C N_D K_1 K_2 \pi^{5/2}}{\varepsilon_1\varepsilon_2(\varepsilon_1+\varepsilon_2)^{1/2}} F_0\left(\frac{\overline{PQ}^2\varepsilon_1\varepsilon_2}{\varepsilon_1+\varepsilon_2}\right) \qquad (5.81)$$

The other type of electron repulsion integral may also be deduced after some algebraic manipulation.

More recently it became clear that the quadrature process in numerical integration[28] was of particular interest for fast computation of two-electron integrals involving high angular momenta.[29] Let us go back to equations (5.78) and (5.79):

$$\mathcal{R} = \frac{N_A N_B N_C N_D K_1 K_2}{\sqrt{\pi}} \int_0^\infty d\rho\, \rho^{-1/2} U_x^{(R)} U_y^{(R)} U_z^{(R)}$$

If we apply twice more the gaussian product theorem by introducing the following quantities, first for the term in h_2:

$$\eta_2 = \varepsilon_2 + \rho$$

$$\bar{h}_2 = \frac{\varepsilon_2 Q_h + \rho h_1}{\varepsilon_2 + \rho}$$

second for the term in h_1:

$$\eta_1 = \varepsilon_1 + \frac{\varepsilon_2 \rho}{\varepsilon_2 + \rho}$$

$$\bar{h}_1 = \frac{\varepsilon_1 P_h + (\varepsilon_2 \rho/(\varepsilon_2 + \rho)) Q_h}{\varepsilon_1 + \varepsilon_2 \rho/(\varepsilon_2 + \rho)}$$

and to shorten the notation:

$$\theta = \frac{\varepsilon_1 \varepsilon_2}{\varepsilon_1 + \varepsilon_2}$$

$$v_h = \eta_1 (h_1 - \bar{h}_1)^2 + \eta_2 (h_2 - \bar{h}_2)^2$$

then we can write:

$$\mathcal{R} = \frac{N_A N_B N_C N_D K_1 K_2}{\sqrt{\pi}} \int_0^{\infty} d\rho\, \rho^{-1/2} \exp\left(-\theta \overline{PC}^2 \frac{\rho}{\theta + \rho}\right)$$

$$\times \prod_{h=x,y,z} \int_{-\infty}^{+\infty} dh_1 h_A^{k_h} h_B^{l_h} \int dh_2 h_C^{m_h} h_D^{n_h} \exp(-v_h) \tag{5.82}$$

Defining the new variable t such as:

$$t^2 = \frac{\rho}{\theta + \rho}$$

we find:

$$\mathcal{R} = \frac{2 N_A N_B N_C N_D K_1 K_2}{\sqrt{\pi} \theta (\varepsilon_1 + \varepsilon_2)^{3/2}} \int_0^1 dt \exp(-\theta \overline{PC}^2 t^2) I_x^{(R)} I_y^{(R)} I_z^{(R)} \tag{5.83}$$

$$I_h^{(R)} = (\eta_1 \eta_2)^{1/2} \int_{-\infty}^{+\infty} dh_1 h_A^{k_h} h_B^{l_h} \int_{-\infty}^{+\infty} dh_2 h_C^{m_h} h_D^{n_h} \exp[-v_h(t^2)] \tag{5.84}$$

where $\eta_1 \eta_2 = \varepsilon_1 \varepsilon_2 / (1 - t^2)$. Thus $I_h^{(R)}$ is an even polynomial in t[30]. This integral can be calculated by the Rys quadrature method with:

$$P_L(t^2) = I_x^{(R)} I_y^{(R)} I_z^{(R)}$$

$$\mathcal{R} = \frac{2 N_A N_B N_C N_D K_1 K_2}{\theta \sqrt{\pi}} \sum_{a=1}^{k} I_x^{(R)}(t_a) I_y^{(R)}(t_a) I_z^{(R)}(t_a) w_a \tag{5.85}$$

where $k > L/2$ and t_a and w_a are the zero and the weight of the kth Rys polynomial. Defining now the smearing transformation:

$$\Gamma(m_h n_h) = \sqrt{\eta_2} \int_{-\infty}^{+\infty} dh_2 h_C^{m_h} h_D^{n_h} \exp[-\eta_2 (h_2 - \bar{h}_2)^2]$$

we can write after some manipulations:

$$\Gamma(m_h n_h) = P_{10}^r r_{01}^s$$

where:

$$P_{01} = \frac{(h_1 - \bar{h}_{m_h})\rho}{\eta_2}$$

and

$$P_{10} = \frac{(h_1 - \bar{h}_{n_h})\rho}{\eta_2}$$

with:

$$\bar{h}_{m_h} = \frac{(\delta + \rho)C_h - \delta D_h}{\rho}$$

and

$$\bar{h}_{n_h} = \frac{(\gamma + \rho)D_h - \gamma C_h}{\rho}$$

After the smearing transformation, the general expression is a simple linear combination of the polynomials P_{10} and P_{01} raised to an integer power. The $I_h^{(R)}$ integral then reduces to a sum of integrals of the type:

$$\sqrt{\eta_1}\int_{-\infty}^{+\infty} dh_1 h_A^{k_h} h_B^{l_h} P_{10}^r P_{01}^s \exp\left[-\eta_1(h_1 - \bar{h}_1)^2\right]$$

Those integrals may be solved by Gauss–Hermite quadrature and collected together to form the $I_h^{(R)}$ value.

5.2.3 Some Technical Features about Integral Computation

The integral computation may be time-consuming. Therefore it is important to work carefully when we write a program. Two principles must be kept in mind: the first one concerns the *operation saving*—we must avoid computing a value twice (a table index as well as an arithmetic operation); the second leads us to *work by shell* (and so it is closely related to the first one). All integrals between gaussian-type functions with the same gaussian exponents have to be computed at the same time. This batching process is nevertheless applied separately for one and two-electron integrals. Thus all the one-electron integrals are computed at once; the general expression is (see also 5.43):

$$O_{pq} = \langle p | O_1 | q \rangle = \langle q | O_1 | p \rangle$$

For symmetry reasons we have only to compute the O_{pq} for which q remains lower or equal to p. Their number is then given by:

$$m_1 = \frac{m(m+1)}{2} \tag{5.86}$$

where m stands for the number of atomic orbitals.

Table 5.7 Two-electron integral number for some small molecules

Molecule	Atomic basis set size m	Two-electron integral number m_2	Non-zero integrals
H_2	2	6	6(100%)
LiH	6	231	83(36%)
H_2O	7	406	206(51%)
H_2O_2	12	3,081	1,740(56%)
CH_4	17	11,781	9,549(81%)

The number of two-electron integrals (see 5.44) increases more rapidly. The symmetry relationship enables us to write:

$$\langle pq \mid rs\rangle = \langle qp \mid rs\rangle = \langle qp \mid sr\rangle = \langle pq \mid sr\rangle = \langle rs \mid pq\rangle$$
$$\langle rs \mid qp\rangle = \langle sr \mid qp\rangle = \langle sr \mid pq\rangle$$

Then the number of two-electron integrals is given by:

$$m_2 = \frac{m_1(m_1+1)}{2} = \tfrac{1}{8}(m^4 + 2m^3 + 3m^2 + 2m) \qquad (5.87)$$

When m increases, the storage requirements rapidly become enormous. As the number of zero integrals (or lower than a given threshold $\simeq 10^{-8}$) is also rapidly increasing (see Table 5.7) one often prefers to use a random storage order instead of the so-called canonical order. The requirements for both cases are now given. For the canonical order:

(a) We have to store all the integrals.
(b) The integral $\langle pq \mid rs\rangle$ stands in place k on tape if k equals:

$$k = \frac{a(a-1)}{2} + b \qquad \text{for } a \geqslant b$$

with

$$a = \frac{p(p-1)}{2} + q \qquad \text{for } p \geqslant q \qquad (5.88)$$

$$b = \frac{r(r-1)}{2} + s \qquad \text{for } r \geqslant s$$

For the random order:[31]

(a) We only want to store the non-zero integrals.
(b) We join the label to any integral. For storage convenience we pack the integral label in one word. One other word is used to store the integral value (in an integer form):

$\lvert p \rvert\, q\, \lvert r \rvert\, s\, \lvert 10^8 * \langle pq \mid rs\rangle \rvert$

5.2.4 The Molecular Integrals

When we want to change the basis set from, for example, the atomic to the molecular basis set according to:

$$\varphi_i = \sum_{p=1}^{m} C_{pi}\chi_p \tag{5.89}$$

we have to reexpress the integral in the new frame. The problem is not tedious for the one-electron integrals which are now:

$$(i \mid O_1 \mid j) = \sum_p \sum_q C_{pi}C_{qj}\langle p \mid O_1 \mid q\rangle \tag{5.90}$$

but it becomes more difficult to perform in a non-expansive way for the two-electron integrals:[32]

$$(ij \mid kl) = \sum_p \sum_q \sum_r \sum_s C_{pi}C_{qj}C_{rk}C_{sl}\langle pq \mid rs\rangle \tag{5.91}$$

Usually we make the transformation twice; first we compute the so-called half-transform:

$$\langle pq \mid kl\rangle = \sum_r \sum_s C_{rk}C_{sl}\langle pq \mid rs\rangle \tag{5.92}$$

which later becomes fully transformed:

$$(ij \mid kl) = \sum_p \sum_q C_{pi}C_{qj}\langle pq \mid kl\rangle \tag{5.93}$$

The following algorithm may be used:

1. The basic integral list is read through and the number of non-zero integrals (m_a) is calculated for each pair pq; notice that the integral $\langle pq \mid rs\rangle$ contributes both to m_a and m_b for $pq \neq rs$ (a and b are in fact the ordering numbers associated with the pq and rs pair index in canonical order; see 5.88).

   Let us suppose that it is possible to store K integrals simultaneously on the direct access memory and N words are available in the core storage; we can then define the integers β and γ:

$$\beta = \frac{N}{m_1} = \frac{2N}{m(m+1)}$$

$$\gamma = \frac{a_{\max} - a_{\min}}{\beta} + 1$$

   where $a_{\max}$ is:

$$\sum_{a=a_{\min}}^{a_{\max}+1} m_a > K$$

2. The integrals are read into the core identified with respect to the pair index pq (packed in the a integer: see 5.88) and distributed in the

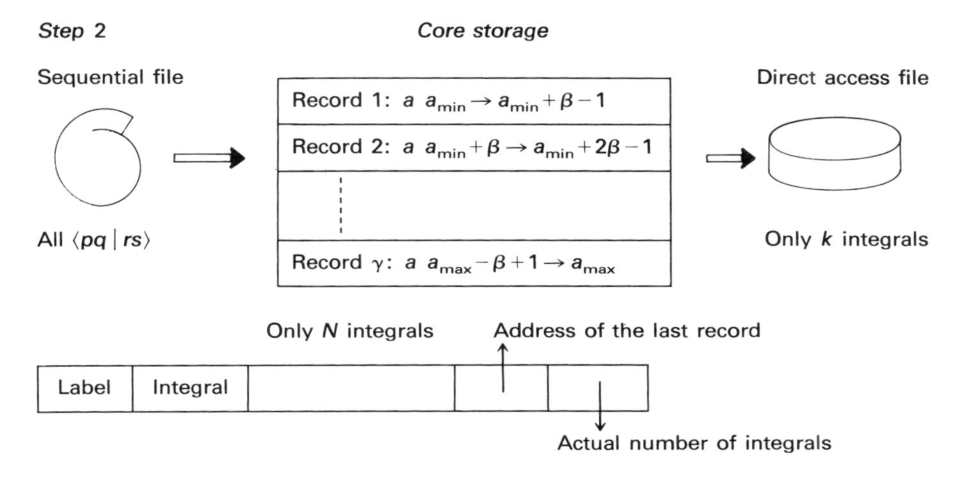

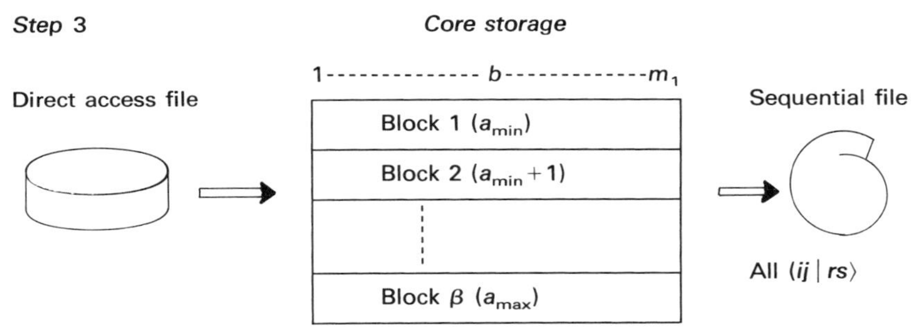

Figure 5.9 An economical strategy for computing the molecular two-electron integrals

appropriate record ($1 \to \gamma$) (see Figure 5.9). When a record is full it is written on the direct access device.

3. After grouping the integrals together with the pair indices in the range $[a_{\min}, a_{\max}]$ we read them into the core once more using a backchaining process in order to collect all integrals belonging to the same group. The integrals are now stored in β blocks of indices between $a_{\min}$ and $a_{\max}$; inside each block, the integrals are ordered according to the pair index rs (with the associated b value; see 5.88). The half-transform can now be calculated as the next matricial product:

$$\langle pq \mid kl \rangle = (\cdots C'_{rk} \cdots)(\langle pq \mid \cdots \rangle) \begin{pmatrix} \vdots \\ C_{sl} \\ \vdots \end{pmatrix}$$

The non-zero integrals are written on a sequential file. This product is performed β times (once for each block) and the overall step 3 must be repeated γ times (once for each record backchained on the direct access memory) (see Figure 5.9).

4. Having now obtained the half-transform we repeat the whole procedure (steps 1 to 3) to compute the integrals $(ij \mid kl)$.

This process requires a minimum of operations and a minimum amount of input/output time. It may already be available on most computers.

5.3 THE SELF-CONSISTENT FIELD PROCESS

5.3.1 The Self-consistent Field Problem

The problem is now to find an explicit solution to the Hartree–Fock equations (see Chapter 4) by expanding the molecular orbitals (φ) as linear combinations of atomic functions (χ):

$$\varphi_i = \sum_p C_{pi}\chi_p$$

We must determine the expansion coefficients ($\mathbf{C}$) by solving the pseudoeigenvalue problem (4.20) or (4.28) which are respectively the RHF[33] and the UHF[34] equations. In both cases the expression is of the form:

$$\sum_q \langle \chi_p | \, h^{\mathrm{HF}} \, | \chi_q \rangle C_{qi} = \sum_q \varepsilon_i \langle \chi_p \mid \chi_q \rangle C_{qi}$$

In matricial notation we write:

$$\mathbf{h}^{\mathrm{HF}}\mathbf{C} = \mathbf{S}\mathbf{C}\boldsymbol{\varepsilon} \tag{5.94}$$

As the Hartree–Fock operator depends on the LCAO coefficients ($h^{\mathrm{HF}} = h^{\mathrm{HF}}(\mathbf{C})$), the eigenvalue problem can only be solved in an iterative fashion. Moreover, it requires some initial guess ($\mathbf{C}(0)$) to be made. Schematically the self-consistent field process may be described as:

$$\text{Choose } \mathbf{C}(0) \rightarrow \text{build } h^{\mathrm{HF}}(\mathbf{C}(0)) \rightarrow \text{solve } h^{\mathrm{HF}}(\mathbf{C}(0))\mathbf{C}(1) = \boldsymbol{\varepsilon}(1)\mathbf{S}\mathbf{C}(1) \rightarrow \tag{5.95}$$

$$\text{if } \mathbf{C}(1) \neq \mathbf{C}(0) \text{ replace } \mathbf{C}(0) \text{ by } \mathbf{C}(1) \text{ and try again; else stop}$$

The set of equations (5.94) cannot be solved directly. Nevertheless transformations exist which enable us to replace (5.94) by another pseudoeigenvalue problem:

$$\mathbf{h}^{\mathrm{HF}\perp}\mathbf{C}^{\perp} = \boldsymbol{\varepsilon}\mathbf{C}^{\perp} \tag{5.96}$$

Löwdin's canonical orthonormalization is the most commonly used procedure[1] (see Section 5.1.5). If $\mathbf{T}$ is a unitary matrix which diagonalizes the overlap $\mathbf{S}$ matrix:

$$\mathbf{T}^{+}\mathbf{S}\mathbf{T} = \mathbf{S}_0 \tag{5.97}$$

we can define a new set of LCAO coefficients ($\mathbf{C}^{\perp}$) related to the old one ($\mathbf{C}$)

by the transformation matrix $\mathbf{V}$:

$$\mathbf{C} = \mathbf{V}^{+}\mathbf{C}^{\perp} \tag{5.98}$$

where:

$$\mathbf{V} = \mathbf{S}_0^{-1/2}\mathbf{T}^{+}$$

Let us now introduce (5.98) in (5.94) and premultiply the result by $\mathbf{V}$; then it appears that:

$$\mathbf{V}\mathbf{h}^{\mathrm{HF}}\mathbf{V}^{+}\mathbf{C}^{\perp} = \boldsymbol{\varepsilon}\mathbf{V}\mathbf{S}\mathbf{V}^{+}\mathbf{C}^{\perp} \tag{5.99}$$

This last relation is simply the one proposed in (5.96) while $\mathbf{VSV}^{+} = \mathbf{E}$ (the unit matrix). We can further write:

$$\mathbf{h}^{\mathrm{HF}\perp} = \mathbf{V}\mathbf{h}^{\mathrm{HF}}\mathbf{V}^{+} \tag{5.100}$$

Löwdin's orthonormalization procedure does not affect the expected value of the total energy. It only corresponds to transformation of the old atomic basis set ($\boldsymbol{\chi}$) into a new one ($\boldsymbol{\chi}^{\perp}$) with:

$$\boldsymbol{\chi}^{\perp} = \boldsymbol{\chi}\mathbf{V}^{+} \tag{5.101}$$

According to the previous definition the new basis set does not remain of an atomic nature. The functions $\chi^{\perp}$ are now delocalized over the whole molecule. Even if the overlap matrix between the χ's is of a general square symmetrical form, the new $\mathbf{S}$ matrix ($\mathbf{S}^{\perp}$) defined in the $\chi^{\perp}$'s frame becomes the unit matrix. In such a case the pseudoeigenvalue problem (5.96) may be solved by using the usual diagonalization algorithms. Moreover, equation (5.100) avoids transformation of the two-electron integrals in the orthogonal basis set; both Hartree–Fock matrices are simply related by a double matricial product. In Table 5.8 we summarize the relations which connect the χ and $\chi^{\perp}$ frames.

From Table 5.8, it is interesting to note that the density matrices of SCF calculations have all the properties of projectors when they are expressed in terms of the orthogonal basis set.[35] So $\mathbf{R}_1$ (defined in Table 5.8) corresponds to the projector into the subspace of the occupied orbitals. Similarly $\mathbf{R}_2$ is the projector into the subspace of the virtual orbitals. Those projectors are idempotent ($\mathbf{R}_1^2 = \mathbf{R}_1$ and $\mathbf{R}_2^2 = \mathbf{R}_2$), the associated subspaces are not overlapping ($\mathbf{R}_1\mathbf{R}_2 = \mathbf{R}_2\mathbf{R}_1 = \mathbf{0}$) and they are complementary to each other ($\mathbf{R}_1 + \mathbf{R}_2 = \mathbf{E}$, the resolution of the identity).

5.3.2 Relation between RHF-like and UHF Solutions for Open-shell Cases

We can use the projector formalism to find a way to connect an unrestricted Hartree–Fock solution to a restricted one—which means to project the UHF solution into a corresponding restricted-type subspace. At least three

Table 5.8 Correspondence between some quantities in terms of orthogonal and non-orthogonal basis sets

	Non-orthogonal basis set (usually the atomic basis set)	Orthonormal basis set
Atomic basis set (line vector)	$\boldsymbol{\chi}$	$\boldsymbol{\chi}^{\perp}=\boldsymbol{\chi}\mathbf{V}^{+}$
Overlap matrix	$\mathbf{S}=\begin{pmatrix} 1 & \neq 0 \\ \neq 0 & 1 \end{pmatrix}$	$\mathbf{S}^{\perp}=\mathbf{V}\mathbf{S}\mathbf{V}^{+}=\mathbf{E}$
Molecular orbitals (line vector)	$\boldsymbol{\varphi}=\boldsymbol{\chi}\mathbf{C}$	$\boldsymbol{\varphi}=\boldsymbol{\chi}^{\perp}\mathbf{C}^{\perp}$
LCAO coefficients	$\mathbf{C}=\mathbf{V}^{+}\mathbf{C}^{\perp}$	$\mathbf{C}^{\perp}=\mathbf{S}_0\mathbf{V}\mathbf{C}=\mathbf{V}\mathbf{S}\mathbf{C}$
SCF density matrices:†	$\mathbf{D}=\mathbf{C}\boldsymbol{\eta}\mathbf{C}^{+}$	$\mathbf{R}=\mathbf{C}^{\perp}\boldsymbol{\eta}\mathbf{C}^{\perp+}$
general expression	$\mathbf{D}=\mathbf{V}^{+}\mathbf{R}\mathbf{V}$	$\mathbf{R}=\mathbf{S}_0\mathbf{V}\mathbf{D}\mathbf{V}^{+}\mathbf{S}_0=\mathbf{V}\mathbf{S}\mathbf{D}\mathbf{S}\mathbf{V}^{+}$
for occupied orbitals	$\mathbf{D}_1=\mathbf{C}\boldsymbol{\eta}_{\text{occ}}\mathbf{C}^{+}$	$\mathbf{R}_1=\mathbf{C}^{\perp}\boldsymbol{\eta}_{\text{occ}}\mathbf{C}^{\perp+}$
for vacant orbitals	$\mathbf{D}_2=\mathbf{C}\boldsymbol{\eta}_{\text{vac}}\mathbf{C}^{+}$	$\mathbf{R}_2=\mathbf{C}^{\perp}\boldsymbol{\eta}_{\text{vac}}\mathbf{C}^{\perp+}$
Projector idempotency constraint	$\mathbf{D}_1=\mathbf{D}_1\mathbf{S}\mathbf{D}_1$ $\mathbf{D}_2=\mathbf{D}_2\mathbf{S}\mathbf{D}_2$ $\mathbf{0}=\mathbf{D}_1\mathbf{S}\mathbf{D}_2=\mathbf{D}_2\mathbf{S}\mathbf{D}_1$ $\mathbf{S}^{-1}=\mathbf{D}_1+\mathbf{D}_2$	$\mathbf{R}_1=\mathbf{R}_1^2$ $\mathbf{R}_2=\mathbf{R}_2^2$ $\mathbf{0}=\mathbf{R}_1\mathbf{R}_2=\mathbf{R}_2\mathbf{R}_1$ $\mathbf{E}=\mathbf{R}_1+\mathbf{R}_2$ $\mathrm{tr}\,(\mathbf{R}_1)=\mathrm{tr}\,(\boldsymbol{\eta}_{\text{occ}})=n$ $\mathrm{tr}\,(\mathbf{R}_2)=\mathrm{tr}\,(\boldsymbol{\eta}_{\text{vac}})=m-n$
Hartree–Fock matrix	$\mathbf{h}^{\text{HF}}$	$\mathbf{h}^{\text{HF}\perp}=\mathbf{V}\mathbf{h}^{\text{HF}}\mathbf{V}^{+}$
Pseudoeigenvalue problem	$\mathbf{h}^{\text{HF}}\mathbf{C}=\boldsymbol{\varepsilon}\mathbf{S}\mathbf{C}$	$\mathbf{h}^{\text{HF}\perp}\mathbf{C}^{\perp}=\boldsymbol{\varepsilon}\mathbf{C}^{\perp}$
Stationary condition	$\mathbf{h}^{\text{HF}}\mathbf{D}\mathbf{S}-\mathbf{S}\mathbf{D}\mathbf{h}^{\text{HF}}=\mathbf{0}$	$\mathbf{h}^{\text{HF}\perp}\mathbf{R}-\mathbf{R}\mathbf{h}^{\text{HF}\perp}=\mathbf{0}$
Orthogonalization matrix	$\mathbf{T}^{+}\mathbf{S}\mathbf{T}=\mathbf{S}_0$ $\mathbf{V}=\mathbf{S}_0^{-1/2}\mathbf{T}^{+}=(\mathbf{S}^{-1/2})^{+}$ $\mathbf{V}\mathbf{V}^{+}=\mathbf{S}_0^{-1}$ $\mathbf{V}^{+}\mathbf{V}=\mathbf{S}^{-1}$	

† $\boldsymbol{\eta}$ is a diagonal square matrix. Any diagonal terms are 0 or 1 according to whether we wish to include in the density matrix the corresponding orbital or not.

reasons exist which can justify this procedure:

(a) There is no guarantee that an UHF wave function which is a pure spin state (eigenfunction of the S^2 and S_Z operators) can be found. On the other hand, we know that 'a single Slater determinant with n_α electrons of $+\frac{1}{2}$ spin and n_β electrons of $-\frac{1}{2}$ spin, such that $n_\alpha \geqslant n_\beta$, represents a pure spin state if, and only if, the number of doubly filled orbitals equals n_β.'[36] This corresponds to a RHF-type wave function; the associated quantum number S equals $\frac{1}{2}(n_\alpha - n_\beta)$.
(b) It is also much simpler to build a configuration interaction calculation if we dispose of a unique set of molecular orbitals used for both α and β electrons, which corresponds to a pure spin state wave function.
(c) It remains nevertheless useful at the SCF level to perform UHF-type calculations instead of RHF ones. The UHF calculations are generally

recognized as being better for studying potential energy surfaces of molecules outside the regions of equilibrium.[37]. It is well known that the RHF calculations dissociate the molecules into wrong electronic states.

For all these reasons let us build the projectors onto the α and β occupied molecular orbitals subspace:

$$\mathbf{R}_\alpha = \mathbf{C}^\perp \boldsymbol{\eta}_\alpha \mathbf{C}^{\perp +}$$
$$\mathbf{R}_\beta = \mathbf{C}^\perp \boldsymbol{\eta}_\beta \mathbf{C}^{\perp +} \qquad (5.102)$$

where $\boldsymbol{\eta}_\alpha$ and $\boldsymbol{\eta}_\beta$ are diagonal square matrices. All diagonal terms are 0 or 1 according to the fact that the corresponding orbital is filled with an electron of the right spin (α for $\boldsymbol{\eta}_\alpha$ and β for $\boldsymbol{\eta}_\beta$) or not. This means that $\mathbf{R}_\alpha$ is built up from the α occupied orbitals and $\mathbf{R}_\beta$ from the β's. According to (4.63) the first-order spinless density matrix of the UHF calculation is defined by:

$$\rho^{\mathrm{UHF}}(r_1 r_1') = \sum_{pq} \chi_p(r_1)(\mathbf{R}_\alpha + \mathbf{R}_\beta)_{pq} \chi_q^*(r_1')^+ \qquad (5.103)$$

Similarly, for the natural orbital obtention we can find a transformation matrix $\mathbf{U}$ which diagonalizes $\mathbf{R}_\alpha + \mathbf{R}_\beta$:

$$\mathbf{U}^+(\mathbf{R}_\alpha + \mathbf{R}_\beta)\mathbf{U} = \boldsymbol{\eta}$$

where $0 \leqslant \eta_{ii} \leqslant 2$ and $\eta_{ij} = 0$ for $i \neq j$. So it appears that we can define a new molecular orbital basis set by:

$$\boldsymbol{\varphi}^{\mathrm{RHF}} = \boldsymbol{\chi}^\perp \mathbf{U} = \boldsymbol{\chi} \mathbf{V}^+ \mathbf{U} = \boldsymbol{\chi} \mathbf{C}^{\mathrm{RHF}} \qquad (5.104)$$

This transformation does not change the first-order density matrix:

$$\rho^{\mathrm{UHF}}(r_1 r_1') = \sum_{pq} \chi_p (\mathbf{C}^{\mathrm{RHF}} \boldsymbol{\eta} \mathbf{C}^{\mathrm{RHF}'})_{pq} \chi_q^* \qquad (5.105)$$

where η represents the occupation number associated to any of the new molecular orbitals. Although the η's are not integers it appears that they are often close to zero, one or two. By rounding off the content of the diagonal $\boldsymbol{\eta}$ matrix n_β times to two, $n_\alpha - n_\beta$ times to one and $m - n_\alpha$ times (if m is the number of atomic basis functions in use) to zero we construct a new wave function of restricted type. The total associated energy will be somewhat increased with respect to the UHF solution. Which RHF Slater determinant we build up depends on which orbitals are occupied twice, once or are empty.

5.3.3 The Convergence Capability of the SCF Procedure

Till now we have discarded the convergence process itself. The logic defined by (5.95) is in fact not necessarily easy to apply and normally leads to a converged result for the electronic ground state of the molecule. However, it happens that the orbital changes from one iteration to the next are quite large (e.g. the symmetry of the final occupied orbitals may be different from the starting one). Moreover the expression (5.95) has absolutely no convergence guarantee. Sometimes the classical Hartree–Fock procedure fails to converge and divergent or oscillatory behaviour sets in. Nevertheless, some attempt to ensure energy descent with a low to high rate of convergence exists.[38–40] We give here the leading ideas of Seeger and Pople's work[39] which generalize the work of Hillier and Saunders.[40]

Starting an iteration in the SCF procedure (see 5.95), it is desirable to carry out a search of the complete spin orbital space for achieving a minimum energy value along some path which contains the starting point (this is usually called a univariate search). Given some trial spin orbital functions in terms of the atomic basis set:

$$\boldsymbol{\varphi}^{\mathrm{o}} = \boldsymbol{\chi}\mathbf{C}^{\mathrm{o}} \tag{5.106}$$

we can construct the corresponding Hartree–Fock matrix ($\mathbf{h}^{\mathrm{HF}}(\mathbf{C}^{\mathrm{o}})$). This matrix may be expressed in terms of the trial spin orbital basis set instead of the atomic one:

$$\mathbf{h}^{\mathrm{HFo}}(\mathbf{C}^{\mathrm{o}}) = \mathbf{C}^{\mathrm{o}+}\mathbf{h}^{\mathrm{HF}}(\mathbf{C}^{\mathrm{o}})\mathbf{C}^{\mathrm{o}} \tag{5.107}$$

and partitioned between occupied (labelled 1) and virtual (labelled 2) spin orbitals:

$$\mathbf{h}^{\mathrm{HFo}}(\mathbf{C}^{\mathrm{o}}) = \begin{pmatrix} \mathbf{h}^{\mathrm{o}}_{11} & \mathbf{h}^{\mathrm{o}}_{12} \\ \mathbf{h}^{\mathrm{o}}_{21} & \mathbf{h}^{\mathrm{o}}_{22} \end{pmatrix}$$

This partitioning enables us to define the pseudocanonical spin orbitals (PCSO) ($\boldsymbol{\varphi}^{\mathrm{p}}$); if $\mathbf{Q}_1$ and $\mathbf{Q}_2$ are unitary transformations which diagonalize $\mathbf{h}^{\mathrm{o}}_{11}$ and $\mathbf{h}^{0}_{22}$ respectively (with $\boldsymbol{\varepsilon}_1$ and $\boldsymbol{\varepsilon}_2$ as eigenvalues), then:

$$\boldsymbol{\varphi}^{\mathrm{p}} = \boldsymbol{\varphi}^{\mathrm{o}} \begin{pmatrix} \mathbf{Q}_1 & \mathbf{o} \\ \mathbf{o} & \mathbf{Q}_2 \end{pmatrix} = \boldsymbol{\varphi}^{\mathrm{o}}\mathbf{Q}$$

The corresponding Hartree–Fock matrix becomes:

$$\mathbf{h}^{\mathrm{HFp}}(\mathbf{C}^{\mathrm{o}}) = \mathbf{Q}'\mathbf{h}^{\mathrm{o}}(\mathbf{C}^{\mathrm{o}})\mathbf{Q} = \begin{pmatrix} \boldsymbol{\varepsilon}_1 & \mathbf{A}^+ \\ \mathbf{A} & \boldsymbol{\varepsilon}_2 \end{pmatrix} = \boldsymbol{\varepsilon} + \begin{pmatrix} \mathbf{o} & \mathbf{A}^+ \\ \mathbf{A} & \mathbf{o} \end{pmatrix} \tag{5.108}$$

where:

$$\mathbf{A} = \mathbf{Q}_2^+\mathbf{h}^{\mathrm{o}}_{21}\mathbf{Q}_1$$

Let us now apply a unitary transformation $\mathbf{V}(\lambda)$ to the PCSO. The elements of this matrix are supposed to vary with a dimensionless path parameter λ. The new molecular spin orbital set will be:

$$\boldsymbol{\varphi}(\lambda) = \boldsymbol{\varphi}^{\mathrm{p}}\mathbf{V}(\lambda) \tag{5.109}$$

and the associated LCAO coefficients are given by:

$$\mathbf{C}(\lambda) = \mathbf{C}^{\mathrm{o}}\mathbf{Q}\mathbf{V}(\lambda) \tag{5.110}$$

These may be used further to solve the pseudoeigenvalue equation (5.94) of the current iteration step. It only remains to select an appropriate $\mathbf{V}(\lambda)$ transformation matrix. It is generated by diagonalizing a general Hartree–Fock matrix $\mathbf{h}^{\mathrm{HF}\lambda}$:

$$\mathbf{h}^{\mathrm{HF}\lambda} = \boldsymbol{\varepsilon} + \lambda\begin{pmatrix} \mathbf{0} & \mathbf{J}^+ \\ \mathbf{J} & \mathbf{0} \end{pmatrix} \tag{5.111}$$

The corresponding initial derivative of the electronic energy with respect to λ is given by:

$$\left(\frac{\mathrm{d}\varepsilon(\lambda)}{\mathrm{d}\lambda}\right)_{\lambda=0} = -\sum_{i}^{\mathrm{occ}}\sum_{k}^{\mathrm{vac}} \frac{J_{ki}A_{ki}^{+} + J_{ki}^{+}A_{ki}}{\varepsilon_k - \varepsilon_i} \tag{5.112}$$

Instead of $\mathbf{A}$ a more general choice is desirable for $\mathbf{J}$ which contains a weighting factor depending on the energy separation between virtual and occupied orbitals $(\varepsilon_k - \varepsilon_i)$ in order to avoid replacement of an occupied orbital by a virtual one:

$$J_{ki} = A_{ki}\left(\frac{\varepsilon_k - \varepsilon_i}{\Delta}\right)^q \tag{5.113}$$

where Δ insures that J_{ki} has the same dimension as A_{ki}. A mean energy difference may be used:

$$\Delta^q = \frac{\sum_{i}^{\mathrm{occ}}\sum_{k}^{\mathrm{virt}} (\varepsilon_k - \varepsilon_i)^q}{(m-n)n} \tag{5.114}$$

where q is a free real parameter that may be chosen to optimize the energy expectation value and m and n stand for the atomic basis set size and the electron number respectively. Experience shows that the best energy occurs for the q parameter close to zero. This only requires about 66 per cent. of the usual iteration number but the Hartree–Fock matrix must be evaluated twice at each iteration.

5.3.4 Extrapolation Procedure

When the iterative process goes well but has too small a rate of convergence, it is permissible to try an extrapolation procedure. This is an attempt to find a better approximation from results of a few successive iterations. Two extrapolation schemes[41] are commonly used which employ three sets of vectors or density matrices (labelled $i-2$, $i-1$ and i) coming from two successive iterations.

The first method is the Aitken method which assumes a geometric decrease of error. The extrapolated solution is given as follows:

$$D_1(\mathrm{ext}) = \frac{D_1(i)D_1(i-2) - D_1^2(i-1)}{D_1(i) - 2D_1(i-1) + D_1(i-2)} \tag{5.115}$$

where additions and multiplications are performed term by term.

In the second method, it is assumed that the end point of the density matrix describes a spiral in a plane centred around the SCF solution. Let us define the displacements of the density matrix between successive iterations by the following vectors (see Figure 5.10a):

$$\begin{aligned}\mathbf{V}_0 &= D_1(i) - D_1(i-1)\\ \mathbf{V}_1 &= D_1(i-1) - D_1(i-2)\\ \mathbf{V}_2 &= D_1(i-2) - D_1(i-3)\end{aligned} \tag{5.116}$$

An obvious precaution before extrapolating is to insure that the three vectors $\mathbf{V}_0$, $\mathbf{V}_1$ and $\mathbf{V}_2$ are nearly coplanar. According to Figure 5.10(b), the angle ψ between $\mathbf{V}_2$ and the plane $\mathbf{V}_0$, $\mathbf{V}_1$ is given by:

$$\cos\psi = \frac{|\mathbf{u}|}{|\mathbf{V}_2|} = \left[\frac{a^2(\mathbf{V}_0\cdot\mathbf{V}_0) + b^2(\mathbf{V}_1\cdot\mathbf{V}_1) + 2ab(\mathbf{V}_0\cdot\mathbf{V}_1)}{(\mathbf{V}_2\cdot\mathbf{V}_2)}\right]^{1/2} \tag{5.117}$$

where:

$$a = \frac{(\mathbf{V}_2\cdot\mathbf{V}_0)(\mathbf{V}_1\cdot\mathbf{V}_1) - (\mathbf{V}_0\cdot\mathbf{V}_1)(\mathbf{V}_1\cdot\mathbf{V}_2)}{(\mathbf{V}_0\cdot\mathbf{V}_0)(\mathbf{V}_1\cdot\mathbf{V}_1) - (\mathbf{V}_0\cdot\mathbf{V}_1)^2}$$

$$b = \frac{(\mathbf{V}_2\cdot\mathbf{V}_1)(\mathbf{V}_0\cdot\mathbf{V}_0) - (\mathbf{V}_0\cdot\mathbf{V}_1)(\mathbf{V}_0\cdot\mathbf{V}_2)}{(\mathbf{V}_0\cdot\mathbf{V}_0)(\mathbf{V}_1\cdot\mathbf{V}_1) - (\mathbf{V}_0\cdot\mathbf{V}_1)^2}$$

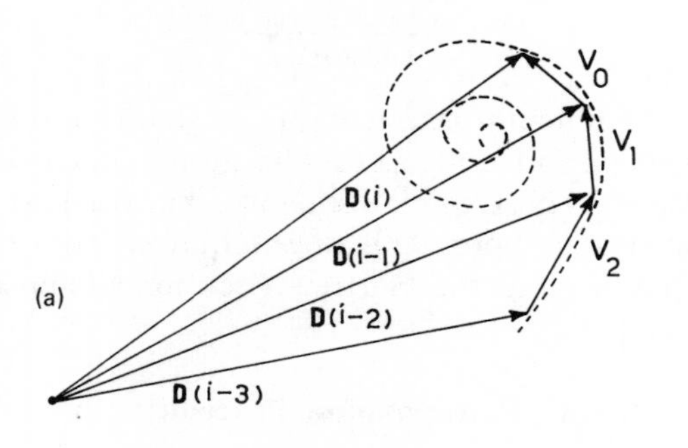

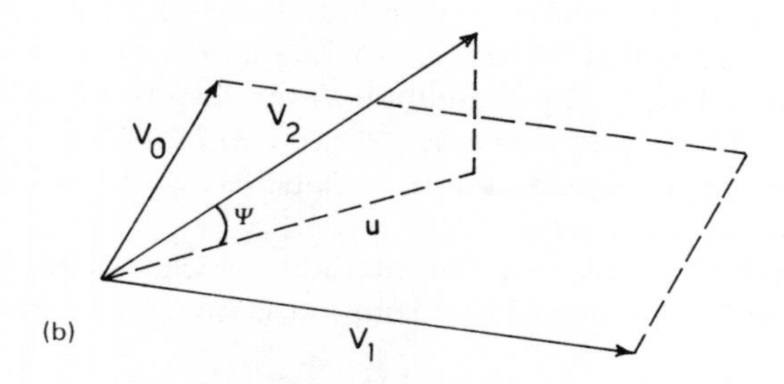

Figure 5.10 Spiralled convergence on the density matrix (a) in density matrix space and (b) in density matrix displacement space

We usually request that $\cos\psi$ be greater than 0.99 for extrapolation. In such a case the extrapolated solution is:

$$D_1(\text{ext}) = \frac{\alpha D_1(i) + \beta D_1(i-1) + \gamma D_1(i-2)}{\alpha + \beta + \gamma} \tag{5.118}$$

where:

$$\alpha = \mathbf{V}_1 \cdot \mathbf{V}_1$$
$$\beta = -2\mathbf{V}_0 \cdot \mathbf{V}_1$$
$$\gamma = \mathbf{V}_0 \cdot \mathbf{V}_0$$

This last equation may be transformed into a four-point extrapolation if we replace the α, β and γ parameters by other ones defined as follows:

$$\alpha = a, \qquad \beta = b \qquad \text{and} \qquad \gamma = -1$$

where a and b are the same as in (5.117).

Although these extrapolation schemes may be useful they nevertheless fail when we are too far from the exact solution.

5.3.5 Corresponding Orbital Transformation

For computations involving a minimal basis set, the extended Hückel method often provides a sufficiently accurate set of trial vectors. In such cases, we expect to have no troubles with the convergence process. This is quite different when we use larger basis sets such as split valence, double zeta or polarized basis sets. To bypass this problem the idea has been given[42] to express the wave function first obtained using a minimal basis set in the larger atomic function set we want to employ. Suppose the wave function for the chemical system of interest is known for a small basis ($\boldsymbol{\chi}^s$), the associated occupied molecular orbitals ($\boldsymbol{\varphi}_1^s$) being:

$$\boldsymbol{\varphi}_1^s = \boldsymbol{\chi}^s \cdot \mathbf{C}_1^s \tag{5.119}$$

How can we generate a new set of orbitals ($\boldsymbol{\varphi}_1^l$) in a larger atomic basis set ($\boldsymbol{\chi}^l$)? It can be obtained simply by maximizing the overlap between both sets:

$$\boldsymbol{\varphi}_1^l = \boldsymbol{\chi}^l \mathbf{C}_1^l \tag{5.120}$$

such that:

$$\langle \boldsymbol{\varphi}_1^s \mid \boldsymbol{\varphi}_1^l \rangle$$

is as large as possible.

The orthonormal basis set corresponding to $\boldsymbol{\chi}^l$ is:

$$\boldsymbol{\chi}^{l\perp} = \boldsymbol{\chi}^l (\mathbf{V}^l)^+$$

Its overlap with the occupied molecular orbital set is:

$$\begin{aligned} \mathbf{P} &= \langle \boldsymbol{\varphi}_1^s \mid \boldsymbol{\chi}^{l\perp} \rangle \\ &= (\mathbf{C}_1^s)^+ \langle \boldsymbol{\chi}^s \mid \boldsymbol{\chi}^l \rangle (\mathbf{V}^l)^+ \end{aligned} \tag{5.121}$$

Let $\mathbf{U}$ consist of orthonormal eigenvectors of the hermitian matrix $\mathbf{P}^{+}\mathbf{P}$:

$$(\mathbf{P}^{+}\mathbf{P})\mathbf{U} = \mathbf{U}\boldsymbol{\lambda} \qquad (5.122)$$

The eigenvectors which correspond to zero λ eigenvalues belong to the space spanned by the virtual orbitals. The others belong to the space spanned by the occupied orbitals. Let $\mathbf{U}_1$ be the part of $\mathbf{U}$ which contains the last eigenvectors. This matrix is nothing more than the matrix of LCAO coefficients we want to obtain expressed in the orthonormal basis set $(\boldsymbol{\chi}^{\text{LL}})$. Then we find:

$$\mathbf{C}_1^{\text{l}} = (\boldsymbol{V}^{\text{l}})^{+}\mathbf{U}_1 \qquad (5.123)$$

This achieves the evaluation of the expression (5.120).

5.3.6 Survey of the SCF Logic

In Figure 5.11 we have reported the flow chart of a SCF program. First of all the Löwdin orthogonalization matrix ($\boldsymbol{V}$) is computed. Then a set of trial vectors must be generated. Either we perform a Hückel-type calculation building and diagonalizing an approximate Hartree–Fock matrix (sometimes this matrix is replaced by the core hamiltonian one, but results are poor), or we search the corresponding orbitals from the wave function obtained in another calculation (i.e. using minimal orbitals), or we load trial vectors corresponding, for example, to molecular orbitals of a previous computation performed with a slightly different geometry. We then decide between the closed- and open-shell type of calculation according to the occupation vectors $\boldsymbol{\eta}_\alpha$ and $\boldsymbol{\eta}_\beta$. The density matrices are computed and the SCF procedure is started.

After evaluation of the Hartree–Fock matrix we can, if requested, perform an univariate search in the pseudocanonical spin orbital frame in order to improve the current density matrices. The iteration may be achieved and the total energy computed. At the end of each iteration we verify whether the convergence threshold is reached. In this way we can compare the total energy of the current iteration with the previous one. This is useful mainly to avoid the propagation of a divergent solution. Decision concerning the convergence itself is generally set up by comparing the two last density matrices:

$$\frac{\sum\limits_{p}\sum\limits_{q\leq p}[D_{pq}(i) - D_{pq}(i-1)]^2}{m(m-1)/2} = \text{r.m.s.} \leq \varepsilon^2 \qquad (5.124)$$

Let us note that the coefficient matrices may be used instead of the density matrices. In such a case we must first insure that both sets of vectors in use—$C(i-1)$ and $C(i)$—have the same relative phase. The relative sign of the vectors obtained at successive iterations may be at random. This nevertheless does not affect the wave function or the density matrices. So the residual mean square (r.m.s.) must be lower than a given threshold (ε^2).

This criterion insures the stationarity of the energy as well as the stationarity of the density matrix (this point is important if we are interested in evaluating other molecular properties). If the inequality (5.124) is not satisfied we must continue the iterative procedure. Before starting a new step it is usual to verify whether the maximum iteration number is reached or not. Finally, at this point we are able to perform an extrapolation as long as we are able to dispose of enough density matrices (from the last extrapolation) and as long as the convergent process is running properly.

5.3.7 SCF Orbitals and Their Localizability

Now that we have resolved the SCF process, molecular orbitals are at our disposal. These orbitals, solution of equation (5.94), are known as the canonical orbitals. Nevertheless one should recall that the general form of the Hartree–Fock equations is given by (5.14), i.e.:

$$h^{\mathrm{HF}}\phi_i = \sum_j \varepsilon_{ij}\phi_j$$

or

$$h^{\mathrm{HF}}\boldsymbol{\phi} = \boldsymbol{\phi}\boldsymbol{\varepsilon} \tag{5.125}$$

where $\boldsymbol{\varepsilon}$ is the matrix of the lagrangian multipliers introduced to ensure the orthogonality during the extremization process.

Let us define a unitary transformation matrix $\mathbf{T}$ in such a way that:

$$\boldsymbol{\phi} = \boldsymbol{\theta}\mathbf{T} \tag{5.126}$$

Then:

$$h^{\mathrm{HF}}\boldsymbol{\theta} = \boldsymbol{\theta}\mathbf{T}\boldsymbol{\varepsilon}\mathbf{T}^{+} = \boldsymbol{\theta}\boldsymbol{\varepsilon}' \tag{5.127}$$

There thus exists an infinity of solutions differing only by a unitary transformation. If we have n orbitals to determine, there are n^2 lagrangian multipliers. The orthonormality condition introduces $n(n+1)/2$ constraints, leaving $n(n-1)/2$ arbitrary values. If we put these remaining parameters to zero, we suppress the off-diagonal elements and have uniquely defined the system of equations in its canonical form (4.17); this leads to the canonical orbital representation of particular interest as its solution is obtained through the well-known diagonalization algorithms. It is interesting now to consider how orbitals, localized on the chemical bonds, lone pairs and atomic cores rather than on the delocalized canonical orbitals, can be obtained, as the former representation is closer to chemical intuition. A good choice of the remaining lagrangian multipliers should lead to those orbitals. Unfortunately such an *a priori* choice is hardly possible. The only way out is to go back to relation (5.127) and try to find an adequate $\mathbf{T}$ transformation matrix to apply on previously obtained canonical orbitals. Localized orbitals $\boldsymbol{\theta}$ can be obtained by the relation:

$$\boldsymbol{\theta} = \boldsymbol{\phi}\mathbf{T} \tag{5.128}$$

but $\mathbf{T}$ must be evaluated with respect to some physical criterion.

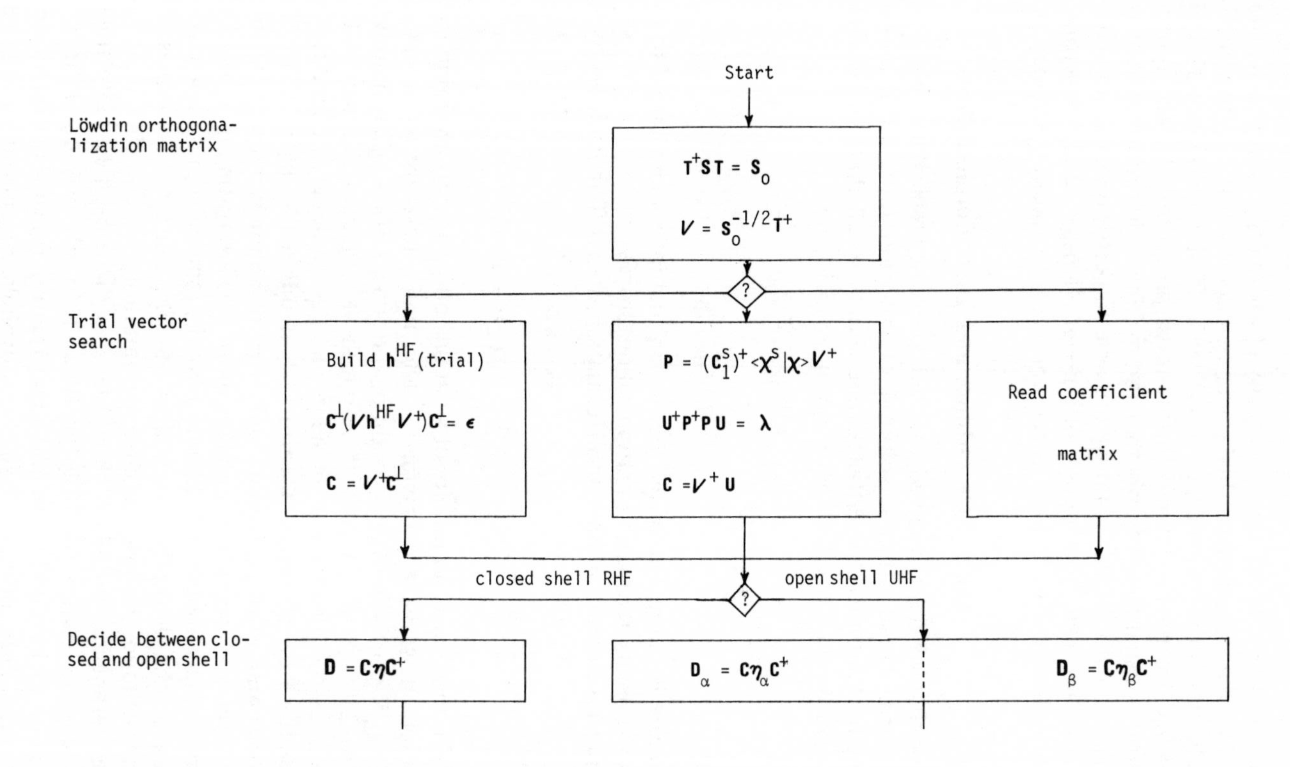

Start
Löwdin orthogona-
lization matrix
$\mathbf{T}^+\mathbf{S}\mathbf{T} = \mathbf{S}_0$
$V = \mathbf{S}_0^{-1/2}\mathbf{T}^+$
?
Trial vector
search
Build $\mathbf{h}^{HF}$(trial)
$\mathbf{C}^{\perp}(V\mathbf{h}^{HF}V^+)\mathbf{C}^{\perp} = \epsilon$
$\mathbf{C} = V^+\mathbf{C}^{\perp}$
$\mathbf{P} = (\mathbf{C}_1^S)^+ \langle \chi^S | \chi \rangle V^+$
$\mathbf{U}^+\mathbf{P}^+\mathbf{P}\mathbf{U} = \lambda$
$\mathbf{C} = V^+\mathbf{U}$
Read coefficient
matrix
closed shell RHF
?
open shell UHF
Decide between clo-
sed and open shell
$\mathbf{D} = \mathbf{C}\eta\mathbf{C}^+$
$\mathbf{D}_\alpha = \mathbf{C}\eta_\alpha\mathbf{C}^+$
$\mathbf{D}_\beta = \mathbf{C}\eta_\beta\mathbf{C}^+$

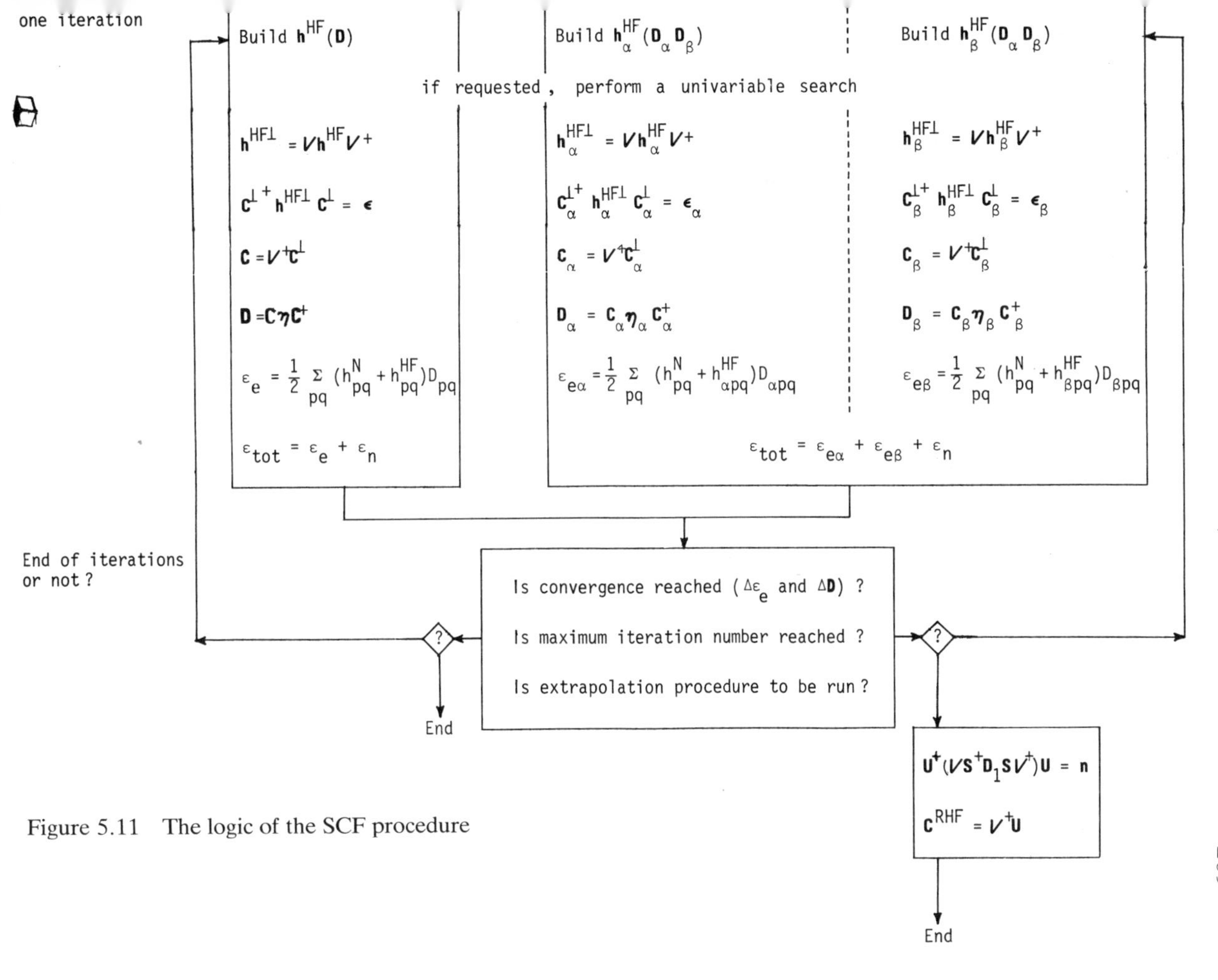

Figure 5.11 The logic of the SCF procedure

5.3.8 Localization Criteria

These criteria are obtained due to the invariance of physical properties against unitary transformations of molecular orbitals. This is easy to show, as any monoelectronic property is obtained in the orbital frame by:

$$\langle G \rangle = \sum_i \langle \varphi_i | G_{op} | \varphi_i \rangle$$

If:

$$\theta_i = \sum_j T_{ij} \varphi_j$$

Then:

$$\sum_i \langle \theta_i | G_{op} | \theta_i \rangle = \sum_k \sum_l \sum_i T^*_{ik} T_{il} \langle \varphi_k | G_{op} | \varphi_l \rangle$$

As **T** is a unitary transformation:

$$\sum_i T^*_{ik} T_{ip} = \delta_{kl}$$

and

$$\begin{aligned} \sum_i \langle \theta_i | G_{op} | \theta_i \rangle &= \sum_k \sum_l \delta_{kl} \langle \varphi_k | G_{op} | \varphi_l \rangle \\ &= \sum_l \langle \varphi_l | G_{op} | \varphi_l \rangle \\ &= \langle G \rangle \end{aligned}$$

This demonstration is easily extended to bielectronic properties and thus to energetic ones (see Section 4.2). It must be noticed that although globally the property stays invariant, its orbital components are subject to variations depending on the nature of **T**. For example, the total electron density, i.e. the sum of the orbital densities, is invariant, but these orbital densities may change considerably according to the type of orbital used. This variation will allow us to define some localization criteria. One distinguishes two types of localization criteria:

(a) The internal criteria, which autodefine the orbitals in the sense that the user does not need to predefine the nature of the orbital he wants to obtain.
(b) The external criteria, which need intervention of the user in the sense that the user defines a correspondence between the orbitals and the regions of space where he wants to localize them (i.e. mainly couples of bonded atoms).

We shall now summarize some of these criteria, differing by their localization function but leading to very similar results.

5.3.8.1 The Edminston–Ruedenberg internal criterion[43]

A good localization of the orbitals also implies an increase in the repulsion between the two electrons associated with those orbitals, as the electrons should 'come closer to each other'. We may thus define a localization function J which must be maximized:

$$J = \sum_i J_{ii} \tag{5.129}$$

The invariant property is here the bielectronic energy:

$$E_2 = \sum_i J_{ii} + \sum_i \sum_{j \neq i} (2J_{ij} - K_{ij}) \tag{5.130}$$

As the first sum must be maximized, the second will be minimized, i.e. we try to obtain the best energetical separation between orbitals. The most important problem related to this criterion is the use of many bielectronic integrals expressed over the molecular basis set, which is quite time-consuming (see Section 5.2.4).

5.3.8.2 The von Niessen internal criterion[44]

Following an idea of Ruedenberg stating that good criteria should be of a bielectronic nature, von Niessen proposed replacing the r_{12}^{-1} operator used in the Edminston–Ruedenberg procedure by a δ function. This leads to the definition of charge overlap integrals $[i^2j^2]$:

$$[i^2j^2] = \int \theta_i^2(x)\theta_j^2(x)\,\mathrm{d}x \tag{5.131}$$

A good localization will be attained by minimizing the charge overlap between pairs, or in other words by maximizing the charge overlap of an orbital with itself. The invariant property is here the electronic density, or, better, its square. Indeed:

$$\rho^2(x) = \sum_i \sum_j \theta_i^2(x)\theta_j^2(x) \tag{5.132}$$

$$\int \rho^2(x)\,\mathrm{d}x = \sum_i \sum_j [i^2j^2] = \sum_i [i^2i^2] + \sum_i \sum_{j \neq i} [i^2j^2] \tag{5.133}$$

The second sum constitutes the localization function which must be minimized. Subsequently the first sum must be maximized. This method introduces a new kind of integral which has to be evaluated.

5.3.8.3 Boys internal criterion[45]

Originally, in this method, the invariance of the barycentre of negative charges is used to find a localization function. Thus:

$$\langle G_e \rangle = \sum_i \langle \theta_i | \, r \, | \theta_i \rangle \tag{5.134}$$

where $\langle\theta_i|\,r\,|\theta_i\rangle$ are components of the electronic dipole moment and are called the 'centroids of charge'. The first idea of Boys was to separate as much as possible those centroids of charge. Later on, following Ruedenberg's idea, he introduced the bielectronic operator r_{12}^2 and suggested the minimization of:

$$L = \sum_i \langle\theta_i^2|\,r_{12}^2\,|\theta_i^2\rangle \tag{5.135}$$

As:

$$\bar{r}_{12}^2 = \bar{r}_1^2 + \bar{r}_2^2 - 2\bar{r}_1\bar{r}_2$$

one finds that:

$$L = \sum_i 2(\langle\theta_i|\,r^2\,|\theta_i\rangle - \langle\theta_i|\,r\,|\theta_i\rangle^2) \tag{5.136}$$

As the first sum is invariant, minimization of L implies maximization of the second sum which refers to the centroids of charge previously defined.

5.3.8.4 The Magnasco–Perico external criterion[46]

This criterion will be the only external one analysed in this book. Like all external criteria, it rests on the LCAO approximation and more precisely on the invariance of the Mulliken population analysis. Indeed:

$$\sum_i \sum_p \sum_q 2C_{ip}C_{iq}S_{pq} = 2n \tag{5.137}$$

or

$$\sum_i \sum_A \sum_B P^i_{AB} = 2n \tag{5.138}$$

where:

$$P^i_{AB} = \sum_p^{\in A} \sum_q^{\in B} 2C_{ip}C_{iq}S_{pq} \tag{5.139}$$

The total contribution is invariant but P^i_{AB}, and any subset of it, is subject to variations. First, and following our chemical intuition, let us define some molecular subspaces corresponding to bonds, lone pairs and atomic cores and associate to those subspaces a set $\boldsymbol{\gamma}$ of concerned atomic orbitals (i.e. 1s atomic orbitals for cores, valence atomic orbitals of a pair of bonded atoms, valence atomic orbitals for a lone pair). To each subspace we shall associate a molecular orbital and to each set $\boldsymbol{\gamma}$ a 'local population' P_i such that:

$$P_i = \sum_p^{\in\gamma} \sum_q^{\in\gamma} 2C_{ip}C_{iq}S_{pq} \tag{5.140}$$

The localization function is simply:

$$P = \sum_i P_i \tag{5.141}$$

and must be maximized, or in other words as many electrons as possible must be associated to the local population of a molecular orbital.

5.3.8.5 The localization process

It still remains to explain how, starting from canonical orbitals having chosen a localization procedure, we can obtain the transformation matrix **T**. As this matrix has the dimension of the molecular orbital set, it may be quite important, and we shall try to put it in the simplest form. The simplest way to treat the problem is to transform only two orbitals and make them satisfy our criterion. Afterwards we can repeat this transformation over all possible couples of orbitals until the localization function is extremized. Let us take two orbitals a and b out of the set of n orbitals $\{\varphi\}$ and apply to these a rotation by an angle α defined by the chosen criterion:

$$\{\varphi\} = (\varphi_1 \varphi_2 \cdots \varphi_a \cdots \varphi_b \cdots \varphi_n)$$

and

$$\begin{aligned} \theta_a &= \varphi_a \cos\alpha + \varphi_b \sin\alpha \\ \theta_b &= -\varphi_a \sin\alpha + \varphi_b \cos\alpha \end{aligned} \tag{5.142}$$

After this elementary process we have:

$$(\varphi_1 \cdots \varphi_2 \cdots \theta_a \cdots \theta_b \cdots \varphi_n) = (\varphi_1 \cdots \varphi_a \cdots \varphi_b \cdots \varphi_n)$$

$$\times \begin{bmatrix} 1 & & & & & & \\ & 1 & & & 0 & & \\ & & 1 & & & & \\ a \cdots\cdots & & (\cos\alpha) & \cdots & (\sin\alpha) & & \\ & 0 & & 1 & & & \\ b \cdots\cdots & & (\sin\alpha) & \cdots & (\cos\alpha) & & \\ & & & & & 1 & \\ & & & & & & 1 \end{bmatrix}$$

This procedure is repeated iteratively until stabilization of the localization function. The final transformation matrix is thus the product of all the elementary transformations realized before convergence. We shall briefly illustrate the obtention of the rotation angle, taking the Boys procedure as an example:

$$\begin{aligned} L_{\text{init}} &= \langle \varphi_a | \, \boldsymbol{r} \, | \varphi_a \rangle^2 + \langle \varphi_b | \, \boldsymbol{r} \, | \varphi_b \rangle^2 + \text{constant} \\ L_{\text{fin}} &= \langle \theta_a | \, \boldsymbol{r} \, | \theta_a \rangle^2 + \langle \theta_b | \, \boldsymbol{r} \, | \theta_b \rangle^2 + \text{constant} \\ &= \langle \cos\alpha\varphi_a + \sin\alpha\varphi_b | \, \boldsymbol{r} \, | \cos\alpha\varphi_a + \sin\alpha\varphi_b \rangle^2 \\ &\quad + \langle -\sin\alpha\varphi_a + \cos\alpha\varphi_b | \, \boldsymbol{r} \, | -\sin\alpha\varphi_a + \cos\alpha\varphi_b \rangle^2 + \text{constant} \end{aligned}$$

Putting:

$$R_a = \langle \varphi_a | \, r \, | \varphi_a \rangle$$
$$R_b = \langle \varphi_b | \, r \, | \varphi_b \rangle$$
$$R_{ab} = \langle \varphi_a | \, r \, | \varphi_b \rangle$$

and using the trigonometric formulas, namely:

$$A \cos^2 \alpha + B \sin^2 \alpha = \tfrac{1}{2}[(A+B)+(A-B) \cos 2\alpha]$$

one finds:

$$L_{\text{fin}} = \left(\frac{R_a - R_b}{2} \cos 2\alpha + \frac{R_a + R_b}{2} + R_{ab} \sin 2\alpha \right)^2$$
$$+ \left(\frac{R_b - R_a}{2} \cos 2\alpha + \frac{R_a + R_b}{2} - R_{ab} \sin 2\alpha \right)^2$$
$$= [\tfrac{1}{4}(R_a - R_b)^2 - R_{ab}^2] \cos 4\alpha + [(R_a - R_b) R_{ab}] \sin 4\alpha + \text{constant}$$

The extremum conditions imply:

$$\frac{\partial L}{\partial \alpha} = 0 \qquad \text{or} \qquad \operatorname{cotg} 4\alpha = \frac{[\tfrac{1}{4}(R_a - R_b)^2 - R_{ab}^2]}{(R_a - R_b) R_{ab}} \tag{5.143}$$
$$\frac{\partial^2 L}{\partial \alpha^2} < 0$$

Combining these two relations, one finds the value of 4α and thus α leading to a maximum. Subsequently, θ_α, θ_b, $\mathbf{T}_{\text{elem}}$ are known.

It must be pointed out here that, after this elementary transformation, we are in a new molecular basis set and that the molecular integrals must be expressed in this new basis set before continuing the process. This procedure is repeated on all pairs of orbitals as long as the rotation angles are significantly different from zero. Finally we obtain:

$$\boldsymbol{\theta} = \boldsymbol{\varphi} \mathbf{T} \tag{5.144}$$

or in the LCAO approximation:

$$\mathbf{C}^{\text{OL}} = \mathbf{C}^{\text{OM}} \cdot \mathbf{T} \tag{5.145}$$

It is a matter of fact that we do not really need the transformation matrix at each step, but the new orbitals. Thus, after convergence we may obtain $\mathbf{T}$ at once by using the relation:

$$\mathbf{T} = \mathbf{C}^{\text{OM}} \mathbf{S} \mathbf{C}^{\text{OL}} \tag{5.146}$$

Knowing $\mathbf{T}$, we may now discuss any property in terms of localized orbitals, especially the Hartree–Fock matrix $\boldsymbol{\varepsilon}$ obtained from relation (5.127), or any other property G computed from the relation:

$$\mathbf{G}^{\text{OL}} = \mathbf{T} \mathbf{G}^{\text{OM}} \mathbf{T}^{+} \tag{5.147}$$

Some examples of localized properties will be presented and discussed later on in Section 6.2.2.

5.4 SOME PRACTICAL EXAMPLES

We intend now to give some simple illustrations of what has been presented until now. Some of these are simple enough to be done on a desk calculator; others need the aid of a computer.

We shall start with the hydrogen atom, treated at various levels of sophistication. Then we shall treat the hydrogen molecule with an elementary basis set first and with a more accurate one afterwards. Some properties will be examined. Finally, we shall present a detailed SCF process for the water molecule and briefly discuss the hydroxyl radical.

5.4.1 The Hydrogen Atom

The wave functions for the hydrogen atom are of course well known, its fundamental state being described by:

$$\psi = \frac{1}{\sqrt{\pi}} \exp(-r) \tag{5.148}$$

of energy -0.5 a.u. Now let us suppose for a while that we did not know the solution. We could approximate it by any spherical function. A normalized gaussian, for example, is:

$$\chi_1 = \frac{2\sqrt{2}}{\pi^{3/2}} \exp(-r^2)$$

The associated energy is given by:

$$E_1 = \langle \chi_1 | H | \chi_1 \rangle \tag{5.149}$$

As we have here a one-electron problem we need only to compute the kinetic and nuclear attraction integrals which were presented in Section 5.2 (Tables 5.3 and 5.5):

$$T = \tfrac{3}{2} \qquad \text{and} \qquad V = -2\sqrt{\left(\frac{2}{\pi}\right)}$$

$$E_1 = -0.09577 \text{ a.u.}$$

This energy is very bad, so we shall have to optimize our gaussian function by introducing a variational parameter. Let us define the function by:

$$\chi_1 = N_\alpha \exp(-\alpha r^2) \tag{5.150}$$

This allows the gaussian to vary its shape depending on the value of α. Referring to (5.149) and Section 5.2 (Tables 5.3 and 5.5), we find:

$$E_1 = \frac{3\alpha}{2} - 2\sqrt{\left(\frac{2\alpha}{\pi}\right)}$$

$$\chi_1 = \left(\frac{2\alpha}{\pi}\right)^{3/4} \exp(-\alpha r^2) \tag{5.151}$$

The best possible energy obtained with one gaussian will be obtained by:

$$\frac{\partial E_1}{\partial \alpha} = 0 = \tfrac{3}{2} - \frac{\sqrt{2}}{\sqrt{\pi}} * \alpha^{-1/2} \tag{5.152}$$

or

$$\alpha = 0.283$$

Then:

$$E_{1.\text{opt}} = -0.42441 \text{ a.u.}$$
$$\chi_{1.\text{opt}} = 0.2765 \exp(-0.283 r^2)$$

This improves our result considerably as can easily be understood by looking at Figure 5.12, which compares the fit obtained by the two χ_1 functions. The optimization improves the behaviour of the approximate function in the middle-range extension. Nevertheless, this behaviour is quite poor for the short range and long range. The best way to improve this approximate function further consists of adding new functions at places where it behaves poorly. We shall thus define:

$$\begin{aligned} \chi_2 &= C_1 N_\alpha \exp(-\alpha r^2) + C_2 N_\beta \exp(-\beta r^2) \\ &= C_1 G_\alpha + C_2 G_\beta \end{aligned} \tag{5.153}$$

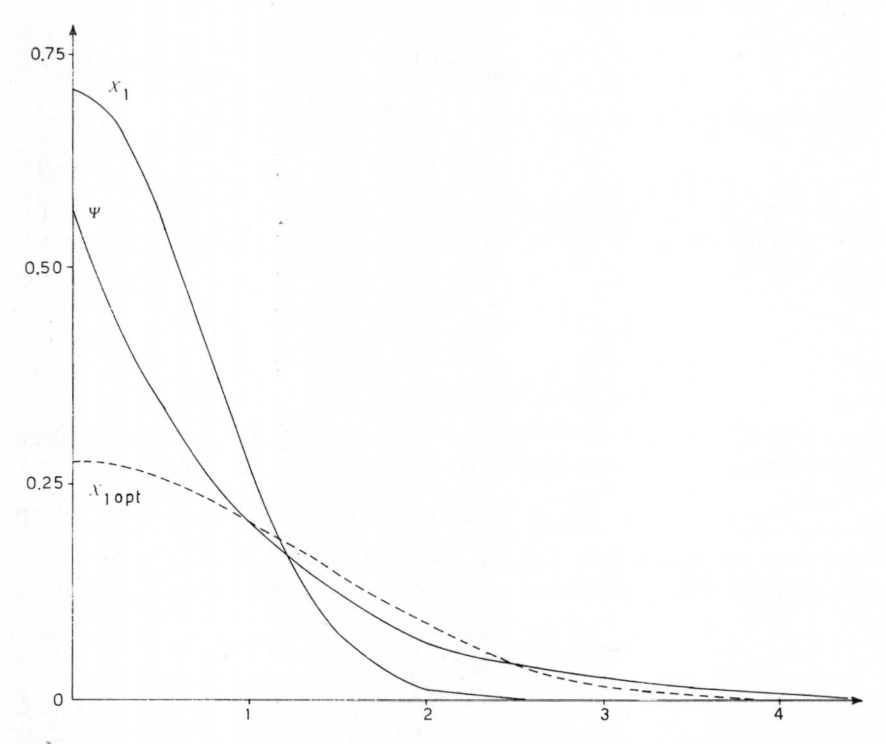

Figure 5.12 Comparison between ψ and a gaussian function

where C_1 and C_2 are the expansion coefficients, α and β being chosen. We shall improve the behaviour by adding a sharp gaussian G_α, the second one being not too different from $G_{\alpha_{opt}}$. Let us take 0.8 and 0.2. The associated energy may be found by the classical resolution of:

$$H\chi_2 = E_2\chi_2 \tag{5.154}$$

and the coefficients obtained by minimization of the energy. We then find as the result:

$$E_2 = -0.47540 \text{ a.u.}$$

which is already better than $E_{1.opt}$.

Of course our choice of α and β is arbitrary and we may try to improve it by searching for the couple α, β giving the lowest energy. This has already been done by Ditchfield, Hehre and Pople.[47] They found:

$$E_2 = -0.48581 \text{ a.u.}$$

with:

$$\alpha = 1.33248 \qquad \beta = 0.2015287$$
$$C_1 = 0.27441 \qquad C_2 = 0.821225$$

As Figure 5.12 showed a poor behaviour for short- and long-range behaviour, it is admitted that three gaussians in the expansion lead to a good compromise between sophistication and accuracy. We shall thus write:

$$\psi \simeq \chi_3 = C_1 G_\alpha + C_2 G_\beta + C_3 G_\gamma \tag{5.155}$$

The optimization of χ_3 may be performed in two ways:

(a) As previously, we search for the triplet (α, β, γ) giving the lowest energy.
(b) As ψ is known, we perform a least square fit on (5.155) and look for the triplet (α, β, γ) giving the best fit.

Both approaches have been realized.[9,47] Figure 5.13 reports the evolution of χ_i towards ψ while Table 5.9 summarizes the results obtained for the different approaches. The best fit of ψ gives coefficients C_i for the exponents α, β, γ, leading to an energy, but this is not the best energy for those coefficients. It is tempting to optimize the coefficients C_i holding the same exponents. This leads to a function which is only slightly better than the previous one, which confirms what has been discussed in Sections 5.1.3 and 5.1.4 about the utility of contracting the atomic basis sets.

5.4.2 The Hydrogen Molecule: A Simple Example

We perform the following calculation in the same sequence as the available gaussian-type programs.

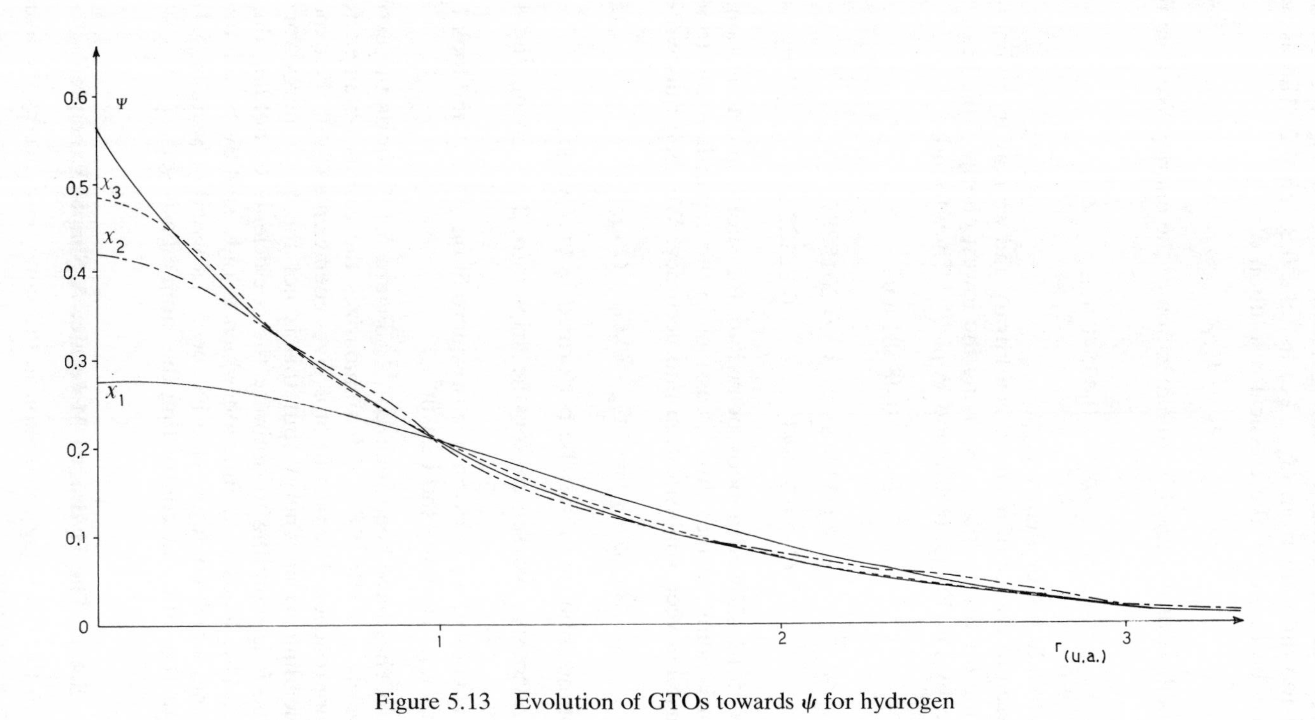

Figure 5.13 Evolution of GTOs towards ψ for hydrogen

Table 5.9 Approximate hydrogen functions

Function	Energy	$\{\alpha\}$	$\{C\}$
1. χ_1 (energy opt)	−0.42441	0.282942	1.0
2. χ_2 (energy opt)	−0.48581	1.332480	0.274408
		0.201529	0.821225
3. χ_3 (energy opt)	−0.49698	4.500225	0.070479
		0.681275	0.407889
		0.151375	0.647669
4. χ_3 (ψopt)	−0.49491	2.227660	0.154329
		0.405771	0.535328
		0.109818	0.444635
5. χ_3 (energy opt) with $\{\alpha\}$ of 4	−0.49501	2.227660	0.147251
		0.405771	0.532657
		0.109818	0.452325

5.4.2.1 The nuclear geometry

We have first to decide what internuclear distance to choose in order to perform the *ab initio* calculation (in the BO frame). Having no more information we take the usual d_{HH} experimental value of 0.725 Å (1.37 a.u.). We dispose the two atoms along the x cartesian axis as shown in Figure 5.14. The actual value does not necessarily correspond to the equilibrium structure but we can repeat the calculations for different interatomic distances in order to find the nuclear configuration corresponding to the lowest total energy.

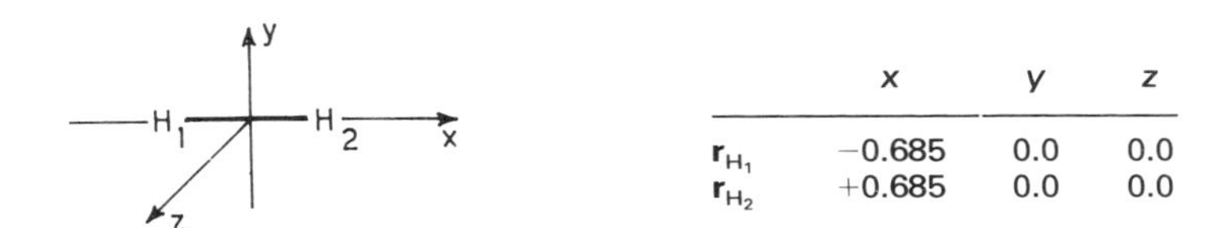

Figure 5.14 H_2 in the cartesian coordinate space

5.4.2.2 The atomic basis set

For computational convenience we decide to use a minimal basis set containing only one gaussian function of the s type. The choice of the best exponent is now not as simple as in the case of the hydrogen atom, so we shall use the gaussian exponent found for the hydrogen atom ($\alpha = 0.283$). This value may be improved by computing the total energy of the H_2 molecule for different values of α. To find the best energy of H_2 (with an atomic basis set which contains only one gaussian function of the s type) we solve a problem with two degrees of freedom, one corresponding to the H_2

interatomic distance and one corresponding to the gaussian exponent. In the present work we will choose the next basis set:

$$\chi_{H_1} = 0.2765 \exp[-0.283(\mathbf{r}-\mathbf{r}_{H_1})^2]$$
$$\chi_{H_2} = 0.2765 \exp[-0.283(\mathbf{r}-\mathbf{r}_{H_2})^2]$$

5.4.2.3 The atomic integrals

The integrals depend only on the nuclear geometry and on the atomic basis set. They can be computed only once before any other calculations:

The overlap integrals
From Table 5.2 we find:

$$S_{pq} = \langle \chi_{H_p} \mid \chi_{H_q} \rangle = N_p N_q K_{pq} \left(\frac{\pi}{\alpha_p + \alpha_q} \right)^{3/2} \tag{5.156}$$

where:

$$K_{pq} = \exp\left[-\frac{\alpha_p \alpha_q}{\alpha_p + \alpha_q} (\mathbf{r}_{H_p} - \mathbf{r}_{H_q})^2 \right]$$

By using matricial notation we write the results as:

$$\mathbf{S} = \begin{pmatrix} 1.0 & 0.76676 \\ 0.76676 & 1.0 \end{pmatrix}$$

The kinetic integrals
From Table 5.3 we have:

$$T_{pq} = \langle \chi_{H_p} | \, T \, | \chi_{H_q} \rangle$$
$$= N_p N_q \left(\frac{\pi}{\alpha_p + \alpha_q} \right)^{3/2} K_{pq} \frac{\alpha_p \alpha_q}{\alpha_p + \alpha_q} \left[3 - \frac{2\alpha_p \alpha_q}{\alpha_p + \alpha_q} (\mathbf{r}_{H_p} - \mathbf{r}_{H_q})^2 \right] \tag{5.157}$$

which gives:

$$\mathbf{T} = \begin{pmatrix} 0.424500 & 0.267859 \\ 0.267859 & 0.424500 \end{pmatrix}$$

The nuclear attraction integrals
From equation (5.65) and Table 5.5 we can write:

$$V_{pq} = \langle \chi_{H_p} | \, V \, | \chi_{H_q} \rangle = -\langle \chi_{H_p} | \frac{1}{r_{H_1}} | \chi_{H_q} \rangle - \langle \chi_{H_p} | \frac{1}{r_{H_2}} | \chi_{H_q} \rangle$$
$$= -N_p N_q (V_{1,pq} + V_{2,pq}) \tag{5.158}$$

where:

$$V_{k,pq} = \frac{2K_{pq}\pi}{\alpha_p + \alpha_q} F_0([\alpha_p + \alpha_q]\overline{PC}_k^2)$$

if

$$\mathbf{PC}_k = \frac{\alpha_p \mathbf{r}_{H_p} + \alpha_q \mathbf{r}_{H_q}}{\alpha_p + \alpha_q} - \mathbf{r}_{H_k}$$

The above equations have the solution:

$$V_{1,11} = 11.101034 F_0(0.0) = 11.101034$$
$$V_{2,11} = 11.101034 F_0(1.062325) = 8.161558$$
$$V_{1,12} = V_{2,12} = 8.511830 F_0(0.265581) = 7.814728$$

The function F_0 has been evaluated according to the expression:

$$F_0(W) = \sum_{i=0}^{k} \frac{(-W)^i}{i!\,(2i+1)}$$

the k value being selected according to the requested precision of 10^{-5} (see Table 5.10).

Table 5.10 The $F_0(W)$ function for some values of W

	$(-W)^i/[i!\,(2i+1)]$			
i	$W = 1.062325$	$W = 0.531163$	$W = 0.265581$	$W = 0.132791$
0	1.0	1.0	1.0	1.0
1	−0.354109	−0.177054	−0.088527	−0.044264
2	0.112854	0.028213	0.007053	0.001763
3	−0.028545	−0.003568	−0.000446	−0.000056
4	0.005896	0.000369	0.000023	0.000001
5	−0.001025	−0.000032	−0.000001	
6	0.000154	0.000002		
7	−0.000020			
8	0.000002			
$F_0(W)$	0.735207	0.847930	0.918102	0.957445

The two-electron repulsion integrals

From equation (5.8) we have:

$$\langle pq \mid rs \rangle = \frac{2N_pN_qN_rN_sK_{pq}K_{rs}\pi^{5/2}}{(\alpha_p + \alpha_q)(\alpha_r + \alpha_s)\sqrt{(\alpha_p + \alpha_q + \alpha_r + \alpha_s)}} F_0\left(\frac{(\alpha_p + \alpha_q)(\alpha_r + \alpha_s)}{\alpha_p + \alpha_q + \alpha_r + \alpha_s} \overline{\mathrm{PQ}}^2\right) \tag{5.159}$$

where:

$$\mathbf{P} = \frac{\alpha_p \mathbf{r}_{H_p} + \alpha_q \mathbf{r}_{H_q}}{\alpha_p + \alpha_q}$$

$$\mathbf{Q} = \frac{\alpha_r \mathbf{r}_{H_r} + \alpha_s r_{H_s}}{\alpha_r + \alpha_s}$$

$$\mathbf{PQ} = \mathbf{P} - \mathbf{Q}$$

We obtain from this expression and according to Table 5.10:

$$\langle 11 \mid 11\rangle = 0.600272 F_0(0.0) = 0.600272$$
$$\langle 11 \mid 22\rangle = 0.600272 F_0(0.531163) = 0.508989$$
$$\langle 12 \mid 12\rangle = 0.352913 F_0(0.0) = 0.352913$$
$$\langle 11 \mid 12\rangle = 0.460265 F_0(0.132791) = 0.440678$$

and by symmetry:

$$\langle 22 \mid 22\rangle = \langle 11 \mid 11\rangle$$
$$\langle 21 \mid 12\rangle = \langle 21 \mid 21\rangle = \langle 12 \mid 21\rangle = \langle 12 \mid 12\rangle$$
$$\langle 11 \mid 21\rangle = \langle 12 \mid 11\rangle = \langle 21 \mid 11\rangle = \langle 11 \mid 12\rangle$$
$$= \langle 21 \mid 22\rangle = \langle 22 \mid 21\rangle = \langle 22 \mid 12\rangle = \langle 12 \mid 22\rangle$$

The first moment integrals

If we want to compute the dipole moment, we have to evaluate the next integrals (see Table 5.4):

$$M_{pq}(x) = N_p N_q K_{pq} \left(\frac{\pi}{\alpha_p + \alpha_q}\right)^{3/2} P_x \tag{5.160}$$

where:

$$P_x = \frac{\alpha_p x_{H_p} + \alpha_q x_{H_q}}{\alpha_p + \alpha_q}$$

Then we find:

$$\mathbf{M}(x) = \begin{pmatrix} -0.685 & 0.0 \\ 0.0 & +0.685 \end{pmatrix}$$

Similiarly we further find that $\mathbf{M}(y) = \mathbf{M}(z) = \mathbf{0}$.

The second moment integrals

The quadrupole moment depends on the following integrals available from (5.61):

$$M_{pq}(x^2) = N_p N_q K_{pq} \left(\frac{\pi}{\alpha_p + \alpha_q}\right)^{3/2} \left[\frac{1}{2(\alpha_p + \alpha_q)} + P_x^2\right] \tag{5.161a}$$

$$M_{pq}(xy) = N_p N_q K_{pq} \left(\frac{\pi}{\alpha_p + \alpha_q}\right)^{3/2} P_x P_y \tag{5.161b}$$

$$\mathbf{M}(x^2) = \begin{pmatrix} 1.352617 & 0.677350 \\ 0.677350 & 1.352617 \end{pmatrix}$$

$$\mathbf{M}(y^2) = \mathbf{M}(z^2) = \begin{pmatrix} 0.883392 & 0.677350 \\ 0.677350 & 0.883392 \end{pmatrix}$$

$$\mathbf{M}(xy) = \mathbf{M}(xz) = \mathbf{M}(yz) = \mathbf{0}$$

5.4.2.4 The Löwdin orthogonalization process

We want to pass from the current atomic basis set $(\chi_1\chi_2)$ which has an overlap matrix of the form:

$$\mathbf{S}=\begin{pmatrix}1 & s\\ s & 1\end{pmatrix}$$

to an orthogonal molecular basis set related to the old one by equation (5.98):

$$\boldsymbol{\chi}^{\perp}=(\chi_1^{\perp}\chi_2^{\perp})=\boldsymbol{\chi}\mathbf{V}^{+}=(\chi_1\chi_2)\begin{pmatrix}v_{11} & v_{21}\\ v_{12} & v_{22}\end{pmatrix}$$

According to Löwdin's orthogonalization procedure we write the eigenvectors ($\mathbf{t}$) and the eigenvalues ($\mathbf{S}_0$) of the overlap matrix (see equation 5.97) as follows:

$$\mathbf{t}=\frac{1}{\sqrt{2}}\begin{pmatrix}1 & 1\\ 1 & -1\end{pmatrix} \tag{5.162a}$$

$$\mathbf{S}_0=\begin{pmatrix}1+s & 0\\ 0 & 1-s\end{pmatrix}=\begin{pmatrix}1.76676 & 0.0\\ 0.0 & 0.23324\end{pmatrix} \tag{5.162b}$$

The $\boldsymbol{V}$ orthogonalization matrix is then of the form:

$$\mathbf{V}=\mathbf{S}_0^{-1/2}\mathbf{t}^{+}=\begin{pmatrix}\sqrt{[1/2(1+s)]} & \sqrt{[1/2(1+s)]}\\ \sqrt{[1/2(1-s)]} & -\sqrt{[1/2(1-s)]}\end{pmatrix}=\begin{pmatrix}0.53198 & 0.53198\\ 1.46414 & -1.46414\end{pmatrix} \tag{5.162c}$$

The new orthogonal basis set corresponds to $\boldsymbol{\chi}^{\perp}=\boldsymbol{\chi}\mathbf{V}^{+}$. Then:

$$\chi_1^{\perp}=0.53198(\chi_1+\chi_2)$$

and

$$\chi_2^{\perp}=1.46414(\chi_1-\chi_2) \tag{5.162d}$$

5.4.2.5 The trial vectors for the SCF process

Before starting the SCF process we want to dispose of trial vectors. They determine approximate molecular orbitals which are introduced in the iterative process:

$$\boldsymbol{\varphi}=(\varphi_1\varphi_2)=\boldsymbol{\chi}\mathbf{C}=(\chi_1\chi_2)\begin{pmatrix}C_{11} & C_{12}\\ C_{21} & C_{22}\end{pmatrix}$$

We know that without electron repulsion the problem could be solved by diagonalizating the $\mathbf{T}+\mathbf{V}$ matrix instead of the Hartree–Fock matrix:

$$\mathbf{h}^{\mathrm{HF}}(\text{trial})=\mathbf{h}^{\mathrm{N}}=\mathbf{T}+\mathbf{V}=\begin{pmatrix}-1.0485 & -0.9273\\ -0.9273 & -1.0485\end{pmatrix}$$

This matrix becomes, in the orthogonal basis set (5.100):

$$\mathbf{h}^{\mathrm{HF}\perp}(\text{trial}) = \mathbf{h}^{\mathrm{N}\perp} = \boldsymbol{V}\mathbf{h}^{\mathrm{N}}\boldsymbol{V}^{+} = \begin{pmatrix} \dfrac{h_{11}^{\mathrm{N}}+h_{12}^{\mathrm{N}}}{1+s} & 0 \\ 0 & \dfrac{h_{11}^{\mathrm{N}}-h_{12}^{\mathrm{N}}}{1-s} \end{pmatrix} = \begin{pmatrix} -1.1183 & 0.0 \\ 0.0 & -0.5196 \end{pmatrix} \tag{5.163}$$

As this matrix is already in diagonal form we conclude that the associated eigenvectors are:

In orthogonal form:

$$\mathbf{C}^{\perp} = \begin{pmatrix} 1 & 0 \\ 0 & 1 \end{pmatrix}$$

In non-orthogonal form:

$$\mathbf{C} = \boldsymbol{V}^{+}\mathbf{C}^{\perp} = \begin{pmatrix} 0.53198 & 1.46414 \\ 0.53198 & -1.46414 \end{pmatrix}$$

This set of eigenvectors ($\mathbf{C}$) will be further used as trial vectors in order to compute the first Hartree–Fock matrix.

5.4.2.6 The self-consistent field process

Using the former $\mathbf{C}$ vectors we compute the Hartree–Fock matrix by the relation (see 4.22):

$$h_{pq}^{\mathrm{HF}} = h_{pq}^{\mathrm{N}} + \sum_{i=1}^{n}\sum_{r=1}^{m}\sum_{s=1}^{m} C_{ri}C_{si}[2\langle pq \mid rs\rangle - \langle pr \mid sq\rangle] \tag{5.164}$$

where n stands for the number of occupied orbitals, here $n=1$, and m stands for the number of atomic orbitals, here $m=2$. Then:

$$\begin{aligned} h_{pq}^{\mathrm{HF}} &= h_{pq}^{\mathrm{N}} + C_{11}^{2}[2\langle pq| \, 11\rangle - \langle p1 \mid q1\rangle] + C_{11}C_{21}[2\langle pq \mid 12\rangle - \langle p1 \mid q2\rangle] \\ &\quad + C_{21}C_{11}[2\langle pq \mid 21\rangle - \langle p2 \mid q1\rangle] + C_{21}^{2}[2\langle pq \mid 22\rangle - \langle p2 \mid q2\rangle] \end{aligned}$$

$$h_{11}^{\mathrm{HF}} = h_{11}^{\mathrm{N}} + (0.53198)^2(\langle 11 \mid 11\rangle + 2\langle 11 \mid 12\rangle - \langle 12 \mid 12\rangle + 2\langle 11 \mid 22\rangle) = h_{22}^{\mathrm{HF}}$$

$$h_{12}^{\mathrm{HF}} = h_{12}^{\mathrm{N}} + (0.53198)^2(\langle 11 \mid 12\rangle + 3\langle 12 \mid 12\rangle - \langle 11 \mid 22\rangle + \langle 12 \mid 22\rangle) = h_{21}^{\mathrm{HF}}$$

$$\mathbf{h}^{\mathrm{HF}} = \begin{pmatrix} -0.441016 & -0.522343 \\ -0.522343 & -0.441016 \end{pmatrix}$$

In the orthogonal basis set the Hartree–Fock matrix becomes:

$$\mathbf{h}^{\mathrm{HF}\perp} = \boldsymbol{V}\mathbf{h}^{\mathrm{HF}}\boldsymbol{V}^{+} = \begin{pmatrix} -0.5453 & 0.0 \\ 0.0 & +0.3487 \end{pmatrix} \tag{5.165}$$

This matrix has a diagonal form and the associated eigenvectors are:

$$\mathbf{C}^{\perp} = \begin{pmatrix} 1.0 & 0.0 \\ 0.0 & 1.0 \end{pmatrix} \quad \text{and} \quad \mathbf{C} = \boldsymbol{V}^{+}\mathbf{C}^{\perp} = \begin{pmatrix} 0.53198 & 1.46414 \\ 0.53198 & -1.46414 \end{pmatrix}$$

As the final eigenvectors are identical to the starting ones we conclude that the convergence is reached. The electronic and nuclear energies are:

$$\begin{aligned} E_e &= \sum_{i=1}^{n} \sum_{p=1}^{m} \sum_{q=1}^{m} C_{pi} C_{qi} (h_{pq}^{HF} + h_{pq}^{N}) \\ &= 2[C_{11}^2 (h_{11}^N + h_{11}^{HF}) + C_{11} C_{21} (h_{12}^N + h_{12}^{HF})] \\ &= -1.6636 \text{ a.u.} \end{aligned} \tag{5.166}$$

$$R_N = \frac{Z_{H_1} Z_{H_2}}{d_{H_1 H_2}} = \frac{1}{1.37} = 0.7299 \tag{5.167}$$

The total energy is now given by:

$$E_t = E_e + R_N = -0.9337 \text{ a.u.}$$

5.4.2.7 The integral transform

Before starting the CI computation we have to compute the molecular form of the one- and two-electron integrals (see 5.90 and 5.91):

$$\begin{aligned} h_{ij}^N &= \langle \varphi_i | \, h^N \, | \varphi_j \rangle \\ &= \sum_p \sum_q C_{pi} C_{qj} h_{pq}^N \end{aligned} \tag{5.168}$$

$$\begin{aligned} h_{ij}^{HF} &= \langle \varphi_i | \, h^{HF} \, | \varphi_j \rangle \\ &= \sum_p \sum_q C_{pi} C_{qj} h_{pq}^{HF} \end{aligned} \tag{5.169}$$

Thus we find:

$$\mathbf{h}^N = \begin{pmatrix} -1.1183 & 0.0 \\ 0.0 & -0.5196 \end{pmatrix}$$

$$\mathbf{h}^{HF} = \begin{pmatrix} \varepsilon_1 & 0 \\ 0 & \varepsilon_2 \end{pmatrix} = \begin{pmatrix} -0.5453 & 0.0 \\ 0.0 & 0.3487 \end{pmatrix}$$

Similarly, for the two-electron integrals we write:

$$(ij \mid kl) = \sum_p \sum_q \sum_r \sum_s C_{pi} C_{qj} C_{rk} C_{sl} \langle pq \mid rs \rangle \tag{5.170}$$

For simplicity let us replace $C_{11} = C_{21}$ by a and $C_{12} = -C_{22}$ by b. Then we find:

$$\begin{aligned} (11 \mid 11) &= a^4 (2\langle 11 \mid 11 \rangle + 2\langle 11 \mid 22 \rangle + 4\langle 12 \mid 12 \rangle + 8\langle 11 \mid 12 \rangle) = 0.573101 \\ (22 \mid 22) &= b^4 (2\langle 11 \mid 11 \rangle + 2\langle 11 \mid 22 \rangle + 4\langle 12 \mid 12 \rangle - 8\langle 11 \mid 12 \rangle) = 0.481242 \\ (11 \mid 22) &= a^2 b^2 (2\langle 11 \mid 11 \rangle + 2\langle 11 \mid 22 \rangle - 4\langle 12 \mid 12 \rangle) = 0.489518 \\ (12 \mid 12) &= a^2 b^2 (2\langle 11 \mid 11 \rangle - 2\langle 11 \mid 22 \rangle) = 0.110755 \\ (11 \mid 12) &= a^3 b (0) = 0 \\ (12 \mid 22) &= a b^3 (0) = 0 \end{aligned}$$

5.4.2.8 The configuration generation

We are to dispose of two electrons which are to be placed into two molecular orbitals. There exist six different ways to distribute them, four corresponding to the occupation of two different molecular orbitals ($\Phi_3\Phi_4\Phi_5\Phi_6$) and two where the electrons occupy the same orbital (Φ_1 and Φ_2):

$\varepsilon_2 = +0.3487(\varphi_2)$	—	↓	↑	↑↓	↑	↓
$\varepsilon_1 = -0.5453(\varphi_1)$	↑↓	↑	↓	—	↑	↓
Associated determinant:	Φ_1	Φ_3	Φ_4	Φ_2	Φ_5	Φ_6

The analytical form of the determinant (or configuration function CF) are:

$$\begin{aligned}
&\Phi_1 = a(\varphi_1(1)\bar{\varphi}_1(2)) && \text{Fundamental state} \\
&\left.\begin{aligned}\Phi_3 = a(\varphi_1(1)\bar{\varphi}_2(2))\\ \Phi_4 = a(\bar{\varphi}_1(1)\varphi_2(2))\end{aligned}\right\} && \text{Monoexcited CF} \\
&\Phi_2 = a(\varphi_2(1)\bar{\varphi}_2(2)) && \text{Diexcited CF} \\
&\left.\begin{aligned}\Phi_5 = a(\varphi_1(1)\varphi_2(2))\\ \Phi_6 = a(\bar{\varphi}_1(1)\bar{\varphi}_2(2))\end{aligned}\right\} && \text{Monoexcited CF with one change of spin}
\end{aligned} \tag{5.171}$$

where $\varphi_i(\mathrm{j})$ and $\bar{\varphi}_i(\mathrm{j})$ are spin orbitals of the α and β types respectively for an electron labelled j:

$$\varphi_i(\mathrm{j}) = \varphi_i\alpha$$
$$\bar{\varphi}_i(\mathrm{j}) = \varphi_i\beta$$

We have now to construct from this basis of CF a new one, usually called a configuration state function (CSF), where each vector is the eigenvector of S^2 and S_Z operators:

$$S^2\, {}^S_M\Phi = S(S+1)\, {}^S_M\Phi$$
$$S_Z\, {}^S_M\Phi = M\, {}^S_M\Phi$$

We can write for the singlet ($S=0$) and triplet ($S=1$) states:

$$\begin{aligned}
&{}^1_0\Phi_0 = \Phi_1 && (S=0,\ M=0) \\
&{}^1_0\Phi_1 = \frac{1}{\sqrt{2}}(\Phi_3 - \Phi_4) && (S=0,\ M=0) \\
&{}^1_0\Phi_2 = \Phi_2 && (S=0,\ M=0) \\
&{}^3_0\Phi_3 = \frac{1}{\sqrt{2}}(\Phi_3 + \Phi_4) && (S=1,\ M=0) \\
&{}^3_1\Phi_4 = \Phi_5 && (S=1,\ M=1) \\
&{}^3_{-1}\Phi_5 = \Phi_6 && (S=1,\ M=-1)
\end{aligned} \tag{5.172}$$

So we find three singlet and three triplet configuration state functions. The CI may then be split in two, according to the orthogonality of the spin part function between singlet and triplet states. The singlet CSF leads to a 3×3 CI; the triplet CSF being degenerate (they differ only by the spin part), a 1×1 CI is enough to know the triplet energy.

5.4.2.9 The CI calculation

We have only to build and to diagonalize the CI matrix; its elements may be computed as:

$$H_{IJ}=\langle\Phi_I|\, H\,|\Phi_J\rangle \tag{5.173}$$

According to McWeeny,[48] we write:

If Φ_I and Φ_J contain the same occupied spin orbitals (called R or S):

$$H_{IJ}=\sum_{R}\varepsilon^{\mathrm{N}}_{RR}+\tfrac{1}{2}\sum_{\substack{RS\\R\neq S}}[(RR\mid SS)-(RS\mid RS)] \tag{5.173a}$$

If Φ_I and Φ_J differ by only one spin orbital $R\in\Phi_I$ and $R'\in\Phi_J$:

$$H_{IJ}=\varepsilon^{\mathrm{N}}_{RR'}+\sum_{S}[(RR'\mid SS)-(SR'\mid RS)] \tag{5.173b}$$

If Φ_I and Φ_J differ by two spin orbitals: R, $S\in\Phi_I$ and R', $S'\in\Phi_J$:

$$H_{IJ}=(RR'\mid SS')-(RS'\mid SR') \tag{5.173c}$$

In all other cases:

$$H_{IJ}=0 \tag{5.173d}$$

For convenience the above expression must be integrated over space as well as over spin. Then:

$$\langle R_iR_j\mid R_kR_l\rangle=(ij\mid kl)\int_{\mathrm{spin}}\omega_i\omega_j\,\mathrm{d}\sigma_1\int_{\mathrm{spin}}\omega_k\omega_l\,\mathrm{d}\sigma_2 \tag{5.174a}$$

$$\varepsilon^{\mathrm{N}}_{ij}=h^{\mathrm{N}}_{ij}\int_{\mathrm{spin}}\omega_i\omega_j\,\mathrm{d}\sigma_1 \tag{5.174b}$$

We can now give a more explicit form to the CI matrix elements. Over the configuration functions and over the CSF respectively:

$$\begin{aligned}\langle\Phi_1|\,H\,|\Phi_1\rangle&=\varepsilon^{\mathrm{N}}_{11}+\varepsilon^{\mathrm{N}}_{\bar{1}\bar{1}}+\tfrac{1}{2}[(11\mid\bar{1}\bar{1})-(1\bar{1}\mid\bar{1}1)+(\bar{1}\bar{1}\mid 11)-(\bar{1}1\mid 1\bar{1})]\\&=2h^{\mathrm{N}}_{11}+(11\mid 11)\\&=-1.6636=\langle{}^1\Phi_0|\,H\,|{}^1\Phi_0\rangle=H_{00}\end{aligned}$$

$$\begin{aligned}\langle\Phi_2|\,H\,|\Phi_2\rangle&=2h^{\mathrm{N}}_{22}+(22\mid 22)\\&=-0.5579=\langle{}^1\Phi_2|\,H\,|{}^1\Phi_2\rangle=H_{22}\end{aligned}$$

$$\langle\Phi_3|\,H\,|\Phi_3\rangle = \varepsilon_{11}^{N} + \varepsilon_{\bar{2}\bar{2}}^{N} + \tfrac{1}{2}[(11\,|\,\bar{2}\bar{2}) + (\bar{2}\bar{2}\,|\,11)]$$
$$= h_{11}^{N} + h_{22}^{N} + (11\,|\,22)$$
$$= -1.1484$$
$$\langle\Phi_4|\,H\,|\Phi_4\rangle = \langle\Phi_3|\,H\,|\Phi_3\rangle$$
$$\langle\Phi_1|\,H\,|\Phi_2\rangle = (12\,|\,\bar{1}\bar{2}) - (1\bar{2}\,|\,\bar{1}2) = (12\,|\,12)$$
$$= +0.1107 = \langle{}^1\Phi_0|\,H\,|{}^1\Phi_2\rangle = H_{02}$$
$$\langle\Phi_1|\,H\,|\Phi_3\rangle = \varepsilon_{\bar{1}\bar{2}}^{N} + (\bar{1}\bar{2}\,|\,11) - (1\bar{2}\,|\,\bar{1}1) = h_{12}^{N} + (12\,|\,11)$$
$$= 0.0$$
$$\langle\Phi_1|\,H\,|\Phi_4\rangle = -\langle\Phi_1|\,H\,|\Phi_3\rangle$$
$$\langle\Phi_2|\,H\,|\Phi_3\rangle = \varepsilon_{21}^{N} + (21\,|\,\bar{2}\bar{2}) - (\bar{2}1\,|\,2\bar{2}) = h_{21}^{N} + (21\,|\,22)$$
$$= 0.0$$
$$\langle\Phi_2|\,H\,|\Phi_4\rangle = -\langle\Phi_2|\,H\,|\Phi_3\rangle$$
$$\langle\Phi_3|\,H\,|\Phi_4\rangle = (1\bar{1}\,|\,2\bar{2}) - (12\,|\,\bar{2}\bar{1}) = -(12\,|\,12)$$
$$= -0.1107$$
$$\langle\Phi_5|\,H\,|\Phi_5\rangle = \varepsilon_{11}^{N} + \varepsilon_{22}^{N} + (11\,|\,22) - (12\,|\,12)$$
$$= -1.2592 = \langle{}^3\Phi_4|\,H\,|{}^3\Phi_4\rangle = H_{44}$$
$$\langle\Phi_6|\,H\,|\Phi_6\rangle = \langle\Phi_5|\,H\,|\Phi_5\rangle = \langle{}^3\Phi_5|\,H\,|{}^3\Phi_5\rangle = H_{55}$$
$$\langle\Phi_5|\,H\,|\Phi_6\rangle = (1\bar{1}\,|\,2\bar{2}) - (2\bar{1}\,|\,1\bar{2})$$
$$= 0.0 = \langle{}^3\Phi_4|\,H\,|{}^3\Phi_5\rangle = H_{45}$$
$$\langle\Phi_3|\,H\,|\Phi_5\rangle = \varepsilon_{\bar{2}2}^{N} + (\bar{2}2\,|\,11) - (12\,|\,\bar{2}1)$$
$$= 0.0$$
$$\langle\Phi_3|\,H\,|\Phi_6\rangle = \varepsilon_{1\bar{1}}^{N} + (1\bar{1}\,|\,\bar{2}\bar{2}) - (\bar{2}\bar{1}\,|\,1\bar{2})$$
$$= 0.0$$

and

$$\langle\Phi_4|\,H\,|\Phi_5\rangle = \langle\Phi_4|\,H\,|\Phi_6\rangle = 0.0$$

For the remaining CSF we find:

$$H_{11} = \langle{}^1\Phi_1|\,H\,|{}^1\Phi_1\rangle = \tfrac{1}{2}\langle\Phi_3 - \Phi_4|\,H\,|\Phi_3 - \Phi_4\rangle$$
$$= \tfrac{1}{2}(\langle\Phi_3|\,H\,|\Phi_3\rangle + \langle\Phi_4|\,H\,|\Phi_4\rangle - 2\langle\Phi_3|\,H\,|\Phi_4\rangle)$$
$$= -1.0377$$
$$H_{01} = \langle{}^1\Phi_0|\,H\,|{}^1\Phi_1\rangle = \frac{1}{\sqrt{2}}(\langle\Phi_1|\,H\,|\Phi_3\rangle - \langle\Phi_1|\,H\,|\Phi_4\rangle)$$
$$= 0.0$$
$$H_{12} = \langle{}^0\Phi_1|\,H\,|{}^1\Phi_2\rangle = \frac{1}{\sqrt{2}}(\langle\Phi_2|\,H\,|\Phi_3\rangle - \langle\Phi_2|\,H\,|\Phi_4\rangle)$$
$$= 0.0$$

$$H_{34} = \langle {}^3\Phi_3 | H | {}^3\Phi_4 \rangle = \frac{1}{\sqrt{2}} (\langle \Phi_3 | H | \Phi_5 \rangle + \langle \Phi_4 | H | \Phi_5 \rangle)$$
$$= 0.0$$

$$H_{35} = \langle {}^3\Phi_3 | H | {}^3\Phi_5 \rangle = \frac{1}{\sqrt{2}} (\langle \Phi_3 | H | \Phi_6 \rangle + \langle \Phi_4 | H | \Phi_6 \rangle)$$
$$= 0.0$$

$$H_{33} = \langle {}^3\Phi_3 | H | {}^3\Phi_3 \rangle = \frac{1}{2} (\langle \Phi_3 | H | \Phi_3 \rangle + \langle \Phi_4 | H | \Phi_4 \rangle + 2\langle \Phi_3 | H | \Phi_4 \rangle)$$
$$= -1.2592$$

The singlet part of the CI matrix is then:

$$^1\mathbf{H} = \begin{bmatrix} 2h_{11}^{\mathrm{N}} + (11 \mid 11) & & \\ \sqrt{2}\ \ (11 \mid 12) & h_{11}^{\mathrm{N}} + h_{22}^{\mathrm{N}} + (11 \mid 22) + (12 \mid 12) & \\ (12 \mid 12) & \sqrt{2}\ \ (12 \mid 22) & 2h_{22}^{\mathrm{N}} + (22 \mid 22) \end{bmatrix}$$
$$= \begin{bmatrix} -1.663637 & & \\ 0.0 & -1.03769 & \\ 0.110755 & 0.0 & -0.557945 \end{bmatrix} \quad (5.175)$$

and for the triplet part (if $A = h_{11}^{\mathrm{N}} + h_{22}^{\mathrm{N}} + (11 \mid 22) - (12 \mid 12)$), we have:

$$^3\mathbf{H} = \begin{bmatrix} A & & \\ 0 & A & \\ 0 & 0 & A \end{bmatrix} = \begin{bmatrix} -1.2592 & & \\ 0.0 & -1.2592 & \\ 0.0 & 0.0 & -1.2592 \end{bmatrix} \quad (5.176)$$

By diagonalization of the two CI matrices we are able to find precisely the electronic energy levels (${}^S E_e$) and the associated eigenvectors (${}^S\mathbf{C}$) (the results are reported in Table 5.11):

$$ {}^S_M\Psi_i = \sum_j^{\mathrm{CSF}} {}^S C_{ji}\, {}^S_M\Phi_j \quad (5.177)$$

Table 5.11 Eigenvalues and eigenvectors of the CI calculation for the H_2 molecule (energy in atomic units)

		Ψ_0 $S=0;\ i=1$	Ψ_1 $S=0;\ i=2$	Ψ_2 $S=0;\ i=3$	Ψ_3 to Ψ_5 $S=1;\ i=4$, 5 or 6
${}^S E_e(i)$		−1.6746	−1.0377	−0.5470	−1.2592
${}^S E_e(i)$		−0.9447	−0.3078	−0.1830	−0.5297
${}^S C_{ij}$	$j=0$	0.9951	0.0	0.0987	0.0
	$j=1$	0.0	1.0	0.0	0.0
	$j=2$	0.0987	0.0	0.99512	0.0
$j=3$, 4 or 5		0.0	0.0	0.0	1.0

The total energy is also given in the same table according to:

$$^{S}E_{T}(i) = {}^{S}E_{e}(i) + R_{N}$$

where:

$$R_{N} = \sum_{A<B} \frac{Z_{A}Z_{B}}{r_{AB}} = \frac{1}{1.37}$$

5.4.2.10 The natural orbitals

To construct the natural orbitals we have first to compute the so-called density matrix (ρ) which is the representation of the first-order reduced density matrix $\rho_1(x_1'x_1)$ in the spin orbital space:

$$\rho_1(x_1'x_1) = \sum_{ij} \rho_{ij}\varphi_i(x_1')\varphi_j(x_1) \tag{5.178}$$

If $\Phi_\mu \Phi_\nu$ differ at most by one spin orbital the ρ_{ij} element corresponds to:[49]

$$\rho_{ii} = \sum_{\mu \ni i} C_\mu^2 \tag{5.179a}$$

$$\rho_{ij} = \sum_{\mu \ni i} \sum_{\nu \ni j} C_\mu C_\nu (-1)^P \tag{5.179b}$$

where $\mu \ni i$ means that the summation runs over all the configuration state functions $^{S}\Phi_\mu$ which contain the molecular spin orbital φ_i; and P is the parity of the permutation which puts equal orbitals into equal positions in Φ_μ and Φ_ν.

We consider only the two singlet states described by a mixture of CSF (Ψ_0 and Ψ_2); in all other cases (Ψ_1 and Ψ_3 to Ψ_5) the canonical orbitals (**C**) correspond to the natural ones ($\mathbf{C}^{NO}$) and the occupation numbers (η_i) are equal to 1 for the occupied spin orbitals and zero elsewhere. For wave functions of the form:

$$\Psi = {}^{1}C_0{}^{1}\Phi_0 + {}^{1}C_2{}^{1}\Phi_2$$

we write:

$$\boldsymbol{\rho} = \left[\begin{array}{cc|cc} {}^{1}C_0^2 & 0 & & \\ 0 & {}^{1}C_2^2 & \multicolumn{2}{c}{\mathbf{0}} \\ \hline & & {}^{1}C_0^2 & 0 \\ \multicolumn{2}{c|}{\mathbf{0}} & 0 & {}^{1}C_2^2 \end{array}\right] = \left[\begin{array}{c|c} \boldsymbol{\rho}^{(\alpha)} & \mathbf{0} \\ \hline \mathbf{0} & \boldsymbol{\rho}^{(\beta)} \end{array}\right]$$

$$\begin{array}{cccc} 1 & \bar{1} & 2 & \bar{2} \end{array}$$

In the case of the fundamental state we find:

$$\boldsymbol{\rho}_0^{(\alpha)} = \boldsymbol{\rho}_0^{(\beta)} = \begin{pmatrix} 0.990 & 0.0 \\ 0.0 & 0.010 \end{pmatrix} \tag{5.180}$$

As these matrices are diagonal the natural orbitals correspond to the canonical ones (this fact is particular to the specific example we treat here). The electron occupation number (of the α and β spins) associated with each natural orbital (η_i) is given by the diagonal terms of $\boldsymbol{\rho}$ matrices in their diagonal form ($\boldsymbol{\eta}_\alpha \boldsymbol{\eta}_\beta$). Then the total number of electrons (n) of the considered system is:

$$n = n_\alpha + n_\beta = \text{trace}\,(\boldsymbol{\rho}^{(\alpha)} + \boldsymbol{\rho}^{(\beta)}) = \text{trace}\,(\boldsymbol{\eta}_\alpha + \boldsymbol{\eta}_\beta) = 2 \qquad (5.181)$$

Finally, for the first diexcited state we have:

$$\boldsymbol{\rho}_2^{(\alpha)} = \boldsymbol{\rho}_2^{(\beta)} = \begin{pmatrix} 0.010 & 0.0 \\ 0.0 & 0.990 \end{pmatrix}$$

5.4.2.11 Some electronic properties of the H_2 molecule

The electronic properties of a molecule may be expressed in terms of the atomic natural density matrix ($\mathbf{D}^{\text{NO}}$). The method enables us to include CI results in a nice way. This matrix (numerically equivalent to the $\mathbf{D}^{\text{CI}}$ matrix previously defined in 4.69) is:

$$D_{pq}^{\text{NO}} = \sum_{i}^{\text{all orbitals}} \eta_i C_{pi}^{\text{NO}} C_{qi}^{\text{NO}} \qquad (5.182)$$

We further consider the fundamental state (Ψ_0) in order to keep our illustration in the short range. Nevertheless, the reader can extend the next computation to any other state (Ψ_1 to Ψ_5). For the fundamental state the $\mathbf{D}^{\text{NO}}$ matrix corresponds to:

$$\mathbf{D}_0^{\text{NO}} = 2\begin{pmatrix} 0.3016 & 0.2587 \\ 0.2587 & 0.3016 \end{pmatrix}$$

Let us now compute the Mulliken population analysis, the dipole and quadrupole moments, the electronic and difference densities and the electrostatic potential.

The Mulliken population analysis

(a) The overlap population (4.79):

$$P_{\text{AB}} = \sum_{p \in A} \sum_{q \in B} D_{pq}^{\text{NO}} S_{pq} = 0.3968$$

This can be considered as a measure of the bond population between the atoms A and B.

(b) The atomic population:

$$Q_{\text{A}} = \sum_{p \in A} Q_p$$

where:

$$Q_p = \sum_q D^{NO}_{pq} S_{pq}$$

$$Q_{H_1} = Q_{H_2} = 1.0$$

This tells us that the net charge (4.83) of any hydrogen atom in H_2 is zero.

The dipole moment

From (4.98), (4.100) and (5.160) we write:

$$\mu(x) = -\sum_p \sum_q D^{NO}_{pq} M_{pq}(x) + \sum_A Z_A x_A$$

$$= 0.0 + 0.0 \qquad (5.183)$$

Similar results occur for $\mu(y)$ and $\mu(z)$. Thus the overall dipole moment of H_2 is zero.

The quadrupole moment

The quadrupole tensor is the expectation value of the one-electron operator:[50]

$$\theta_{\alpha\beta} = \sum_{\substack{\text{electrons}\\ \text{and nuclei}}} \frac{Z}{2}(3\alpha\beta - r^2\delta_{\alpha\beta}) \qquad (5.184)$$

where α, β stand for x, y or z. Using the equation (4.44) jointly with the second moment integral previously computed (5.160) we write:

$$Q_{xx} = \tfrac{1}{2}[-2M(x^2) + M(y^2) + M(z^2)] + \tfrac{1}{2}\sum_A 2Z_A X_A^2$$

$$= -0.0968$$

$$Q_{yy} = \tfrac{1}{2}[-2M(y^2) + M(x^2) + M(z^2)] - \tfrac{1}{2}\sum_A Z_A X_A^2$$

$$= 0.0484$$

$$Q_{zz} = Q_{yy}$$

$$Q_{xy} = Q_{xz} = Q_{yz} = 0$$

where:

$$M(\alpha\beta) = \sum_{pq} D^{NO}_{pq} M_{pq}(\alpha\beta)$$

The electronic density

According to (4.74) we write:

$$\rho(M) = \sum_{pq} D^{NO}_{pq} \chi_p(M)\chi_q(M)$$

$$= D^{NO}_{11}\chi_1(M)^2 + D^{NO}_{22}\chi_2(M)^2 + 2D^{NO}_{12}\chi_1(M)\chi_2(M) \qquad (5.185a)$$

where M stands for any point of the cartesian space.

Table 5.12 Electronic density along the H_2 bond (x in a.u. and ρ in e^-/a.u.^{3})

x(M)	ρ_0^0(M)	ρ_1^0(M)	ρ_2^0(M)	ρ_0(M)
0.000	0.1327	0.0663	0.0000	0.1314
0.228	0.1299	0.0687	0.0096	0.1287
0.457	0.1277	0.0780	0.0283	0.1208
0.685(H)	0.1091	0.0824	0.0557	0.1086
0.913	0.0936	0.0877	0.0819	0.0935
1.142	0.0767	0.0886	0.1006	0.0769
1.370	0.0601	0.0838	0.1076	0.0606
1.800	0.0333	0.0624	0.0916	0.0339
3.000	0.0025	0.0077	0.0129	0.0026
5.000	0.0000	0.0000	0.0000	0.0000

Another available expression which is equivalent to the preceding one is:

$$\rho(\text{M}) = C_0^2\rho_0^0(\text{M}) + C_2^2\rho_2^0(\text{M}) \tag{5.185b}$$

where:

$$\rho_0^0(\text{M}) = 2\varphi_1(\text{M})^2$$

and

$$\rho_2^0(\text{M}) = 2\varphi_2(\text{M})^2$$

ρ_0^0(M) being the electronic density of the SCF solution. In Table 5.12 we show the evolution of the electronic density along the H_2 bond axis. We remark, by comparing ρ_0^0(M) and ρ_0(M), that the CI electron density is somewhat lower in the interatomic region and larger outside than the SCF electron density. In Figure 5.15 we represent the CI electron density for the three singlet state wave functions. We note that the Ψ_2 state leads to a vanishing electron density for $x = 0$ and so we consider this state to be a purely non-bonding state.

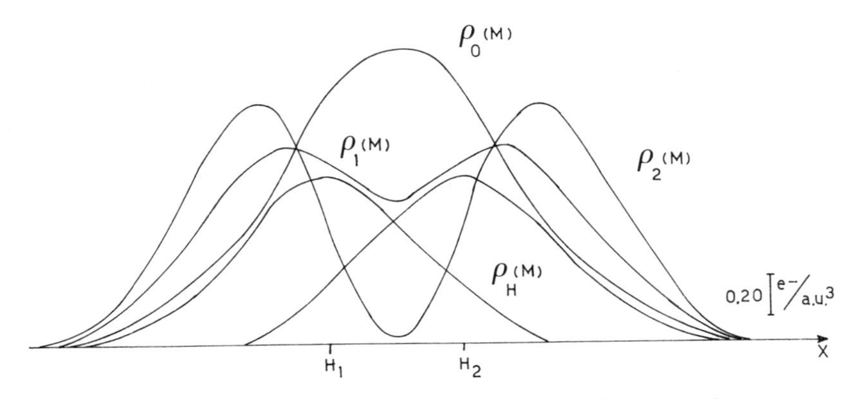

Figure 5.15 Electronic density of H_2 along its bond axis

Table 5.13 The $\delta(M)$ function corresponding to the fundamental state

$x(M)$	$\rho_{H_1}(M)$	$\rho_{H_2}(M)$	$\delta_0(M)$
0.0	0.0586	0.0586	0.0142
0.228	0.0477	0.0679	0.0131
0.457	0.0365	0.0743	0.0100
0.685(H)	0.0264	0.0765	0.0057
0.913	0.0180	0.0743	0.0012
1.142	0.0116	0.0679	−0.0026
1.370	0.0070	0.0586	−0.0050
1.800	0.0023	0.0378	−0.0062
3.000	0.0	0.0037	−0.0011
5.000	0.0	0.0	0.0

The difference density

Having at our disposal the wave function for H_2 as well as for H (5.150), we can compute the difference density function Δ defined by (4.84):

$$\delta(M) = \rho_{H_2}(M) - \rho_{H_1}(M) - \rho_{H_2}(M) \tag{5.186}$$

The results obtained for the fundamental state are given in Table 5.13. The corresponding graphical representation is reported in Figure 5.16. This figure shows that the state described by Ψ_0 is the only one which presents positive values for δ in the bonding region (between the two hydrogen atoms of H_2). Therefore we can consider that this state corresponds to a real bonded state.

The electrostatic potential

General formulation of the electrostatic potential is (4.105):

$$V(M) = -\sum_{pq} D_{pq}^{NO} \langle \chi_p(M) | \frac{1}{r(M)} | \chi_q(M) \rangle + \sum_A \frac{Z_A}{r_A(M)} \tag{5.187}$$

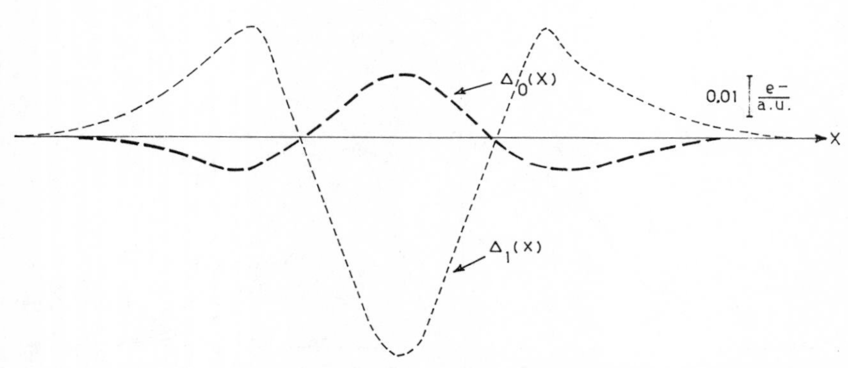

Figure 5.16 The $\delta(M)$ function for fundamental and first excited singlet states

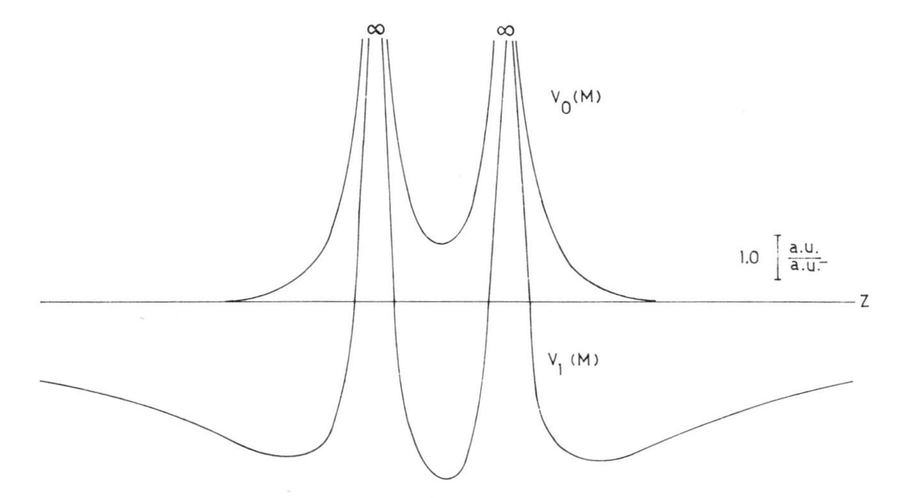

Figure 5.17 The electrostatic potential for H_2 in its fundamental and its first excited state

where, from Table 5.5, we have:

$$\langle \chi_p(\mathrm{M})| \frac{1}{r(M)} |\chi_q(\mathrm{M})\rangle = N_p N_q K_{pq} \pi \frac{2}{\alpha_p + \alpha_q} F_0([\alpha_p + \alpha_q]\overline{PM}^2)$$

if:

$$\mathbf{PM} = \frac{\alpha_p \mathbf{r}_{H_1} + \alpha_q \mathbf{r}_{H_2}}{\alpha_p + \alpha_q} - \mathbf{M}$$

The computed functions along the HH bond axis are reported in Figure 5.17 for the fundamental and the first excited state. Each function presents a discontinuity at the hydrogen atoms as $Z_A/r_A(\mathrm{M})$ becomes infinite when r_A tends to zero.

5.4.2.12 Some energetical comparisons

By using this quite simple basis set the energetical results we have found can be summarized as follows:

H_2	Ψ_0	singlet	−0.9447
2H	—	—	−0.8488
H_2	Ψ_3	triplet	−0.5297
H_2	Ψ_1	singlet	−0.3078
H_2	Ψ_2	singlet	−0.1830
$2H^+ + 2e^-$	—	—	0.0

60.2 kcal mol^{-1}

260.3 kcal mol^{-1} | 200.1 kcal mol^{-1}

399.5 kcal mol^{-1}

477.8 kcal mol^{-1}

592.6 kcal mol^{-1}

Compared to the SCF results we find an energy lowering of 6 kcal mol^{-1} for the CI Ψ_0 wave function. Nevertheless our dissociation energy (60.2 kcal mol^{-1}) remains too weak when compared to the experimental one. This can be improved slightly by optimizing the α gaussian exponent of our s-type basis function and by looking for an equilibrium interatomic distance consistent with the used atomic basis set. Let us note also that no CI can be performed on the H isolated atom which contains only one electron. Finally, we find that except for the fundamental state, all other states of the H_2 molecule are of a higher energy than the dissociated system.

5.4.3 The H_2 Molecule: Some More about Optimization

If we want to improve our calculation on the H_2 molecule, we have to introduce more gaussian functions to fit the wave function. This may be done in several ways, but we shall restrict ourselves to a commonly used approach.

In Section 5.4.1, we discussed the improvement obtained for a hydrogen atom by introducing two or three gaussians into the approximate function. Figure 5.13 showed that with three gaussians the fit obtained was quite good and Table 5.9 showed that the energy also was not too bad. Of course, a basis set optimized for the atom is not optimized for a molecule, and we have to reoptimize the GTOs if we desire a better function. Any of the three types of χ_3 would do, but we shall follow the idea of Hehre, Stewart and Pople[9] of contracting the GTOs which fit the STOs as well as possible (i.e. in our case the wave function of H) and scaling this fit in order to introduce some flexibility into the function (i.e. the ζ exponent of general STOs). We shall thus look for a wave function of H_2 such that:

$$\Psi(1,2) \simeq \Phi_0 = (\varphi_1(1)\bar{\varphi}_1(2))$$

with:

$$\varphi_1 = C_1\chi_{H_1}(\zeta) + C_2\chi_{H_2}(\zeta)$$

where:

$$\chi(1) = \sum_{k=1}^{3} C_k G_k(\alpha, r) \qquad \text{(Table 5.9, function 4)}$$

Φ_0 may be obtained easily by a SCF process and, if desired, a CI computation may be done afterwards (as in Section 5.4.2).

5.4.3.1 The optimization itself

We now have two variables in our problem: first the internuclear distance and second the scaling factor ζ. We shall thus vary both and tabulate the results in order to find the equilibrium distance r_e for the optimized ζ. This may be done for any chosen methodology (RHF or CI) and for any chosen state (fundamental singlet, excited singlet, triplet, etc.). For our purpose, we have studied the fundamental RHF state, the fundamental CI state and the triplet state. Table 5.14 gives the results obtained at several internuclear

Table 5.14 Full optimization of H_2 in a minimal basis set

		r							r
	ζ	1.0 a.u.	1.2 a.u.	1.4 a.u.	1.6 a.u.	1.8 a.u.	2.0 a.u.	ζ	50.0 a.u.
RHF	1.05	−1.00425	−1.07210	−1.09895	−1.10288	−1.09381	−1.07750	0.90	−0.67869
CI–singlet		−1.01521	−1.08552	−1.11538	−1.12295	−1.11821	−1.10701		−0.95980
CI–triplet		−0.26500	−0.46625	−0.60653	−0.70618	−0.77794	−0.83033		−0.95980
RHF	1.10	−1.02790	−1.08951	−1.11089	−1.11011	−1.09705	−1.07728	0.95	−0.67055
CI–singlet		−1.03936	−1.10366	−1.12834	−1.13156	−1.12327	−1.10917		−0.96728
CI–triplet		−0.24170	−0.44919	−0.59337	−0.69536	−0.76857	−0.82196		−0.96728
RHF	1.15	−1.04620	−1.10160	−1.11757	−1.11217	−1.09571	−1.07191	1.00	−0.65747
CI–singlet		−1.05819	−1.11652	−1.13609	−1.13507	−1.12330	−1.10631		−0.96981
CI–triplet		−0.21339	−0.42736	−0.57552	−0.67988	−0.75461	−0.80912		−0.96981
RHF	1.20	−1.05926	−1.10850	−1.11911	−1.10916	−1.08820	−1.06137	1.05	−0.63944
CI–singlet		−1.07178	−1.12420	−1.13875	−1.13357	−1.11836	−1.09843		−0.96740
CI–triplet		−0.18020	−0.40088	−0.55305	−0.65982	−0.73618	−0.79188		−0.96740
RHF	1.25	−1.06717	−1.11028	−1.11561	−1.10113	−1.07616	−1.04558	—	—
CI–singlet		−1.08025	−1.12681	−1.13641	−1.12713	−1.10846	−1.08548		—
CI–triplet		−0.14231	−0.36985	−0.52604	−0.63528	−0.71336	−0.77028		—
RHF	1.30	−1.07003	−1.10704	−1.10713	−1.08813	—	—	—	—
CI–singlet		−1.08369	−1.12443	−1.12914	−1.11579	—	—		—
CI–triplet		−0.09986	−0.33434	−0.49455	−0.60634	—	—		—

distances and for different values of ζ around the minimum. A simple analysis of this table leads to very important conclusions.

For the fundamental state

(a) The minima for RHF and for CI lie in the same region for r, as well as for ζ (i.e. $\zeta \pm 1.2$; $r \pm 1.4$).
(b) Near equilibrium the scaling factors for RHF and CI, although they change with the distance, present close values.
(c) The scaling factor shows a contraction of the basis functions in the molecule (i.e. $\zeta > 1$).
(d) An increase of the distance corresponds to a lowering of the scaling factor; however, as the distance increases, discrepancies appear between CI and RHF.
(e) This discrepancy is very important at 50.0 a.u. The scaling factor for RHF decreases, introducing diffusion in the basis functions. This reflects the bad behaviour of RHF in dissociation processes. As RHF improperly introduces ionic states, the diffusion has to appear.
(f) The scaling factor for CI tends towards 1, which is the optimized value for the separated atoms.

For the excited state

(a) The scaling factor is very different for the singlet and triplet states. It is lower than 1 and shows the well-known diffuse behaviour of excited states.
(b) When the distance increases, the scaling factor increases, also tending to the optimized value for separated atoms, and rejoins the results obtained for the singlet state.

The analysis we have performed here is important as it shows the behaviour of the molecular wave functions compared to isolated atoms for a wide range of internuclear distances. Generally such an optimization is hardly possible but it has shown that the most important thing to enable attainment of enough flexibility is to add more or less diffuse functions in the basis set. We would have obtained the same conclusion if, instead of working with a minimal basis set, we had used a double, triple zeta or even bigger basis set. This latter approach is nowadays the most commonly used, as already mentioned in Section 5.1.

5.4.3.2 The potential energy curves

As we have performed our optimization we may now try to find the best dissociation curve. Table 5.15 shows the results obtained for the optimization of the ground state, and allows us to look for the rotational and vibrational levels as well as for thermochemical properties (see Sections 4.2.5 and 4.2.6). From Table 5.15, we may deduce the equilibrium distance and its energy in the harmonical approximation; moreover, a local potential

Table 5.15 Optimized results for the H_2 fundamental state

	RHF		CI	
Distance	ζ_{opt}	E_{opt}	ζ_{opt}	E_{opt}
1.0	1.3035	−1.07004	1.3095	−1.08378
1.2	1.2425	−1.11034	1.2510	−1.12682
1.4	1.1900	−1.11921	1.2015	−1.13875
1.6	1.1450	−1.11220	1.1600	−1.13517
1.8	1.1065	−1.09709	1.1255	−1.12391

function may be proposed:

At the RHF level:

$$r_e = 1.412 \text{ a.u.}$$
$$E_e = -1.11924$$

and

$$E = 0.19860r^2 - 0.560732r - 0.72345$$

or

$$E = 0.19860(r - 1.412)^2 - 1.11924$$

At the CI level:

$$r_e = 1.454 \text{ a.u.}$$
$$E_e = -1.139313$$

and

$$E = 0.193913r^2 - 0.563828r - 0.72946$$

or

$$E = 0.193913(r - 1.454)^2 - 1.139313$$

The force constants are obtained by relation (4.155) and are respectively:

$$k_{RHF} = 0.3972 \text{ Hartrees/Bohr}^2$$
$$k_{CI} = 0.3878 \text{ Hartrees/Bohr}^2$$

5.4.3.3 Rotations and vibrations of the system

We may now look for the rotational and vibrational levels of H_2 using the harmonic potential (relation 4.156) or the Morse potential (relations 4.158 to 4.161) as the dissociation energy is known. As these potentials are electronic potentials the same analysis holds for any isotopic compound.

Table 5.16 gives some typical results for H_2, HD and D_2 for both potentials in the RHF and CI approximation while Table 5.17 gives information about some transitions obtained at the CI level for the Morse potential. It can be seen that the isotopic effects appear clearly in the transition energies and subsequently on the wave numbers related to the

Table 5.16 Rotation–vibration of H_2, HD and D_2

Quantum state		Harmonic potential		Morse potential	
v	k	RHF	CI	RHF	CI
		H_2			
0	0	−1.108839	−1.129036	−1.107048	−1.129212
	1	−1.108294	−1.128522	−1.108519	−1.128711
	2	−1.107206	−1.127496	−1.107464	−1.127710
1	0	−1.088038	−1.108481	−1.089919	−1.110071
	1	−1.087493	−1.107967	−1.089422	−1.109595
	2	−1.086406	−1.106941	−1.088431	−1.108645
2	0	−1.067238	−1.087928	−1.072462	−1.092343
	1	−1.066692	−1.087414	−1.071998	−1.091893
	2	−1.065605	−1.088388	−1.071072	−1.090994
		HD			
0	0	−1.110232	−1.130413	−1.110389	−1.130546
	1	−1.109822	−1.130027	−1.109989	−1.130168
	2	−1.109006	−1.129256	−1.109194	−1.129414
1	0	−1.092218	−1.112613	−1.093629	−1.113805
	1	−1.091808	−1.112226	−1.093250	−1.113443
	2	−1.090992	−1.111456	−1.092497	−1.112722
2	0	−1.074204	−1.094812	−1.078122	−1.098124
	1	−1.073794	−1.094426	−1.077765	−1.097778
	2	−1.072978	−1.093656	−1.077053	−1.097091
		D_2			
0	0	−1.111886	−1.132046	−1.111991	−1.132134
	1	−1.111612	−1.131788	−1.111723	−1.131882
	2	−1.111068	−1.131274	−1.111190	−1.131376
1	0	−1.097178	−1.117512	−1.098118	−1.118306
	1	−1.096904	−1.117254	−1.097861	−1.118062
	2	−1.096359	−1.116740	−1.097351	−1.117575
2	0	−1.082469	−1.102978	−1.085081	−1.105186
	1	−1.082195	−1.102720	−1.084836	−1.104951
	2	−1.081651	−1.102206	−1.084349	−1.104482

transition. The domain of vibrational excitations (IR) and of rotational excitations (microwave) appear clearly in the table.

5.4.3.4 Thermochemical data

Knowing the spectrum of the molecules we may now try to find some elementary properties related to these compounds. We may analyse, for example, the dissociation reactions:

$$H_2 \rightarrow 2H \quad \text{(a)}$$

$$HD \rightarrow D + H \quad \text{(b)}$$

$$D_2 \rightarrow 2D \quad \text{(c)}$$

Table 5.17 Some typical transitions

Transition		Energy a.u.	Energy kcal mol^{-1}	Wave number (cm^{-1})
$v=0\rightarrow 1$	H_2	0.01914	12.00	4201
$k=0$	HD	0.01674	10.50	3674
	D_2	0.01383	8.67	3035
$v=1\rightarrow 2$	H_2	0.01773	11.12	3891
$k=0$	HD	0.01568	9.84	3441
	D_2	0.01312	8.23	2880
$k=0\rightarrow 1$	H_2	0.00050	0.314	110
$v=0$	HD	0.00038	0.237	83
	D_2	0.00025	0.158	55
$k=1\rightarrow 2$	H_2	0.00100	0.628	219
$v=0$	HD	0.00075	0.473	165
	D_2	0.00051	0.317	112
$k=0\rightarrow 1$	H_2	0.00048	0.299	105
$v=1$	HD	0.00036	0.227	79
	D_2	0.00024	0.153	53

The total energy for the atoms is known (Table 5.9, function 4). From Section 4.2.6 it can be seen that the reaction energy at zero degree is simply the difference between the total energies of products and reactants in their fundamental states. As the temperature rises, thermal corrections must be introduced. Reaction enthalpies may now be obtained. Table 5.18 summarizes such information for each reaction:

$$\Delta U(0) = 2E_{\mathrm{H}} - E_{x_2}\begin{pmatrix} v=0 \\ k=0 \end{pmatrix}$$

$$\Delta U(300) = \Delta U(0) + \frac{RT}{2} - \frac{Nh\nu}{e^{h\nu/kT}-1}$$

$$\Delta H(300) = \Delta U(300) + RT$$

Moreover, we may look at the properties of the isotopic exchange reaction:

$$H_2 + D_2 \rightleftharpoons 2HD$$

for which:

$$\Delta U^0(0) = 0.16 \text{ kcal mol}^{-1}$$

Table 5.18 Thermochemical properties (kcal mol^{-1})

Property	(a)	(b)	(c)
$\Delta U^0(0)$	87.45	88.28	89.28
$\Delta U^0(300)$	87.75	88.58	89.58
$\Delta H^0(300)$	88.34	89.17	90.17

and

$$\Delta H(300) = \Delta U^{0}(300) = 0.16 \text{ kcal mol}^{-1}$$

as thermal corrections are negligible.

We may further determine the reaction entropy by the well-known statistical mechanics relations and so calculate the equilibrium constant of the exchange reaction:

$$\Delta S^{0} = 2S^{0}_{HD} - S^{0}_{H_2} - S^{0}_{D_2}$$

where:

$$S = S_{tr} + S_{rot} + S_{vib} + S_{e}$$

and

$$S_{tr} = R[1.5 \ln \mathrm{M}\,(\mathrm{g\ mol}^{-1}) + 2.5 \ln T - 1.1650]$$

$$S_{rot} = R\left(\ln \frac{IT}{\sigma} - 2.188618\right)$$

$$S_{vib} = R\left[\frac{\theta/T}{e^{\theta/T} - 1} - \ln\left(1 - e^{-\theta/T}\right)\right]$$

$$S_{e} = R \ln m$$

if:

I = inertia tensor
σ = number of C_{2v} symmetry axis
$\theta = \frac{h\nu}{k}$ = vibrational characteristic temperature
m = 1 for singlet states, 2 for doublet states

Thus:

$$\Delta S^{0} = 1.5R \ln \tfrac{9}{8} + R \ln \tfrac{8}{9} + (-1.10^{-6}) + 0.0 = 0.12 \text{ e.u.}$$

$$\Delta G^{0}(300) = 0.16 - 300 * 0.12 * 10^{-3} = +0.124 \text{ kcal}$$

$$\ln K = \frac{-0.124}{0.594} = -0.2087$$

$$K = 0.8116$$

5.4.4 A more advanced example of the SCF procedure

Lastly we propose an illustration of the SCF logic which has been previously exposed in Figure 5.11. Two examples will be described. The first is a closed-shell system computed using the minimal STO-6G basis set of Pople:[9] the water molecule. The second is an open-shell radical in its doublet state calculated in the 6-31G frame of Pople:[51] the $OH^{\cdot}$ compound. The UHF equations have been solved and the results are projected into a RHF subspace.

5.4.4.1 The water molecule

We choose the experimental water geometry to perform the SCF calculation. The cartesian coordinates employed are reported in Figure 5.18. The

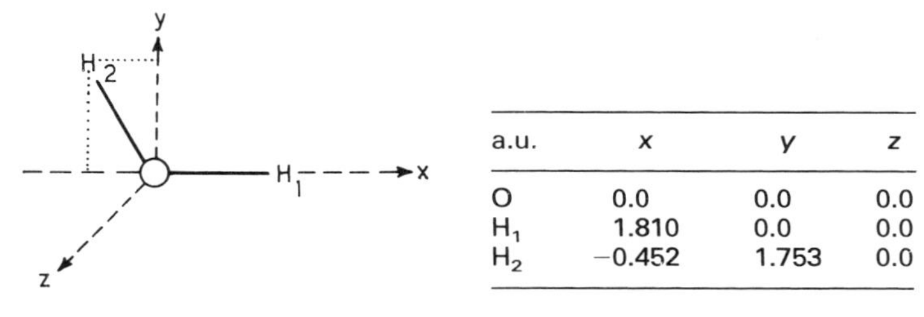

a.u.	x	y	z
O	0.0	0.0	0.0
H_1	1.810	0.0	0.0
H_2	−0.452	1.753	0.0

Figure 5.18 The water molecule in the chosen cartesian coordinate frame

atomic basis set used is defined according to:

$$\chi(H_2O) = 1s(O)2s(O)2p_x(O)2p_y(O)2p_z(O)1s(H_1)1s(H_2)$$

The Löwdin orthogonalization process is given in Table 5.19. We dispose of the unsymmetrical square matrix $\mathbf{V}$ to express the pseudoeigenvalue problem in its orthogonal form (equation 5.96).

Table 5.19 The Löwdin orthogonalization procedure of the H_2O molecule

Overlap matrix (**S**)						
1.000						
0.231	1.000					
0.0	0.0	1.000				
0.0	0.0	0.0	1.000			
0.0	0.0	0.0	0.0	1.000		
0.055	0.476	0.392	0.0	0.0	1.000	
0.055	0.476	−0.098	0.380	0.0	0.0	1.000
Eigenvalues ($\mathbf{S}_0$) *and eigenvectors* (**t**) *of the overlap*						
1.932	1.331	1.090	1.000	0.888	0.417	0.342
0.207	0.0	−0.642	0.0	−0.718	−0.0	0.171
0.584	0.0	−0.310	0.0	0.281	0.0	−0.696
0.165	0.631	0.415	0.0	−0.371	0.476	−0.196
0.213	−0.489	0.535	0.0	−0.478	−0.369	−0.253
0.0	0.0	0.0	1.000	0.0	0.0	0.0
0.521	0.426	0.127	0.0	0.141	−0.565	0.439
0.521	−0.426	0.127	0.0	0.141	0.565	0.439
The orthogonalization matrix (**V**)						
0.149	0.420	0.119	0.153	0.0	0.375	0.375
0.0	0.0	0.547	−0.424	0.0	0.369	−0.369
−0.615	−0.297	0.398	0.513	0.0	0.121	0.121
0.0	0.0	0.0	0.0	1.000	0.0	0.0
−0.762	0.298	−0.393	−0.507	0.0	0.150	0.150
0.0	0.0	0.736	−0.570	0.0	−0.874	0.874
0.292	−1.188	−0.336	−0.433	0.0	0.750	0.750

Table 5.20 The first iteration in the H_2O run

(0-a) $\mathbf{h}^{N} = \mathbf{T} + \mathbf{V}$ *is in use as trial Hartree–Fock matrix* ($\mathbf{h}^{HF}$ (*trial*))

−33.035						
−7.303	−9.248					
−0.012	−0.137	−7.608				
−0.016	−0.177	0.030	−7.593			
0.0	0.0	0.0	0.0	−7.475		
−1.760	−3.732	−2.604	−0.058	0.0	−5.077	
−1.760	−3.732	0.594	−2.536	0.0	−1.607	−5.077

(0-*b*) *Trial Hartree-Fock matrix in orthogonal basis set* :$\mathbf{h}^{HF\perp}$ (*trial*)

−8.706						
0.0	−6.670					
5.113	0.0	−18.732				
0.0	0.0	0.0	−7.475			
5.163	0.0	−11.509	0.0	−19.269		
0.0	−1.625	0.0	0.0	0.0	−5.331	
1.125	0.0	−2.415	0.0	2.431	0.0	−6.421

(0-*c*) *Eigenvalues* ($\boldsymbol{\varepsilon}$) *and eigenvectors* ($\mathbf{C}^{\perp}$) *of* $\mathbf{h}^{HF\perp}$

−33.042	−8.367	−7.758	−7.489	−7.475	−4.243	−4.231
−0.280	0.733	0.0	0.005	0.0	0.0	0.620
0.0	0.0	0.831	0.0	0.0	−0.556	0.0
0.667	0.082	0.0	0.714	0.0	0.0	0.198
0.0	0.0	0.0	0.0	1.000	0.0	0.0
0.682	0.112	0.0	0.700	0.0	0.0	0.180
0.0	0.0	0.556	0.0	0.0	0.831	0.0
−0.112	−0.666	0.0	0.024	0.0	0.0	0.737

(0-*d*) *Canonical coefficients* ($\mathbf{C}$) *in atomic basis set*

−1.004	−0.221	0.0	−0.088	0.0	0.0	0.049
0.019	1.108	0.0	0.390	0.0	0.0	−0.621
0.001	0.299	0.864	−0.567	0.0	0.307	−0.166
0.001	0.386	−0.669	−0.732	0.0	−0.238	−0.214
0.0	0.0	0.0	0.0	1.000	0.0	0.0
0.005	−0.197	−0.179	0.035	0.0	−0.930	0.837
0.005	−0.197	0.179	0.035	0.0	0.932	0.837

(0-e) *The atomic density matrix* ($\mathbf{D}_1$)

2.129						
−0.587	2.761					
−0.035	0.220	2.317				
−0.045	0.284	−0.095	2.268			
0.0	0.0	0.0	0.0	2.000		
0.091	−0.410	−0.468	0.037	0.0	0.145	
0.091	−0.410	0.152	−0.444	0.0	0.016	0.145

(0-f) *Electronic energy calculation*

$E_e = -83.134002$ a.u.

(1-a) *Next iteration, build new Hartree–Fock matrix*

−19.130						
−4.702	−1.790					
−0.010	−0.140	0.089				
−0.013	−0.182	0.031	0.105			
0.0	0.0	0.0	0.0	0.224		
−1.134	−0.759	−0.305	−0.060	0.0	−0.262	
−1.334	−0.759	0.018	0.310	0.0	−0.299	−0.262

We start this SCF process using the core hamiltonian (h^N) as the Hartree–Fock trial matrix. That means we neglect the two electron repulsions in the first Hartree–Fock matrix. In general this is not a good choice but may suffice for our purpose. In Table 5.20 the first iteration (label '0') is detailed. If we compare the trial Fock matrix (0-a) with the corresponding matrix built at the beginning of the second step (1-a), we can see the large evolution which appears.

To shorten this process we follow the SCF procedure shown in Table 5.21. A rapid energy lowering may be observed. We also note that the density matrix reaches its stationary value a few iterations after the energy.

Table 5.21 Summarizing the convergence procedure in the H_2O case

Iteration number	Electronic energy	Convergence on energy	Convergence on density matrix
1	−84.848599	−1.714600	3.08065
2	−84.865265	−0.016666	0.34207
3	−84.866009	−0.000744	0.06080
4	−84.866105	−0.000096	0.02321
5	−84.866122	−0.000017	0.00907
Extrapolation	-84.866124_8	−0.000003	—
7	-84.866125_3	5×10^{-7}	0.00161
8	-84.866125_5	2×10^{-7}	0.00069
9	-84.866125_6	3×10^{-7}	0.00030
10	-84.866125_6	2×10^{-7}	0.00013
Extrapolation	-84.866125_6	0	—
11	-84.866125_6	0	0.00002

This illustrates our purpose of Section 5.3.6. Table 5.22 gives the final results. From step 11-d we recognize the oxygen core function in the first molecular orbital and a pure p_z function perpendicular to the molecular plane in the last occupied molecular orbital (φ_5).

We also report the projectors R (defined in Table 5.8). The trace of $R_1(\alpha)$ or $R_1(\beta)$ corresponds to the number of electrons of associated spin (5α and 5β electrons). Similarly, the trace for $R_2(\alpha)$ or $R_2(\beta)$ corresponds to the number of unoccupied orbitals of α and β spin. It is also trivial to verify the projector properties of those matrices ($R^2=R$; $R_1R_2=0$; $R_1+R_2=1$).

5.4.4.2 The hydroxyl radical

The hydroxyl radical in its neutral form contains nine electrons. They are divided into five electrons of α spin and four of β spin. We use the experimental geometry for the OH radical ($d_{OH}=1.884$ a.u.). The OH bond is oriented along the cartesian z axis. The 6-31G basis set leads to the

Table 5.22 The last iteration of the H_2O run

(11-a) *The final Hartree–Fock matrix*

−20.501						
−4.964	−2.343					
−0.018	−0.078	−0.339				
−0.023	−0.101	−0.015	−0.347			
0.0	0.0	0.0	0.0	−0.397		
−1.196	−1.011	−0.538	−0.049	0.0	−0.579	
−1.196	−1.011	0.087	−0.533	0.0	−0.393	−0.579

(11-d) *The canonical LCAO coefficients* (**C**) *and the associated eigenvalues* ($\boldsymbol{\varepsilon}$)

−20.504	−1.276	−0.621	−0.459	−0.398	0.597	0.729
−0.997	−0.222	0.0	−0.096	0.0	0.121	0.0
−0.016	0.838	0.0	0.534	0.0	−0.882	0.0
−0.002	0.079	0.482	−0.476	0.0	−0.454	0.780
−0.003	0.103	−0.374	−0.614	0.0	−0.586	−0.605
0.0	0.0	0.0	0.0	1.000	0.0	0.0
0.004	0.154	0.442	−0.279	0.0	0.799	−0.839
0.004	0.154	0.442	−0.279	0.0	0.799	0.839

(11-e) *The atomic density matrix* ($\mathbf{D}_1$)

1.052						
−0.221	0.987					
0.030	−0.187	0.466				
0.039	−0.242	0.120	0.528			
0.0	0.0	0.0	0.0	1.000		
−0.011	−0.020	0.358	0.022	0.0	0.297	
−0.011	−0.020	−0.068	0.352	0.0	−0.094	0.297

(11-g) *The projectors into the occupied* ($\mathbf{R}_1$) *and unoccupied* ($\mathbf{R}_2$) *subspace*:

$\mathbf{R}_1(\alpha)=\mathbf{R}_1(\beta)=\frac{1}{2}\mathbf{R}_1(\alpha+\beta)$

0.961						
0.0	0.993					
0.022	0.0	0.988				
0.0	0.0	0.0	1.000			
−0.063	0.0	0.034	0.0	0.898		
0.0	−0.085	0.0	0.0	0.007		
−0.183	0.0	0.100	0.0	−0.294	0.0	0.153

$$\mathrm{tr}\,(\mathbf{R}_1(\alpha))=\mathrm{tr}\,(\mathbf{R}_1(\beta))=5$$

$\mathbf{R}_2(\alpha)=\mathbf{R}_2(\beta)=\frac{1}{2}\mathbf{R}_2(\alpha+\beta)$

0.039						
0.0	0.007					
−0.022	0.0	0.012				
0.0	0.0	0.0	0.0			
0.063	0.0	−0.035	0.0	0.102		
0.0	0.085	0.0	0.0	0.0	0.993	
0.183	0.0	−0.100	0.0	0.294	0.0	0.847

$$\mathrm{tr}\,(\mathbf{R}_2(\alpha))=\mathrm{tr}\,(\mathbf{R}_2(\beta))=2$$

following atomic functions:

$$\chi(\mathrm{OH})=1s(\mathrm{O})2s(\mathrm{O})2p_x(\mathrm{O})2p_y(\mathrm{O})2p_z(\mathrm{O})2s'(\mathrm{O})$$
$$2p_x'(\mathrm{O})2p_y'(\mathrm{O})2p_z'(\mathrm{O})1s(\mathrm{H})1s'(\mathrm{H})$$

In Table 5.23 we list the UHF molecular orbitals obtained after convergence of the SCF procedure. We note that the α and β molecular spin

Table 5.23 The eight first molecular orbitals (after SCF convergence) of the hydroxyl radical at the UHF level and the corresponding orbitals projected into the RHF subspace

The eigenvalues ($\boldsymbol{\varepsilon}_\alpha$) and eigenvectors ($\mathbf{C}_\alpha$) of the α Hartree–Fock operator

−20.641	−1.380	−0.661	−0.641	−0.556	0.204	1.060	1.069
0.996	0.224	0.006	0.0	0.0	−0.006	0.003	0.0
0.002	−0.521	−0.154	0.0	0.0	0.010	−0.009	0.0
0.0	0.0	0.0	−0.489	−0.474	0.0	0.0	−0.655
0.0	0.0	0.0	0.489	−0.474	0.0	0.0	0.655
0.0	−0.009	0.549	0.0	0.0	0.295	−0.418	0.0
0.0	−0.506	−0.257	0.0	0.0	0.839	0.006	0.0
0.0	0.0	0.0	−0.322	−0.338	0.0	0.0	0.751
0.0	0.0	0.0	0.322	−0.338	0.0	0.0	−0.751
0.0	−0.006	0.326	0.0	0.0	0.592	−0.170	0.0
0.0	−0.137	0.284	0.0	0.0	−0.010	1.348	0.0
0.0	0.002	0.139	0.0	0.0	−1.524	−0.882	0.0

The eigenvalues ($\boldsymbol{\varepsilon}_\beta$) and eigenvectors ($\mathbf{C}_\beta$) of the β Hartree–Fock operator

−20.600	−1.222	−0.610	−0.503	0.127	0.219	1.063	1.154
0.996	0.214	−0.008	0.0	0.0	0.006	−0.004	0.0
0.002	−0.482	0.177	0.0	0.0	−0.009	0.113	0.0
0.0	0.0	0.0	0.460	0.349	0.0	0.0	0.676
0.0	0.0	0.0	0.460	−0.349	0.0	0.0	0.676
0.0	−0.107	−0.508	0.0	0.0	−0.302	0.384	0.0
0.0	−0.492	0.321	0.0	0.0	−0.874	−0.004	0.0
0.0	0.0	0.0	0.354	0.464	0.0	0.0	−0.737
0.0	0.0	0.0	0.354	−0.464	0.0	0.0	−0.737
0.0	−0.005	−0.321	0.0	0.0	−0.623	0.259	0.0
0.0	−0.187	−0.297	0.0	0.0	0.007	−1.360	0.0
0.0	−0.001	−0.175	0.0	0.0	1.543	0.845	0.0

The molecular orbitals projected in the RHF subspace

Doubly filled orbitals ($\alpha+\beta$)				Singly filled orbital (α)	Virtual unoccupied orbitals		
−1.004	0.0	−0.173	−0.009	0.0	0.007	0.008	0.0
0.0	0.0	0.479	0.223	0.0	−0.285	−1.123	0.0
0.0	−0.467	0.0	0.0	−0.489	0.0	0.0	−0.671
0.0	−0.467	0.0	0.0	0.489	0.0	0.0	−0.671
0.0	0.0	−0.289	0.454	0.0	−0.632	0.517	0.0
0.003	0.0	0.562	0.132	0.0	−0.405	1.213	0.0
0.0	−0.346	0.0	0.0	−0.322	0.0	0.0	0.741
0.0	−0.346	0.0	0.0	0.322	0.0	0.0	0.741
0.0	0.0	−0.182	0.273	0.0	−0.298	−0.658	0.0
0.0	0.0	−0.008	0.323	0.0	0.664	0.423	0.0
0.0	0.0	−0.113	0.110	0.0	0.674	−0.477	0.0

orbitals have not the same spatial part as in the RHF results. They remain, nevertheless, orthogonal by the spin part. We can characterize the molecular orbitals as follows:

(a) The first is the oxygen core orbital; we find $\varphi_1(\alpha) \simeq \varphi_1(\beta)$.
(b) The orbitals 2 and 3 are the σ(OH) bonding and antibonding orbitals.

Table 5.24 The evolution of the occupation number from UHF to RHF

Open-shell run type:		UHF		NO associated to UHF	RHF C^{RHF} Round off $\eta_{\alpha+\beta}$ to the closest integer
Canonical orbital set:		$C^{UHF}(\alpha) \neq C^{UHF}(\beta)$		$V'U = C^{RHF}$	
Occupation vector		η_α	η_β	$\eta_{\alpha+\beta}$	
	1	1.0	1.0	2.00000	2.0
	2	1.0	1.0	1.99978	2.0
	3	1.0	1.0	1.99975	2.0
	4	1.0	1.0	1.99833	2.0
	5	1.0	0.0	1.00000	1.0
Orbital:	6	0.0	0.0	0.00167	0.0
	7	0.0	0.0	0.00025	0.0
	8	0.0	0.0	0.00022	0.0
	9	0.0	0.0	0.0	0.0
	10	0.0	0.0	0.0	0.0
	11	0.0	0.0	0.0	0.0
Number of electrons:		5 +	4	9.00000	9.0
$\langle E_t \rangle$ a.u.)		−75.36239		−75.36239	−75.36102
S		0.502		0.502	0.500
$\langle S^2 \rangle$		0.754		0.754	0.750

(c) The orbitals 4 and 5 are both π(OH) bonding orbitals which contain the 3π electrons of the radical, 2 of α spin ($\varphi_4(\alpha)$ and $\varphi_5(\alpha)$) and 1 of β spin ($\varphi_4(\beta)$).

(d) The remaining orbitals are all unoccupied.

Let us now transform this double set of molecular orbitals by projection into the RHF subspace (see Section 5.3.2). In Table 5.24 we report the evolution of the occupation number along the procedure. The first column refers to the UHF wave function where the occupation numbers are zero or one. We also report the S and $\langle S^2 \rangle$ expected values and note that the UHF wave function is not a pure spin state, but is not far from it. The transformation into the natural orbital frame modifies the occupation numbers; they no longer remain as integer values. The expectation values for any observable are not affected by the transformation. If we now round off the NO occupation numbers to the closest integer (this must be done in such a way to preserve the total number of electrons) the energy is slightly increased. However the wave function now becomes a pure spin state, as shown by S and $\langle S^2 \rangle$. Table 5.23 lists the RHF type of molecular orbital which remain quite similar to the corresponding UHF orbitals.

REFERENCES

1. P. O. Löwdin, *Adv. Phys.*, **5,** 1 (1956).
2. F. R. Burden and R. M. Wilson, *Adv. Phys.*, **21** (94), 825 (1972).

3. I. Shavitt and M. Karplus, *J. Chem. Phys.*, **36,** 550 (1962).
4. E. Steiner and S. Sykes, *Mol. Phys.*, **23,** 643 (1972).
5. S. F. Boys, *Proc. Roy. Soc.* (*London*), **A200,** 542 (1950).
6. H. Preuss, *Z. Naturf.*, **11,** 823 (1956); J. L. Withen and L. C. Allen, *J. Chem. Phys.*, **43,** 3165 (1965).
7. M. Roche and J. C. Simon, *Theoret. Chim. Acta*, **27,** 165 (1972).
8. A. A. Frost, in *Modern Theoretical Chemistry* (Ed. H. F. Schaeffer III), Vol. 3, Plenum Press, New York, 1977.
9. W. J. Hehre, R. F. Stewart and J. A. Pople, *J. Chem. Phys.*, **51,** 2657 (1969).
10. T. H. Dunning Jr. and J. P. Hay, in *Modern Theoretical Chemistry* (Ed. H. F. Schaeffer III), Vol. 3, Plenum Press, New York, 1977.
11. G. Page and G. Ludwig, *J. Chem. Phys.*, **56,** 5626 (1972).
12. J. A. Pople *et al.*, *J. Chem. Phys.*, **51,** 2657 (1969), **52,** 2769 (1970), **53,** 932 (1970), **54,** 7241 (1971); *Mol. Phys.*, **27,** 209 (1974); *J. Amer. Chem. Soc.*, **102,** 139 (1980).
13. B. Roos and P. Sieghbahn, *Theoret. Chim. Acta*, **17,** 199, 209 (1969).
14. A. Veillard and J. Demuynck, in *Modern Theoretical Chemistry* (Ed. H. F. Schaeffer III), Vol. 4, Plenum Press, New York, 1977.
15. G. Maroulis, Ph.D. Thesis, UCL, 1981; G. Maroulis, M. Sana and G. Leroy, *Internat. J. Quantum Chem.*, **19,** 43 (1981).
16. W. J. Hehre, R. Ditchfield, R. F. Stewart and J. A. Pople, *J. Chem. Phys.*, **52,** 2769 (1970).
17. S. Huzinaga, *J. Chem. Phys.*, **42,** 1293 (1965); T. H. Dunning Jr., *J. Chem. Phys.*, **53,** 2823 (1970).
18. S. Huzinaga and C. Arnau, *J. Chem. Phys.*, **52,** 2224 (1970).
19. T. H. Dunning Jr., *J. Chem. Phys.*, **55,** 716 (1971).
20. P. O. Löwdin, *J. Chem. Phys.*, **18,** 365 (1950).
21. L. Pauling, *J. Amer. Chem. Soc.*, **53,** 1367 (1931); J. C. Slater, *Phys. Rev.*, **37,** 481 (1931).
22. K. R. Roby, *Mol. Phys.*, **28,** 1441 (1974).
23. V. R. Saunders, in *Computational Techniques in Quantum Chemistry and Molecular Physics* (Eds. G. H. F. Diercksen, B. T. Sutcliffe and A. Veillard), Reidel, 1975, p. 347.
24. F. Oberheittinger and L. Baddii, *Tables of Laplace Transform*, Springer-Verlag, 1973.
25. H. S. Wall (Ed.), *Analytic Theory of Continuous Fractions*, Wiley, 1948, pp. 15, 350, 356; A. N. Khovanskii, *The Application of Continued Fractions and Their Generalizations to Problems in Approximation Theory* (Translated by P. Wynn), Noordhoff, 1963, pp. 2, 144; P. J. Davis, in *Handbook of Mathematical Functions* (Eds. M. Abramowitz and I. A. Stegun), Dover, 1964, p. 263.
26. I. Shavitt, *Methods in Computational Physics*, Vol. 2, Academic Press, 1963, p. 1; V. R. Saunders, *Computational Techniques in Quantum Chemistry and Molecular Physics*, Series C, Reidel, 1975, p. 366.
27. G. Szegö, *Orthogonal Polynomials*, American Mathematics Society, New York, 1959; H. F. King and M. Dupuis, *J. Comp. Phys.*, **21,** 144 (1976).
28. M. Dupuis, J. Rys and H. F. King, *J. Chem. Phys.*, **65,** 111 (1965).
29. M. Dupuis, J. Rys and H. F. King, *HONDO*, QCPE, no. 338, 1976.
30. H. F. King and M. Dupuis, *J. Comp. Phys.*, **21,** 144 (1976).
31. W. J. Hehre, W. A. Lathan, R. Ditchfield, M. D. Newton and J. A. Pople, *GAUSSIAN*-70, QCPE, no. 236, 1970.
32. M. Yoshimine, IBM Corp. Technical Report RJ 555, 1973; G. H. F. Diercksen, *Theoret. Chim. Acta*, **33,** 1 (1974); B. Roos, in *Computational Techniques in Quantum Chemistry and Molecular Physics* (Eds. G. H. F. Diercksen, B. T. Sutcliffe and A. Veillard), Reidel, 1975, p. 251.

33. C. C. J. Roothaan, *Rev. Mod. Phys.*, **23,** 69 (1951), **32,** 179 (1960); A. Veillard, in *Computational Techniques in Quantum Chemistry and Molecular Physics* (Eds. G. H. F. Diercksen, B. T. Sutcliffe and A. Veillard), Reidel, 1975, p. 201.
34. J. A. Pople and R. K. Nesbet, *Chem. Phys. Lett.*, **22,** 571 (1954).
35. R. McWeeny, *Rev. Mod. Phys.*, **32,** 335 (1960); R. McWeeny, in *Computational Techniques in Quantum Chemistry and Molecular Physics* (Eds. G. H. F. Diercksen, B. T. Sutcliffe and A. Veillard), Reidel, 1975, p. 505.
36. P. O. Löwdin, *Adv. Chem. Phys.*, **2,** 207 (1959).
37. H. Fukutome, *Prog. Theoret. Phys.*, **47,** 1156 (1972); T. Takabe and H. Fukutome, *Prog. Theoret. Phys.*, **56,** 689 (1976).
38. R. McWeeny, *Proc. R. Soc. London,* **A235,** 496 (1956); *Rev. Phys.*, **32,** 335 (1960); A. Veillard: in *Computational Techniques in Quantum Chemistry and Molecular Physics* (Eds. G. H. F. Diercksen, B. T. Sutcliffe and A. Veillard), Reidel, 1975, pp. 226–233; P. de Mongolfier and A. Hoareau, *J. Chem. Phys.*, **65,** 2477 (1976).
39. R. Seeger and J. A. Pople, *J. Chem. Phys.*, **65,** 265 (1976).
40. I. H. Hillier and V. R. Saunders, *Internat. J. Quantum Chem.*, **4,** 503 (1970); *Proc. R. Soc. London,* **A320,** 161 (1970); *Internat. J. Quantum Chem.*, **7,** 699 (1973).
41. C. C. J. Roothaan and P. S. Bagus, in *Methods in Computational Physics* (Eds. B. Alder, S. Fernbach and M. Rotenberg), Vol. 2, Academic Press, 1963, p. 47.
42. H. F. King, R. E. Stanton, H. Kim, R. E. Wyatt and R. G. Parr, *J. Chem. Phys.*, **47,** 1906 (1967).
43. C. E. Edminston and K. Ruedenberg, *Rev. Mod. Phys.*, **35,** 457 (1963).
44. W. von Niessen, *J. Chem. Phys.*, **56,** 4249 (1972); *Theoret. Chim. Acta*, **27,** 9 (1972).
45. S. F. Boys, *Rev. Mod. Phys.*, **32,** 296, 300 (1960); S. F. Boys, in *Quantum Theory in Atoms, Molecules and Solid State* (Ed. P. O. Löwdin), Academic Press, 1966, p. 253.
46. V. Magnasco and A. Perico, *J. Chem. Phys.*, **48,** 800 (1968).
47. R. Ditchfield, W. J. Hehre and J. A. Pople, *J. Chem. Phys.*, **52,** 5001 (1970).
48. R. McWeeny, *Rev. Mod. Phys.*, **32,** 335 (1960).
49. B. Roos, *Computational Techniques in Quantum Chemistry and Molecular Physics* (Eds. G. H. F. Diercksen, B. T. Sutcliffe and A. Veillard, Reidel, 1975, p. 251.
50. P. Swanstrom and F. Hegelund, *Computational Techniques in Quantum Chemistry and Molecular Physics* (Eds. G. H. F. Diercksen, B. T. Sutcliffe and A. Veillard), Reidel, 1975, p. 299.
51. R. Ditchfield, W. J. Hehre and J. A. Pople, *J. Chem. Phys.*, **54,** 724 (1971).

CHAPTER 6

Some Applications of Quantum Chemistry: From Computations to Concepts

6.1 CONFIGURATIONAL AND CONFORMATIONAL ANALYSIS

6.1.1 The Concept of Equilibrium Structure

In the previous chapter, we have seen how we could realize a computation for a molecule. It appeared that for a configuration, i.e. a given arrangement of the constituting nuclei, the wave function and related energy could be obtained within the Born–Oppenheimer frame by resolution of the Hartree–Fock equations, further approximated by the LCAO approach, followed when desired by a CI calculation. This energy is thus a function of the method used, of a chosen basis set (namely through the non-linear parameters $\{\bar{\zeta}\}$) and of some internal molecular parameters $\{\bar{R}\}$ defining the geometry of the system.

When we have chosen a method and a basis set (i.e. $\{\bar{\zeta}\}$ is frozen) only the internal parameters remain as variables. This set of $3N-6$ components ($3N-5$ for linear molecules) defines configurations and generates a hypersurface in a $(3N-6)$-dimensional world. The continuous hypersurface will inform us not only on all possible arrangements but also on the possible motions of the nuclei in the molecular landscape. There may be smooth valleys and abrupt ravines or unsurmountable mountains. All of these constrain the molecular reality in some well-defined regions corresponding to some chemical reality. It is a matter of fact that the detailed analysis of such hypersurfaces is quite difficult. We do not intend to perform this analysis here as the third part of this book will be devoted to it; however, from now on we must become familiar with the hypersurface and its relationship with the molecule in order to extract some important concepts.

It is clear for mechanical reasons that the most convenient situation for a molecular system will be at the bottom of a well; nevertheless, one should keep the fact in mind that the hypersurface describes the potential energy of the nuclei and that the total energy of the molecule is slightly higher, as some corrections (i.e. ZPE, thermal, etc.) must be applied to obtain some 'observable' value. Mathematically, admitting that we have an analytical function describing the hypersurface, the most probable configuration of the system turns out to be a point on this surface presenting a zero value for any first derivative and positive second derivatives for all variables of space. Such a point is known to represent an *equilibrium structure*.

On such a broken surface many points correspond to this situation and many minima will appear corresponding to stable arrangements of nuclei. Therefore we shall find one *absolute minimum* corresponding to the most stable arrangement and other *relative minima* corresponding to other possible arrangements (isomers, conformers, reaction products, complexes, etc.). These different minima are related to the absolute minimum which we take as reference, by different kinds of *chemical processes*, following a *reaction pathway*, defined as the way of lowest energy on the surface connecting the minima of interest and passing through a *transition state* on the surface which is the maximum along the reaction path. The reaction pathway can thus be discussed in terms of nuclear motion corresponding to a concerted evolution of the internal molecular parameters describing the *reaction coordinate*. The local energy maximum along the reaction coordinate (i.e. the transition state energy) is separated from the minima by some *energy barrier*, currently referred to as the activation barrier of the process. This transition state has some mathematical properties as its first derivatives are zero and its second derivatives are positive, except for one value corresponding to the maximum along the reaction coordinate. The problems arising from the analysis of such peculiar points are somewhat mistreated here as they are explicitly studied in the third part.

We can clarify our assertions by presenting some examples, starting from a brute chemical formula such as C_4H_8 or CH_5N:

(a) For C_4H_8 there are some 30 variables defining the hypersurface. Chemists know that many molecules correspond to that formula; one can think of butene-1, butene-2 *cis*, butene-2 *trans*, methyl-2 propene, cyclobutane and methyl cyclopropane.

All these molecules are quite different but lie on the same hypersurface, correspond to minima in different portions of the surface and are known as isomers. One of them is the most stable; of the others some are a few kilocalories per mole and some more than 20 kcal mol^{-1} higher depending on the isomerization energies. Moreover, butene-1, for example, although it corresponds to a unique molecule, presents more than one minimum describing different ‘conformers’. One of these is the most stable and is related to the others by rotations around the CC bond of methyl and vinyl groups passing through rotational barriers along a rotation path.

(b) For CH_5N, corresponding to methylamine, there is more than one absolute minimum as by symmetry some conformers correspond to the same energy. Here we have the possible rotation of a methyl group around the CN bond leading to a rotation barrier, but also a possible inversion of the amine group through an inversion barrier leading after rotation to another equivalent minimum. It is a matter of fact that rotation and inversion are quite different chemical processes involving different reaction coordinates and energies on the same surface. Moreover, many other minima could be found on the border of this hypersurface. Any chemical reaction starting from, or leading to, methylamine may be analysed on the same hypersurface. In our example we could think of:

$$
\begin{aligned}
CH_3NH_2 \rightarrow\ & \dot{C}H_2\text{—}NH_2 + H \\
& CH_3\text{—}\dot{N}H + H \\
& CH_2\text{=}NH + H_2 \\
& \dot{C}H_3 + \dot{N}H_2 \\
& C + N + 6H
\end{aligned}
$$

each leading to a reaction energy, a reaction path along some coordinates, crossing or not some activation barriers. As this turns out to be chemical reactivity, we shall leave it to further sections and continue this section with more practical considerations on the equilibrium structure search and conformational analysis, following throughout the methylamine example for the numerical application.

6.1.2 The Practice of Equilibrium Structure Search

Generally when a chemist looks for an equilibrium structure, he already has an idea of the isomer, and even conformer, he is interested in. He has then to propose an initial guess on its structure and compute that point of the hypersurface. As his chemical intuition is often good enough, the equilibrium structure will not be very far from his guess and the obtention of this structure may be easy as we are in a quadratic region of the hypersurface.

First of all we have to build a molecular geometry. The initial parameters will be found in some tables of atomic distances[1] if the experimental structure is known, or we shall refer to some standard geometry.[2] At this point, we have to transform the internal parameters in cartesian coordinates. Many programs exist in the QCPE catalog,[3] but the Z-matrix technique of Pople and coworkers[4] is very commonly used. By this technique, the $3N$ cartesian coordinates are reduced to $3N-6$ variables by an appropriate choice of the cartesian axis. The usual choice is:

First atom at origin	: $x_1 = y_1 = z_1 = 0$
Second atom along the Z axis	: $x_2 = y_2 = 0$
Third atom in the XZ plane	: $y_3 = 0$

The origin is generally an atom but it could also be a point specified by some local symmetry, in which case we shall call this point a *dummy atom.*

Sometimes dummy atoms may be introduced to discuss some interesting parameter such as, for example, an elevation angle of an atom with respect to a plane. In this case the number of internal parameters may increase artificially. These artificial parameters may be frozen at the beginning of a search as they are of no chemical interest.

The structure optimization needs the $3N-6$ parameters: more parameters introduce useless optimization parameters and linear dependencies which may slow the search, while less parameters correspond to a limited optimization which will not lead to an equilibrium structure but to a point on the hypersurface optimized under certain constraints (this is the case in the optimization of a reaction path).

Table 6.1 gives an example of the methylamine molecule using the Z-matrix technique on a standard geometry. The first column gives the bond

Table 6.1 Standard geometry Z-matrix for methylamine

	Bond length		Bond angle		Dihedral angle	
C						
N	CN	1.47				
D_1	ND_1	1.00	CND_1	126.00		
D_2	ND_2	1.00	D_1ND_2	90.00	CND_2; D_1ND_2	0.0
H_1	CH_1	1.09	NCH_1	109.47	H_1CN; D_2NC	0.0
H_2	CH_2	1.09	NCH_2	109.47	H_2CN; H_1CN	120.0
H_3	CH_3	1.09	NCH_3	109.47	H_3CN; H_1CN	240.0
H_4	NH_4	1.01	D_1NH_4	54.735	H_4ND_1; D_2ND_1	90.0
H_5	NH_5	1.01	D_1NH_5	54.735	H_5ND_1; D_2ND_1	270.0

Optimization parameter (- - -).

distances, the second column gives the bond angles and the last column gives dihedral angles between the planes defined by the bonds. Figure 6.1 gives the corresponding configuration more explicitly.

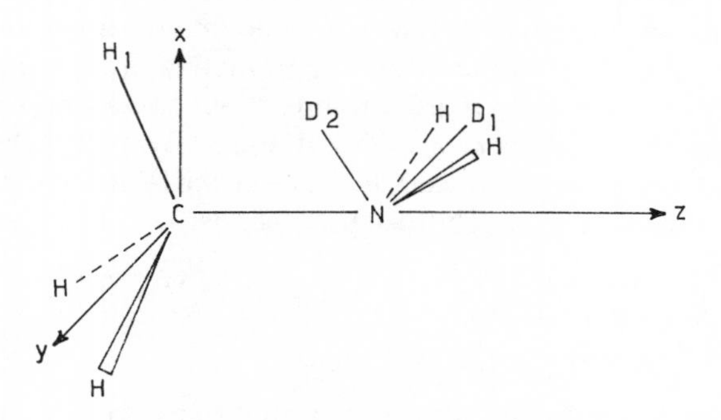

Figure 6.1 Methylamine configuration

As later on we shall discuss the rotation of the methyl group and the inversion of the amine group, two dummy atoms were introduced to define the elevation angle of the amine group with respect to the CN bond.

From now on we may start the equilibrium structure search using the gradient or force technique. This method rests on the fact that for the equilibrium structure the first derivative of the potential hypersurface vanishes. Or, in more physical terms, as the gradient of a potential is a force, the equilibrium structure is the one for which residual forces acting on the nuclei are zero. Starting at the initial guess, one may compute the energy and its derivative. The gradient will point towards the minimum and we may modify our guess moving along the force direction by a certain step. Starting over again, we shall after a few steps converge and attain the local minimum. A more explicit description of this method will be given in the third part, but to illustrate the technique in our example Table 6.2 gives the

Table 6.2 Forces computed in the STO-3G basis set on the $3N-6$ significant parameters of methyl amine (atomic units)

Total energy:	−94.0299759				
CN	0.08704				
		CND_1	−0.02860		
CH_1	0.03133	NCH_1	0.10733	H_1CN; D_2CN	0.0
CH_2	−0.01270	NCH_2	0.01541	H_2CN; H_1CN	−0.60883
CH_3	−0.01270	NCH_3	0.01541	H_3CN; H_1CN	−0.60883
NH_4	0.16006	D_1NH_4	−0.05090		
NH_5	0.16006	D_1NH_5	−0.05090		

forces computed for our initial guess on the $3N-6$ parameters of interest. The final values obtained at the STO-3G level are reported in Table 6.3. The optimized results are obtained with a gain in energy of some 1.8 kcal mol^{-1} and reveal the behaviour of the internal parameter (i.e. in respect of the symmetry plane only).

6.1.3 A Conformational Analysis

It is now tempting to look at the chemical behaviour of the methylamine molecule. Does the methyl group rotate easily? Is the inversion of the amine group possible? To answer these questions we shall have to explore the hypersurface near the equilibrium structure.

As the force technique seems very relevant, we shall use it to optimize the structures near the equilibrium, but we shall have to introduce some constraints on our technique, otherwise we shall irremediably come back to our initial structure. We shall thus define two parameters: α, the elevation angle of the amine group, and θ, the rotation angle of the methyl group. The optimized geometry reveals that the methyl group does not exactly present a

Table 6.3 Final optimization results of CH_3NH_2

Z-matrix (angström and degree)					
CN	1.4855				
ND_1		180-α	119.08		
ND_2					
CH_1	1.0931	NCH_1	113.74	θ	0.0
CH_2	1.0888	NCH_2	109.14	H_2CN; H_1CN	120.93
CH_3	1.0888	NCH_3	109.14	H_3CN; H_1CN	239.07
NH_4	1.0333	D_1NH_4	52.20		
NH_5	1.0333	D_1NH_5	52.20		
Energy STO-3G: −94.0328587 (atomic units)					
Forces (atomic units)					
CN	−0.00007				
		180-α	−0.00007		
				θ	0.0
CH_1	0.00030	NCH_1	−0.00058	H_2CN; H_1CN	0.00125
CH_2	−0.00051	NCH_2	−0.00005	H_3CN; H_1CN	−0.00125
CH_3	−0.00051	NCH_3	−0.00005		
NH_4	0.00050	DNH_4	0.00001		
NH_4	0.00050	DNH_5	0.00001		

threefold axis, but as its deviation is meaningless, we shall refer to it in what follows, i.e. after rotation of 120° we have found the same structure as the fully optimized one. Table 6.4 gives some relative energy values for fully optimized points of the hypersurface, represented in Figure 6.2.

Table 6.4. Relative energies of the methylamine rotation inversion process (in kilocalories per mole)

α \ θ	0	30	60	90	120
60	0.0	1.25	2.79	1.25	0.0
30	5.98	6.82	7.67	6.82	5.98
0	10.47	10.47	10.47	10.47	10.47

It appears that the rotation needs some 2.79 kcal mol^{-1} to cross the barrier while the inversion needs 10.47 kcal mol^{-1}. In its planar structure, methylamine rotates freely without any activation energy. These computations can be a refined little, e.g. by allowing α to relax while θ changes, but generally, as well as in our example, α changes less than 1° and the fully optimized barrier becomes 2.77 kcal mol^{-1} instead of 2.79 kcal mol^{-1}.

During the rotation, the geometry does not change significantly, but the CN bond and angles rearrange smoothly to go from one value to the other.

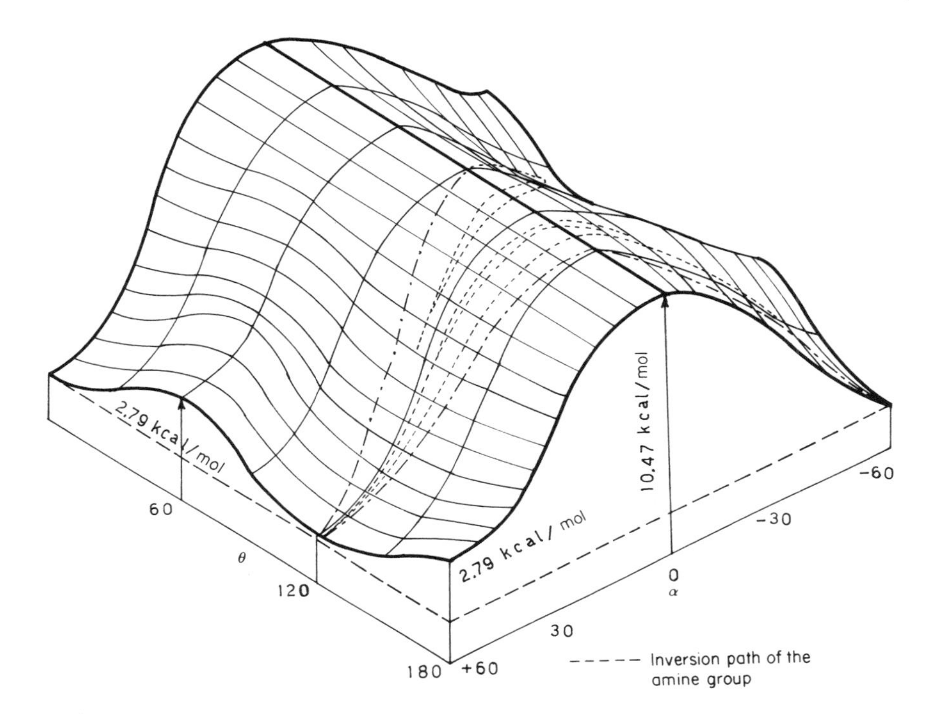

Figure 6.2 Hypersurface for rotation and inversion of methylamine

The situation is slightly different for the inversion. As in the planar structure of the amine group, the hybridization of the nitrogen atom becomes of the sp^2 type, the CN bond length shortens by some 0.035 Å, the HNH angle goes up to 120.9°, while the NH bond length shortens by 0.025 Å.

These variations show important electronic rearrangements which need more comment and criticism on the reliability of our results.

6.1.4 Comparison with Experiment

When we compare our results to some experimental values, we should take care to check the reliability of our results. Some errors may indeed appear due to approximate calculations.

6.1.4.1 For the equilibrium structure

Although the basis set is scaled and gives fairly good structures, its small size could induce a false absolute conformation due to the fact that certain orientations complete the limited basis set better than others. Extension of the basis set to a larger one would reveal this fact.

6.1.4.2 For energy results

The origin of the errors rests in two approximations: first, the limitation of the basis set leading to a basis error; second, the correlation error due to the Hartree–Fock approximation. As we are mainly interested in differences between energies, cancellation of errors may appear under certain conditions. It is admitted that when the number and nature of electronic pairs are conserved, the correlation error remains constant. This is the case when the rotation of a methyl group is considered, so the rotation barriers should be correctly computed. In the case of inversion, as the nitrogen changes its hybridization there will be more important electronic reorganizations and a non-negligible error will be introduced. The geometry of the planar structure clearly reveals this fact. Thus, in order to obtain good results correlation must be introduced.

The discussion above criticizes the reliability of the potential energy surface, but in order to compare the results with experimental values one has to add the ZPE and thermal corrections which are admitted to be constant when we compare energy barriers to experimental values. This is not necessarily true as, for instance, the planar structure describes a transition state which presents the loss of one vibrator by becoming imaginary. Thus special care should be taken with inversion energies, as many sources of error may appear. For rotational barriers the results are often very encouraging, suggesting that the error cancels along the rotation of the groups. Moreover, the results obtained, without full optimization, using standard geometries are currently good enough to be compared with the experimental values.[5]

6.2 THE ELECTRONIC STRUCTURE OF MOLECULES

This section is devoted to a static description of the electronic structure of molecules. The topic is obviously related to the problem of the chemical bond discussed in several sections of Part I from the viewpoint of loge theory. We will not study explicitly the nature of the chemical bond as many excellent papers and textbooks have deeply analysed this subject.[6–8]

The main object of this section is in fact to show that the localized orbital approach not only gives some support to the early electronic theories of bonding but also enables generalization of loge theory for large molecules. We shall first describe briefly the Linnett theory of the electronic structure of molecules based on the double quartet rule.

6.2.1 The Linnett Theory

In the Lewis theory of valence the covalent bonds between atoms are formed by sharing pairs of electrons. This pairing of electrons leading to stable octets is identified as the driving force for bond formation. In Lewis structures of molecules, shared electrons are represented by dots between the bonded atoms. Furthermore, after the discovery of the spin, paired electrons were recognized to have opposite spins. Thus, for example, the electronic structure of the methane molecule can be represented by formulas such as:

$$\begin{array}{c} \mathrm{H} \\ \bullet \\ \mathrm{H}\bullet\bullet\mathrm{C}\bullet\bullet\mathrm{H} \\ \bullet \\ \mathrm{H} \end{array} \qquad \begin{array}{c} \mathrm{H} \\ \uparrow\downarrow \\ \mathrm{H}\uparrow\downarrow\mathrm{C}\uparrow\downarrow\mathrm{H} \\ \uparrow\downarrow \\ \mathrm{H} \end{array} \qquad \begin{array}{c} \mathrm{H} \\ | \\ \mathrm{H}-\mathrm{C}-\mathrm{H} \\ | \\ \mathrm{H} \end{array}$$

the latter being the modern representation commonly used by chemists. Completed by the theory of resonance,[6,9] the Lewis electronic theory of bonding has been very successful in giving a static description of the mean positions of electron pairs in most closed-shell systems. We show here that the Linnett approach[7] gives a more detailed description of electronic structures, particularly in the case of open-shell systems. This is based on the concept of charge and spin correlation of electrons, a meaning which is obvious in terms of Coulomb and Pauli interactions between particles of the same charge and the same spin. More explicitly, electrons having the same spin tend to avoid one another and to keep apart more than electrons of opposite spin which influence each other only by virtue of their negative charge. This is a qualitative justification of the concept of electron pair which is the basis of the modern theory of valence.

We summarize now the main results of the Linnett approach. Let us first consider a collection of uncharged particles having spins which are confined to a circle. Explicit calculations show that it is satisfactory to consider two sets of particles according to their spin. The most probable disposition of the particles is that in which the particles of each set are as far as possible from each other. If the two sets contain the same number of particles, the most probable configuration is that in which the particles are paired. If the two sets contain unequal numbers of particles, then the probability does not depend on the mutual disposition of the two sets. For electrons (negatively charged particles) of the same spin the effects of charge correlation will obviously add to the effects of spin correlation, but for electrons of opposite spins the effects of charge correlation will operate against the effects of spin correlation. Thus, in closed-shell systems, the charge correlation tends to reduce the probability of pair formation. In order to be used for atoms, Linnett's theory must be generalized in three dimensions. In this case, we shall look for the most probable disposition of electrons on a sphere. For atoms having an L valence shell, the maximum number of electrons of each spin will be four. Assuming that electrons are approximately equidistant from the nucleus, it is easy to anticipate the most probable configurations of electrons of the same spin:

Number of electrons	Most probable arrangement
2	Linear
3	Equilateral triangle
4	Tetrahedron

These results suggest the following arrangement for the eight electrons of the external shell of neon. Those four having parallel spins will be disposed at the corners of a regular tetrahedron and the other four having spins opposite to those of the first set will have a similar arrangement. The correlation between the electrons of each set is expected to be very strong but the correlation between the two tetrahedral sets will be relatively small. The most probable configuration of the eight electrons suggested by these considerations is shown in Figure 6.3(a). The fluoride anion should have the same electronic structure.

Thus Linnett's theory replaces the Lewis octet corresponding to four pairs of electrons of opposite spins by a double quartet corresponding to two sets of four electrons having the same spin within a given set. The comparison of the two models is given in Figure 6.3(b).

The electronic structure of molecules can be anticipated using the double quartet rule, which states that each atomic core must be surrounded by two sets of four electrons weakly correlated. However, it must be stressed that in some circumstances the two sets may be drawn together if this leads to a lowering in the potential energy of the system.

The Linnett procedure for finding the electronic structure of a molecule consists of localizing the valence electrons around the atomic cores taking account of the charge and spin correlation effects and of lowering the potential in the regions lying between the cores. As an application of this procedure we shall look for the electronic structure of fluorohydric acid HF. For this purpose, let us analyse what happens when a proton approaches a fluoride anion. Two electrons of opposite spin will be attracted in the region of low potential lying between H^+ and the fluorine core. In other terms, two

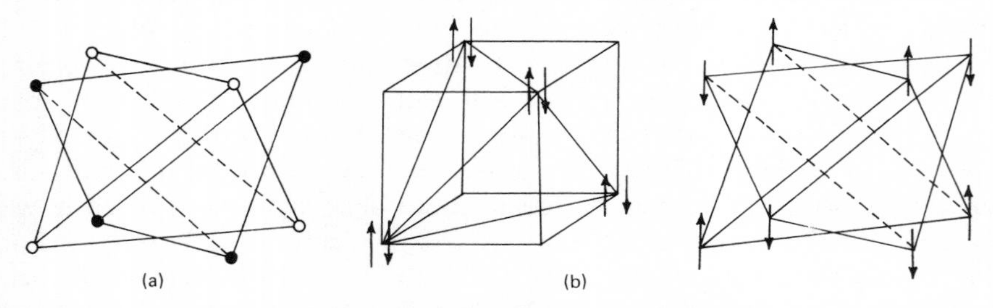

Figure 6.3 (a) Fluoride anion electronic structure. (b) Comparison between the Linnett and Lewis models for the fluoride anion. Reproduced by permission of C.N.R.S. Gauthier–Villars

corners of the two tetrahedral sets will coincide in this region, the other corners always being separated because of the Coulomb repulsions between the electrons. Thus the electronic structure of HF may be described such as:

The cross (×) and the circle (○) represent respectively electrons with α and β spins. So the HF bond is formed by a pair of electrons of opposite spin shared by the two atoms. According to Linnett, this is a spatial pair represented by a heavy line because the two electrons occupy the same spatial (bond) orbital. The other couples of electrons of opposite spins (represented by light lines) would not occupy the same spatial orbital. They would therefore not be considered as true lone pairs in the Lewis sense.

Thus, in general, the octet rule will be replaced by the double quartet rule, according to which each atomic core tends to be surrounded by two sets of four electrons disposed at the corners of two tetrahedra which may eventually be drawn together. In molecules, the number of electrons of a given spin is generally higher than four and one has to use the data of Table 6.5, which gives the most stable arrangements of sets of two to seven

Table 6.5 Most probable arrangements of different sets of electrons

Number	Arrangement
2	
3	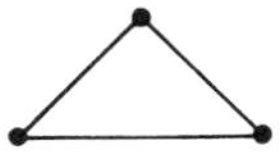
4	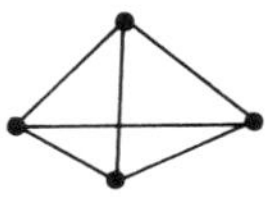
5	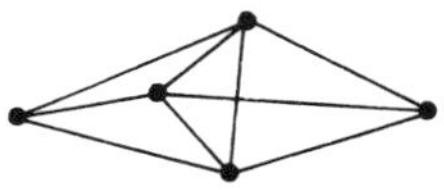
6	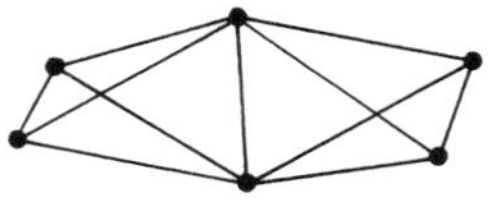
7	

Table 6.6 Linnett's structure of diatomic compounds

Number of valence electrons	Compound	Linnett's structure	More conventional structure
9	CN, N_2^+, CO^+	×○ ×A×○B×○ ×○	× A≡B |
10	N_2, CN^-, CO, NO^+, C_2^{2-}	×○ ×○A×○B×○ ×○	|A≡B|
11	NO, O_2^+	○ ○×A×○B×○ ×○×	$\|\overset{\times\ \circ\ \times}{A=B}\|$
12	O_2, NO^-	× ○ × ○×A×○B×○ × ○ ×	$\|\overset{\times\ \circ\ \times}{\underset{\times\ \circ\ \times}{A-B}}\|$
13	O_2^-	○× ×○ ○×A×○B×○ ○× ×○	$\|\overset{\times\ \circ\ \times}{\underline{A}-\underline{B}}\|$
14	F_2	○× ×○ ○×A×○B×○ ○× ×○	$\|\underline{\bar{A}}-\underline{\bar{B}}\|$

electrons. We describe below some characteristic results obtained by using the Linnett procedure considering only covalent compounds. In some cases several structures can be proposed. The choice of the most probable one is often determined by means of empirical arguments.

Electronic structures of a series of diatomic systems are schematized in Table 6.6. Further information on these results may be obtained from Linnett's book, especially concerning the problem of the choice of the best structure(s).[7]

In order to clarify the results of Table 6.6 we shall deal with a specific example, i.e. the fluorine molecule. This contains 14 valence electrons, seven having a α spin ($\uparrow$ or $\times$) and seven a β spin ($\downarrow$ or $\circ$). The most probable arrangement of each set is given below:

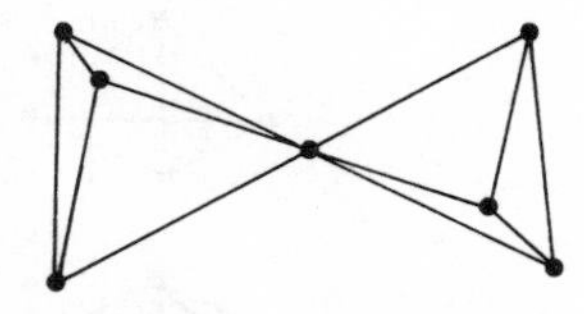

The corners (× and ○) are led to coincidence in the low potential region between the two fluorine cores. The corresponding electrons form a spatial pair. The other corners of the two sets do not coincide but are disposed in a starred arrangement. Thus there are no true lone pairs around the two fluorine cores. We have then the following representations of the electronic structure of F_2:

×○ ×○
○×F×○F×○
○× ○×   or $|\bar{F}-\bar{F}|$

The double quartet rule is obeyed in this molecule.

In all the compounds of Table 6.6 (those with nine valence electrons excepted), each atom has an octet of electrons made up of four electrons of one spin and four of the other. Linnett's theory is very useful for understanding the existence, the stability and even the reactivity of diatomic compounds. This may be very easily extended to triatomic and tetratomic species but not to larger systems for which its use becomes generally very intuitive and its results as ambiguous as those of resonance theory.

6.2.2. The Localized Orbital Approach

6.2.2.1 Survey of the Boys localization procedure

It is well known that the SCF molecular orbitals of the Hartree–Fock–Roothaan procedure are not uniquely defined. Canonical orbitals which diagonalize the Hartree–Fock matrix are delocalized on the whole system. They are particularly useful when studying excited and ionized states of the molecule. It is usually possible by an appropriate unitary transformation to relocalize these orbitals in certain regions of the molecular space corresponding to the classical concepts of atomic core, bond and lone pair. Thus, one will obtain a description of electronic structures which is very close to the usual one and particularly interesting to describe the chemical bonds in molecular systems. It must be stressed that the localized and delocalized approaches are completely equivalent as they lead to the same wave function and molecular properties.

As shown in Chapter 5, localized molecular orbitals may be obtained by using a criterion of spatial localization proposed by Boys[10] in which a unitary transformation maximizes the sum of the squares of the orbital centroid distances from an arbitrarily defined origin of the molecular coordinate system. The Boys method provides the shape of localized molecular orbitals (LMO) and interesting information such as the position of the centroid of charge of each LMO and the dispersion about this centroid.

The Boys procedure allows us to relocalize not only the doubly occupied molecular orbitals of closed-shell systems but also the singly occupied orbitals obtained by the UHF method.[11] In this case, the centroids of charge give exactly the same description of electronic structures as in the Linnett approach. Thus it can be pointed out that the Boys localization procedure is a mathematification of the qualitative Linnett theory. In this way the correct meaning of the Linnett symbols ($\times$ and $\circ$) can be found. We may assume that they represent not localized electrons but centroids of charge of localized orbitals. Within this interpretation Linnett's approach agrees much better with the ideas of quantum mechanics such as the wave character of electrons which prevents their localization. The localized orbital approach allows us to interpret, to justify and to generalize the Linnett theory. Figure 6.4 demonstrates the perfect equivalence of the descriptions

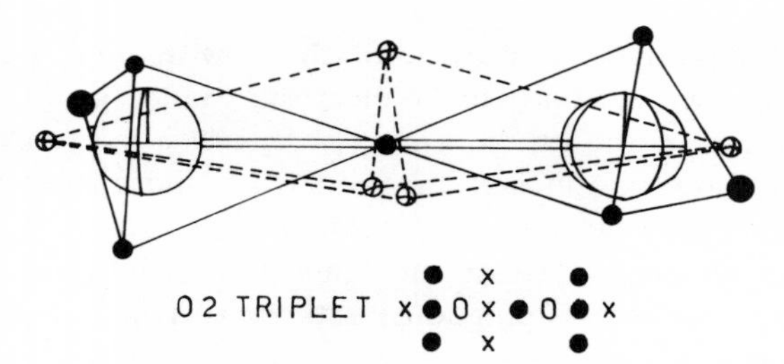

Figure 6.4 Electronic structure of $O_2(^3\Sigma_g^-)$. Reproduced by permission of C.N.R.S. Gauthier-Villars

of electronic structures in terms of centroids of charge and of Linnett symbols.

6.2.2.2 Some illustrations of the LMO approach

Open-shell systems

Electronic structures of a series of atoms, radicals and diradicals are described in Table 6.7. The centroids of charge of core orbitals which are localized on the nuclei are not indicated. Obviously the spatial arrangement of centroids of a given spin (α or β) depends on their number (see also Table 6.5). The distance between α (or β) centroids is larger than the distance between centroids corresponding respectively to α and β spin orbitals. This result is an illustration of the basic idea of Linnett's theory, according to which the correlation between electrons of the same spin is stronger than the correlation between electrons of opposite spins.

In the oxygen atom, the distribution of the centroids does not indicate the existence of spatial pairs. In the CH system, the presence of two centroids (α and β) near the H core suggests the existence of a covalent bond between the two atoms. Electronic structures of $\dot{C}N$, NO and O_2 are identical to those of Linnett's theory but for the $\dot{C}N$ radical another structure was also proposed by this author. In homonuclear diatomic compounds the centres of the two polyhedra coincide with the centre of gravity of the negative charges of the system. In heteronuclear diatomic molecules the centres of the polyhedra are separated and the centroids are drawn nearer to the most positive core.

Other examples of electronic structures obtained by the LMO approach are given in Figures 6.5, 6.6 and 6.7 for diatomic systems and in Table 6.8 for some polyatomic open-shell compounds. These results show the general usefulness of the LMO method.

Closed-shell systems

For closed-shell systems RHF and UHF methods give identical results and molecular orbitals are then doubly occupied. The Boys localization procedure will now provide the positions of the centroids of charge associated with the different electron pairs of the system.

Table 6.7 Electronic structure of open-shell systems. Reproduced by permission of C.N.R.S. Gauthier-Villars

System	Number of centroids α	β	Electronic structure	Representation
Carbon (^3p)	3	1	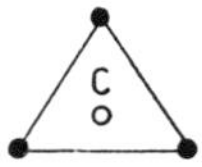	
Carbon (^5s)	4	0	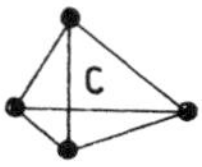	
Nitrogen	4	1	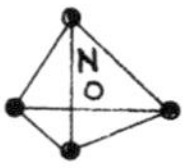	
Oxygen	4	2	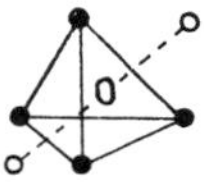	
Fluorine	4	3	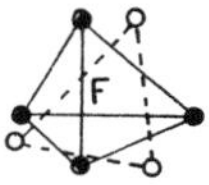	
CH (quadruplet)	4	1	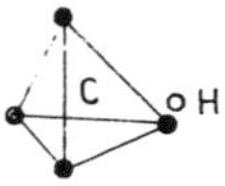	 :C•×H
B_2 (triplet)	4	2	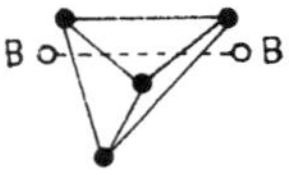	 •× •× B••B
CN	5	4	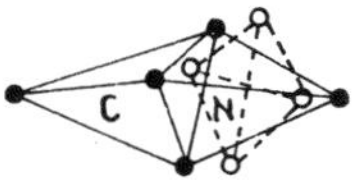	 •C•×N×•
NO	6	5	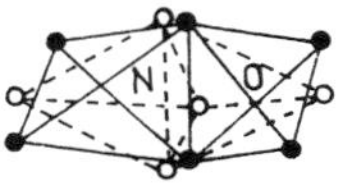	 ×:N×•O:×
O_2 (triplet)	7	5	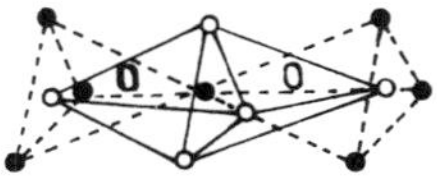	 ×•O×•O•×

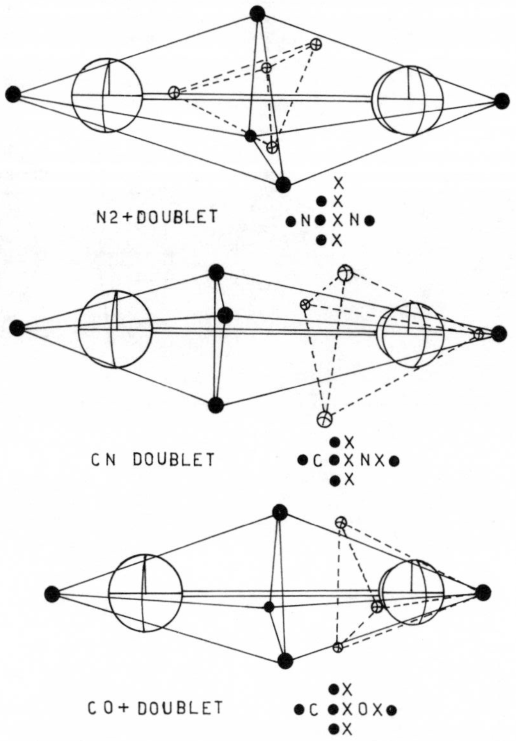

N2+DOUBLET
CN DOUBLET
CO+DOUBLET

Figure 6.5 Electronic structure of nine-valence electrons in open-shell systems. Reproduced by permission of C.N.R.S. Gauthier-Villars

CO- DOUBLET
NO DOUBLET
O2 + DOUBLET

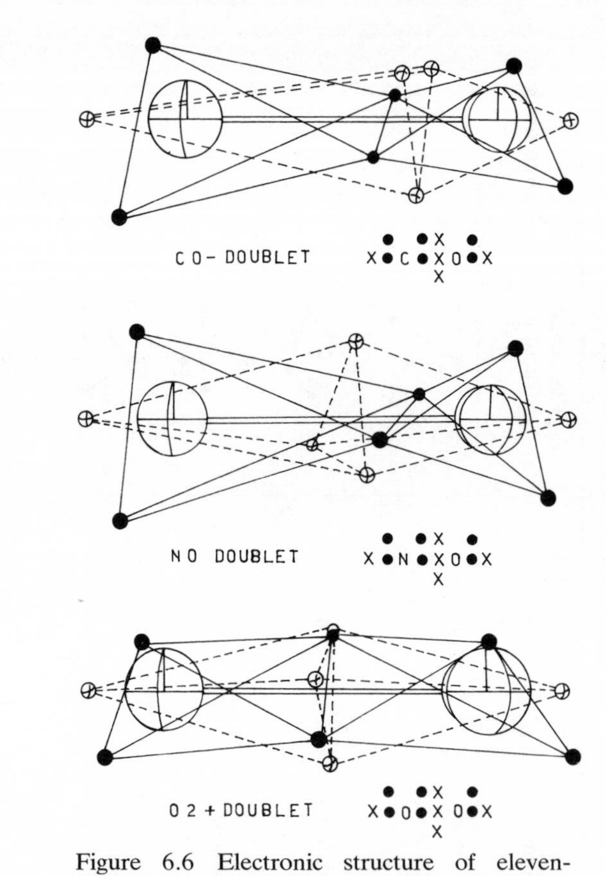

Figure 6.6 Electronic structure of eleven-valence electrons in open-shell systems. Reproduced by permission of C.N.R.S. Gauthier-Villars

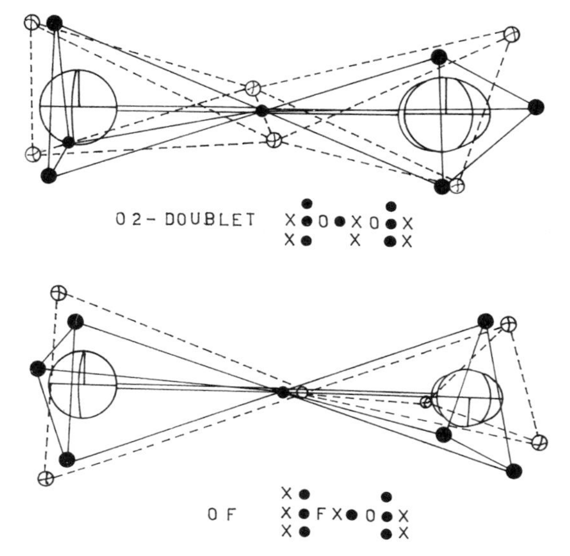

Figure 6.7 Electronic structure of O_2^- and FO (thirteen-valence electrons). Reproduced by permission of C.N.R.S. Gauthier-Villars

Table 6.8 Electronic structure of some polyatomic open-shell systems. Reproduced by permission of C.N.R.S. Gauthier-Villars

System	Number of α centroids	Number of β centroids	Notation
HCO	6	5	H×•C:×O:×
CO_2^+	8	7	×•O:×C×:O•×
NCO	8	7	×•N:×C×:O:×
NO_2	9	8	:×O:×N:×O×:
ClO_2	10	9	×:O×•Cl×•O:×
NO_3	12	11	:×O×:N×: (×•O×:) (×•O:×)

Table 6.9 Electronic structure of closed-shell diatomic species. Reproduced by permission of C.N.R.S. Gauthier-Villars

Molecule	Valence electrons	Centroids	Structure	Notation	Similar cases
C_2	8	4	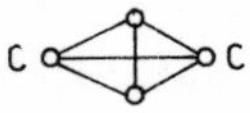	$\|C{=}C\|$	CN^+
N_2	10	5	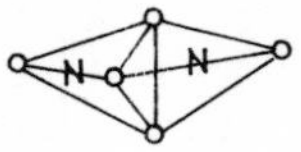	$\|N{\equiv}N\|$	C_2^{2-}, CN^-, NO^+, CO, HCN
HNO	12	6	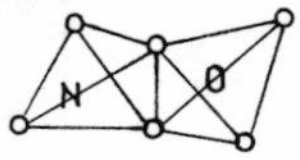	$H{-}\bar{N}{=}O\rangle$	N_2H_2
O_2^{2-}	14	7	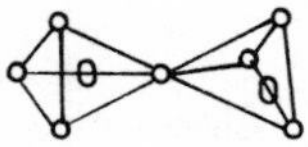	$\|\bar{\underline{O}}{-}\bar{\underline{O}}\|$	N_2H_4

Therefore it is no more possible to obtain electronic structures of the Linnett type. These will be in fact of the Lewis type, each centroid corresponding to a pair of electrons of opposite spins. Electronic structures of some diatomic closed-shell molecules and of some polyatomic systems are described respectively in Tables 6.9 and 6.10.

Remarks

(a) Some electronic structures obtained in the LMO approach may suggest the existence of bonds formed by more than six electrons. As a consequence the Lewis octet rule may seem to be violated. Some typical examples of this situation are given below:

N_2^+ 

C_2 $C{\equiv}C$

N_3^- $|N{\equiv}N{\equiv}N|$

According to these formulas, N_2^+ would be more stable than N_2, which is in contradiction with experimental facts, and the central nitrogen atom of N_3^- would be surrounded by 12 electrons. It fact, the presence of a centroid in a certain molecular region does not necessarily imply that a given number of electrons (one or two) are confined in the same region because the orbitals are not strictly localized. In order to obtain quantitative information about the number of electrons as-

Table 6.10 Electronic structure of closed-shell polyatomic species. Reproduced by permission of C.N.R.S. Gauthier-Villars

Number of cores	Number of centroids	Structure	Notation	Examples
3	8	(a) Symmetrical case	$\lvert A\equiv B\equiv C\rvert$	NO_2^+, CO_2, N_3^-, CN_2^{2-}
		(b) Dissymmetrical case	$\lvert A\equiv B-\underline{\overline{C}}\rvert$	CNO^-, N_2O NCO^-
3	9	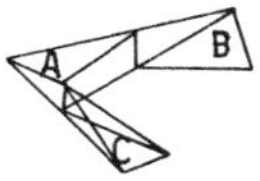	$\lvert\underline{B}=\overline{A}=\underline{C}\rvert$	NO_2^-, O_3
4	10	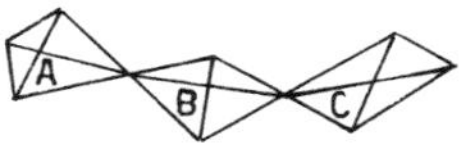	$\lvert\underline{\overline{A}}-\underline{\overline{B}}-\underline{\overline{C}}\rvert$	F_2O, HOClO
4	11	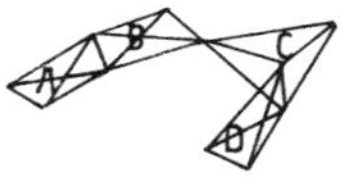	$\langle A=\overline{B}-\overline{C}=D\rangle$	N_2O_2
4	12	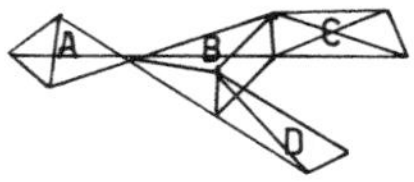	$\lvert\underline{\overline{A}}-B(=C\rangle)(=D\rangle)$	$HONO_2$, H_2NNO_2
4	13	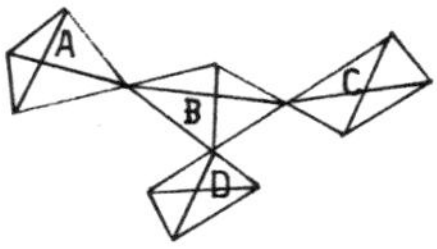	$\lvert\underline{\overline{A}}-\overline{B}(-\lvert\underline{D}\rvert)-\underline{\overline{C}}\rvert$	$HOSO_2$, $HOClO_2$

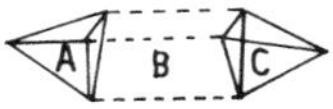

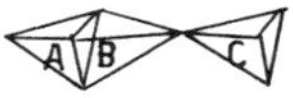

sociated with bonds and lone pairs, we have to integrate the electronic density in well-defined regions of the molecular space such as the loges or the pseudologes (see Section 6.4.3). Such integrations have been performed, for N_2^+, C_2 and N_3^-, in volumes defined by planes and half-spheres which correspond approximately to bond and lone pair loges. The results are shown in Figure 6.8. One finds that in N_2^+ the chemical bond is intermediate between the double and triple bonds. In C_2 it appears to be a double bond but in N_3^- there remain 10 electrons

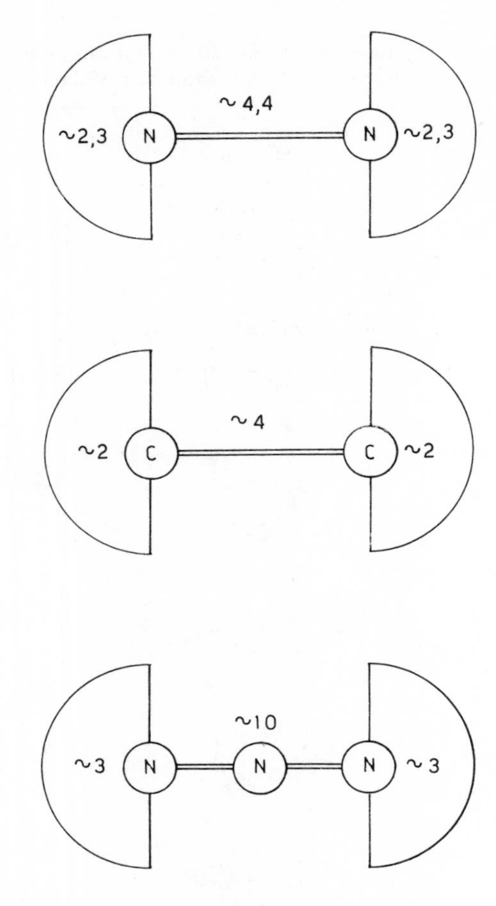

Figure 6.8 Electronic structure of C_2, N_2^+ and N_3^-

around the central nitrogen atom. More refined calculations should be made to confirm these preliminary results.

(b) Although electronic structures in terms of centroids may occasionally seem confused and have then to be used with caution, in most cases they are in perfect agreement with those suggested by experimental data or other theoretical approaches. A typical illustration of this remark is given by electronic structures of molecular oxygen and its positive and negative ions:

O_2	$\lvert O=O \rvert$	'Double bond'
O_2^+	$\lvert O\equiv O \rvert$	Five-electron bond
O_2^-	$\lvert O-O \rvert$	Three-electron bond
O_2^{2-}	$\lvert \bar{O}-\bar{O} \rvert$	Single bond

These LMO results lead to the same conclusions as the shell model based on the energetic diagram of the symmetry orbitals of diatomic compounds: the positive ion O_2^+ is more stable than the molecule O_2, which is itself more stable than its negative ions O_2^- and O_2^{2-}. This is also suggested by experimental facts.

6.2.3 The Pseudologe Approach

As we have shown in several sections of Part I, the loge theory of Daudel is the most rigorous approach to the electronic structure of atoms and molecules and also to the concept of the chemical bond. Using criteria such as the probability of occurrence of electronic events, the missing information function or the fluctuation of the number of electrons, this theory leads to partitions of the molecular space into fragments associated predominantly with cores, bonds and lone pairs and generally containing a definite number of electrons with a certain organization of their spins.

Unfortunately, the loge theory requires sophisticated calculations which are as yet impossible to perform on large molecules. We shall describe here a much simpler but obviously less rigorous method to partition the molecular space into loges. It is based on the intuitive assumption that the domains in which molecular orbitals are best localized are approximate loges. This is the basic idea of the pseudologe approach.[12] We shall call the pseudologe a region of the molecular space in which the contribution of a given localized orbital to the total electronic density is much larger than the contribution of any other LMO. The frontiers of these pseudologes are simply the collection of points in which the isodensity contour lines of the LMOs, taken two by two, intersect. For any pseudologe Ω, the following properties can be calculated:

(a) The mean number of electrons:

$$N = \int_{\Omega} \rho(\mathrm{M})\, \mathrm{d}v \tag{6.1}$$

(b) The volume:

$$V = \int_{\Omega} \mathrm{d}v \tag{6.2}$$

(c) The position of the centre of charge (centroid):

$$\langle u \rangle = \frac{1}{N} \int_{\Omega} u\rho(\mathrm{M})\, \mathrm{d}v \qquad \text{for} \qquad u = (x, y, z) \tag{6.3}$$

(d) The dispersion of the electronic charge (quadratic or second moment):

$$\langle uv \rangle = \int_{\Omega} uv\rho(\mathrm{M})\, dv \qquad \text{for} \qquad (u, v) = (x, y, z) \tag{6.4}$$

Another way to define this dispersion is to consider the expectation value of the spherical quadratic moment operator with its origin at the centroid of charge:

$$\langle r^2 \rangle = \langle x^2 \rangle + \langle y^2 \rangle + \langle z^2 \rangle \tag{6.5}$$

This is often assumed to be a measure of the 'size' of the corresponding domain (of the LMO or, in this case, of the pseudologe).[13] We shall now give some typical results of the pseudologe approach.[14]

First, Figure 6.9 shows tridimensional representations of the BH molecule and its valence pseudologes. The external density contour corresponds to $\rho = 0.001\ \mathrm{e\ (a.u.)}^{-3}$. In the following we shall not consider individual lone

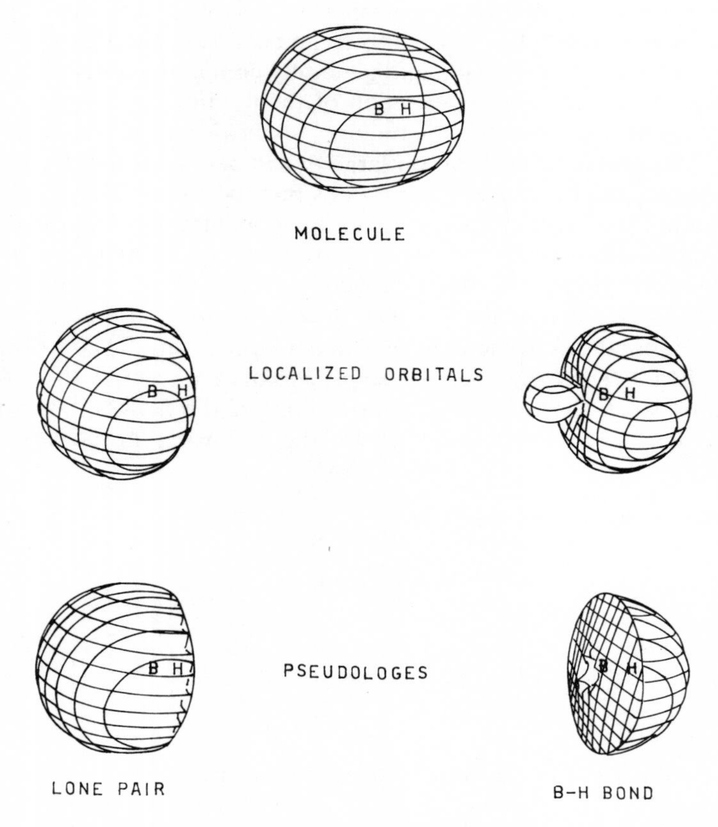

Figure 6.9 Tridimensional representation of the molecule BH and its valence pseudologes

Table 6.11 Mean number of electrons (the number of electrons inside the contour $\rho = 0.001$ being given in parentheses)

Molecule	XH_n XH_3YH_n	Core X	Core Y	Bond XY	Bond XH	Bond YH	Lone pair(s) X, Y
LiH	4.0 (3.837)	2.0	—	—	2.000	—	—
BeH_2	6.0 (5.868)	2.0	—	—	2.000	—	—
BH	6.0 (5.916)	2.0	—	—	1.999	—	2.001
CH_4	10.0 (9.905)	2.0	—	—	2.000	—	—
NH_3	10.0 (9.925)	2.0	—	—	1.988	—	2.036
H_2O	10.0 (9.953)	2.0	—	—	1.937	—	4.126
HF	10.0 (9.960)	2.0	—	—	1.876	—	6.124
C_2H_2	14.0 (13.906)	2.0	2.0	5.710	2.145	2.145	—
C_2H_4	16.0 (15.885)	2.0	2.0	3.790	2.052	2.052	—
C_2H_6	18.0 (17.870)	2.0	2.0	1.971	2.004	2.004	—
CH_3NH_2	18.0 (17.878)	2.0	2.0	1.966	2.010	2.004	2.003
CH_3OH	18.0 (17.889)	2.0	2.0	1.902	2.013	1.959	4.100
CH_3F	18.0 (17.911)	2.0	2.0	1.829	2.011	—	6.139
BH_3NH_3	18.0 (17.851)	2.0	2.0	2.011	1.998	2.000	—

pair pseudologes but preferably an 'unshared' pseudologe which is simply the sum of the domains determined from the lone pair localized orbitals (see, for example, F in HF and O in H_2O). In the same way we shall consider only one CC bond pseudologe in C_2H_4 and C_2H_2. In order to avoid the error due to truncation of the 'molecular volume', the mean electronic populations of valence pseudologes have been normalized. Their sum is equal to the total number of electrons of the molecule, the populations of the core pseudologes being taken equal to two. The mean number of electrons in each pseudologe of a series of molecules of growing complexity is given in Table 6.11.

This result is always found to be close to two or a multiple of two, which allows us to assume *a posteriori* that the pseudologes are very similar to the loges as they were initially defined by Daudel using the probability criterion. However, in some cases, pseudologe populations are significantly different

Table 6.12 Volumes of molecules and pseudologes (in cubic atomic units)

Molecule	XH_n XH_3YH_n	Core X	Core Y	Bond XY	Bond XH	YH	Lone pair(s) X, Y
LiH	218.47	10.40	—	—	209.02	—	—
BeH_2	275.81	3.09	—	—	136.36	—	—
BH	202.51	1.25	—	—	93.43	—	107.84
CH_4	256.30	0.65	—	—	63.79	—	—
NH_3	197.95	0.36	—	—	54.86	—	33.43
H_2O	147.55	0.22	—	—	50.81	—	46.06
HF	103.51	0.14	—	—	50.59	—	52.78
C_2H_2	284.57	0.63	0.63	131.05	76.14	76.14	—
C_2H_4	343.10	0.63	0.63	70.46	67.83	67.83	—
C_2H_6	401.31	0.65	0.65	19.06	63.45	63.45	—
CH_3NH_2	350.76	0.65	0.36	15.75	64.14	54.68	32.38
CH_3OH	306.89	0.65	0.22	13.74	65.14	50.60	45.98
CH_3F	268.21	0.66	0.14	13.87	66.50	—	53.87
BH_3NH_3	414.61	1.25	0.37	19.97	78.52	53.52	—

Table 6.13 Centroids of charge (distance atom–centroid in atomic units), the corresponding values for localized orbitals being given in parentheses

Molecule	XH_n XH_3YH_n d(XC)	Core X d(XC)	Core Y d(YC)	Bond XY d(XC)	Bond XH d(XC)	Bond YH d(YC)	Lone pair(s) X,Y d(X, YC)
LiH	1.193	0.011	—	—	2.479	—	—
		(0.007)			(2.469)		
BeH_2	0.0	0.0	—	—	1.991	—	—
		(0.0)			(1.979)		
BH	0.322	0.0	—	—	1.841	—	0.866
		(0.003)			(1.768)		(0.794)
CH_4	0.0	0.0	—	—	1.569	—	—
		(0.0)			(1.412)		
NH_3	0.144	0.0	—	—	1.408	—	0.893
		(0.002)			(1.223)		(0.673)
H_2O	0.151	0.0	—	—	1.303	—	0.397
		(0.001)			(1.089)		(0.291)
HF	0.117	0.0	—	—	1.220	—	0.184
		(0.0)			(0.984)		(0.124)
C_2H_2	1.138	0.0	0.0	1.138	1.413	1.413	—
		(0.0)	(0.0)	(1.138)	(1.354)	(1.354)	
C_2H_4	1.259	0.0	0.0	1.259	1.529	1.529	—
		(0.0)	(0.0)	(1.259)	(1.406)	(1.406)	
C_2H_6	1.449	0.0	0.0	1.449	1.573	1.573	—
		(0.0)	(0.0)	(1.449)	(1.417)	(1.417)	
CH_3NH_2	1.370	0.0	0.0	1.487	1.566	1.404	0.894
		(0.0)	(0.003)	(1.528)	(1.418)	(1.222)	(0.647)
CH_3OH	1.290	0.0	0.0	1.478	1.564	1.297	0.400
		(0.0)	(0.003)	(1.566)	(1.417)	(1.084)	(0.282)
CH_3F	1.222	0.0	0.0	1.466	1.568	—	0.175
		(0.0)	(0.0)	(1.596)	(1.431)		(0.102)
BH_3NH_3	1.518	0.0	0.0	1.912	1.850	1.392	—
		(0.0)	(0.0)	(2.028)	(1.742)	(1.196)	

from two, four or six. The volumes of molecule and pseudologes are collected in Table 6.12. Their variations seem to be essentially dependent on the electronegativities of the corresponding atoms. With the exception of the BH molecule, it can be seen that the volume of a lone pair pseudologe is always smaller than the volume of the nearby bond pseudologe.

Table 6.13 gives the positions of the electronic charge centroids of the different domains and also of the corresponding localized orbitals. In general the two series of values are quite similar.

The expectation value of the spherical quadratic moment operator is given in Table 6.14 for each domain of the various systems. It can be seen that the variations of $\langle r^2 \rangle$ are not always parallel to those of the corresponding volumes.

Table 6.14 Expectation values of the spherical quadratic moment operator (in atomic units)

	XH_n XH_3YH_n	Core X	Core Y	Bond XY	Bond XH	Bond YH	Lone pair(s) X,Y
LiH	3.690	0.618	—	—	1.956	—	—
BeH_2	5.141	0.439	—	—	1.665	—	—
BH	4.076	0.336	—	—	1.440	—	1.569
CH_4	5.717	0.279	—	—	1.286	—	—
NH_3	4.869	0.234	—	—	1.190	—	1.027
H_2O	4.117	0.201	—	—	1.143	—	1.112
HF	3.413	0.176	—	—	1.077	—	1.039
C_2H_2	7.501	0.277	0.277	1.654	1.344	1.344	—
C_2H_4	8.743	0.278	0.278	1.519	1.306	1.306	—
C_2H_6	10.179	0.279	0.279	1.144	1.288	1.288	—
CH_3NH_2	9.515	0.282	0.234	1.071	1.290	1.196	1.025
CH_3OH	8.941	0.279	0.201	1.006	1.293	1.126	1.115
CH_3F	8.480	0.279	0.176	0.965	1.300	—	1.045
BH_3NH_3	10.366	0.341	0.235	1.055	1.393	1.201	—

We may conclude that the pseudologe approach leads to a very detailed description of the electronic structure of chemical systems and provides original information on the nature of the chemical bonds.

6.2.4. Other Approaches

6.2.4.1 The shell model for diatomic systems

Without going into much detail we will recall some characteristic features of the shell model for diatomic compounds. Electronic configurations of diatomic systems may be obtained by using the energetic diagram of 'symmetry orbitals' and the building-up principle. In the series:

$$H_2, He_2, Li_2, Be_2, B_2, C_2, CN, N_2^+, CO, BO, CN^-$$

the following energetic diagram is generally assumed:

$\sigma_u^* 2p$	——
$\pi_g^* 2p$	—— ——
$\sigma_g 2p$	——
$\pi_u 2p$	—— ——
$\sigma_u^* 2p$	——
$\sigma_g 2s$	——
$\sigma_u^* 1s$	——
$\sigma_g 1s$	——

However, for the series:

NO, NF, O_2, OF, F_2, Ne_2 and the corresponding ions

another diagram is adopted where $\pi_u 2p$ and $\sigma_g 2p$ levels are permuted.

We shall describe some typical results of this model, considering a few examples of the two series of compounds. electronic configurations of C_2; CN and N_2^+; N_2, CN^- and CO are shown in Figure 6.10. The number of bonding electrons and the corresponding bond order introduced by Herzberg[(15)] are also given with the chemical formulas suggested. For closed-shell systems, these results are very similar (although less detailed) to those obtained by the Linnett and LMO methods. This is no longer true in the case of free radicals. The Herzberg model gives no information about the localization of the centroid of charge of the unpaired electron and leads to erroneous electronic structures for CN and N_2^+ although the bond orders seem to be plausible when compared with the number of bonding electrons calculated by integration of electronic density in an approximate bond loge (4,4 in N_2^+ and 4,3 in CN).

The electronic structures of NO and O_2^+, NF and O_2, OF and O_2^-, F_2 and O_2^{2-} are also described in Figures 6.10. In Herzberg's model NO and CN would have the same structure, which is completely wrong if one refers to the Linnett and LMO results and to the experimental properties of these radicals, such as their electron affinity and their reactivity. In fact, Herzberg's and Linnett's approaches are complementary in the case of diatomic compounds but the best information on these systems remains that obtained by a loge or pseudologe partitioning.

6.2.4.2 The density approach

As we have shown previously, the electronic structure of a chemical system may be described in terms of the electronic density $\rho(M)$ (Sections 3.1 and 4.2). This is commonly represented by contour maps showing the lines of isodensity. Such maps do not clearly show the effect of binding on the electronic distribution in a molecule. However, a representation of the electronic cloud using dots whose darkening is related to the value of

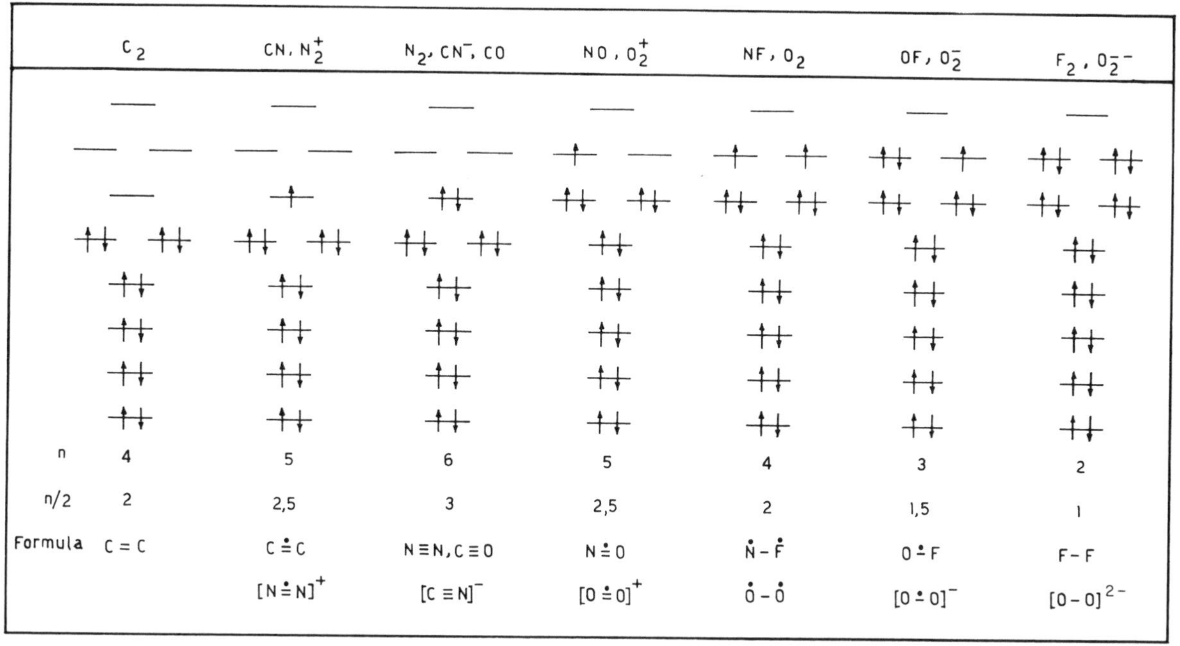

	C_2	CN, N_2^+	N_2, CN^-, CO	NO, O_2^+	NF, O_2	OF, O_2^-	F_2, O_2^{--}
n	4	5	6	5	4	3	2
n/2	2	2,5	3	2,5	2	1,5	1
Formula	$C = C$	$C \dot{\equiv} C$	$N \equiv N$, $C \equiv O$	$N \dot{\equiv} O$	$\dot{N} - \dot{F}$	$O \dot{=} F$	$F - F$
		$[N \dot{\equiv} N]^+$	$[C \equiv N]^-$	$[O \dot{\equiv} O]^+$	$\dot{O} - \dot{O}$	$[O \dot{=} O]^-$	$[O - O]^{2-}$

Figure 6.10 Electronic configurations of diatomic systems

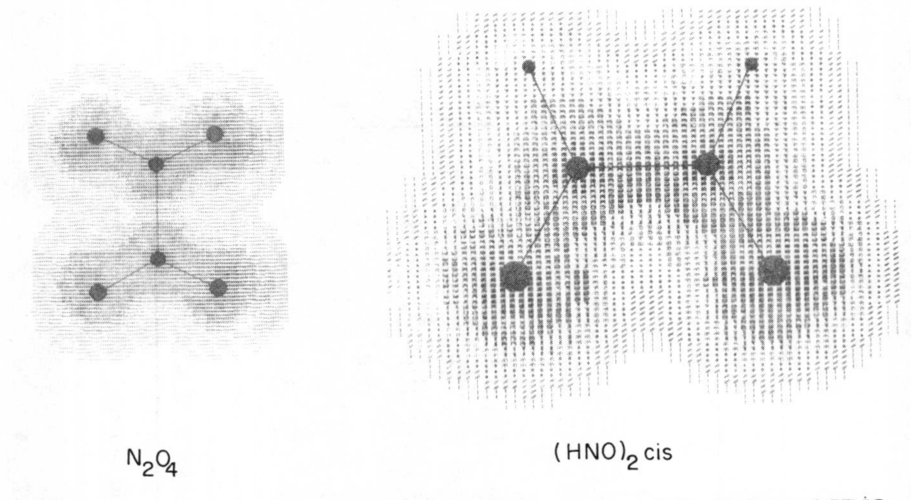

Figure 6.11 Electronic density in the molecular plane of the N_2O_4 and HNO dimer. Reproduced by permission of the Journal de Chimie Physique

electronic density at the corresponding point may provide qualitative information on the strength of the chemical bonds. This is shown in Figure 6.11 which represents $\rho(M)$ in N_2O_4 and $(HNO)_2$ molecules. It can readily be seen that the NN bond in very weak in N_2O_4, which resembles two isolated NO_2 groups. On the other hand, the very important electronic density between the two nitrogens in the HNO dimer may be related to the double bond character of the NN bond.

Density difference functions give much more detailed information on the nature of chemical bonds. They can be defined in different ways. In general, $\delta(M)$ may be written as:

$$\delta(M) = \rho(M) - \rho^f(M) \tag{6.6}$$

In Daudel's definition,[16] the fictitious density ρ^f results from the addition of the densities in the free atoms. Therefore at a point where δ is positive, the bonding has led to a decrease in the electronic density. This is why the δ function is also called the bond density function. If ρ^f is defined as a density corresponding to normal covalent bonds:

$$\rho^f = \frac{1}{2}(\rho_{AA} + \rho_{BB}) \qquad \text{(diatomic case)} \tag{6.7}$$

then the δ function provides information on the polarity of the bond. Typical examples of this type of δ function are given in Figure 6.12. These maps clearly show the opposite signs of the hydrogen atom charges in LiH and HF.

It is preferable to analyse the electronic structure of free radicals in terms of spin densities. As shown previously (Section 4.2), the spin density of an

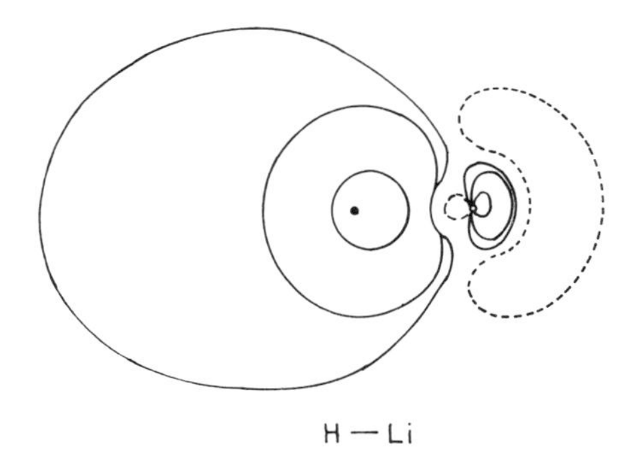

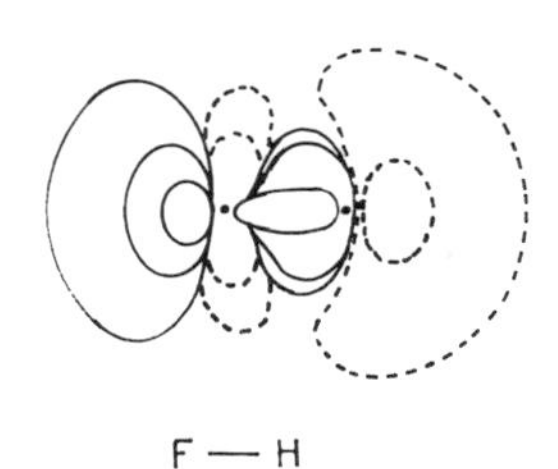

Figure 6.12 Typical difference electron distributions

open-shell system may be defined as:

$$\rho^S(M) = \rho^\alpha(M) - \rho^\beta(M) \tag{6.8}$$

The spin density can be either positive, negative or equal to zero and its value at a given nucleus $\rho^S(N)$ is connected with the corresponding hyperfine splitting constant measured in ESR spectroscopy.[17]

Spin densities of free radicals give interesting information on the electronic structures of such systems. These are not only complementary to the results obtained by the LMO approach but also generally support them, as shown by the following examples:

Cyanide radical, $\dot{C}N$

The spin density of $\dot{C}N$ calculated at the 4-31G level is represented in Figure 6.13. One finds that the carbon atom is lying in a region of largely positive spin density (an excess of α electrons with respect to β electrons). The distribution of the charge centroids agrees perfectly with this spin density map:

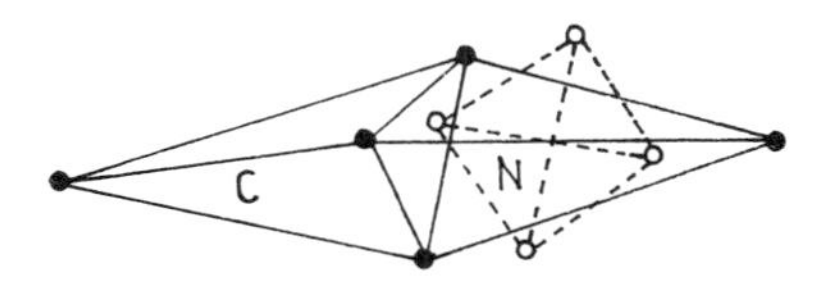

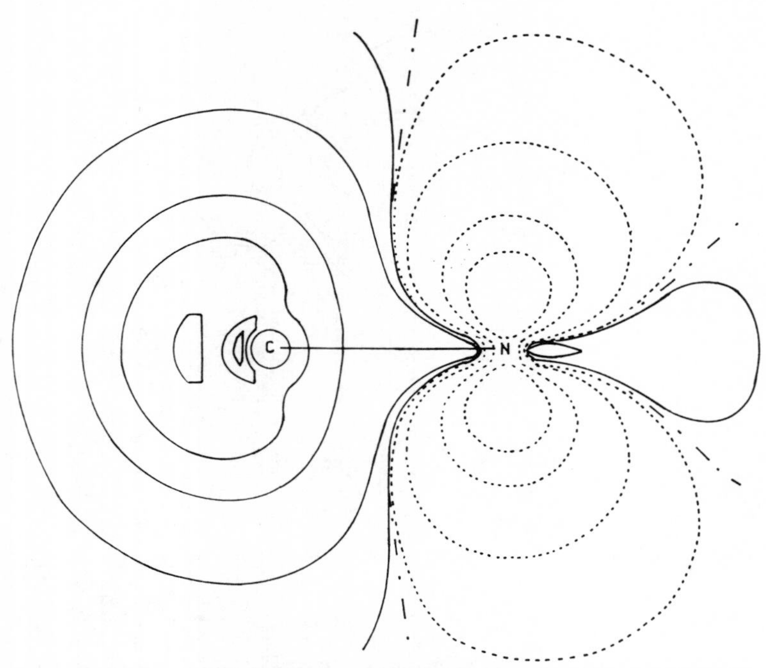

Figure 6.13 Spin density of the cyanide radical. Reproduced by permission of D. Reidel Publishing Company, Dordrecht, Holland

These results suggest the following representation of electronic structure of the cyanide radical:

$$\bullet C \equiv N| \quad \text{or} \quad \bullet \dot{C} \overset{\backslash \, /}{=} N|$$

They allow us to interpret the large ^{13}C hyperfine splitting constant observed in the ESR spectrum of this radical:[18]

$$a(^{13}C) = 210 \text{ G}$$

Nitrogen radical cation, N_2^+ (4-31G results)

The spin density represented in Figure 6.14 shows the delocalization of the unpaired electron on the whole system as anticipated by the centroid of charge distribution:

N N

Thus, the following electronic structure may be proposed:

$$N \overset{\backslash \, /}{=} N$$

Compared to $\dot{C}N$ which is a localized (carbon-centred) radical disobeying the Lewis octet rule (and the Linnett double quartet rule), N_2^+ is a delocalized radical where the Lewis octet rule (but not Linnett's double quartet rule) is satisfied.

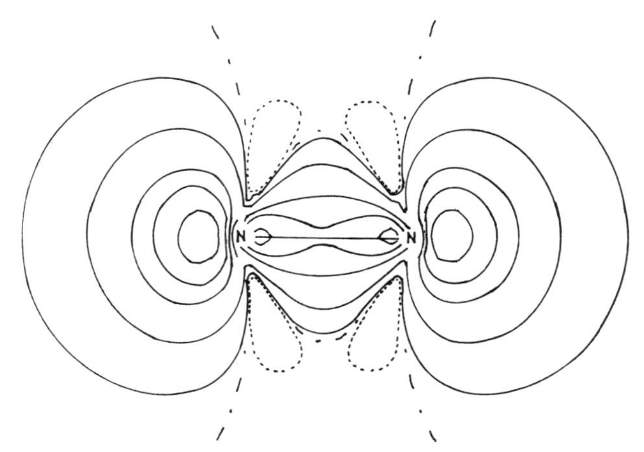

Figure 6.14 Spin density of the N_2^+ radical cation. Reproduced by permission of D. Reidel Publishing Company, Dordrecht, Holland

Nitric oxide molecule, NO (4-31G results)

The centroid of the charge distribution of the NO molecule (see below) and its spin density (Figure 6.15) lead to the electronic structure:

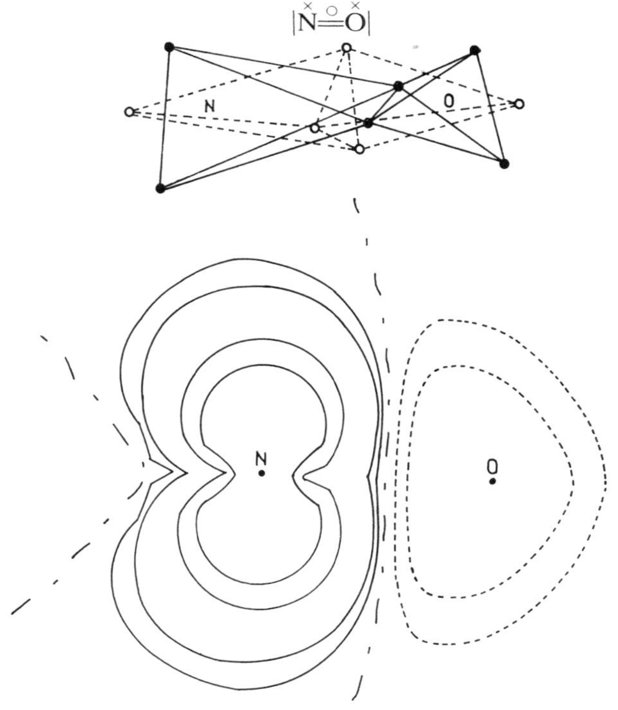

Figure 6.15 Spin density of the NO molecule. Reproduced by permission of D. Reidel Publishing Company, Dordrecht, Holland

The octet and double quartet rules are satisfied for the two atoms of this delocalized radical.

FȮ radical (4-31G results)
LMO results and spin density show that the FȮ radical which is isoelectronic to O_2^- has a quite different electronic structure (see the centroid of charge distribution below and also Figure 6.16). FȮ is an atom-centered radical:

$$|\underline{\bar{F}}—\underline{\bar{O}}^{\cdot}$$

where the oxygen atom has only seven electrons in its valence shell.

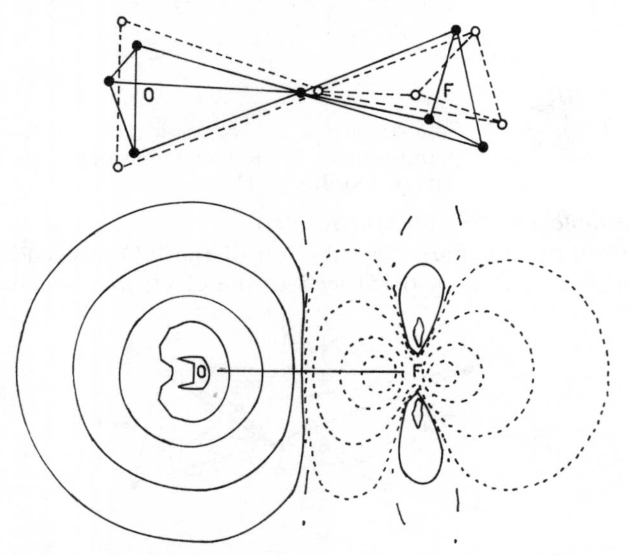

Figure 6.16 Spin density of the FȮ radical. Reproduced by permission of D. Reidel Publishing Company, Dordrecht, Holland

Model radicals

Let us now consider some of these model radicals: hydroxyl, ȮH, prototype of alkoxy radicals; $\dot{N}H_2$, prototype of $R_2\dot{N}$ radicals; and $\dot{C}H_3$, prototype of alkyl radicals. They have a very similar electronic structure, being all atom-centered radicals. $H_2N\dot{O}$ is a good model of nitroxide radicals. Its centroid of charge distribution and spin density (Figure 6.17) reveal its delocalized character. The NO bond is a 'three-electron' bond as

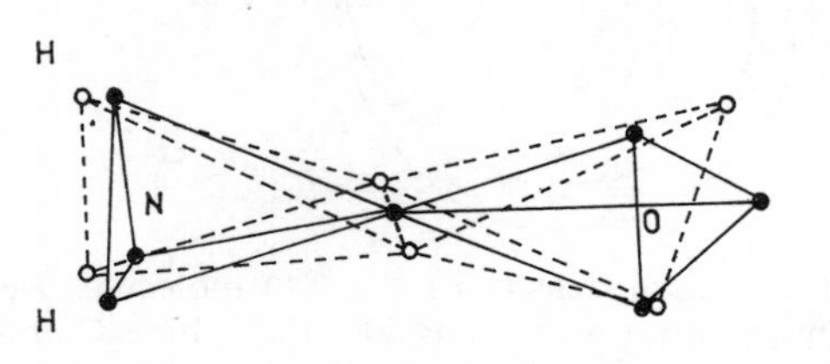

shown in the following formula:

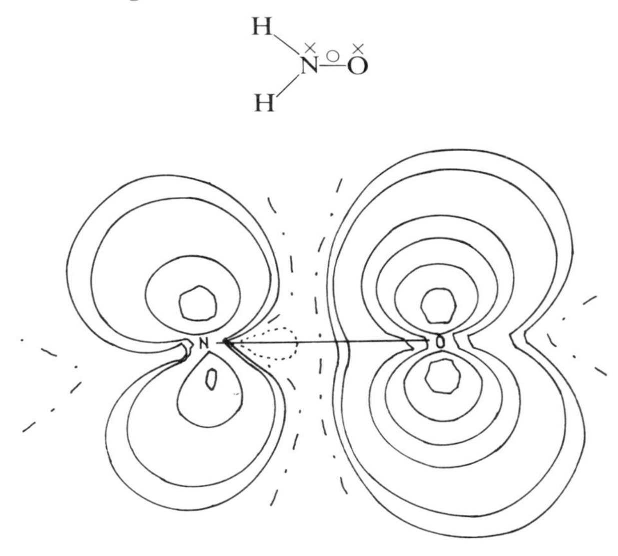

Figure 6.17 Spin density of $H_2N\dot{O}$ (perpendicular plane). Reproduced by permission of D. Reidel Publishing Company, Dordrecht, Holland

Note that $H_2N\dot{O}$ is very similar to the isoelectronic system $NH_2\dot{N}H$ (see Figure 6.18) prototype of the hydrazyl radicals as revealed by the spin density maps. This has the following electronic structure:

H
N—N
H H

It is interesting to note that the sign of the spin density in the different regions of molecular space is generally in perfect agreement with the predictions based on the models of spin polarization[19] and hyperconjugation[20] Therefore, the α hydrogen atoms of an alkyl radical are always in regions of negative spin density and the β hydrogen atoms in regions of positive spin density, as shown in Figure 6.19.

6.3 THE THEORETICAL APPROACH OF THERMOCHEMICAL DATA

6.3.1 The Theoretical Heats of Reaction

Heats of reaction may be calculated directly from the enthalpies of the constituents by using the following equation, if one refers to standard state

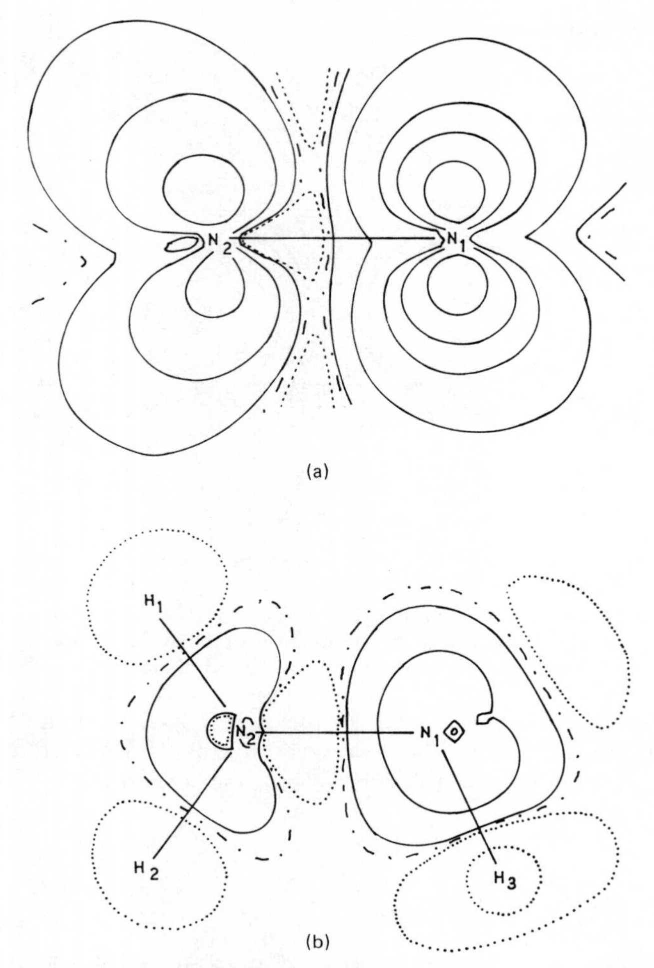

Figure 6.18 Spin density of $NH_2\dot{N}H$ in (a) the perpendicular plane and (b) the molecular plane. Reproduced by permission of D. Reidel Publishing Company, Dordrecht, Holland

(pressure of 1 atm) at 298.15 K

$$\Delta H^0(298.15) = \sum_i k_i N_i H_i^0(298.15) \tag{6.9}$$

For a linear polyatomic molecule, $H^0(298.15)$ may be written more explicitly:

$$H^0(298.15) = Nu_0(s_e) + \frac{R}{2}\sum_{j=1}^{n} \theta_j + [H^0(298.15) - H^0(0)] \tag{6.10}$$

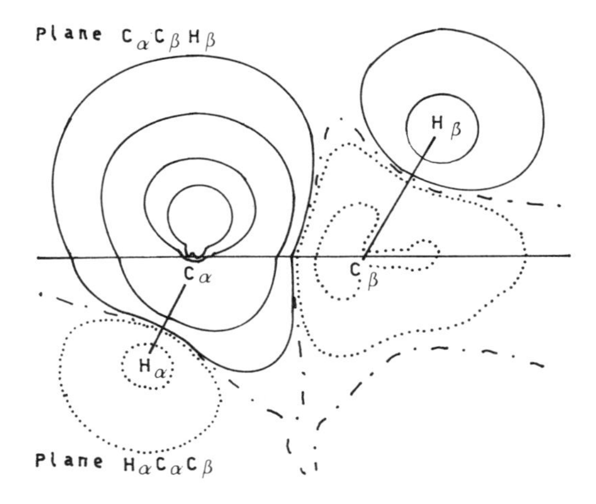

Figure 6.19 Spin density of the ethyl radical in two perpendicular planes. Reproduced by permission of D. Reidel Publishing Company, Dordrecht, Holland

where:

$$\theta_j = \frac{hc}{k}\,\tilde{\nu}_j \tag{6.11}$$

$\tilde{\nu}_j$ being the fundamental frequency of the jth normal mode of vibration of the molecule, and n is the number of normal modes.

The other notations have the same meaning as in expression (4.173) of which (6.10) is a trivial generalization. One will also remember that:

$$H^0(298.15) - H^0(0) = \frac{7}{2}RT + \sum_{j=1}^{n} \frac{R\theta_j}{e^{\theta_j/T} - 1} \tag{6.12}$$

We will show in Part III how to calculate the fundamental frequencies $\tilde{\nu}_j$ of a polyatomic compound.

It is therefore possible to obtain theoretical heats of reaction at any temperature. Unfortunately the results obtained in the Hartree–Fock–Roothaan approximation are not very reliable due to the correlation errors on total energies which usually do not neutralize each other. To illustrate this remark we shall consider a specific example, i.e. the recombination reaction of $\dot{C}N$ radicals:[21]

$$2\,CN \rightleftharpoons (CN)_2$$

The total and relative energies of the two constituents calculated at the 6-31G level for the equilibrium geometries are given in Table 6.15.
One finds that the heat of reaction at 0 K is equal to:

$$\Delta H^0(0) = -109.08 \text{ kcal mol}^{-1} \tag{6.13}$$

Table 6.15 Total and relative energies of the equilibrium structures of (2) CN and $(CN)_2$

System	$u_0(s_e)$ (a.u.)	Δu_0 (kcal mol^{-1})
2 $\dot{C}N$	−184.326472	0.0
NC—CN	−184.500352	−109.08

to be compared to the experimental value at 298.15 K:[22]

$$\Delta H^0(298.15) = -134.13 \pm 5.43 \text{ kcal mol}^{-1} \tag{6.14}$$

The theoretical values of the $\tilde{\nu}_j$ are collected in Table 6.16 with the corresponding experimental values. They allow us to calculate the thermal and ZPE corrections and so to obtain the theoretical heat of reaction at 298.15 K or the experimental heat of reaction at 0 K. The results are given in Table 6.17. They show that the agreement between theory and experiment is far from being satisfactory.

Table 6.16 Fundamental frequencies of the normal modes of vibration (per centimetre) for $\dot{C}N$ and NCCN

$\tilde{\nu}_j$	NCCN(^{12}C; ^{14}N) Theory	NCCN(^{12}C; ^{14}N) Experiment[23]	$\tilde{\nu}_j$	$\dot{C}N$(^{12}C; ^{14}N) Theory	$\dot{C}N$(^{12}C; ^{14}N) Experiment[24]
$\tilde{\nu}_1$	2,710	2,329	$\tilde{\nu}_1$	1,910	2,068.7
$\tilde{\nu}_2$	2,464	2,159		—	—
$\tilde{\nu}_3$	942	851		—	—
$\tilde{\nu}_4 = \tilde{\nu}_5$	605	507		—	—
$\tilde{\nu}_6 = \tilde{\nu}_7$	258	235		—	—

Table 6.17 Theoretical and experimental heat of reaction for the recombination of CN radicals (in kilocalories per mole)

	$\Delta H^0(0)$	Thermal and ZPE corrections	$\Delta H^0(298.15)$
Theory	−109.08	4.47	−104.61
Experiment	−136.81	2.68	−134.13

6.3.2. The Semiempirical Heats of Reaction

Some years ago, Pople introduced the interesting concept of isodesmic reaction.[25] By definition this is a reaction in which all the bonds are conserved (in number and nature). The corresponding theoretical heat of reaction is generally reliable as the correlation errors on the total energies of reactants and products approximately cancel. As an example, let us consider the following isodesmic reaction:

$$\dot{C}N + CH_4 \rightarrow HCN + CH_3 \tag{6.15}$$

Using the data of Tables 6.15, 6.16 and 6.18 one obtains:

$$\Delta H^0(0) = -19.14 \text{ kcal mol}^{-1} \tag{6.16}$$

which compares pretty well with the experimental value taken at 0 K: $-19.95 \text{ kcal mol}^{-1}$. Thus, semiempirical heats of formation may be calculated using the theoretical ΔH^0 of the isodesmic reaction and experimental heats of formation of the reference compounds. If the necessary data are available, ΔH^0 would be evaluated at 298.15 K, but satisfactory results can also be obtained using $\Delta H^0(0)$ and empirical heats of formation at 298.15 K.

Table 6.18 The 6-31G level total energies (in atomic units), ZPE and thermal corrections (in kilocalories per mole) and heats of formation (in kilocalories per mole) of some reference systems

System†	$u_0(s_e)$	Corrections (298.15)	ΔH_f (298.15)
HCN	−92.82763[26]	11.97[24]	32.30[27]
CH_4	−40.18055[28]	29.48[24]	−17.89[29]
C_2H_6	−79.19748[28]	48.04[24]	−20.24[29]
CH_3	−39.54666[30]	23.50[30]	34.82[31]

† ZPE and thermal corrections for CN and its dimer are respectively 5.03 and 12.74 kcal mol^{-1}.[22]

The following examples illustrate the reliability of this procedure.

ĊN radical

Isodesmic reaction:

$$\begin{gathered} \dot{C}N + CH_4 \rightarrow HCN + CH_3 \\ \Delta H^0(0) = -19.14 \text{ kcal mol}^{-1} \end{gathered} \tag{6.17}$$

$$\text{Empirical ZPE and thermal corrections} = 0.96$$

$$\Delta H^0(298.15) = -18.18 \text{ kcal mol}^{-1}$$

One may write:

$$\Delta H^0(298.15) = \Delta H_f^0(\text{HCN}) + \Delta H_f^0(\text{CH}_3) - \Delta H_f^0(\dot{\text{C}}\text{N}) - \Delta H_f^0(\text{CH}_4)$$

$$-18.18 = 32.3 + 34.82 - x + 17.89 \tag{6.18}$$

$$x = \Delta H_f^0(\dot{\text{C}}\text{N}) = 103.19 \text{ kcal mol}^{-1} \tag{6.19}$$

Neglecting the ZPE and thermal corrections one obtains:

$$\begin{gathered} -19.14 = 32.3 + 34.82 - x + 17.89 \\ x = \Delta H_f^0(\dot{\text{C}}\text{N}) = 104.15 \text{ kcal mol}^{-1} \end{gathered} \tag{6.20}$$

The dimer NCCN

Isodesmic reaction:

$$\mathrm{NCCN + 2CH_4 \rightarrow 2HCN + CH_3CH_3} \qquad (6.21)$$

$$\Delta H^0(0) = 5.47 \text{ kcal mol}^{-1}$$

Empirical ZPE and thermal corrections = 0.28

$$\Delta H^0(298.15) = 5.75 \text{ kcal mol}^{-1}$$

Thus:

$$\Delta H^0(298.15) = 2\Delta H_f^0(\mathrm{HCN}) + \Delta H_f^0(\mathrm{CH_3CH_3}) - \Delta H_f^0(\mathrm{NCCN}) - 2\Delta H_f^0(\mathrm{CH_4}) \qquad (6.22)$$

$$5.75 = 2\times 32.3 - 20.24 - x + 2\times 17.89$$

$$x = \Delta H_f^0(\mathrm{NCCN}) = 74.40 \text{ kcal mol}^{-1} \qquad (6.23)$$

Without the ZPE and thermal corrections one finds:

$$5.47 = 2\times 32.3 - 20.24 - x + 2\times 17.89$$

$$x = \Delta H_f^0(\mathrm{NCCN}) = 74.68 \text{ kcal mol}^{-1} \qquad (6.24)$$

These results are collected in Table 6.19 and compared with experimental data. The general agreement between theory and experiment allows one to adopt this procedure for calculating unknown heats of formation with or without ZPE and thermal corrections. Furthermore, semiempirical heats of formation may be utilized for estimating semiempirical heats of reaction, which are generally much better than those computed from the total energies. This is illustrated by the results of Table 6.20 which allows

Table 6.19 Heats of isodesmic reactions; heats of formation of the $\dot{\mathrm{C}}\mathrm{N}$ radical and its dimer (in kilocalories per mole)

Compound		$\Delta H^0(0)$	$\Delta H_f^0(298.15)^\dagger$	$\Delta H^0(298.15)$	$\Delta H_f^0(298.15)$
$\dot{\mathrm{C}}\mathrm{N}$	Theory	−19.14	104.15	−18.18	103.19
	Experiment	−19.95	—	−18.99	104.00 ± 2.50[31]
NCCN	Theory	5.47	74.68	5.75	74.40
	Experiment	6.02	—	6.30	73.84 ± 0.43[29]

† Calculated directly assuming $\Delta E^0(0) = \Delta H^0(298.15)$.

Table 6.20 Semiempirical and experimental thermodynamical parameters for the recombination reaction of the $\dot{\mathrm{C}}\mathrm{N}$ radical (kcal mol^{-1}, Gibbs; 298.15 K)

Reaction		ΔH^0	ΔC_p^0	ΔS^0	ΔG^0
$\mathrm{2CN \rightleftharpoons NCCN}$	Theory	−131.98	−5.1	−39.9	−120.08
	Experiment	−134.16	−4.4	−38.9	−122.56

comparison of the semiempirical and experimental thermodynamical parameters of the dimerization reaction of the $\dot{C}N$ radical.

The entropy and Gibbs free energy changes and the ΔC_p^0 values are obtained using the well-known formulas of statistical thermodynamics with the data of Table 6.16. One can conclude that the semiempirical approach based on the utilization of isodesmic reactions leads to quite satisfactory results.

6.3.3 The Semiempirical Bond Dissociation Energies

Semiempirical heats of formation may also be used in the calculation of interesting properties such as bond dissociation energies and related quantities as electron affinities of free radicals. The dissociation energy of a bond R—X is nothing more than the heat of the following reaction:

$$R\text{—}X \rightarrow R^{\cdot} + X^{\cdot} \tag{6.25}$$

leading to unpaired electron species, i.e. a free radical and an atom. Thus we can write:

$$\text{BDE}(R\text{—}X) = \Delta H^0 = \Delta H_f^0(R^{\cdot}) + \Delta H_f(X^{\cdot}) - \Delta H_f(RX) \tag{6.26}$$

Taking semiempirical values for $\Delta H_f^0(RX)$ and $\Delta H_f^0(R)$ and the experimental value for $\Delta H_f^0(X^{\cdot})$ one easily obtains BDE(RX). If RX is a reference compound of an isodesmic reaction, its empirical heat of formation will be used, the semiempirical one being unattainable by the procedure previously described. Some typical results of bond dissociation energy calculations are given in Table 6.21. The semiempirical BDEs are compared with the

Table 6.21 Bond dissociation energies of some compounds and electron affinities of the corresponding radicals (kcal mol^{-1}; 298.15 K)[32]

Compound		$\Delta H^0(R^-H^+)$	BDE(RH)	EA($R^{\cdot}$)
HCN	Theory	—	122.99	87.29
	Experiment	349.3	123.80	88.10
HNO	Theory	—	50.47†	2.87
	Experiment	361.2	49.90	2.30
HCF_3	Theory	—	106.74‡	44.74
	Experiment	375.6	106.40	44.40
HCH_2CN	Theory	—	92.59	33.99
	Experiment	372.2	93.00	34.40
$HCH(CN)_2$	Theory	—	82.60	60.20
	Experiment	336.0	—	—

† Obtained from ΔH_f^0 calculated at 6-31G level without ZPE and thermal corrections.
‡ Obtained from ΔH_f^0 calculated at 4-31G level without ZPE and thermal corrections.

corresponding experimental data. Furthermore, one can calculate electron affinities of radicals $R^{\cdot}$ using the following expression which is only valid if $X = H$:

$$A(R) = BDE(RH) + I(H) - \Delta H^0(R^-H^+) \tag{6.27}$$

where $I(H)$ is the ionization potential of the hydrogen atom (313.6 kcal mol^{-1}) and $\Delta H^0(R^-H^+)$ is the heat of the reaction:

$$RH \rightleftharpoons R^- + H^+ \tag{6.28}$$

This is simply the 'acidity' of RH measured in the gaseous phase. Electron affinities of radicals and the data used in the calculations are also given in Table 6.21. For more details on these calculations, the reader may refer to the corresponding paper.[32] One finds that semiempirical BDE(RH) and EA($R^{\cdot}$) values are quite trustworthy.

6.3.4 The Stability Concept

To conclude this section we will give an example of how quantum chemical reasoning can be applied to practical problems. Furthermore, we will show that the joint utilization of theory and experiment helps us to make common concepts clearer without necessarily performing sophisticated calculations. As this section is devoted to thermochemistry we will analyse the concept of stability which is commonly used by the experimentalists, often without explicit definition.

In fact, the stability concept may be defined in different ways. Taken from a kinetic point of view, we may attribute the stability of a compound to its lack of reactivity. This definition has been deeply analysed by Griller and Ingold[33] for the case of carbon-centred radicals so we will not discuss it further here. From a thermodynamical point of view, the stability of a chemical species can be measured, for example, by its heat of atomization, ΔH^0_a. However, this definition is ambiguous as it depends on the choice of an arbitrary reference which is here the isolated atoms. Moreover, with this definition, the stabilities of different systems are only comparable if they contain the same atoms; this would apply to isomers or the reactants and the products of a chemical reaction. Thus, if a reaction is endothermic, the reactants are more stable than the products. In this case there is no absolute definition of the stability concept. However, the 'conventional stabilization energy' as defined by Cox and Pilcher[29] may be considered as an intrinsic characteristic of every compound. This stabilization energy may be written as:

$$SE^0 = \Delta H^0_a - \sum_i \varepsilon^0_l \tag{6.29}$$

where ΔH^0_a represents the heat of atomization of the species, experimentally determined or theoretically estimated, and $\sum_i \varepsilon^0_l$ is the heat of atomization calculated using standard bond energies corresponding to an arbitrary chemical formula of the compound. The formula chosen most often will be that suggested by the valence of the atoms, localizing multiple bonds, if any.

The standard bond energies will be determined within the framework of a given bond energy scheme. It is to be noted that the stabilization energy SE^0 may, *a priori*, contain stabilizing contributions due to electronic effects such as the delocalization of π electrons and also destabilizing terms due to steric effects or to ring strain. In alkanes, SE^0 will measure the joint influence of the electronic and steric effects of the substituents on saturated carbons; in microcycles, SE^0 will be identified with the conventional ring strain energy; in conjugated compounds, SE^0 will essentially represent the stabilization energy due to the delocalization of π electrons which is most often related to mesomeric effects of substituents. However, as pointed out by Dewar and Schmeising,[34] that stabilization energy may contain contributions due to hybridization changes of the atoms related to modifications of bond lengths. Thus, the stabilization energy of a conjugated system cannot be identified with the π delocalization energy which is itself different from the classical resonance energy derived from heats of hydrogenation. Although of composite origin, the conventional stabilization energy proves to be very useful as it may easily be estimated for any kind of compound provided that we dispose of appropriate bond energies. Those obtained by the Laidler bond energy scheme seem to be very convenient as they have been obtained by considering reference compounds where the effects responsible for SE^0 are not present.[35] Before giving some examples of how this approach can be applied we shall describe the formalism to be used for different types of chemical species.

6.3.4.1 The expressions of the stabilization energy

(a) The stabilization energy of neutral systems such as molecules and free radicals can be evaluated by the equation:

$$SE^0(298.15) = \Delta H_a^0(298.15) - \sum_l \varepsilon_l(298.15) \tag{6.30}$$

where:

$$\Delta H_a^0(298.15) = \Delta H_a^0(298.15)(\text{elements}) - \Delta H_f^0(298.15) \tag{6.31}$$

Thus, for calculating $SE^0(298.15)$, we need to know the standard heat of formation of the compound. This can be measured experimentally or estimated semiempirically as shown previously (Section 6.2.2).

As an example we shall calculate the stabilization energies of the $\dot{C}N$ radical and its dimer NCCN using experimental heats of formation of these species.

The $\dot{C}N$ radical:

$$\begin{aligned} SE^0(298.15) &= \Delta H_a^0(298.15) - \varepsilon^0_{C\equiv N} \\ SE^0(298.15) &= 179.9 - 214.0 = -34.1 \text{ kcal mol}^{-1} \end{aligned} \tag{6.32}$$

This compound is much destabilized due to the localization of the unpaired electron on the carbon atom as seen in section 31.

The NCCN dimer:

$$\begin{aligned} SE^{0}(298.15) &= \Delta H^{0}_{a}(298.15) - \varepsilon_{C—C} - 2\varepsilon_{C\equiv N} \\ SE^{0}(298.15) &= 493.86 - 85.48 - 428.0 = -19.62 \text{ kcal mol}^{-1} \end{aligned} \quad (6.33)$$

The dimer is also destabilized—probably due to polar effects.

(b) In the case of carbocations, we have to consider the 'atomization' process which leads to the positive ion C^{+}. We have, in the case of $CH_3{}^{+}$:

$$CH_3{}^{+} \rightarrow C^{+} + 3H \quad (6.34)$$

It is easy to show that the stabilization energy of carbocations may be written as (see Figure 6.20):

$$SE^{0}(R^{+}) = SE(R) - I(R) + I(C) \quad (6.35)$$

where $SE^{0}(R)$ is the stabilization energy of the corresponding free radical, $I(R^{\cdot})$ is the ionization potential of the latter and $I(C)$ the ionization potential of the carbon atom (259.65 kcal mol^{-1}).[36] Thus, for example:

$$\begin{aligned} SE^{0}(CN^{+}) &= SE^{0}(\dot{C}N) - I(\dot{C}N) + I(C) \\ SE^{0}(CN^{+}) &= -34.1 - 327.5^{(36)} + 259.65 = -101.95 \text{ kcal mol}^{-1} \end{aligned} \quad (6.36)$$

The large destabilization energy of CN^{+} is not surprising as the $\dot{C}N$ radical is a strong electronegative species.

(c) Finally, in the case of carbanions, one will consider the 'atomization' process leading to the C^{-} anion as shown for $CH_3{}^{-}$:

$$CH_3{}^{-} \rightarrow C^{-} + 3H \quad (6.37)$$

The stabilization energy of carbanions will then be written as:

$$SE^{0}(R^{-}) = SE^{0}(R) + EA(R) - EA(C) \quad (6.38)$$

where EA(R) and EA(C) represent respectively the electron affinities of

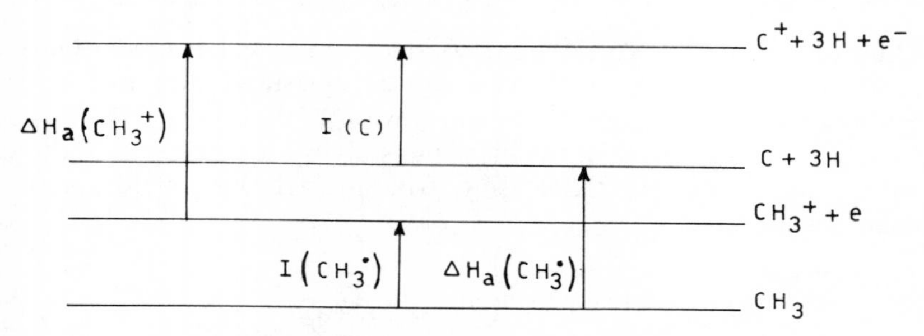

Figure 6.20 Stabilization energy of carbocations

the corresponding radical and carbon atom (29.33 kcal mol^{-1}[37]). Thus, we can evaluate $SE^0(CN^-)$:

$$SE^0(CN^-) = -34.1 + 88.1 - 29.33 = 24.67 \text{ kcal mol}^{-1}$$

As expected, this anion is stabilized.

6.3.4.2 Some applications

Substituted ethanes

The stabilization energies of some substituted ethanes are collected in Table 6.22. These are either positive or negative and are generally relatively small. The results may be qualitatively rationalized in terms of (destabilizing) steric repulsions and (stabilizing or destabilizing) electrostatic interactions between the substituents. The conjugation of polar and steric effects would be responsible for the large destabilization energy of hexachloroethane.

All the compounds of Table 6.22 may be considered as dimers of

Table 6.22 Stabilization energies of substituted ethanes (kcal mol^{-1}; 298.15 K)

Molecule	ΔH_f^0[38]	ΔH_a^0	$\sum_l \varepsilon_l^0$	SE^0
$(CH_3)_2$	−20.24	674.64	664.62	[0.02]†
$(CH_3CH_2)_2$	−30.36	1234.96	1234.66	[0.30]†
$[(CH_3)_2CH]_2$	−42.61	1797.41	1798.34	[−0.93]†
$[(CH_3)_3C]_2$	−54.00	2359.00	2365.78	−6.78
$(CH_2F)_2$	−105.40	693.40	693.06	[0.34]†
$(CHF_2)_2$	−207.90	729.50	731.82	−2.32
$(CF_3)_2$	−320.90	776.10	773.02	3.08
$(CCl_3)_2$	−33.20	548.40	590.26	−41.86
$(CH_2NH_2)_2$	−4.07	988.67	988.24	[0.43]†
$(CH_2OH)_2$	−93.90	867.50	865.22	2.26
$(CH_2CN)_2$	52.58	1065.42	1073.48	−8.06
$[CH(OH)_2]_2$	−181.92	1074.72	1071.30	3.42
$(C_6H_5CH_2)_2$	32.40	3089.60	3087.72	1.88
$(CH_2{=}CH{-}CH_2)_2$	20.11	1526.29	1526.52	[−0.23]†

† Not significant values.

carbon-centered radicals obtained by the recombination reaction:

$$R^{\cdot} + R^{\cdot} \rightarrow R-R \tag{6.39}$$

The dissociation energy of the central CC bond of the substituted ethane is equal to minus the heat of this reaction. It is easy to show that this can be written as:

$$BDE(CC) = \varepsilon^0(CC) + SE^0(RR) - 2SE^0(R^{\cdot}) \tag{6.40}$$

Substituted ethylenes

The stabilization energies of some disubstituted ethylenes are given in Table 6.23.

Table 6.23 Stabilization energies of disubstituted ethylenes (kcal mol^{-1}; 298.15 K)

Olefin	$\Delta H_f^{0(38)}$	ΔH_a^0	$\sum_l \varepsilon_l^0$	SE^0
$CH_2{=}C(CN)_2$	84.79†	929.01	943.52	−14.51
trans-CNCH=CHCN	81.30	932.50	942.20	−9.70
$CH_2{=}C(OH)_2$	−73.65†	743.05	733.08	9.97
trans-HOCH=CHOH	−61.24†	730.64	731.76	−1.12
$CH_2{=}C(OH)(CN)$	8.92†	832.68	838.30	−5.62
trans-HOCH=CHCN	3.33†	838.27	836.98	1.29

† Semiempirical values calculated using total energies of fully optimized structures at 4-31G level.

Using the notation c for the π-captor group (CN) and d for the π-donor group (OH), one finds the following order of growing stabilization:

$$\text{cc gem} < \text{cc trans} \simeq \text{cd gem} < \text{dd trans} < \text{cd trans} < \text{dd gem}$$

Therefore, the stabilization energy is a good measure of mesomeric effects of substituents according to their nature and their position.

Free radicals

The stabilization energy of a large variety of free radicals has been calculated using the same expression as for closed-shell systems, with Laidler parameters. The results obtained are collected in Table 6.24. In the

Table 6.24 Stabilization energies of free radicals (kcal mol^{-1}; 298.15 K)

Radicals	$\Delta H_f^{0(38)}$	ΔH_a^0	$\sum_l \varepsilon_l^0$	SE^0	$\Delta SE^0(\dot{C}H_3)$
$\dot{C}H_3$	34.30	292.90	294.57	−1.67	0.00
$CH_3\dot{C}H_2$	25.60	576.70	574.59	2.11	3.78
$CN\dot{C}H_2$	58.50	500.50	494.00	6.50	8.17
$NH_2\dot{C}H_2$	37.00	455.30	451.38	3.92	5.59
$CH_3NH\dot{C}H_2$	30.00	737.40	725.22	12.18	13.85
$(CH_3)_2N\dot{C}H_2$	26.00	1016.50	1001.78	14.72	16.39
$OH\dot{C}H_2$	−6.20	393.00	389.87	3.13	4.80
$CH_3O\dot{C}H_2$	−2.80	664.70	661.06	3.64	5.31
$F\dot{C}H_2$	−7.90	301.90	303.79	−1.89	−0.22
$(CH_3)_2\dot{C}H$	18.20	859.20	856.63	2.57	4.24
$(CN)_2\dot{C}H$	90.27†	700.53	695.36	5.17	6.84
$(OH)_2\dot{C}H$	−44.46†	490.86	492.91	−2.05	−0.38
$(CN)(NH_2)\dot{C}H$	57.68†	666.42	653.48	12.94	14.61
$(CN)(OH)\dot{C}H$	22.31†	596.29	594.14	2.15	3.82
$(CN)F\dot{C}H$	21.46†	504.34	507.13	−2.79	−1.12
$F_2\dot{C}H$	58.26	319.06	323.17	−4.11	−2.44
$(CH_3)_3\dot{C}$	8.00	1144.50	1140.15	4.35	6.02
$CH_3\dot{C}(CN)_2$	82.97	982.93	979.01	3.92	5.59
$CH_3\dot{C}(OH)_2$	−56.50	777.90	777.75	0.15	1.82
$CH_3\dot{C}(CN)(OH)$	13.06	880.64	878.38	2.26	3.93
$\dot{C}F_3$	−112.00	339.60	343.77	−4.17	−2.50
$\dot{C}Cl_3$	19.00	238.60	252.39	−13.79	−12.12
$CH_2{=}CH{-}\dot{C}H_2$	39.40	733.80	720.52	13.28	14.95
$C_6H_5{-}\dot{C}H_2$	47.80	1513.20	1501.12	12.08	13.75
$CH_2{=}\dot{C}H$	68.35	429.75	435.91	−6.16	−4.49
$HC{\equiv}C^{\cdot}$	122.00	271.90	287.47	−15.57	−13.90
$C_6H_5^{\cdot}$	78.50	1207.40	1217.67	−10.27	−8.60
$C_5H_5^{\cdot}$	60.90	1054.10	1034.86	19.24	20.91
$\dot{N}H_2$	46.00	171.20	183.14	−11.94	—
$\dot{O}H$	9.40	102.30	107.83	−5.53	—
$CH_3O^{\cdot}$	3.83	382.97	379.02	3.95	—
t-BuO$^{\cdot}$	−21.60	1233.70	1231.17	2.53	—
$CH_3OO^{\cdot}$	6.70	439.70	419.06	20.24	—
$\dot{C}N$	101.00	182.90	214.00	−31.10	—
$H\dot{C}O$	9.00	273.60	260.25	13.35	—
NO	21.60	151.00	110.69‡	40.31	—
FO	26.00	52.50	48.12§	4.38	—

† Semiempirical values (4-31G level).
‡ ε(N—O) calculated from ΔH_f^0((HNO).
§ ε(F—O) calculated from ΔH_f^0(HOF).

case of carbon-centred radicals ($\dot{C}XYZ$), the relative stabilization energies with respect to methyl (taken as reference) are given in the last column. As these systems are not always planar, the stabilization energies cannot be rationalized in terms of mesomeric effects alone. It can be seen that carbon-centred radicals are best stabilized by unsaturated groups such as

Table 6.25 Classification of free radicals according to their SE^0 (kcal mol^{-1})

Radicals	SE^0 (kcal mol^{-1})	BDE(RH)[38] (kcal mol^{-1})	Notation	Type
$\dot{C}{\equiv}N$	−31.10	120.80	—	σ, centred on C
$\dot{C}{\equiv}CH$	−15.57	119.76	—	σ, centred on C
$\dot{C}Cl_3$	−13.79	95.70	ddd	σ, centred on C
$\dot{N}H_2$	−11.94	109.10	—	σ, centred on N
$C_6H_5^{\cdot}$	−10.27	110.80	—	σ, centred on C
$\dot{C}H{=}CH_2$	−6.16	108.00	—	σ, centred on C
$\dot{O}H$	−5.53	119.30	—	σ, centred on O
$\dot{C}F_3$	−4.17	106.40	ddd	σ, centered on C
$CN\dot{C}HF$	−2.79	90.40†	cd	π
$OH\dot{C}HOH$	−2.05	101.31†	dd	σ, centred on C
$\dot{C}H_2F$	−1.89	101.00	d	σ, centred on C
$\dot{C}H_3$	−1.67	104.30	—	π
$CH_3\dot{C}H_2$	2.11	98.00	d	π
$CN\dot{C}HOH$	2.15	85.89†	cd	π
$\dot{C}H_2OH$	3.13	93.97	d	σ
$\dot{C}H_2NH_2$	3.92	94.60	d	σ
$\dot{C}H(CN)_2$	5.17	80.29†	cc	π, delocalized
$\dot{C}H_2CN$	6.50	93.00	c	π, delocalized
$C_6H_5\dot{C}H_2$	12.08	87.90	c	π, delocalized
$CH_3NH\dot{C}H_2$	12.12	87.00	d	(π, delocalized)
$CN\dot{C}HNH_2$	12.94	81.84†	cd	π, delocalized
$CH_2{=}CH{-}\dot{C}H_2$	13.28	86.60	—	π, delocalized
$(CH_3)_2N\dot{C}H_2$	14.72	84.00	d	(π, delocalized)
$C_5H_5^{\cdot}$	19.24	80.60	—	π, delocalized
$CH_3OO^{\cdot}$	20.64	—	—	π, delocalized
NO	40.31	49.90	—	π, delocalized

† Semiempirical values (4-31G level).

vinyl or benzyl, by *N*-alkyl groups and also by specific capto-dative substitution[39] (see, for example, $(CN)(NH_2)\dot{C}H$). As shown in Table 6.25, stabilization energies may be used to classify free radicals according to their type, which corresponds to their electronic structure and therefore to their configuration. These results can be summarized in the following way:

SE^0(kcal mol^{-1})	Type	Configuration
<0	σ, atom centered	Pyramidal
<5	σ, atom centered	Pyramidal
	or π	Planar
>5	π, delocalized	Planar

It is well known that the rate of recombination of $\dot{C}N$ radicals ($SE^0 = -31.1$ kcal mol^{-1}) is very high while NO ($SE^0 = 40.31$ kcal mol^{-1}) is a persistent radical. This leads to the belief that in the absence of steric or polar effects the concepts of kinetic stability and thermodynamic stability have the same meaning. In other terms, persistent and *a fortiori* stable radicals should have in most cases a very high stabilization energy.

The heat of the recombination reaction of a given species, or the corresponding bond dissociation energy BDE(CC), can be analysed in terms of stabilization energies. As shown by the following examples, BDE(CC) will be all the more important as the dimer will be more stabilized and the corresponding radical will be less stabilized:

	$F_3C—CF_3$	$\longrightarrow$	$2\,\dot{C}F_3$	
SE^0	3.08		$2\times(-4.17)$	BDE(CC) = 96.90
	Stabilized		Destabilized	Large
	$Cl_3C—CCl_3$	$\rightarrow$	$2\,\dot{C}Cl_3$	
SE^0	−41.86		$2\times(-13.79)$	BDE(CC) = 71.20
	Highly destabilized		Destabilized	Small

(6.41)

It may be concluded that, contrary to the commonly assumed opinion, BDE(CC) is not a good measure of the stability of a carbon-centred free radical.

Carbocations

Stabilization energies of selected carbocations are collected in Table 6.26. The relative stabilities with respect to CH_3^+ are given in the last column. It can be seen that all the carbocations, generally assumed to be planar, are stabilized by means of the mesomeric effect of the substituent(s)—whatever their nature (π-captor or π-donor groups). A recently accepted opinion,[40]

Table 6.26 Stabilization energies of carbocations (kcal mol^{-1}; 298.15 K)

Ion(R^+)	$SE^0(R^\cdot)$	$I(R^\cdot)$[38]	SE^0	$SE^0(CH_3^+)$
CH_3^+	−1.67	227.37	30.61	0.00
$CH_3CH_2^+$	2.11	193.70	68.06	37.45
$(CH_3)_2CH^+$	2.57	172.95	89.27	58.66
$(CH_3)_3C^+$	4.35	159.11	104.89	74.28
FCH_2^+	−1.89	205.23	52.53	21.92
F_2CH^+	−4.11	200.62	54.92	24.31
F_3C^+	−4.17	211.46	44.02	13.47
Cl_3C^+	−13.79	202.47	43.39	12.78
$NH_2CH_2^+$	3.92	142.97	120.60	89.99
$OHCH_2^+$	3.13	175.26	87.52	56.91
$CNCH_2^+$	6.50	250.70	15.45	−15.16
$CH_2{=}CH—CH_2^+$	13.28	187.48	85.45	54.84
$C_6H_5CH_2^+$	12.08	166.03	105.70	75.09
$(CH_3)_2N—CH_2^+$	14.72	131.44	142.93	112.32

according to which a captor group such as CN may stabilize a carbocation by delocalizing the positive charge, is illustrated by the following resonance structures:

$$H_2\overset{+}{C}-C\equiv N \longleftrightarrow H_2C=C=N^+ \tag{6.42}$$

However, taking CH_3^+ as reference, one may still assume that captor groups destabilize carbocations.

The results of Table 6.26 allow us to order the substituents according to their growing stabilizing influence:

$$CN < H < F < CH_3 < OH \simeq CH{=}CH_2 < C_6H_5 < NH_2 < N(Me)_2$$

As expected, the best π-donor groups are the most stabilizing substituents.

Negative ions

Table 6.27 collects stabilization energies of a series of negative ions and the relative stabilities of these species with respect to CH_3^- (last column). It can be seen that all substituents stabilize CH_3^-, which turns out to be the most destabilized anion. Mesomeric effects of the various groups do not allow rationalization of these results as the carbanions are generally not planar. The relatively important stabilization energy of CF_3^- may certainly be interpreted in terms of inductive effects of fluorine atoms which permit the delocalization of the negative charge on the whole system.

We may retain the following order of stabilizing influence of substituents:

$$H < CH_3 < CH_2{=}CH < C_6H_5 < CN$$

Finally, the high stabilization energy of cyclopentadienyl anion (comparable

Table 6.27 Stabilization energies of negative ions (kcal mol^{-1}; 298.15 K)

Ion(R^-)	$SE^0(R^\cdot)$	$A(R^\cdot)$[38]	SE^0	$SE^0(CH_3^-)$
CH_3^-	−1.67	1.80	−29.20	0.00
$CH_3CH_2^-$	2.11	<7.80	>−19.42	<9.78
F_3C^-	−4.17	46.40	12.90	42.10
Cl_3C^-	−13.79	33.20	−9.92	19.28
$CNCH_2^-$	6.50	34.70	11.87	41.07
$(CN)_2CH^-$	5.17	60.20	36.04	65.24
$CN{-}CHOH^-$	2.15	30.30	3.12	32.32
$CH{=}CH{-}CH_2^-$	13.28	12.70	−3.35	25.85
$C_6H_5CH_2^-$	12.08	20.30	3.05	32.25
$C_5H_5^-$	19.24	41.20	31.11	60.31
$C_6H_5^-$	−10.26	50.73	11.74	40.34
CN^-	−31.10	88.10	27.67	56.87

with the stabilization energy of benzene: 45.8 kcal mol^{-1}) demonstrates the aromatic character of this species in agreement with the $4n+2$ Hückel rule.

6.4 QUANTUM THEORY OF CHEMICAL REACTIVITY

6.4.1 General Considerations

A fundamental goal of quantum molecular physics is the rationalization and prediction of the reactivity of chemical species. In Figure 6.21 we summarize the main theoretical approaches to this problem. As the experiments are always performed on a collection of molecules, we start with the corresponding wave function $\Psi(a\{x, X, t\})$, where a stands for the number of molecules studied at time t and x, X represent formally the cartesian coordinates associated respectively with electrons and nuclei. The time variable t is requested as the chemical process is effectively time-dependent and not a stationary process. In most theoretical studies the dimension of

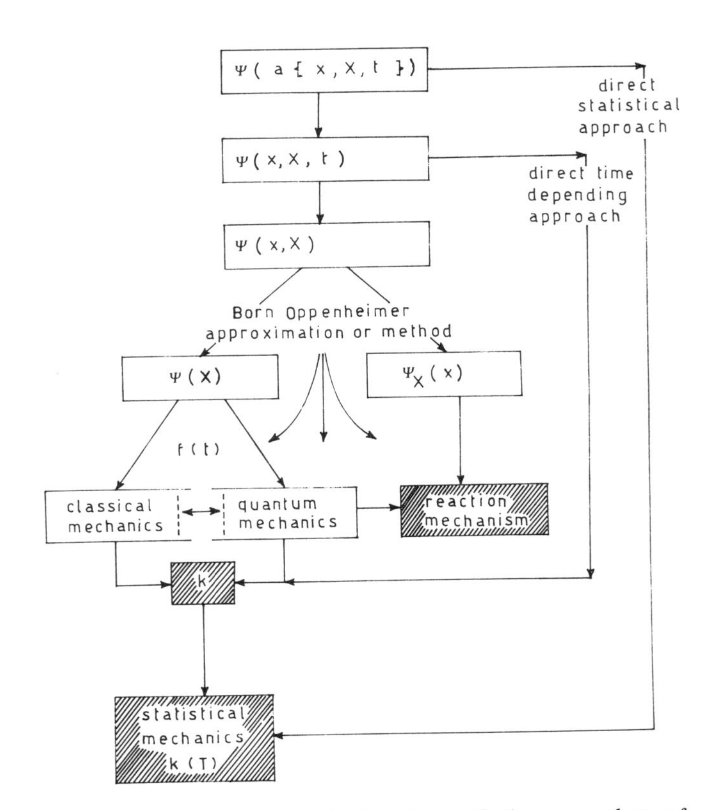

Figure 6.21 Summary of the theoretical approaches of chemical reactivity

the problem is reduced by considering the elementary reaction involving only one or two molecules. Thus we have now to calculate the wave function $\Psi(x, X, t)$. A further simplification will be introduced by assuming that the reacting system may be described by a series of stationary solutions of the time-independent Schrödinger equation. The searched wave function then becomes $\Psi(x, X)$. Finally, the Born–Oppenheimer approximation leads to the separation of electronic and nuclear motions described respectively by the corresponding wave functions $\Psi(X)$ and $\Psi_X(x)$, the latter being calculated for any given nuclear configuration. The Born–Oppenheimer method may also be used for systems where the separation of electronic and nuclear motions is not allowed. The energy associated with the electronic wave function is commonly called the potential energy of the reacting system. It leads to the concept of potential energy hypersurface which is related to the force field applied to any nuclear configuration. Depending on their accuracy, the potential surfaces may be used either to determine the reaction pathway and the corresponding reaction mechanism to solve the nuclear equation. These two aspects will be analysed later on in this section.

The resolution of the nuclear equation allows us to introduce time as an explicit variable. For very simple systems, it may be performed by using the tools developed in quantum mechanics. One may also use a classical approach where nuclei are no longer considered as quantum particles but as classical ones moving on the potential energy hypersurface. Both approaches lead to the concept of reaction probability which is related to the reaction rate constant by the following equation deduced in statistical mechanics:

$$k \text{ (statistical)} = \int \text{probabilities} \times \text{density of state}$$

In this way results describing the evolution of the whole system may be obtained. Furthermore, the calculated rate constants are temperature-dependent.

In this section, we shall consider both the problem of rate constant computation and the problem of the determination of reaction mechanisms. Historical surveys of the theoretical approach of chemical reactivity may be found in other books.[41]

6.4.2 The Rate Constant Calculation

From among the main methods for calculating the rate constant of a chemical reaction we shall only retain here two of them based on the transition state theory (TST) and on the collision theory. However, we will also make some comments about the quantum and statistical approaches. Let us first recall some aspects of chemical reactions.

6.4.2.1 General remarks concerning chemical reactions

It is possible to distinguish between four modes of production of chemical reactions. The simplest one is simply to heat the reaction mixture. By heating we mean any operation which brings the reaction mixture to a temperature which is higher than absolute zero. Consequently, it is reasonable to call this type of reaction *thermochemical.*

Certain reactions are produced between electrodes with the help of an electric field; such reactions are *electrochemical.* Others are produced under the effect of radiation; these are *radiochemical* reactions. Reactions produced under the effect of UV radiations or visible light are called *photochemical* reactions. Finally, it is now possible to produce reactions by *molecular beam experiments.* This new kind of reaction is especially convenient for theoretical interpretation.

We must also recall that there are a few relations between the mechanism of a reaction and its analytical balance. When we write:

$$H_2 + Cl_2 \rightarrow 2\,HCl$$

for the photochemical reaction between chlorine and hydrogen, we do not give any indication of the mechanism of the reaction, which could much better be suggested by writing:

$$\begin{aligned} Cl_2 + h\nu &\rightarrow Cl + Cl \\ Cl + H_2 &\rightarrow ClH + H \\ H + Cl_2 &\rightarrow Cl \end{aligned}$$

In fact, this chain reaction gives a mean quantum yield of the order of one hundred thousand. This means that a single photon gives rise, on an average, to the formation of one hundred thousand molecules of hydrochloric acid. The mean length of chain initiated by a single photon is therefore about one hundred thousand steps. Each step consisting of a collision between a molecule and an atom or between a molecule and a photon can be called an elementary process (or simply a process) of the complete reaction. The main purpose of this section of the book is to explain how it is possible to calculate the rate constant of such processes.

To complete the analysis of the mechanism of the reaction we can say that the walls of the apparatus containing the reacting molecules play an important role. They permit the recombination of the active species, i.e. of the H and Cl atoms following the processes:

$$\begin{aligned} Cl + Cl + \text{wall} &\rightarrow Cl_2 + \text{wall} \\ H + H + \text{wall} &\rightarrow H_2 + \text{wall} \\ H + Cl + \text{wall} &\rightarrow HCl + \text{wall} \end{aligned}$$

Furthermore, impurities like O_2 can also contribute to the cutting of the chains:

$$Cl + O_2 \rightarrow ClO_2$$

This simple example shows that before using wave mechanics to study a chemical reaction it is necessary to have a certain amount of information about its mechanism, based on a careful interpretation of many experimental results.

6.4.2.2 The transition state theory

Let us consider a bimolecular process:

$$A+B \xrightarrow{k_c} C+D+\cdots$$

taking place in the gaseous phase.

The rate v of the reaction is, by definition:

$$v=\frac{d[C]}{dt} \tag{6.43}$$

where $[C]$ denotes the concentration of molecules C. The kinetic theory leads to the equation:

$$v=k_c[A][B] \tag{6.44}$$

where k_c denotes the rate constant of the bimolecular process, expressed in concentration units.

During the collision between molecules A and B a certain 'activated complex' $M^{\neq}$ is formed. The hypothesis of the transition state[42] rests in the fact that a state of that complex possesses a sufficiently long life to be in thermodynamical equilibrium with the initial products. That state is called the transition state.

One can therefore write:

$$A+B \underset{}{\overset{K_c^{\neq}}{\rightleftarrows}} M^{\neq} \xrightarrow{k^{\neq}} C+D+\cdots$$

If $k^{\neq}$ denotes the rate constant of decomposition of the activated complex (or transition complex), the rate of the reaction can be written as:

$$v=k^{\neq}[M^{\neq}] \tag{6.45}$$

Let $K_c^{\neq}$ be the equilibrium constant of the process leading to the activated complex:

$$K_c^{\neq}=\frac{[M^{\neq}]}{[A][B]} \tag{6.46}$$

By considering the two last equations it is readily seen that:

$$v=k^{\neq}k_c^{\neq}[A][B] \tag{6.47}$$

Taking into account equation (6.44) we obtain the final result:

$$k_c=k^{\neq}K_c^{\neq} \tag{6.48}$$

The calculation of k_c amounts to the calculations of $k^{\neq}$ and $K_c^{\neq}$. We shall discuss these two kinds of calculation separately.

However, before doing so we must add two comments. First, if the reaction is reversible it would be necessary to take the reverse process into account:

$$C + D \rightarrow M^{\neq}$$

but since, in any case, at the beginning of the reaction the concentrations of C and D are zero this process is negligible. Therefore the study we are going to develop applies both in the case where the reaction is not reversible and at its beginning.

Second, we have neglected the perturbation caused by the disappearance of $M^{\neq}$ according to the process:

$$M^{\neq} \rightarrow C + D + \cdots$$

Various authors[43] have shown that this approximation is frequently very convenient.

Eyring[44] has calculated $k^{\neq}$, the rate constant of the decomposition of the activated complex into the final products. He has shown that a first approximation for $k^{\neq}$ is given by the expression:

$$k^{\neq} = \frac{kT}{h} \tag{6.49}$$

where k denotes the Boltzmann constant. (The derivation of this expression and many details about the transition state theory have been given by Daudel.[45] Therefore $k^{\neq}$ is independent of the nature of the activated complex. The specificity of a rate constant k_c lies in the value of $K_c^{\neq}$.

$K_c^{\neq}$ is an equilibrium constant which can be expressed in terms of partition functions. Indeed, it is easy to show that:

$$K_c^{\neq} = \frac{[M^{\neq}]}{[A][B]} = K_p^{\neq}(R'T)^{-\Delta\nu^{\neq}} \tag{6.50}$$

where $K_p^{\neq}$ is the equilibrium constant in pressure units, R' is the ideal gas constant in litre-atmosphere units and $\Delta\nu^{\neq}$ is the mole change in the process leading to the activated complex (here $\Delta\nu = -1$).

Furthermore, as shown in Section 4.2.6, we may write:†

$$K_p^{\neq} = \frac{f_{M^{\neq}}}{f_A f_B} \exp\left[\frac{-\Delta U^{0\neq}(0)}{RT}\right] \tag{6.51}$$

where $\Delta U^{0\neq}(0)$ is the energy of formation of $M^{\neq}$ at 0 K from A and B:

$$\Delta U^{0\neq}(0) = N\Delta\varepsilon^{\neq} = N(\varepsilon_{0M^{\neq}} - \varepsilon_{0A} - \varepsilon_{0B}) \tag{6.52}$$

† In this expression the Avogadro number is implicitly introduced in the translational partition function.

We recall that the fundamental energy level of each species contains the zero point energy:

$$\varepsilon_{0j} = (\text{ZPE})_j + u_{0j} \tag{6.53}$$

Thus it is readily seen that:

$$K_c^{\neq} = \frac{f_{M^{\neq}}}{f_A f_B} \exp\left(-\frac{\Delta\varepsilon^{\neq}}{kT}\right) R'T \tag{6.54}$$

The quantity $\Delta\varepsilon^{\neq}$ is called the potential energy barrier at 0 K. Thus one finally obtains the fundamental equation of the transition state theory:

$$k_c = \frac{kT}{h}\frac{f_{M^{\neq}}}{f_A f_B} \exp\left(-\frac{\Delta\varepsilon^{\neq}}{kT}\right) R'T \tag{6.55}$$

which refers to a standard state of 1 atm. The rate constant k_c may also be written in the more general form:

$$k_c = \frac{kT}{h}\frac{f_{M^{\neq}}}{f_A f_B} \exp\left(-\frac{\Delta\varepsilon^{\neq}}{kT}\right)\left(\frac{R'T}{P}\right)^{m-1} \tag{6.56}$$

where m is the overall order of the reaction and P is the pressure (generally taken as 1 atm). This expression is particularly useful for defining the units of k_c.

It is also easy to show that:

$$k_p = \frac{kT}{h}\frac{f_{M^{\neq}}}{f_A f_B} \exp\left(-\frac{\Delta\varepsilon^{\neq}}{kT}\right)(P)^{1-m} \tag{6.57}$$

or, formally, if $P = 1$ atm:

$$k_p = \frac{kT}{h}\frac{f_{M^{\neq}}}{f_A f_B} \exp\left(-\frac{\Delta\varepsilon^{\neq}}{kT}\right) \tag{6.58}$$

Let us now express the rate constants in terms of $\Delta H^{\neq}$ and $\Delta S^{\neq}$. We may write:

$$k_p = \frac{kT}{h} K_p^{\neq} \tag{6.59}$$

but:

$$K_p^{\neq} = \exp\left(-\frac{\Delta G_p^{0\neq}}{RT}\right) \tag{6.60}$$

Thus, it follows that:

$$k_p = \frac{kT}{h} \exp\left(-\frac{\Delta G_p^{\neq}}{RT}\right) \tag{6.61}$$

and

$$k_c = \frac{kT}{h} \exp\left(-\frac{\Delta G_p^{\neq}}{RT}\right) R'T \tag{6.62}$$

(We have now dropped the superscript ^{0} for convenience, and shall continue to do so.) The explicit form of $\Delta G^{\neq}_{\mathrm{p}}$, $\Delta H^{\neq}_{\mathrm{p}}$ and $\Delta S^{\neq}_{\mathrm{p}}$ is easy to deduce by using (6.51), (6.52) and (6.60):

$$\Delta G^{\neq}_{\mathrm{p}} = -RT \ln k^{\neq}_{\mathrm{p}} \tag{6.63}$$

$$\Delta G^{\neq} = N\Delta\varepsilon^{\neq} - RT \ln \frac{f_{\mathrm{M}^{\neq}}}{f_{\mathrm{A}} f_{\mathrm{B}}} \tag{6.64}$$

On the other hand:

$$\Delta H^{\neq}_{\mathrm{p}} = RT^2 \left(\frac{\partial \ln k_{\mathrm{p}}}{\partial T}\right)_{\mathrm{p}} \tag{6.65}$$

$$\Delta H^{\neq}_{\mathrm{p}} = RT^2 \left[\frac{\partial \ln (f^{\neq}/f_{\mathrm{A}}/f_{\mathrm{B}})}{\partial T}\right]_{\mathrm{p}} + N\Delta\varepsilon^{\neq} \tag{6.66}$$

Finally:

$$\Delta S^{\neq}_{\mathrm{p}} = \frac{\Delta H^{\neq}_{\mathrm{p}} - \Delta G^{\neq}_{\mathrm{p}}}{T} \tag{6.67}$$

and

$$\Delta S^{\neq}_{\mathrm{p}} = R \ln \frac{f^{\neq}}{f_{\mathrm{A}} f_{\mathrm{B}}} + RT \left[\frac{\partial \ln (f^{\neq}/f_{\mathrm{A}} f_{\mathrm{B}})}{\partial T}\right]_{\mathrm{p}} \tag{6.68}$$

Thus, the activation entropy only depends on the partition functions but the activation enthalpy depends mainly on the potential barrier at 0 K.

At this point, the rate constants k_{p} and k_{c} may respectively be written as:

$$k_{\mathrm{p}} = \frac{kT}{h} \exp\left(-\frac{\Delta H^{\neq}_{\mathrm{p}}}{RT}\right) \exp\left(\frac{\Delta S^{\neq}_{\mathrm{p}}}{R}\right) \tag{6.69}$$

and

$$k_{\mathrm{c}} = \frac{kT}{h} \exp\left(-\frac{\Delta H^{\neq}_{\mathrm{p}}}{RT}\right) \exp\left(\frac{\Delta S^{\neq}_{\mathrm{p}}}{R}\right) R'T \tag{6.70}$$

It can easily be shown that:[46]

(a) $\Delta H^{\neq}_{\mathrm{p}} = \Delta H^{\neq}_{\mathrm{c}} + \Delta\nu^{\neq} RT = \Delta U^{\neq}_{\mathrm{c}} + 2\Delta\nu^{\neq} RT$ (6.71)

where $\Delta U^{\neq}_{\mathrm{c}}$ is the activation barrier in concentration units, at temperature T:

$$\Delta U^{\neq}_{\mathrm{c}} = \Delta U^{\neq}(T) = N\Delta\varepsilon^{\neq} + \Delta[U(T) - U(0)]^{\neq} \tag{6.72}$$

(b) $\Delta S^{\neq}_{\mathrm{p}} = \Delta S^{\neq}_{\mathrm{c}} + R\Delta\nu^{\neq} \ln (R'T)$ (6.73)

so, for a second-order reaction:

$$k_{\mathrm{c}} = \frac{kT}{h} e^2 \exp\left(-\frac{\Delta U^{\neq}}{RT}\right) \exp\left(\frac{\Delta S^{\neq}_{\mathrm{c}}}{R}\right) \tag{6.74}$$

Let us now deduce the explicit expressions of the Arrhenius parameters

E_a and A, which are generally given in concentration units:

$$E_a = E_c; \qquad A = A_c$$

Introducing (6.62) in the well-known relation:

$$E_c = RT^2\left(\frac{\partial \ln k_c}{\partial T}\right)_p \tag{6.75}$$

one obtains:

$$E_c = RT^2\left[\frac{1}{T} + \frac{\Delta G_p^{\neq}}{RT^2} - \frac{1}{RT}\left(\frac{\partial \Delta G_p^{\neq}}{\partial T}\right)_p + \frac{1}{T}\right] \tag{6.76}$$

and, as:

$$\left(\frac{\partial \Delta G_p^{\neq}}{\partial T}\right)_p = -\Delta S_p \tag{6.77}$$

equation (6.76) becomes:

$$E_c = 2RT + \Delta H_p^{\neq} \tag{6.78}$$

Finally, using (6.71) with $\Delta\nu^{\neq} = -1$, one finds:

$$E_c = \Delta U_c^{\neq} \qquad \text{or} \qquad E_c = \Delta U^{\neq}(T) \tag{6.79}$$

More explicitly:

$$E_c = N\Delta\varepsilon^{\neq} + \Delta[U(T) - U(0)]^{\neq} \tag{6.80}$$

or

$$E_c = N\Delta u_0^{\neq} + \Delta(\text{ZPE})^{\neq} + \Delta[U(T) - U(0)]^{\neq} \tag{6.81}$$

We can distinguish three kinds of 'activation energy':

(a) The true activation energy, i.e. the experimental Arrhenius activation energy, E_c, which is equal to $\Delta U^{\neq}(T)$,
(b) The activation barrier at 0 K, $\Delta\varepsilon^{\neq}$, which differs from E_c by the thermal correction term and, finally,
(c) The potential barrier at 0 K which does not contain the ZPE correction term.

In pressure units, the Arrhenius activation energy will be written:

$$E_p = E_c - RT \tag{6.82}$$

or, more explicitly:

$$E_p = N\Delta u_0^{\neq} + \Delta(\text{ZPE})^{\neq} + [U(T) - U(0)]^{\neq} - RT \tag{6.83}$$

Of course, one can also define the corresponding activation enthalpies, but they are not used very much. The computational details concerning the evaluation of ZPE and thermal corrections will not be given here. Notice that the thermodynamical properties of the transition state must be computed without considering the imaginary vibrational frequency corresponding to the movement along the reaction coordinate. Finally, an explicit expression of the preexponential factor of the Arrhenius equation may be deduced from the relations (6.74) and (6.79):

$$A_c = \frac{kT}{h} e^2 \exp\left(\frac{\Delta S_c^{\neq}}{R}\right) \tag{6.84}$$

or, using equation (6.73):

$$A_c = \frac{kT^2}{h} R' e^2 \exp\left(\frac{\Delta S_p^{\neq}}{R}\right) \tag{6.85}$$

The corresponding expression for A_p is obtained from (6.69), (6.71) and (6.82):

$$A_p = \frac{kT}{h} e \exp\left(\frac{\Delta S_p^{\neq}}{R}\right) \tag{6.86}$$

Equation (6.81) shows that the activation energy mainly depends on the activation barrier $\Delta u_0^{\neq}$ which can only be computed by the *ab initio* methods of quantum chemistry. This is why the calculation of potential barriers is a central problem in the framework of the transition state theory. Let us consider the classical example of the isotopic exchange reaction:

$$H + D_2 \rightarrow DH + D$$

in which D signifies deuterium. During the collision an activated complex is formed containing three electrons and three nuclei. Let r_1 and r_2 be the respective distances of one of the nuclei, say H, from the two others, D_1 and D_2, and let θ be the angle formed by the vectors $\mathbf{r}_1$ and $\mathbf{r}_2$. In the framework of the Born–Oppenheimer approximation we can calculate the electronic energy u of the system as a function of r_1, r_2 and θ. One may find the appearance of the isoenergetic contours for $\theta = 180°$ by using a method proposed by Eyring which introduces one empirical parameter (further details being given by Daudel[(45)], and see two potential valleys (one for a large value of r_1 and the other for a large value of r_2) separated by a kind of hill. Eyring proposed arguments leading to the conclusion that for other values of θ the hill is higher. Therefore to go classically from the initial products ($H + D_2$) to the final ones ($DH + D$) the electronic energy of the system must go over the top of the hill. During the collision the total energy of the reactants is a constant; thus, when the electronic energy increases, the kinetic energy associated with the nuclei must decrease. The state corresponding to the top of the valley must possess a relatively long life. This is why it is assumed that this top corresponds to the activated complex.

The contribution to the potential barrier of the electronic energy is therefore the height of the hill. As shown previously, to obtain the actual activation energy we must take account of the vibration energy of the nuclei at 0 K and of the thermal corrections for both the transition state and the initial products.

The formula (6.55) can be improved by taking account of the tunnel effect and of the transmission coefficient η. The formula:

$$k^{\neq} = \frac{kT}{h}$$

has been obtained by assuming that $\eta = 1$, that is to say that all collisions leading to the transition state lead also to the final product.

By introducing all these improvements and calculating the various partition functions, Eyring has been able to compute the rate constants for the reactions:

$$H + H_2 \rightarrow H_2 + H$$
$$D + D_2 \rightarrow D_2 + D$$
$$H + HD \rightarrow H_2 + D$$

Table 6.28 compares the results of the calculations with experimental measurements. The agreement is somewhat miraculous and gives confidence to the transition state theory. This is why that theory has been widely used.

Table 6.28 Theoretical and experimental rate constants for some elementary reactions (in cubic centimetres per mole-second)

Reaction		300 K	1000 K
$H + H_2 \rightarrow H_2 + H$	Calculated	7.3×10^7	1.5×10^{12}
	Observed	9×10^7	2×10^{12}
$D + D_2 \rightarrow D_2 + H$	Calculated	3×10^7	0.76×10^{12}
	Observed		1.2×10^{12}
$H + HD \rightarrow H_2 + HD$	Calculated	2.2×10^7	0.52×10^{12}
	Observed		0.68×10^{12}

Before ending this brief description of the transition state theory we shall give two of the various expressions which have been proposed for the 'tunnelling factor' $(1 + t)$. In 1932, Wigner[47] developed a method which is a first approximation applicable to any form of potential curve. Wigner's quantum correction for a vibration with imaginary frequency $i\nu_t^{\neq}$ is given by:

$$(1 + t) = 1 + \frac{(h\nu_t^{\neq}/RT)^2}{24} \tag{6.87}$$

Recent studies[48] confirmed the validity of Wigner's corrections for proton transfer reactions. The Arrhenius parameters become, with the tunnelling corrections:

$$E_a(t) = E_a - RT\frac{(h\nu_t^{\neq}/kT)^2}{12}\left[1 + \frac{(h\nu_t^{\neq}/kT)^2}{24}\right]$$

$$A(t) = -A\left[1 + \frac{(h\nu_t^{\neq}/kT)^2}{24} \exp\left(\frac{E_a(t) - E_a}{RT}\right)\right]$$

A more precise correction has been proposed by Bell[49] for a parabolic potential curve. Christov[50] has given a deduction of Bell's formula:

$$(1+t)=\frac{\theta}{\sin\theta}\exp[\text{cotg }\theta(1-\theta)] \tag{6.88}$$

where:

$$\theta=\frac{\pi T_c}{2T} \tag{6.89}$$

T_c being the critical temperature defined as:

$$T_c=\frac{h\nu_t^{\neq}}{\pi k} \tag{6.90}$$

According to Christov's results, the classical behaviour remains as long as the temperature is greater than twice T_c. A large tunnel effect appears for temperatures lower than $T_c/2$; below $T_c/2$, a complete quantum treatment is necessary.

Goldanskii has introduced an alternative definition of the characteristic tunnelling temperature based not on the imaginary frequency, i.e. the curvature of the reaction pathway at the transition state, but on the height ($\Delta\varepsilon^{\neq}$) and the width (d) of the barrier:[51]

$$T_g=\frac{h}{k\pi d}\sqrt{\left(\frac{\Delta\varepsilon^{\neq}}{2\mu}\right)} \tag{6.91}$$

where μ is the reduced mass of the particle passing through the barrier. The Arrhenius parameters with the tunnelling correction become:

$$E_a(t)=E_a-RT\,(1-\theta\text{ cotg }\theta)$$

$$A(t)=A\frac{\theta}{\sin\theta}\exp[\text{cotg}(\theta-1)]$$

For a strictly parabolic barrier both characteristic temperatures, T_c and T_g, should coincide.

6.4.2.3 Collision theory

Molecular beam experiments

For theoretical purposes the best way to produce a chemical reaction is to generate a collision between a molecule possessing a given velocity characterized by known rotational, vibrational and electronic quantum numbers and another molecule described with the same precision. Molecular beams almost provide these good conditions.[52]

The reaction is produced in a large chamber where a good vacuum is created. The chamber contains two mobile generators of molecular beams in such a way that the angle between the beams can be chosen. When neutral molecules are concerned thermal diffusion is used. For ion beams, ion accelerators are chosen. They present a significant advantage as they produce ionic beams in which all ions possess approximately the same velocity. For neutral molecules, velocity selectors must be added if a monocinetic beam is needed. Various mobile detectors are also introduced in the chamber. They permit measurement of the angular distribution of the products. If ions are produced mass spectrometers can be used to measure the velocity of the products.

For example, to study the reaction:

$$K + Br_2 \rightarrow KBr + Br$$

chemical detectors are needed. A detector of tungsten is able to extract an electron from both K and KBr. On the other hand, a detector made of an alloy of tungsten and platinum containing 8 per cent. of tungsten does not react with KBr but reacts with K. An electron counter related to the tungsten detector will count both K and KBr. An electron counter related to the platinum detector will count only K. By difference, the number of KBr molecules produced will be determined.

Figure 6.22 shows the angular distribution of KBr in a cross-molecular beam experiment in which K and Br_2 beams are produced by thermal

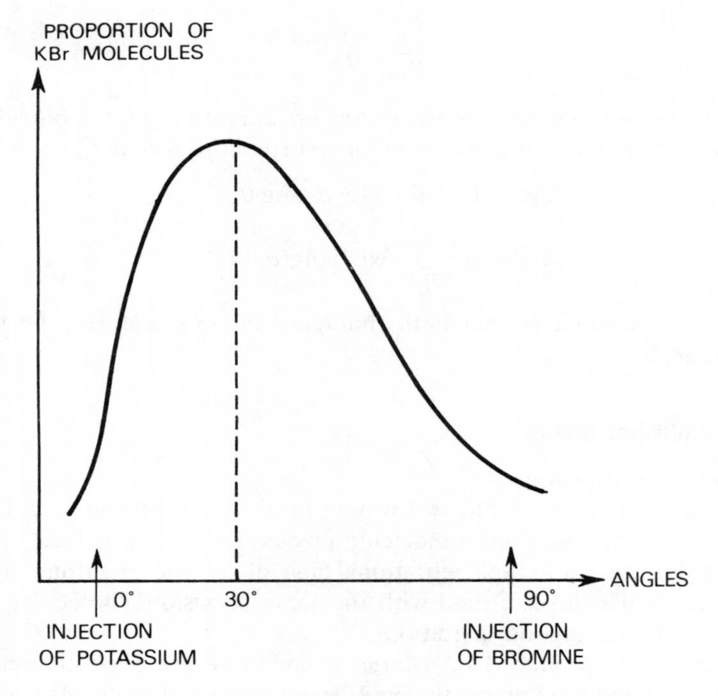

Figure 6.22 Angular distribution of KBr in a cross-molecular beam experiment

diffusion at 686 and 314 K respectively, the angle between the beams being 90°. More details about molecular beams have been given.[53] From such experiments the cross-section of the reaction can be measured for any diffusion angle.

Semiclassical approximation

A complete wave mechanical treatment of a molecular collision remains a very difficult problem. As the nuclei are heavy compared with the electrons, a semiclassical treatment is usually made. It consists in computing the potential energy surface of the colliding molecule considered as a supermolecule. The potential is introduced into the hamiltonian of classical mechanics and the trajectories of the nuclei are calculated as a function of the initial conditions. Such calculations allow the computation of the reaction rate for given experimental conditions.

To illustrate this procedure we shall consider once again the reaction:

$$\mathrm{D}+\mathrm{H}_2 \rightarrow \mathrm{HD}+\mathrm{H}$$

In the first step we calculate the potential energy of the system HDH as a function of three interatomic distances:

$$\mathrm{H}^{(1)}\mathrm{H}^{(2)} = R_1; \qquad \mathrm{H}^{(1)}\mathrm{D} = \mathrm{R}_2; \qquad \mathrm{H}^{(2)}\mathrm{D} = R_3$$

Such calculations can be achieved by using a large number of configurations built on a sufficiently large basis set of atomic functions. Such a basis could be made of $1s, 2s, 2p_x, 2p_y, 2p_z$ type orbitals associated with each nucleus. It is interesting to note that such a calculation confirms the hypothesis made by Eyring (Section 6.4.2.2) which assumes that the saddle point corresponds to a linear structure of the H_2D system.

The second step of the procedure consists in introducing the function $u(R_1, R_2, R_3)$ obtained in the hamiltonian of classical mechanics. In fact the foregoing calculations do not generate the function u itself but a set of numerical values of u for a set of values of the three distances R_1, R_2, R_3. Therefore we need to find an analytical function which fits as well as possible the various numerical values obtained by using the configuration interaction method.

The classical hamiltonian can be written as:

$$H = \frac{1}{2\mu_{\mathrm{HH}}}\sum_{i=1}^{i=3} P_i^2 + \frac{1}{2\mu_{\mathrm{D,HH}}}\sum_{i=4}^{i=6} P_i^2 + \frac{1}{2M}\sum_{i=7}^{i=9} P_i^2 + W(Q_1, Q_2, \ldots, Q_9) \quad (6.92)$$

where Q_1, Q_2, Q_3 denote the relative coordinates of one hydrogen atom with respect to the other, Q_4, Q_5, Q_6 the coordinates of D with respect to the centre of gravity of H_2 and Q_7, Q_8, Q_9 the coordinates of the centre of gravity of the molecule. The μ's represent the reduced masses and the P_i's are the conjugated momenta.

The classical equations are:

$$\frac{dQ_i}{dt} = \frac{\partial H}{\partial P_i}$$
$$\frac{dP_i}{dt} = -\frac{\partial H}{\partial Q_i} = -\frac{\partial u}{\partial Q_i} \tag{6.93}$$

By solving them for a given set of initial conditions one obtains the corresponding trajectories of the nuclei.

Figure 6.23 shows a typical result.[54] The distance R_1 corresponding to H_2 oscillates during the approach of D. Suddenly the HD molecule is formed and the corresponding R_2 distance oscillates. The distances R_1 and R_3 increase, showing that one hydrogen atom is going far from the new HD molecule obtained. This figure therefore corresponds to initial conditions leading to the formation of HD; they are reactive conditions leading to the final products.

Obviously other conditions only lead to an elastic collision, the H_2 molecule being perturbed during a short time but not disturbed. Figure 6.24 shows such a case.

Therefore the third step is to calculate the probability of the reaction by considering all the possible initial conditions and taking account of the experimental conditions.

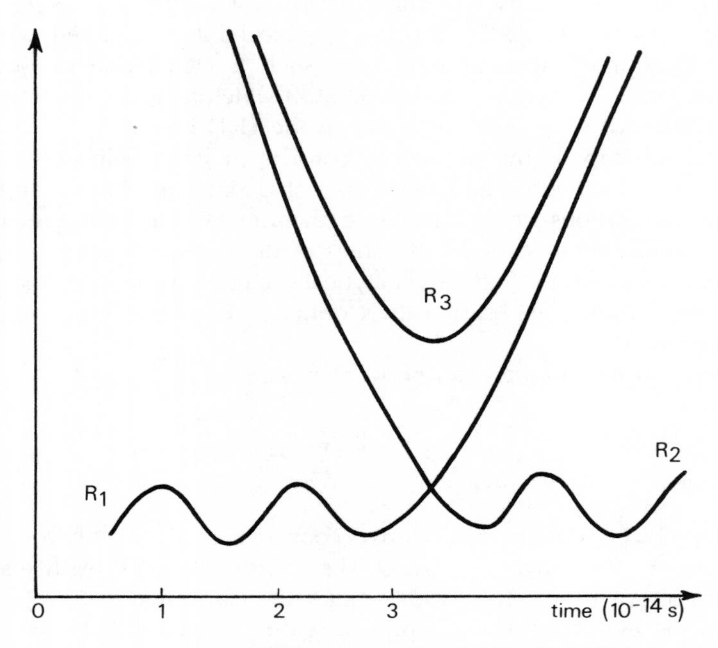

Figure 6.23 Typical trajectory for the $D+H_2$ collision

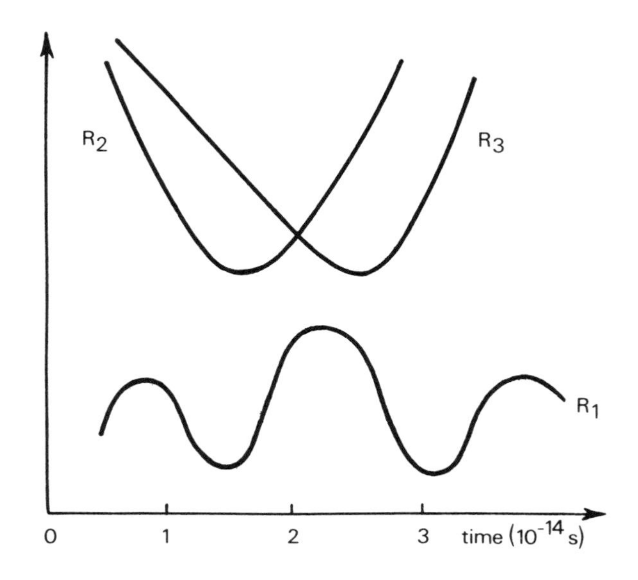

Figure 6.24 Elastic collision of $D+H_2$

However, before discussing this point we must add an important remark. Consider again Figure 6.24. It shows that the reaction takes place in a very short time (of the order of 10^{-14} s). During that time, the activated complex *cannot* be in thermodynamical equilibrium with the initial products. It is impossible to write partition functions for the transition state. Therefore the Eyring and Polanyi hypothesis assuming that a state of that complex possesses a sufficiently long life to be in thermodynamical equilibrium with the initial products is not valid. *The transition state theory is wrong in the case of the reaction considered here.* We shall discuss more fully later on the consequences of this very important remark.

Going back to the third step of the semiclassical treatment, we need to calculate the probability of the reaction. To do that Karplus, Porter and Sharma[(54)] have used the Monte Carlo method. That is to say, they have generated stochastically a lot of initial conditions, calculated the trajectories and computed the ratio of the number of reactive trajectories and of the total number of trajectories. It is convenient to express the reaction probability P_r as a function of the impact parameter b of the collision, the relative velocity V_R and the rotational and vibrational quantum numbers J and v describing the initial state of the hydrogen molecule.

We can write:

$$P_r(V_R, J, v, b) = \lim_{N\to\infty} \frac{N_r(V_R, J, v, b)}{N(V_R, J, v, b)} \tag{6.94}$$

where N denotes the total number of trajectories corresponding to a given choice of the variables V_R, J, v, b and N_r, the number of reactive trajectories. The total reaction cross-section S_r is readily obtained from the equation:

$$S_r(V_R, J, v) = 2\pi \int_0^{b_{max}} P_r(V_R, J, v, b) b \, db \tag{6.95}$$

The final step is to calculate the reaction rate from the reaction cross-section. Obviously the procedure of calculating that rate will depend on the experimental conditions. If thermal diffusion is used without velocity selectors we have to introduce a maxwellian distribution for the velocities and a Boltzmann distribution for the rotation–vibration state. Then the rate constant $k_c(t)$ can be written as:

$$k_c(t) = Q_{Jv}^{-1} \left[\sum_{v,J} f_J(2J+1) \exp\left(-\frac{E_{v,J}}{kT}\right) \times N_A \left(\frac{2}{\pi}\right)^{1/2} \left(\frac{\mu_{DHH}}{kT}\right)^{3/2} \int_0^{\infty} S_r(V_R, J, v) \times \exp\left(-\frac{\mu_{D,HH} V_R^2}{2kT}\right) V_R^3 \, dV_R \right] \tag{6.96}$$

where Q_{Jv} denotes the rotation–vibration partition function, f_J the statistical weight of the state $E_{v,J}$, $S_r(V_R, J, v)$ the reaction cross-section and N_A the Avogadro number which is introduced so that $k_c(T)$ has units of cubic centimetres per mole-second.

Many important results have been obtained by Karplus, Porter and Sharma by following the procedure described above. It is seen that the reaction does not occur under a certain relative energy threshold evaluated to 5.69 kcal mol^{-1}. Furthermore, the computed value of $k_c(T)$ is well represented between 300 and 1000 K by the simple relation:

$$k_c(T) = A_c \exp\left(\frac{-E_c}{RT}\right) \tag{6.97}$$

with:

$$E_c = 7.435 \text{ kcal mol}^{-1}$$

$$A_c = 4.334 \times 10^{13} \text{ cm}^3 \text{ mol}^{-1} \text{ s}^{-1}$$

These results are in very good agreement with experimental results:

$$E_c = 7.5 \pm 1 \text{ kcal mol}^{-1}$$

$$A_c = 5.4 \times 10^{13} \text{ cm}^3 \text{ mol}^{-1} \text{ s}^{-1}$$

Theory and experiment lead to an activation energy of 7.5 kcal mol^{-1}. It is interesting to add that this result corresponds to the use of the potential energy surface for which:

$$N\Delta u_0^{\neq} = 9.13 \text{ kcal mol}^{-1}$$

and

$$N\Delta\varepsilon^{\neq} = 8.85 \text{ kcal mol}^{-1}$$

These results underline the fact that the reaction threshold, potential barriers and activation energy are not identical concepts. Let us also notice that, using the same potential energy surface, the absolute rate theory leads to an activation energy of 8.812 kcal mol^{-1}), which is very close to the barrier corrected for $\Delta(\text{ZPE}^{\neq})$.

Quantum mechanical treatment

We must now go further and consider a more complete quantum mechanical treatment to measure the importance of the semiclassical approximation introduced previously. In fact a complete quantum mechanical calculation for an exchange reaction with the collision taking place in three-dimensional space does not yet exist. Karplus[55] has considered the simplified case of a linear collision on a simplified potential surface. Figure 6.25 shows a comparison between the reaction probabilities calculated for the process:

$$D + H_2 \rightarrow DH + H$$

following this quantum mechanical treatment (curve QM) and the reaction probabilities coming from a semiclassical treatment (curve CM). It is seen that the threshold of the reaction is found for a smaller energy value with the quantum mechanical treatment. This is the well-known tunnel effect. Furthermore, the QM curve shows a minimum value which does not appear on the CM curve and corresponds to a kind of resonance phenomena.

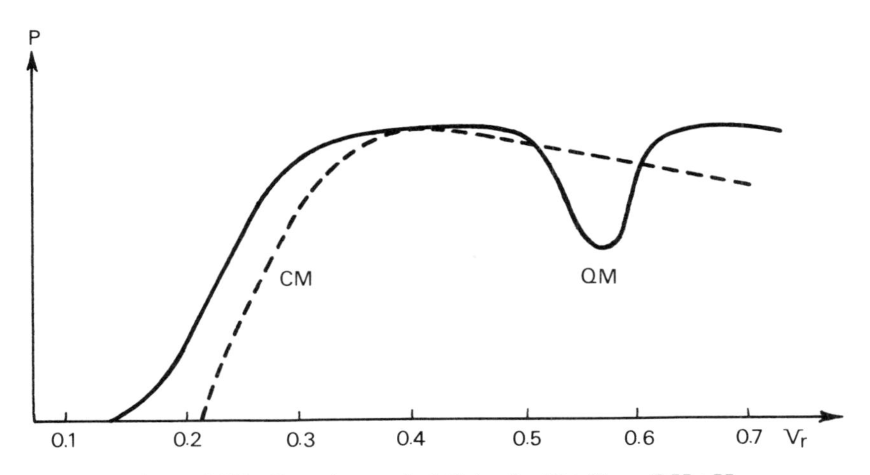

Figure 6.25 Reaction probabilities for $D + H_2 \rightarrow DH + H$

However, the two curves have roughly the same aspect and it turns out that the semiclassical treatment is not too bad. This is because it is the only one which can at present be performed for reactions between molecules of real chemical interest.

Table 6.29 shows the quantitative effect resulting mainly from the tunnel effect. It can be seen that the differences between quantum and semiclassical treatments are not too large, except for the low temperature conditions. For more details on the quantum theory of reactive molecular collisions see, for example, the paper of Michz.[56]

Table 6.29 Quantum and semiclassical H+ H_2 rate constants (in cubic centimetres per mole-second)

T	$k_{class} \times 10^{-11}$	$k_{quant} \times 10^{-11}$
300	0.0025	0.029
400	0.051	0.31
500	0.32	1.54
900	9.56	18

6.4.2.4 On the validity of the transition state theory

It is now time to go back to the discussion on the validity of the transition state theory. The fact that during the collision of two molecules there is no evidence for a long-lived complex seems to be well established. Such a complex can only be obtained when the potential surface shows a sufficiently deep hole, as in the reaction:

$$\mathrm{Na + ClCs \rightarrow ClNa + Cs}$$

Molecular beam experiments provide a tool to confirm theoretical results in this respect. The angular distribution of the products in such an experiment is significantly changed if a long-life complex is formed.

We have seen that the molecular beam experiments for the exchange reaction:

$$\mathrm{D + H_2 \rightarrow DH + H}$$

are in agreement with the collision theory results and confirm that this reaction is mainly an impulsive one.

Conversely, for the alkali atom–alkali halide reaction the distribution in angle and velocity of the products is found to agree well with the predictions obtained from a statistical model of the kind familiar in the theory of unimolecular decay or nuclear fission.[57] The data show that complex formation occurs with high probability at a distance as large as 8 Å.

Qualitatively, the electronic structure of the complex can be visualized as follows:

$$M + X^-M'^+ \rightleftarrows \begin{pmatrix} M \\ M' \end{pmatrix}^+ X^- \rightarrow M' + X^-M^+$$

In the halogen atom–halogen molecule reaction the distribution in angle and translational energy of the products is well accounted for by an oscillating complex model. This employs the same statistical postulates used for the long-lived case, but makes allowance for the decay of the complex with a mean lifetime comparable to the rotational period. In any event, a long-life intermediate complex corresponding to an absolute minimum on the potential surface cannot be confused with an activated complex of the transition state theory which corresponds to a minimax on this surface (see Part III).

The overall picture is that the validity of the transition state theory has not yet been really proved and its success seems to be mysterious. This is why various investigators have tried to build a theory as handy as that of Eyring but in which the existence of a long-lived collision complex is not required. Christov[41] starts from a different quantum mechanical formulation from Eyring, Walter and Kimball.[58] In the framework of this problem, statistical mechanics is used for calculating the distribution of the reactants among the various energy levels (rotational and vibrational), and for each of them it is possible to obtain the probability of a reaction through or over the potential barrier. Finally, the rate constant is written as:

$$k = \frac{\chi T}{h} \kappa_t \eta \frac{\tilde{f}_{A,B}}{f_A f_B} e^{-\Delta\varepsilon^{\neq}/\chi T} \qquad (6.98)$$

where χ denotes the Boltzmann constant, κ_t refers to a tunnel effect and $\tilde{f}_{A,B}$ is a partition function describing the distribution of the reactants among the various degrees of freedom, except for those which are related with the reaction coordinate. Furthermore, $\Delta\varepsilon^{\neq}$ is the height of the saddle point which does not necessarily correspond to a long-lived complex. Therefore this approach is a rehabilitation of the potential barrier which remains important even if there is no activated complex in thermodynamical equilibrium with the initial products.

6.4.2.5 The stochastic approach[59]

Let us consider a brownian molecule confined in a potential well by forces responsible for chemical bonds or by intermolecular forces. The molecule leaks across the potential barrier as the result of white noise forces acting on it. From a mathematical point of view this situation, which is characterized by a diffusion process in a force field, may be described by Langevin's stochastic differential equation.[60] If one assumes that the medium surrounding the molecule is in thermal equilibrium, the velocity (**y**) distribution due to the random collisions is maxwellian, the motion then being described by the following equations:

$$\mathrm{d}\mathbf{x} = \mathbf{y}\,\mathrm{d}t \tag{6.99}$$

$$\mathrm{d}\mathbf{y} = -\beta\mathbf{y}\,\mathrm{d}t - \boldsymbol{\nabla}\phi(\mathrm{X})\,\mathrm{d}t + \left(\frac{2\beta\chi T}{m}\right)^{1/2}\mathrm{d}\mathbf{w} \tag{6.100}$$

In equation (6.100), the first term on the right-hand side comes from the intermolecular long-range forces, $-\beta\mathbf{y}$ being the dynamic viscosity per unit mass which expresses the slowing down rate of molecules by random collisions with the particles of the medium. The second term represents the contribution of the internal force field of the reactant molecules, $\phi(\mathrm{X})$ being the potential function of $V(\mathrm{X})$ per mass unit. The last term comes from the white noise forces due to collisions. The intensity of such white noise forces determines the temperature of the reactants. These internal forces may activate the molecules in such a way that they surmount the potential barrier.

When the molecules are immersed in a fluid, the parameter β can be computed using the Stokes formula:

$$\beta = \frac{6\pi a\eta}{\mu} \tag{6.101}$$

for a monomolecular process, where a stands for the radius of the sphere representing the molecule, η is the usual viscosity coefficient and μ the reduced mass of the molecule. In the framework of this stochastic approach, the final expression of the rate constant:

$$k = -\frac{1}{C(t)}\frac{\mathrm{d}C(t)}{\mathrm{d}t} \tag{6.102}$$

where $C(t)$ is the actual concentration of the reactant molecule, may be written as:

$$k = \frac{1}{2\pi\beta}\frac{\prod_{i}^{n} \nu_i^{(\mathrm{R})}\nu_{\mathrm{t}}^{(\neq)}}{\prod_{i}^{n-1} \nu_i^{(\neq)}}\exp\left(-\frac{\Delta\varepsilon^{\neq}}{\chi T}\right) \tag{6.103}$$

In this expression, the ν_i's are respectively the principal frequencies of the reactant side ($\nu_i^{(\mathrm{R})}$) and the real vibrational frequencies of the transition structure ($\nu_i^{(\neq)}$) The term $\nu_{\mathrm{t}}^{(\neq)}$ is the imaginary frequency of the transition structure. Finally, $\Delta\varepsilon^{\neq}$ is the energy difference between the top and the bottom of the potential well at the transition point.

This expression of k is a generalization of the result obtained by Kramers for a one-dimensional case.[61] The approach is only valuable, however, if the potential barrier is large relative to the thermal energy.

6.4.2.6 Concluding remarks

Before ending this brief survey of the various theories of chemical reactivity, we would like to mention that they all lead to an Arrhenius-like expression of the rate constant:

$$k = A \exp\left(-\frac{E_a}{RT}\right)$$

However, the interpretation of A and E_a and also their temperature dependence may largely differ from one theory to another. Furthermore, all these theories assume that the thermal equilibrium of the reactants is not perturbed by the chemical reaction, and this is certainly not always true.

In spite of their failures these different approaches certainly provide much interesting information on the reactivity of chemical species.

6.4.3 The Reaction Mechanisms

Nowadays, for purely practical reasons, accurate calculations of potential energy hypersurfaces and rate constants can only be performed on small supermolecules.

For 'medium size' supersystems, the hypersurfaces may be computed locally with a good accuracy in the neighbourhood of the transition and equilibrium structures. Therefore the kinetic parameters of the reaction and the corresponding rate constant may be obtained within the framework of the transition state theory. Furthermore, the analysis of the electronic structure of the activated complex gives some information on the reaction mechanism.

For large supermolecules, it is only possible to obtain approximate potential energy hypersurfaces—most often at the SCF level using small basis sets. Therefore one calculates only the portion of the surface which is expected to contain the reaction pathway. This approach provides heats of reaction and activation barriers which are generally not very reliable. Furthermore, as the analytical expression of the surface is unreachable, thermal and ZPE corrections cannot be performed. However, as shown below, a detailed analysis of the geometric and electronic characteristics of the supermolecule along the reaction path may give interesting information on the mechanism of the corresponding chemical transformation. We will give here some results concerning 'medium size' and large supersystems.

6.4.3.1 Hydrogen abstraction reactions

Hydrogen transfer reactions involving methane and five different open-shell species have been studied at the SCF and CI levels.[62] These elementary reactions may be described by the equation:

$$CH_4 + R \rightarrow CH_3 + RH$$

The chosen R and the corresponding RH are collected in Table 6.30. For the SCF calculations, the 6-31G basis set has been used as it gives an excellent ratio of performance/cost. The CI calculations have been done with perturbation selection of configurations using the CIPSI program.[63]

Table 6.30 Hydrogen abstraction reactions

Reaction	R	RH or XH_n
I	H	H_2
II	CH_3	CH_4
III	NH_2	NH_3
IV	OH	OH_2
V	F	FH

The local potential energy hypersurface in the neighbourhood of the equilibrium structure of each polyatomic radical and molecule has been calculated at the SCF level. The full geometry optimization of these systems and their vibrational analysis were performed using methods explicitly described in Part III of this book.

Optimized structures and vibrational analysis

The parameters of the equilibrium structures and the vibrational frequencies calculated at the SCF level are respectively given in Tables 6.31 and 6.32 and are compared with the corresponding experimental data. To study the transition states, the internal coordinate system defined in Figure 6.26 has been adopted. Depending on the nature of X, hydrogen atoms H_7, H_8, H_9 and the corresponding internal coordinates may either exist or not. Distances and angles will normally be referred to according to Table 6.33.

Table 6.31 Equilibrium structures (in angstroms or degrees)

(a) Optimized geometries at SCF level, basis set: 6-31G								
	H_2	CH_3	CH_4	NH_2	NH_3	OH	H_2O	HF
Distances X—H	0.730	1.072	1.082	1.015	0.991	0.967	0.948	0.921
Angles H—X—H	—	120.0	109.5	108.6	116.1	—	111.5	—
(b) Experimental geometries								
	H_2	CH_3	CH_4	NH_2	NH_3	OH	H_2O	HF
Distances X—H	0.742	1.079	1.085	1.024	1.012	0.971	0.957	0.917
Angles H—X—H	—	120.0	109.5	103.4	106.7	—	104.5	—

Table 6.32 Vibrational frequencies (reciprocal centimetres)

Compound	$\bar{\nu}$ Theory(6-31G)	Experiment[64]	Symmetry	Type
H_2	4,643.83	4,405.3	A_1	H—H stretch
CH_3	3,321.02	3,184	E_1	C—H stretch
	3,124.85	3,002	A_1	C—H stretch
	1,469.53	1,383	E_1	H—C—H scissors
	776.47	580	A_1	Umbrella
CH_4	3,372.05	3,018.7	F_2	C—H stretch
	3,226.10	2,916.5	A_1	C—H stretch
	1,718.17	1,534.0	E	H—C—H scissors
	1,532.95	1,306.0	F_2	Umbrella
NH_2	3,676.11	3,220	B_1	N—H stretch
	3,553.77	3,173	A_1	N—H stretch
	1,651.19	1,499	A_1	H—N—H scissors
NH_3	3,985.25	3,444	E	N—H stretch
	3,780.77	3,336	A_1	N—H stretch
	1,814.29	1,627	E	H—N—H scissors
	596.81	950	A_1	Umbrella
OH	3,954.96	3,735.21	A_1	O—H stretch
H_2O	4,143.22	3,755.79	B_1	O—H stretch
	3,986.99	3,657.05	A_1	O—H stretch
	1,678.01	1,594.59	A_1	H—O—H-scissors
HF	4,150.24	4,138.32	A_1	H—F stretch

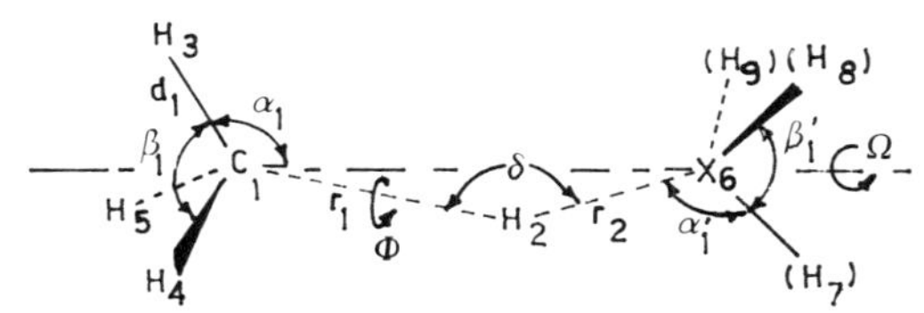

Figure 6.26 Internal coordinates of the transition structures

Table 6.33 Internal coordinate system for $H_3C\cdots H\cdots XH_{n-1}$

C—H_2	r_1
X—H_2	r_2
C—H_3, C—H_4, C—H_5	d_1, d_2, d_3
X—H_7, X—H_8, X—H_9	d'_1, d'_2, d'_3
H_2—C—H_3, H_2—C—H_4, H_2—C—H_5	$\alpha_1, \alpha_2, \alpha_3$
H_2—X—H_7, H_2—X—H_8, H_2—X—H_9	$\alpha'_1, \alpha'_2, \alpha'_3$
H_3—C—H_4, H_3—C—H_5	β_1, β_2
H_7—X—H_8, H_7—X—H_9	β'_1, β'_2
C—H_2—X	δ
H_3—C—H_2—X	Φ
H_3—C—X—H_7	Ω

Note: When C, H and X are collinear, ϕ degenerates into a linear angle perpendicular to δ.

The properties of a transition structure may be obtained by the methods used for the equilibrium structures. However, the initial estimation of its position is generally more tedious. Further details about this problem may be found elsewhere.[62] The fully optimized geometries of the transition structures are described in Table 6.34.

Table 6.34 Geometries of the transition structures (in angstroms or degrees)

	H	CH_3	NH_2	OH	F
d_1	1.077	1.079	1.078	1.078	1.077
d_2	1.077	1.079	1.079	1.077	1.077
d_3	1.077	1.079	1.079	1.077	1.077
d'_1	—	1.079	1.008	0.962	—
d'_2	—	1.079	1.008	—	—
d'_3	—	1.079	—	—	—
r_1	1.3627	1.3571	1.3407	1.3335	1.2816
r_2	0.9344	1.3571	1.2615	1.1920	1.1789
α_1	104.3	105.1	108.3	106.5	104.1
α_2	104.3	105.1	103.4	102.7	104.1
α_3	104.3	105.1	103.4	102.7	104.1
α'_1	—	105.1	104.8	104.6	—
α'_2	—	105.1	104.8	—	—
α'_3		105.1	—	—	—
β_1	114.1	113.5	113.6	114.3	114.3
β_2	114.1	113.5	113.6	114.3	114.3
β'_1	—	113.5	110.4	—	—
β'_2	—	113.5	—	—	—
δ	180.0	180.0	187.7	186.5	180.0
ϕ	180.0	180.0	180.0	180.0	180.0
Ω	—	180.0	58.1	180.0	—
Found symmetry	$C_{3}v$	$D_{3}d$	Cs	Cs	$C_{3}v$

The vibrational frequencies of the activated complexes have also been calculated. The results are given in Figure 6.27 with the symmetry species and the description of the vibrators. Some correlations among the most clearly assignable motions are represented by dotted lines. Imaginary frequencies corresponding to the movement along the reaction coordinate (a non-return oscillation) are given in the bottom part of the diagram. Let us recall that the norm of an imaginary frequency is a measure of the sharpness of the barrier.

It can be seen that geometry optimizations performed at the CI level for isolated species give, on average, only a slightly improvement of the structural parameters. Therefore one may assume that the SCF optimized geometries of the transition structures would be very similar to the CI ones which have been computed. Thus, the discussion on the geometrical changes at the transition state will be based on SCF results. Furthermore, we will only consider in this discussion the main geometrical parameters of the

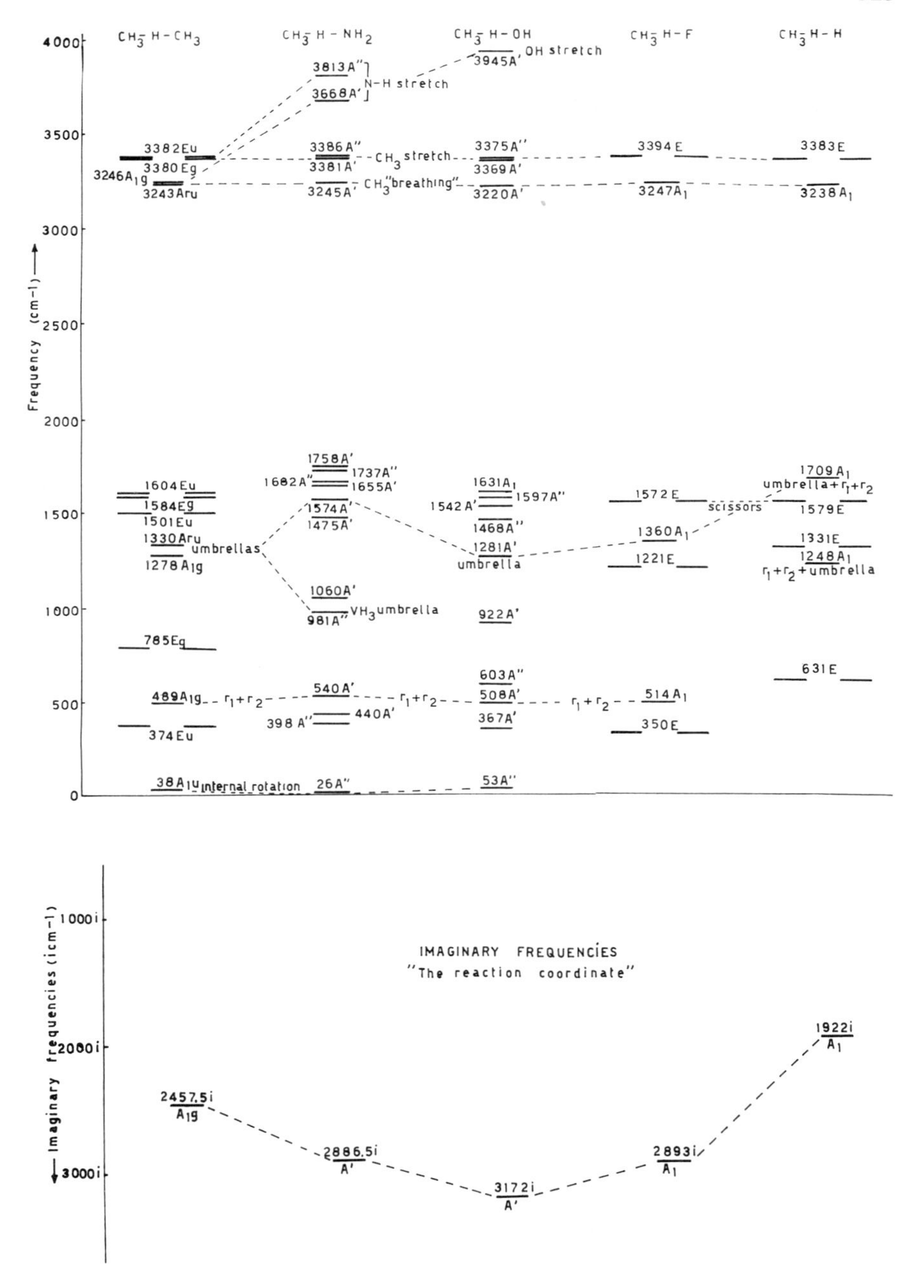

Figure 6.27 Vibrational frequencies (cm^{-1}) of the transition structure $CH_3 \ldots H \ldots R$

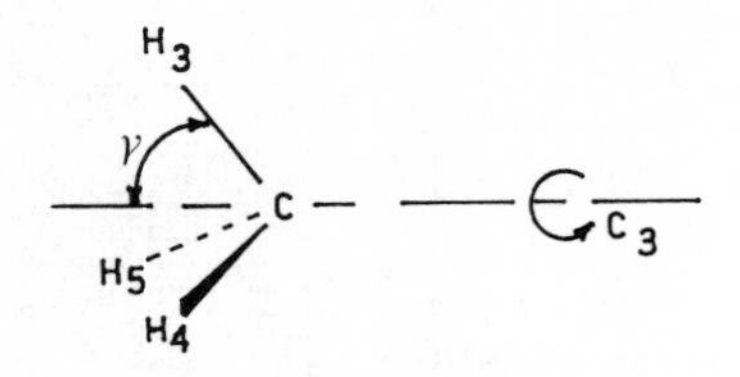

Figure 6.28 The 'umbrella' angle of the methyl group

transition states, i.e. r_1, r_2, C—X and the 'umbrella' angle of the methyl group, γ, defined in Figure 6.28. The values of these parameters are given in Table 6.35.

Table 6.35 Main geometrical parameters of the transition structures (in angstroms or degrees)

	H	CH_3	NH_2	OH	F
r_1	1.363	1.357	1.341	1.334	1.282
r_2	0.934	1.357	1.261	1.192	1.179
C—X	2.297	2.714	2.596	2.521	2.461
γ	75.74	74.92	75.05	75.94	75.97

It is also interesting to define the relative distances $r_{rel.1}$ and $r_{rel.2}$, i.e. r_1 and r_2 divided respectively by the corresponding bond distance in the stable molecule:

$$r_{rel.1} = \frac{r_1}{d(CH_4)}; \qquad r_{rel.2} = \frac{r_2}{d(RH)} \tag{6.104}$$

Their values are collected in Table 6.36.

Table 6.36 Relative distances from migrating hydrogen

	$r_{rel.1}$	$r_{rel.2}$	$r_{rel.1}/r_{rel.2}$
H	1.26	1.28	0.98
CH_3	1.25	1.25	1.00
NH_2	1.24	1.27	0.98
OH	1.23	1.26	0.98
F	1.18	1.28	0.92

As shown in Table 6.35, the elongation of the CH breaking bond at the transition state is always smaller than 0.3 Å. Thus, according to Benson's definition,[65] all the reactions considered here have tight transition states.

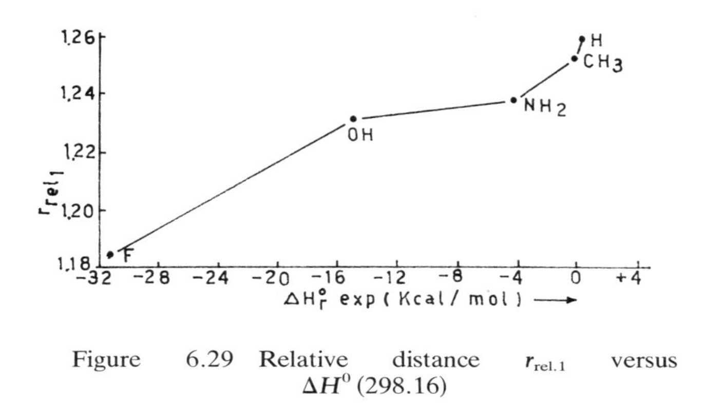

Figure 6.29 Relative distance $r_{rel.1}$ versus ΔH^0 (298.16)

Furthermore, if we except the first reaction (R=H), the relative distance $r_{rel.1}$ decreases with the electronegativity of atom X of the attacking radical. Let us summarize the conclusions which can be deduced from the results of geometry optimizations and vibrational analysis. The most important geometrical changes at the transition state concern the distances r_1 and r_2, which appear to be the only parameters for characterizing the different reactions. More precisely, as shown in Figure 6.29, the relative distance $r_{rel.1}$ is clearly related to the exothermicity of the reaction which is itself partly responsible for the corresponding activation energy according to the Evans–Polanyi relation.[66] In Figure 6.29 the $\Delta H^0(298.16)$ are obtained from $\Delta H_f^0(298.16)$.[64,65]

Geometrically speaking, Hammond's postulate[67] would be able to predict the importance of the deformation of the breaking bond at the transition state. It would also indicate the relative amount of formation of the new bond in the activated complex. Thus, as shown in the last column of Table 6.36, the ratio $r_{rel.1}/r_{rel.2}$ is equal to unity for the strictly thermoneutral hydrogen abstraction but, for the most exothermic one, it has the value 0.92, clearly corresponding to a more reactant-like transition state. In atom abstraction reactions, a reactant-like transition state would be one in which the relative distance $r_{rel.1}$ is small but also one in which the new bond is formed to a lesser extent than the other one is broken.

Thermodynamical properties

The thermodynamical properties of reactants, transition structures and products can be calculated using the formalism described in Section 4.2.6 with the vibrational frequencies computed at the SCF level (see Figure 6.27). In this way, one obtains the thermal and ZPE corrections, the molar heat capacities and the entropies collected in Table 6.37 and compared with the corresponding available experimental data. It can be seen that the theoretical and experimental values correlate quite well. A first-order re-

Table 6.37 Thermodynamical properties at 298.16 K (cal mol^{-1}, Gibbs)†

Species	$H^0(T)-H^0(0)$	ZPE	C_p^0 Theory	C_p^0 Experiment	S^0 Theory	S^0 Experiment
H	1,481.25	0	4.97	4.97	27.39	27.39
CH_3—H—H	2,577.39	28,653.33	10.48	—	51.58	—
H_2	2,073.75	6,641.32	6.96	6.89	31.05	31.21
CH_3	2,430.71	19,273.91	8.81	9.25	46.06	46.38
CH_3—H—CH_3	3,364.77	49,761.55	15.27	—	64.53	—
CH_4	2,380.53	30,560.35	8.22	8.52	44.36	44.48
NH_2	2,371.63	12,696.04	7.99	8.02	46.36	46.50
CH_3—H—NH_2	3,213.88	44,015.92	13.98	—	63.76	—
NH_3	2,473.12	22,839.60	9.04	8.52	46.18	46.03
OH	2,073.75	5,653.85	6.96	7.17	42.53	43.88
CH_3—H—OH	3,166.85	34,069.90	13.61	—	64.23	—
H_2O	2,371.46	14,020.74	7.99	8.03	44.93	45.11
F	1,481.25	0	4.97	5.44	36.15	37.92
CH_3—H—F	2,985.56	26,018.00	12.93	—	59.57	—
HF	2,073.74	5,933.03	6.96	6.96	41.46	41.51

† All experimental values in this table are taken from Ref. 64.

gression fitting gives for C_p^0 and S^0:

$$C_{(p)}^0 = 0.9442 C_p^0(\text{th}) + 0.500 \tag{6.105}$$

$$S^0(\text{exp}) = 0.9904 S^0(\text{th}) + 0.783 \tag{6.106}$$

with correlation coefficients of 0.9896 and 0.9956 respectively. This agreement between theory and experiment is a confirmation of the good degree of accuracy not only of the computed vibrational frequencies but also of the optimized geometries which are used to calculate the rotational entropies.

Thermochemical and kinetic parameters

The total energies computed at the CI level using CI optimized geometries for the isolated species and SCF optimized geometries for the transition structures are given in Table 6.38. These results and the data of Table 6.37 allow us to calculate the standard heats of reaction and the corresponding entropy changes as well as the activation parameters $\Delta U^{\neq}(298.16)$ and $\Delta S^{\neq}(298.16)$ for each hydrogen abstraction reaction. These quantities are collected in Table 6.39. In order to obtain theoretical activation energies comparable to experimental ones it is necessary to take account of the tunnel effect which is known to play a role in proton transfer reactions. According to the theory outlined above, it appears to be important for temperatures lower than $T_c/2$. The characteristic temperatures calculated using the formula (6.90) proposed by Christov[50] are given in Table 6.40. It can be seen that all T_c are far above the standard temperature. This means that we should expect a large fraction of H atoms to pass through the barrier by the tunnel effect at 300 K. Using Wigner's method[47] it is easy to obtain

Table 6.38 Total energies at the CI level (in atomic units)

Species	CI energy	Experimental energy
H	(−0.49823)	−0.5
CH_3—H—H	−40.76502	—
H_2	−1.15152	−1.1744
CH_3	−39.63947	−39.846
CH_3—H—CH_3	−70.90067	—
CH_4	−40.29260	−40.526
NH_2	−55.63231	−55.911
CH_3—H—NH_2	−95.89932	—
NH_3	−56.28750	−56.588
OH	−75.46154	−75.783
CH_3—H—OH	−115.73600	—
H_2O	−76.11736	−76.483
F	−99.44823	−99.809
CH_3—H—F	−139.72906	—
HF	−100.11222	−100.533

Table 6.39 Thermochemical and kinetic parameters at 298.16 K (kcal mol^{-1}, Gibbs)

R	H	CH_3	NH_2	OH	F
ΔH^0(th)	−4.10	0	−2.28	−4.25	−11.52
ΔH^0(exp)	0.62	0	−4.26	−14.52	−31.28
ΔS^0(th)	5.36	0	1.52	4.10	7.01
ΔS^0(exp)	5.72	0	1.43	3.13	5.49
$\Delta U^{\neq}$(th)	14.18	19.35	16.45	9.13	3.13
$\Delta U^{\neq}$(exp)	11.9[68]	14.0[69]	10.3[70]	3.77[71]	1.15[72]
$\Delta S_p^{\neq}$(th)	−20.17	−25.89	−26.96	−22.67	−20.94
$\Delta S_p^{\neq}$(exp)	—	—	—	—	

Table 6.40 Characteristic temperatures T_c (K)

R	H	CH_3	NH_2	OH	F
T_c	880	1125	1322	1453	958

the tunnelling corrections to activation energies given in Table 6.41 with the corresponding $\Delta U^{\neq}$.

As shown in Figure 6.30, theoretical activation energies correlate well with experimental ones. However, they are overestimated by about 30 per cent. over the whole set of reactions. On the other hand, the theoretical enthalpies of reaction are somewhat less accurate than the activation energies. Nevertheless, the order of experimental heats of reaction is reproduced

Table 6.41 Tunnelling corrections to $\Delta U^{\neq}$ and activation energies (kcal mol^{-1}, 298.16 K)

R	Correction	$\Delta U^{\neq}$(th)	$\Delta U^{\neq}$(exp)
H	−1.01	13.17	11.9
CH_3	−1.05	18.30	14.0
NH_2	−1.07	15.38	10.3
OH	−0.95	8.18	3.77
F	−0.93	2.20	1.15

by the theoretical results (see Table 6.39 and also Figure 6.31) even though the *ab initio* value differs from the experimental one by about 5 kcal mol^{-1} for the nearly thermoneutral reaction and is systematically underestimated for the more exothermic processes.

Rate constants from transition state theory

The rate constants of hydrogen abstraction reactions may be computed at different temperatures using equation (6.74). The corresponding Arrhenius-type plots are represented in Figure 6.32. In each case, one obtains a curved line showing a non-Arrhenius behaviour which is more pronounced in the reactions with F and OH and less pronounced in the reaction with CH_3.

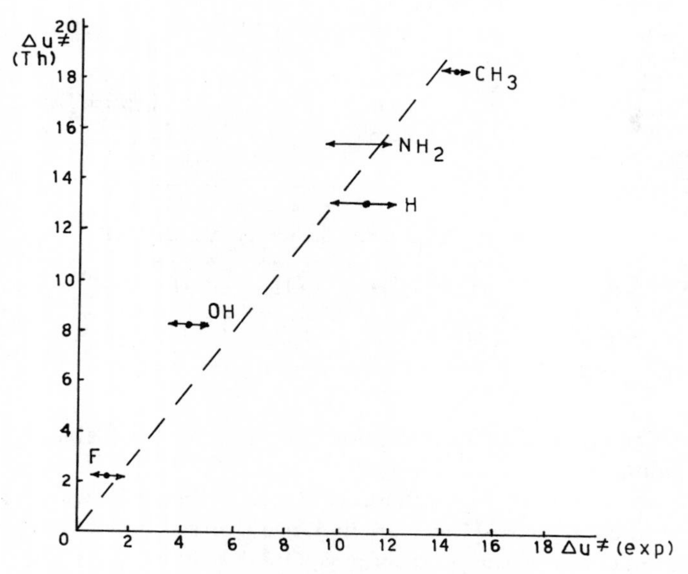

Figure 6.30 Activation energies at 298.16 K. Theory versus experiment

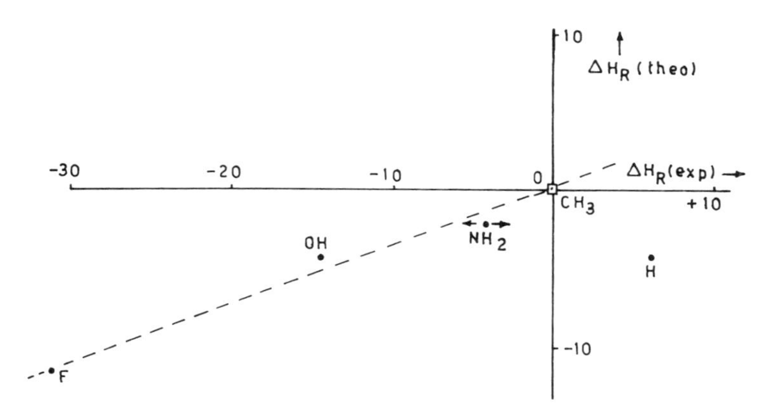

Figure 6.31 Enthalpies of reaction at 298.16 K. Theory versus experiment

Shaw[73] and Zellner[74] have reviewed the data on reactions with H and OH and, in both cases, conclude that experimental results show distinct non-Arrhenius behaviour. Pacey and Purnell[75] have suggested that all Arrhenius plots for hydrogen transfer reactions of alkyl radicals probably exhibit strong temperature dependence of their Arrhenius parameters. However, Kerr and Parsonage[69] have reviewed the reactions with the CH_3 radical and conclude that: 'the general conclusion of curvature in Arrhenius plots on hydrogen transfer reactions should be treated with caution'. The theoretical results seem to confirm the hypothesis of Pacey and Purnell while at the same time they explain the careful attitude of Kerr and Parsonage. A

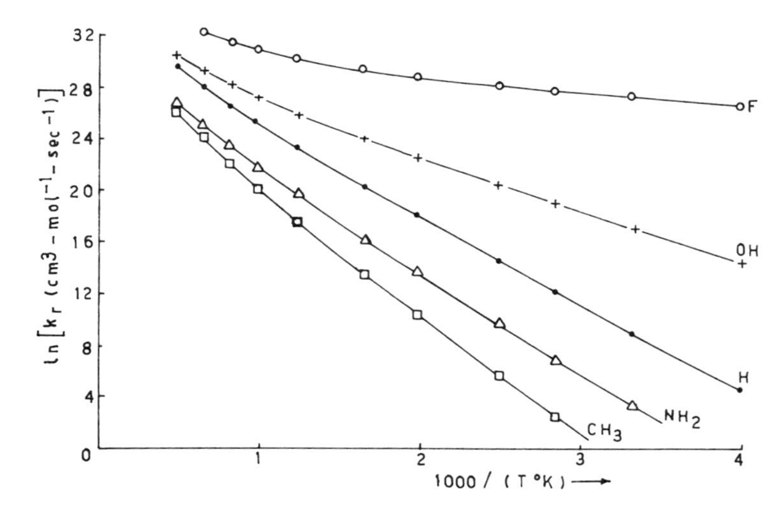

Figure 6.32 Arrhenius type plots

non-Arrhenius behaviour is theoretically obtained in every hydrogen transfer reaction considered but this effect is rather small, particularly in the case of the methyl radical.

For reactions with H and OH, the large amount of kinetical data at various temperatures allows a detailed comparison to be made between theoretical and experimental results. Figures 6.33 and 6.34 show the theoretical Arrhenius plots for these reactions with and without tunnel corrections. Experimental results from various authors are indicated by dots. The critical temperatures of Christov (T_c) and Goldanski (T_g) are also given in these graphs.

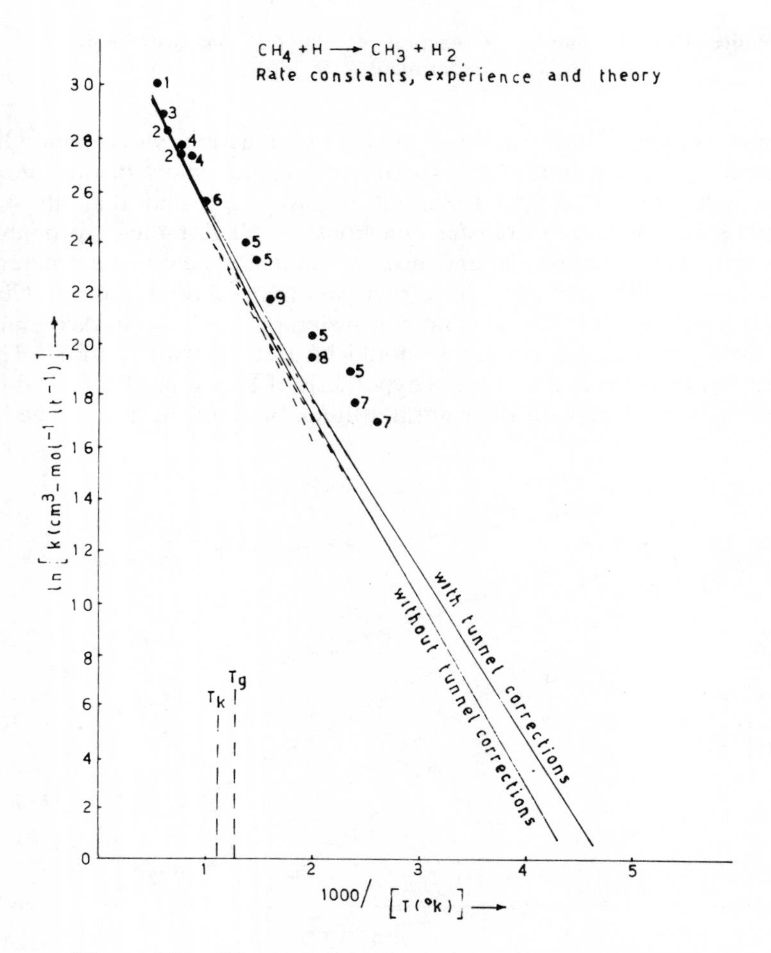

Figure 6.33 Theoretical and experimental rate constants for the reaction $CH_4 + H \rightarrow CH_3 + H_2$

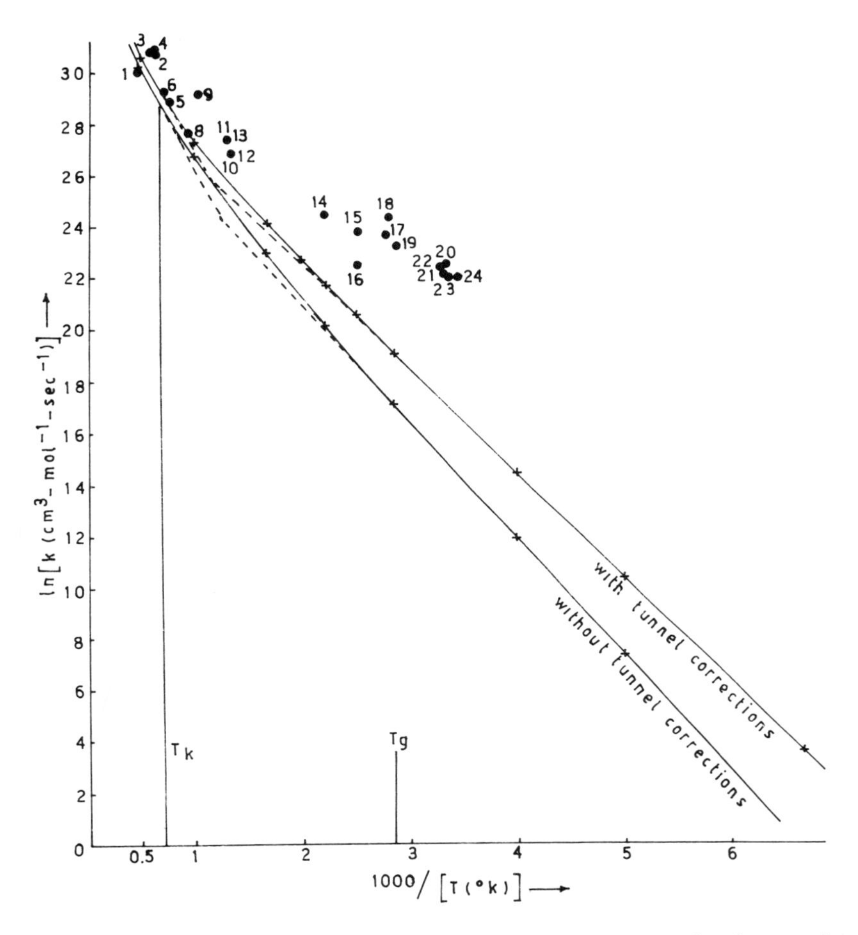

Figure 6.34 Theoretical and experimental rate constants for the reaction $CH_4 + OH \rightarrow CH_3 + H_2O$

It can be seen that the theoretical curves are very close to the experimental points at temperatures near or above Christov's characteristic temperature, but at lower temperatures *ab initio* results are substantially lower than experimental ones. This is partly due to the importance of tunnelling at low temperatures, which is not taken into account very well in Wigner's method. This is not an unexpected result as it is well known that, at temperatures lower than $T_c/2$, a complete quantum treatment is needed to guarantee the accuracy of the results. Nevertheless, at high temperatures the transition state theory seems to give a precision of the order of experimental errors.

It is also interesting to compare the theoretical and experimental values of the preexponential factor A for the different reactions studied. From Table 6.42 it can be seen that the two series of values are well correlated, although

Table 6.42 Arrhenius-like preexponential factors (in litres per mole-second)

	H	CH_3	NH_2	OH	F
T K	667	500	500	300	298
ΔS_p(Gibbs)	−21.16	−26.19	−26.11	−22.67	−20.94
log A_c	11.12	9.78	9.79	10.10	10.47
log $A_c(t)$	11.02	9.81	9.91	10.43	10.75
log A_c (exp)	9.85	8.60	8.70	9.15	11.82
Ref.	(73)	(69)	(70)	(76)	(72)

experiment generally gives lower values than theory. This is presumably due to the fact that the temperatures considered here are much lower than T_c.

Finally, it is to be noticed that *ab initio* results suggest that the main part of the curvature of experimental Arrhenius plots for hydrogen abstraction reactions comes from tunnel effects. Only a more rigorous quantum mechanical treatment could give some support to this hypothesis.

Electronic properties

Some electronic properties of the reactants, transition structures and products of the five reactions studied have been calculated using the basis of natural orbitals provided by the CI treatment. Let us summarize briefly the main results obtained.

Maps of spin density at the transition state are given in Figure 6.35. They show that the five transition complexes are rather similar from the electronic point of view. For example, the spin density is always negative in the region surrounding the migrating hydrogen. More details about the electronic reorganization at the transition state may be obtained by plotting the α and β densities along the reaction axis. As shown in Figure 6.36, both densities increase on the side of the attacking radical with an increasing nuclear charge of the atom (H, C, N, O, F), but simultaneously the β density decreases around the migrating hydrogen. It is also seen that the new bond loge is partly filled with electrons at the transition state at the expense of the bond-breaking loge. Moreover, the electronic changes taking place at the transition state are described fairly well by the derivatives of the different electronic densities represented in Figure 6.37 for the reaction involving hydrogen as the attacking radical. It can clearly be seen that the β density increases around the future H_β atom on the attacking hydrogen side. The α density behaves approximately in the opposite way but the α pattern is more complicated, reflecting the fact that there is an electron 'jumping' from the CH loge to the HX one.

The information provided by the density maps can be summarized as follows. At the transition state of a hydrogen abstraction reaction, the migrating atom confronted with the presence of a radical with an α density

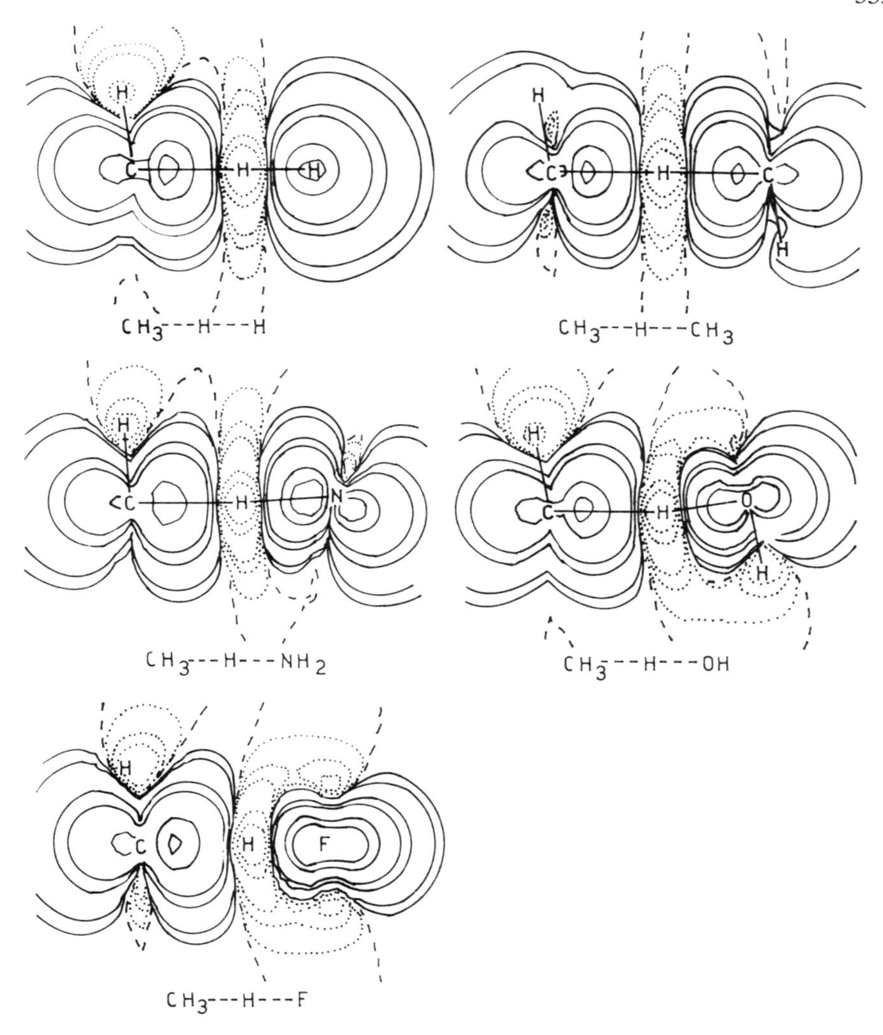

Figure 6.35 Spin density maps of the transition structures

takes a β electron of the CH bond and jumps to the radical. The activated complexes are characterized by the fact that the migrating hydrogen has already begun to share this electron with the radical. During this process there occurs a separation of charges depending on the nature of the radical, as described by the results of the Mulliken population analysis collected in Table 6.43. In this table we retain only the net charges and total spin populations of the three fragments CH_3, H and R.

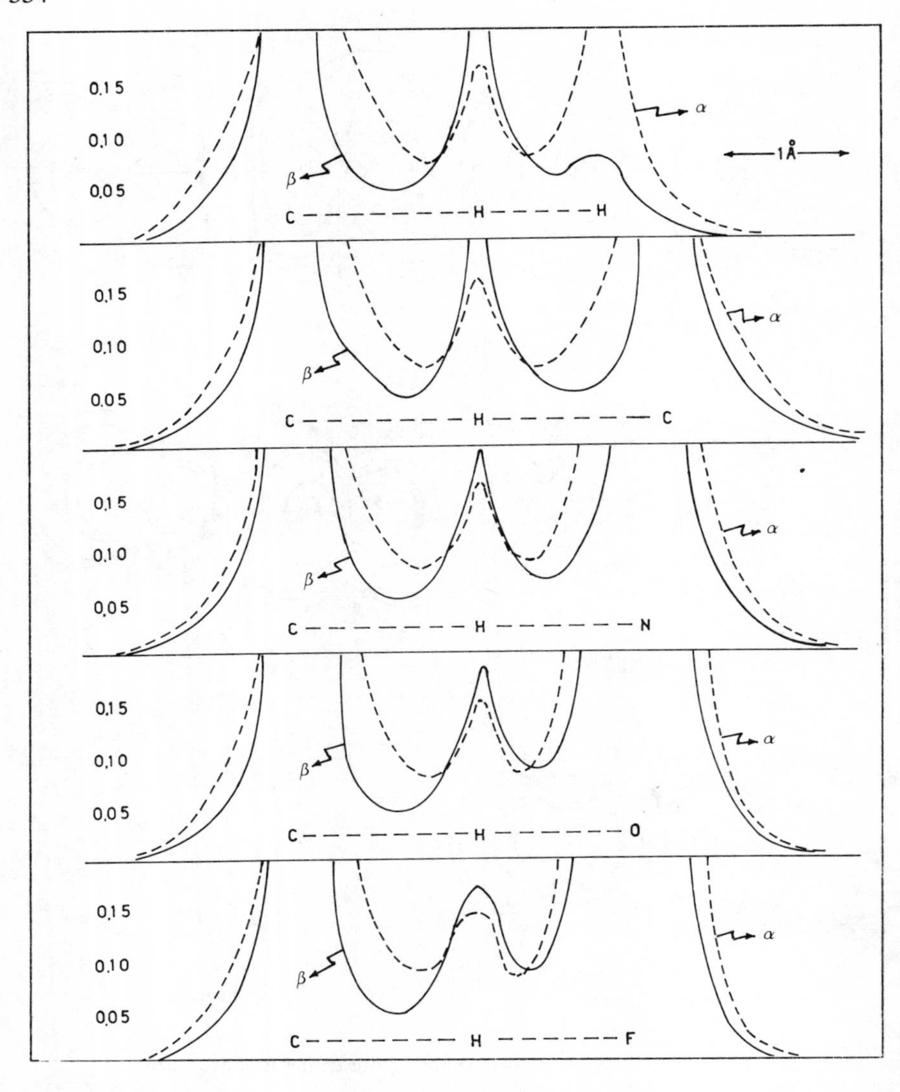

Figure 6.36 Alpha and beta electron densities along the reaction axis C—H—X

Another way of looking at the charge distribution is to analyse the dipole moments listed in the last column of Table 6.43. The negative sign implies that $\boldsymbol{\mu}$ is oriented from X to C. In the isoelectronic radical series the dipole moment, i.e. the charge separation, increases regularly with increasing electronegativity of X. On the other hand, it is worth noticing that for the five reactions studied the dipole moments at the transition state correlate well with the corresponding activation energies.

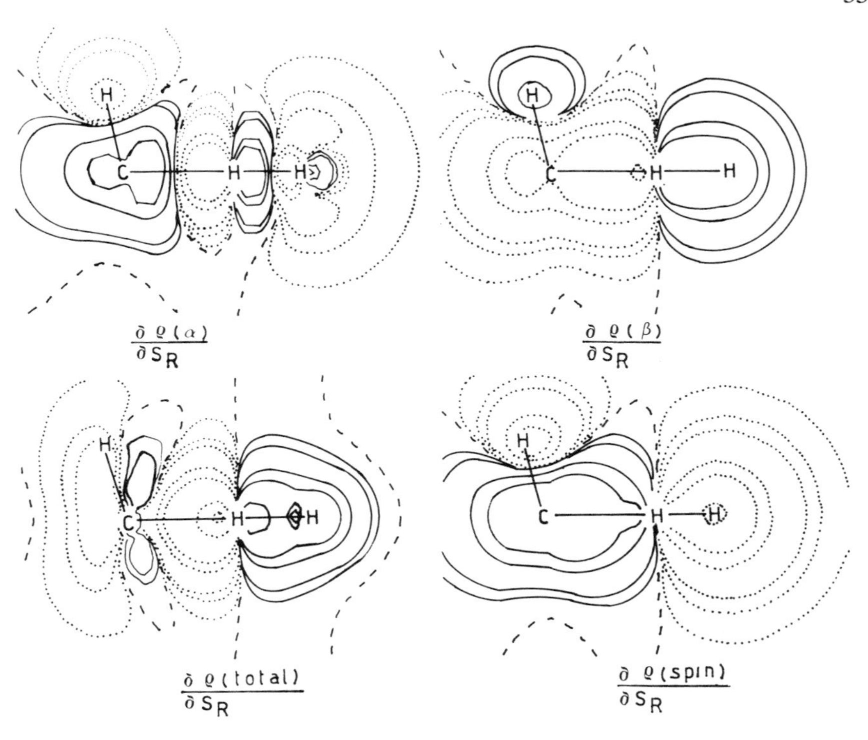

Figure 6.37 Density derivatives at the transition state for $CH_4 + H \rightarrow CH_3 + H_2$

Table 6.43 Net charges and total spin populations of CH_3, H and R; dipole moment (Debye units)

$Q^{\neq}$	0.0	0.0	0.0	$\mu^{\neq}$
$P_T^{s\neq}$	$[CH_3]$---------	H-----------	$[H]$	
	0.7α	0.2β	0.5α	−0.729
$Q^{\neq}$	0.0	0.0	0.0	
$P_T^{s\neq}$	$[CH_3]$---------	H----------	$[CH_3]$	
	0.6α	0.2β	0.6α	0.000
$Q^{\neq}$	0.0	+0.1	−0.1	
$P_T^{s\neq}$	$[CH_3]$---------	H ---------	$[NH_2]$	
	0.5α	0.1β	0.6α	−0.055
$Q^{\neq}$	0.0	+0.3	−0.3	
$P_T^{s\neq}$	$[CH_3]$---------	H----------	$[OH]$	
	0.6α	0.1β	0.5α	−1.508
$Q^{\neq}$	0.0	+0.3	−0.3	
	$[CH_3]$---------	H-----------	$[F]$	
$P_T^{s\neq}$	0.6α	0.1β	0.5α	−2.596

Localized orbitals still provide another description of the electron distribution at the transition state. The centroids of charge and the second moments of UHF molecular orbitals localized according to the Boys criterion are respectively shown in Figures 6.38 and 6.39. In Figure 6.38 the centroids of the α orbitals are represented by small circles and the centroids of the β orbitals by small crosses. In Figure 6.39, each second moment is represented by an ellipse proportional to the 'size' of the corresponding localized orbital. The second moment value is given inside the ellipse. Beta

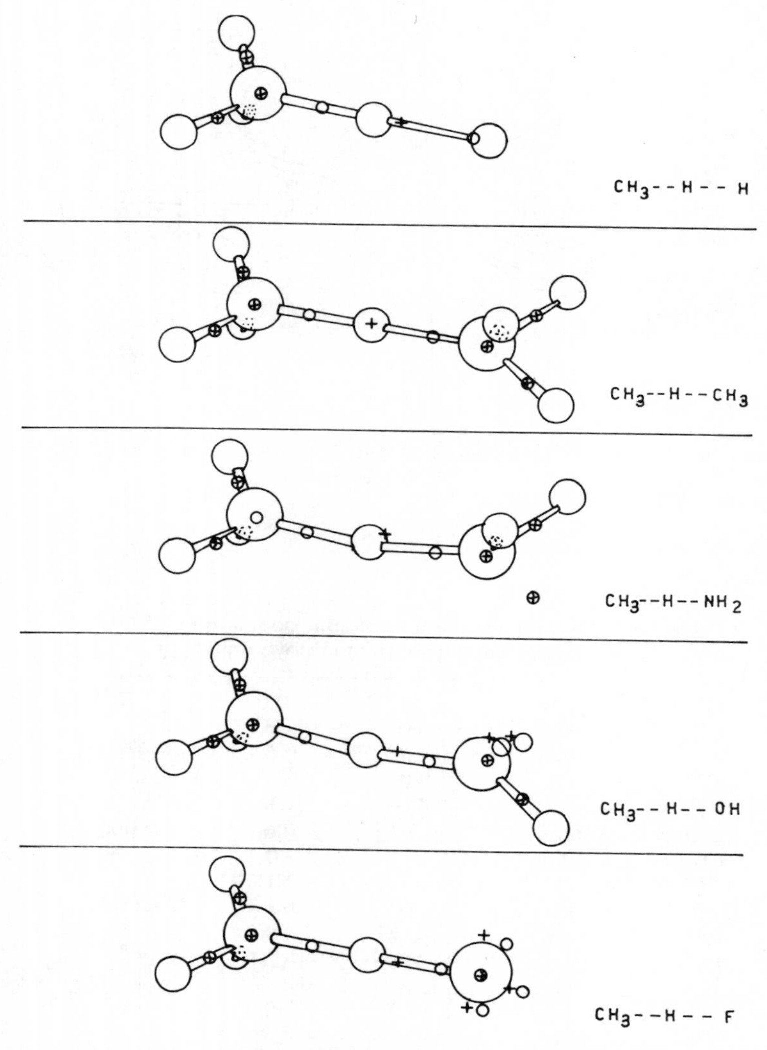

Figure 6.38 Centroids of charge of UHF localized orbitals at the transition states

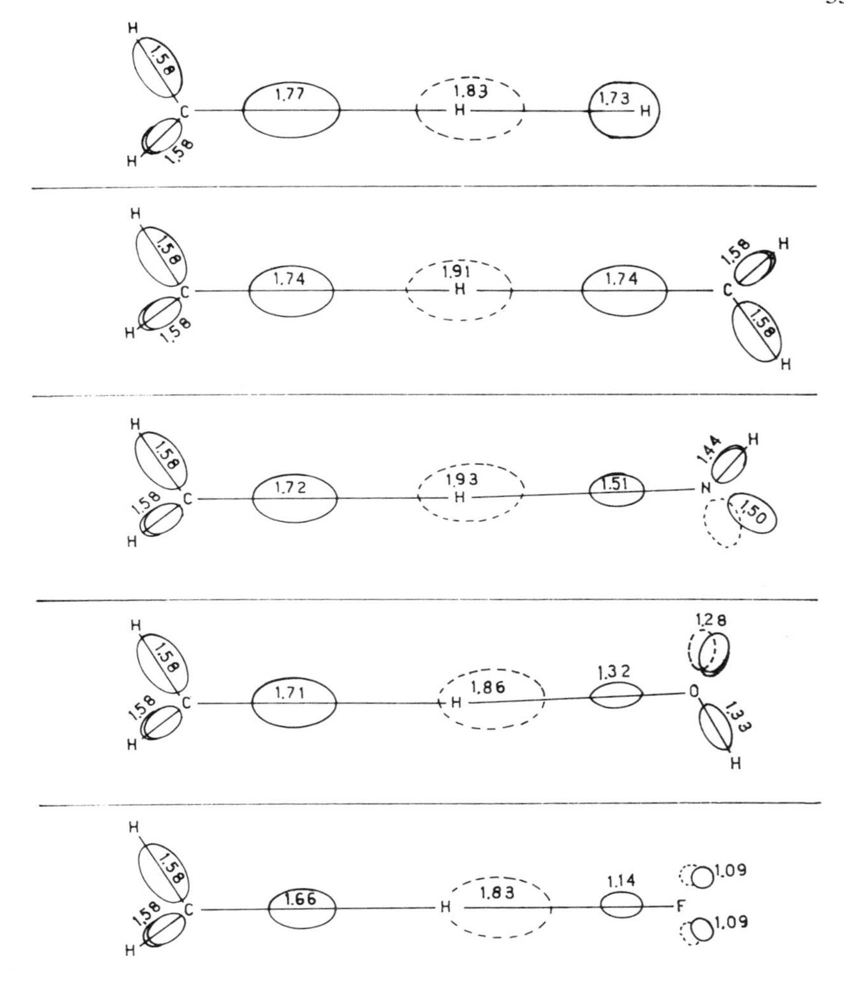

Figure 6.39 Second moments and sizes of UHF localized orbitals at the transition states

orbitals (dotted-line ellipses) are included only when they are significantly different from their alpha partners.

These results confirm the main previous observations concerning the electronic structure of the transition complexes, namely:

(a) The five complexes are very similar from the electronic point of view.
(b) The hydrogen atom associated to a beta electron moves between two 'alpha regions'.
(c) The centroid of this beta electron which is exactly located on H in the symmetrical CH_3—H—CH_3 complex is displaced even towards the attacking radical as the electronegativity of X increases.

Furthermore, it is interesting to note that the size of the orbitals involved in the transfer process is appreciably increased with respect to the size of the corresponding orbitals in stable molecules.

It is debatable whether the theoretical results justify the definition of the nucleophilicity or electrophilicity of a radical commonly used by the experimentalists[77] and which is based on the following resonance formulas:

$$\underset{\text{I}}{R\cdot\dot{H}\cdot X} \qquad \underset{\text{II}}{\bar{R}\!:\ \dot{H}\ \overset{+}{\dot{X}}} \qquad \underset{\text{III}}{\overset{+}{R}\ \dot{H}\ :\ \bar{X}}$$

According to this definition an electrophilic radical (X) would favour structure III while a nucleophilic one would favour structure II.

The various theoretical results concerning the electronic structure of the complexes CH_3—H—R lead to a representation given by the following formulas:

$$CH_3\cdot\dot{H}\cdot R\ (R=H, CH_3); \qquad CH_3\cdot\overset{\delta+}{\dot{H}}:\overset{\delta-}{\dot{R}}\ (R=NH_2, OH, F)$$

Thus, attacking radicals commonly considered as electrophilic, such as OH and F, would effectively favour structure III. Correspondingly, according to the net charges, NH_2 would be less electrophilic than OH and F. On the other hand, it is not possible to discriminate the polarities of H and CH_3 as the net charges are zero in both corresponding complexes. However, if one assumes that the dipole moment of the activated complex is a good measure of the relative polarity of the attacking radical, one would anticipate the following order of growing electrophilicity:

$$CH_3 < NH_2 < H < OH < F$$

which also corresponds to the order of growing electron affinity, as shown in the conclusions.

Conclusions

The results obtained in the theoretical approach of hydrogen abstraction reactions lead to some interesting conclusions concerning both the methodology used and the reactions themselves.

(*a*) *The methodology* It has been shown that UHF procedure is accurate enough to determine optimized geometries and to perform vibrational analysis. However, configuration interaction is needed to calculate the energetic properties of isolated species and activated complexes. Thus one obtains acceptable activation energies but heats of reaction remain very approximate. We can anticipate that the use of larger basis sets, including polarization functions in the initial SCF step and correspondingly the inclusion of a larger number of configurations in the CI subsequent step, could considerably improve the accuracy of the theoretical heats of reaction and perhaps also the description of the transition states.

To handle the problem of selecting configurations in a truncated CI, the diagonalization perturbation iterative technique (CIPSI) within an iterative natural orbital (INO) scheme and with an extrapolation for estimating the energies corresponding to a full CI appears to be a very effective procedure.

We want to now make an important remark. Even though it is highly desirable to be able to reproduce experimental data such as heats of reaction, activation energies and rate constants, this is not the main goal of the theoretical approach of chemical reactivity. In fact, we want rather to rationalize the experimental data by providing original information on properties which are not measurable, such as those characterizing a transition state. Therefore, in this field, theory acts essentially as a complement to experiment.

(*b*) *The hydrogen transfer reactions* In spite of its imperfections this work provides some useful information on hydrogen transfer reactions and, more specially, on their transition states.

Although the exothermicities of the five reactions are quite different the transition structures are very similar from all points of view. Only the properties of the bonds directly involved in the transfer vary appreciably from one reaction to another. Thus, the concept of reactant likeness related to Hammond's postulate may be clarified. Furthermore, all the reactions studied have tight transition states from a geometrical point of view, even though their theoretical A factor is relatively large. It is to be pointed out that the preexponential factor A increases with the exothermicity of the reaction. Therefore the more reactant-like transition state is also the less tight.

Analysis of the electronic properties of activated complexes enables one to propose the following mechanism for hydrogen abstraction reactions from methane:

$$H_3C{\uparrow}{\downarrow}H + {\uparrow}R \longrightarrow \left[H_3C{\uparrow} \quad \overset{\delta+}{H}{\downarrow} \; {\uparrow}\overset{\delta-}{R} \right]^{\neq} \longrightarrow H_3C{\uparrow} + H{\downarrow}{\uparrow}R$$

The driving force of this atom transfer would be the electron affinity of the attacking radical responsible for a low potential zone of electrons in the formation of the bond loge. The electron affinity of a radical (and the corresponding electronegativity) would be a good measure of its electrophilicity related to the charge separation at the transition state (indicated by the corresponding dipole moment). Thus the activation energy of a hydrogen abstraction reaction would decrease with an increase in the attacking radical electron affinity. Correspondingly, the dipole moment of the transition structure would increase as well as the exothermicity of the reaction. The results collected in Table 6.44 seem to confirm this rationalization.

Table 6.44 Relation between the reaction parameters and the properties of the attacking radical

R	CH_3	NH_2	H	OH	F
EA (kcal mol^{-1})	1.8	17.1	17.4	42.1	78.4
IP (kcal mol^{-1})	227.4	258.3	313.6	298.4	401.7
x†	1.91	2.29	2.75	2.83	4.00
$\mu^{\neq}$(D)	0.000	−0.055	−0.729	−1.508	−2.596
ΔH^0_{th}(kcal mol^{-1})	0.00	−2.28	−4.10	−4.25	−11.52
ΔH^0_{exp}(kcal mol^{-1})	0.00	−4.26	0.62	−14.52	−31.28
$\Delta U^{\neq}_{th}$(kcal mol^{-1})	19.35	16.45	14.28	9.13	3.13
$\Delta U^{\neq}_{exp}$(kcal mol^{-1})	14.0	10.3	11.9	3.77	1.15

† Calculated according to Mulliken's electronegativity definition in order to have $x_F = 4.00$.

A critical analysis of the determination of electron affinities (EA) and ionization potentials (IP) of radicals and the corresponding references may be found elsewhere.[32]

The parallelism between electron affinities and either activation energies and transition state dipole moments or reaction enthalpies is clear both from the theoretical and experimental points of view. This not only explains the existence of Evans–Polanyi relations in narrow series of reactions but also suggests other correlations such as between $\Delta U^{\neq}$ and between ΔH^0 and $\mu^{\neq}$.

In order to analyse these relations in more detail, least square calculations have been performed using the following equation:

$$\Delta U^{\neq}(\mathrm{T}) = a_0 + a_1 P_1 \qquad (6.107)$$

where P_1 stands for EA, IP, x, $\mu^{\neq}$ and ΔH^0 (theoretical and/or experimental). The test of the Evans–Polanyi relation is illustrated in Figure 6.40 for theoretical results at different temperatures as well as for experimental data at 298 K.

The results of the regressions and the correlation coefficients r are listed in Table 6.45. We recall that r^2 measures the fraction of the total variation in $\Delta U^{\neq}$ which is explained by the chosen parameter (here ΔH^0). It can be seen that the lowest value of r is obtained for the fixed nuclei case, suggesting that a certain part of the Evans–Polanyi relation comes from the nuclear motions.

The correlation between activation energies and some other properties is shown in Table 6.46. It is interesting to note that the correlation coefficients for regressions made on experimental data are in each case slightly lower than the corresponding coefficients for theoretical values. This may be partly due to experimental incertitudes on ΔH^0 and E_a. Furthermore, it can be seen that electron affinity always gives the best correlation. This property of the attacking radical which defines its polarity seems to be the determining factor responsible for both the activation energy and the reaction enthalpy of hydrogen abstraction reactions.

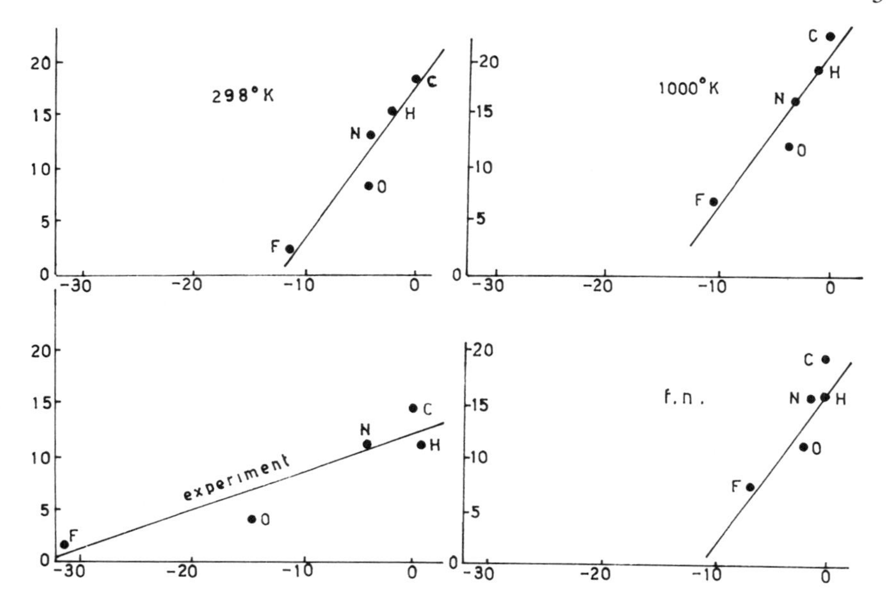

Figure 6.40 Evans–Polanyi plots: $\Delta U^{\neq}$ versus ΔH^0

Table 6.45 Evans–Polanyi regression coefficients

	a_0	a_1	r
fixed nuclei	16.986	1.498	0.883
298 K	17.638	1.398	0.950
1000 K	20.956	1.449	0.950
Exp. (298 K)	12.049	0.373	0.929

Table 6.46 Correlation between activation energies and one independent parameter P (298 K)

E_a	P	a_0	a_1	r
Theory	ΔH^0 (th)	17.638	1.398	0.950
	IP	38.344	−0.090	0.935
	AE	18.003	−0.209	0.987
	x	32.938	−7.798	0.967
Experiment	ΔH^0 (exp)	12.049	0.383	0.949
	IP	28.957	−0.069	0.845
	AE	13.734	−0.176	0.958
	x	24.860	−6.036	0.865

It is important to remember that the statistical analysis performed in this work is based on a very limited number of data and therefore in no way may be considered as conclusive. However, the good correlation coefficients obtained for the electron affinities give some support to the *a priori* rationalization of hydrogen transfer reactions presented above.

Finally, this work demonstrates the potential capability of the transition state theory to provide accurate rate constants of chemical reactions—at least at temperatures at which possible tunnelling plays a negligible role.

6.4.3.2 Reactions involving a large supermolecule

To end this section, we shall describe briefly the theoretical approach of the mechanism to some typical organic reactions involving a large supermolecule. In each of these reactions, called molecular, the number of electron pairs is preserved such that the correlation energy of the supersystem remains approximately constant during the process. Thus, the calculations may be performed at the SCF level without having to apply a configuration interaction systematically. Furthermore, only that portion of the potential energy hypersurface which is needed for determining the reaction path of the corresponding transformation is calculated.

The originality of this work lies in the detailed analysis of the energetic, geometric and electronic properties of the supermolecule along the reaction pathway. Particularly, the evolution of the LMO's centroid of charge distribution allows one to demonstrate the reorganization of the electron pairs during the reaction and therefore to determine its mechanism. The detailed results of this work may be found elsewhere;[78] in this section we shall restrict ourselves to the discussion of the reaction mechanisms.

The 1,3-dipolar cycloadditions

The concept of the 1,3-dipole was first introduced by Huisgen in the beginning of the sixties.[79] According to Huisgen, a 1,3-dipole is a compound which is isoelectronic with the propargyl anion:

$$[HC{\equiv}C{-}\bar{C}H_2]^-$$

or the allyl anion:

$$[H_2C{=}CH{-}\bar{C}H_2]^-$$

having respectively 22 and 24 electrons, four of which are π electrons delocalized on three adjacent centres. Every 1,3-dipole can react with an unsaturated compound, called a dipolarophile, to give a five-membered heterocycle. These cycloadditions which bring in 3+2 centres are described by the symbol $3+2 \rightarrow 5$ or, more explicitly, by the general equation:

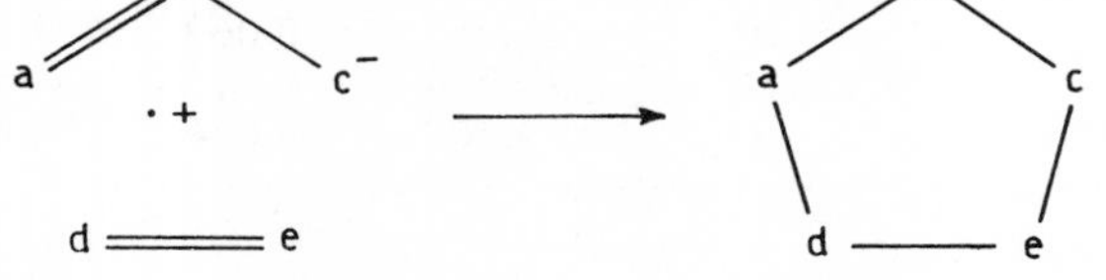

where a, b, c are most often a CH_2 or NH group or an oxygen atom. The propargyl anion-type dipoles which have an 'orthogonal π system' are linear, the allyl anion-type ones being bent dipoles.

Let us summarize the most important experimental results concerning the 1,3-dipolar cycloadditions and the conclusions they suggest. The stereospecificity of the reactions of 1,3-dipoles with alkenes led Huisgen to propose a concerted process for these additions. The two new bonds would be formed simultaneously, in a synchronous or asynchronous manner. This means that the population of one of the bonds will increase more than the other with the decrease of the reaction coordinate. In the same perspective, no intermediate would appear on the reaction path. Then 1,3-dipolar cycloadditions would be geometrically and energetically concerted reactions. It follows that the two-step mechanism calling for a diradical intermediate seems to be most unlikely. Moreover, the small effect of solvent polarity on the reaction rate allows one to question the existence of a zwitterionic intermediate.

From a kinetic point of view, 1,3-dipolar cycloadditions are second-order reactions characterized by relatively small activation energies, 8 to 18 kcal mol^{-1}, and large negative activation entropies, -20 to -40 Gibbs. This last result is also in agreement with a concerted mechanism which involves a highly organized activated complex. In addition, it is interesting to note that activation energies of the reactions of ozone with olefins are near 0 kcal mol^{-1}. According to Huisgen the reactant molecules would approach each other on two parallel planes and, in the case of linear dipoles, the π orbitals of the reactants would interact before the bending of the dipole. In the Woodward–Hoffman terminology, the dipole and the dipolarophile would approach each other in a suprafacial–suprafacial manner and the reaction would be classified as a $[\pi_s^4 + \pi_s^2]$ cycloaddition, with symmetry allowed in the ground state.

Finally, the 1,3-dipolar cycloadditions between asymmetrical molecules can be either regioselective, regiospecific or non-regiospecific, as shown in Figure 6.41.

These experimental results would suggest the possibility to explore that

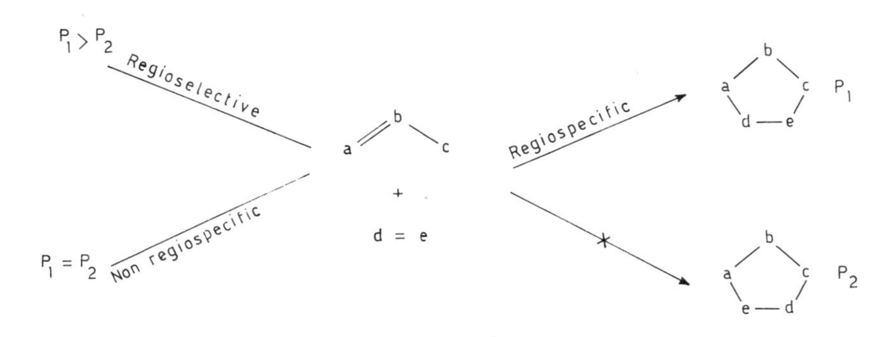

Figure 6.41 Regiochemistry of 1,3-dipolar cycloaddition

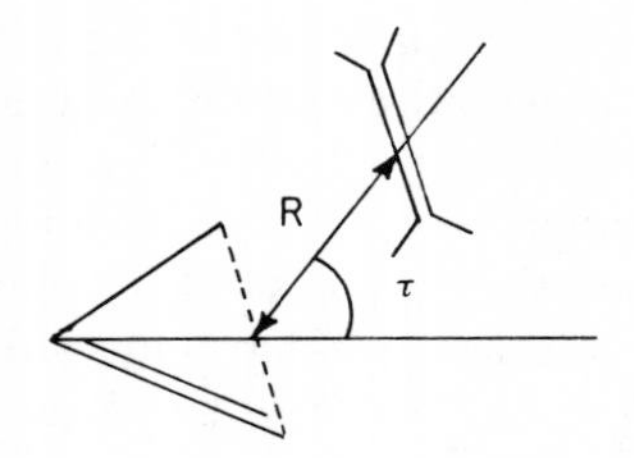

Figure 6.42 Definition of τ (envelope angle)

portion of the potential energy hypersurface which corresponds to the concerted approach only. We summarize below the leading results obtained in the theoretical study of 41 reactions, at the SCF level using the STO-3G basis set with the Gauss-70 program. This basis set has proved to be reliable enough to study reactions between neutral molecules in their ground state, assuming a concerted approach.

(*a*) *Geometrical results* *A priori*, in 1,3-dipolar cycloadditions, the reactants may approach each other either on two planes, parallel or not (P approach), or in the same plane (C approach). The envelope angle τ defined in Figure 6.42 permits these different situations to be distinguished quantitatively. For example, $\tau = 0$ for the coplanar approach and $\tau = 90°$ for the parallel one.

It has been found that the coplanar approach is systematically favoured in the case of linear dipoles and the 'parallel' one with bent dipoles. This result may be explained as follows. The planar structure of the supermolecule corresponding to a linear dipole allows the delocalization of the remaining π electrons leading to the stabilization of the transition state. On the other hand, the coplanar approach is at a great disadvantage in the case of bent dipoles such as nitrone and carbonylylide, as it would require the rotation of the CH_2 group(s) and consequently the rupture of the π bond(s) to obtain the transition state. Moreover, the asynchronism of the concerted approach is always found to be relatively small.

In general, the geometry of the transition structure is rather similar to that of the reactants. This might be an artefact of the chosen methodology which overestimates the heat of reaction. Thus, according to Hammond's postulate, the reactant likeness of the activated complex would also be overestimated.

Finally, the intermolecular parameter R at the transition state lies between 1.85 and 2.35 Å depending on the reaction under consideration. For the different atom couples (ad and ce) they correspond to the following mean values ($\bar{d}$) and standard deviation (S):

C—C	$\bar{d} = 2.28$ Å;	$S = 0.07$ Å
C—N	$\bar{d} = 2.16$ Å;	$S = 0.08$ Å
C—O	$\bar{d} = 2.19$ Å;	$S = 0.06$ Å
N—N	$\bar{d} = 2.05$ Å;	$S = 0.09$ Å
N—O	$\bar{d} = 2.07$ Å;	$S = 0.05$ Å

(*b*) *Electronic results* The electronic reorganization which takes place during the reaction can be described by the net charges of the atoms and by the centroids of charge of the localized molecular orbitals. The charge transfer from one molecule to the other is estimated by the relation:

$$t = -\sum_{A} q_A \tag{6.108}$$

where q_A denotes the net charge of atom A and the sum is taken over all the atoms of the dipolarophile. Thus t is positive when the charge transfer takes place from the dipole towards the dipolarophile.

The value of t is always rather small. It does not exceed $0.2\,(e^-)$ at the transition state and is most often positive. Moreover, it is correlated to the nature of the substituents in the dipole and the dipolarophile, as shown in Table 6.47.

Table 6.47 Influence of substituents on $t^{\neq}$

Reaction	$t^{\neq}$ (e^-)
$C_2H_4 + CH_2N_2$	0.060
$C_2H_4 + CNCHN_2$	0.035
$C_2H_4 + CH_3CHN_2$	0.067
$CNCH{=}CH_2 + CH_2N_2$	0.098
$CH_3CH{=}CH_2 + CH_2N_2$	0.055

The evolution of the centroid of charge distribution at the vicinity of the transition state provides a good description of the electronic reorganization responsible for the chemical reaction.

As a first illustration we shall consider the cycloaddition of diazomethane to ethylene. As shown in Figure 6.43, at the transition state ($R = 2.25$ Å) the new bonds are not yet formed even though there is some displacement of the centroids. Immediately after the transition state ($R = 2.00$ Å), one observes a complete reorganization of the centroids which prefigures those of the cyclic product (1-pyrazoline). It can be seen that electrons of the new CC bond come from the dipole and those of the new CN bond from the dipolarophile. Finally, the lone pair on the former central nitrogen atom of diazomethane results from the migration of a π electron pair from the triple NN bond of the dipole. This cycloaddition is then carried out by a cyclic movement of six π electrons of the two molecules, occurring just after the transition state. In fact, this result is quite general whatever the nature of the dipole may be. This is illustrated by the two specific examples given in Figure 6.44.[80]

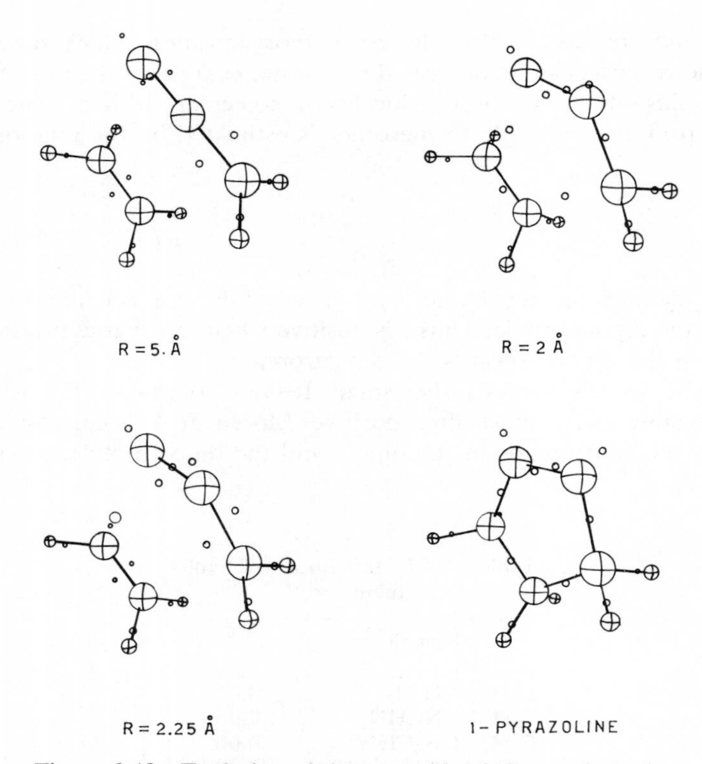

Figure 6.43 Evolution of the centroids of charge along the reaction path of the cycloaddition of diazomethane to ethylene. Reproduced by permission of the Société Scientifique de Bruxelles

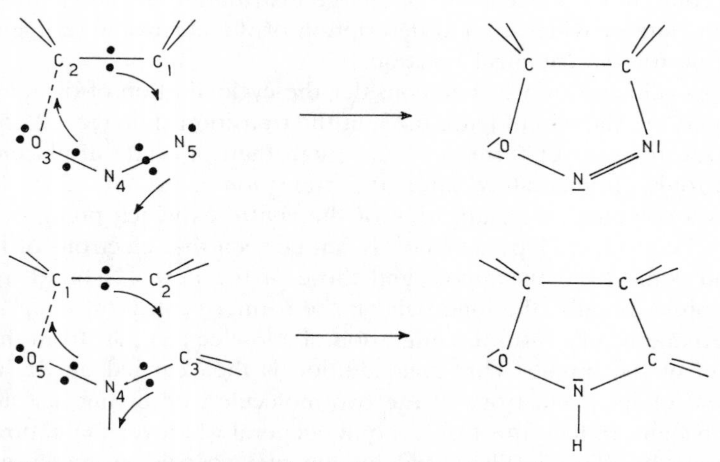

Figure 6.44 Evolution of the centroids of charge in two typical 1,3-dipolar cycloadditions

The electronic mechanism of 1,3-dipolar cycloadditions has been now well established. It may be described in terms of the usual symbols:

$$\mathrm{a{\equiv}b{=}c + d{=}e} \longrightarrow \longrightarrow$$

or

$$\mathrm{a{=}b{=}c + d{=}e} \longrightarrow \longrightarrow$$

The direction of bond migration in some typical reactions is shown in Figure 6.45. The most precocious bond and the net charges of the terminal atoms of the molecules are also given. These results suggest that the direction of the π bond migration is determined by the charges of the terminal atoms of the reactants. Indeed, the new bonds always start from the most negative ends of the molecules.

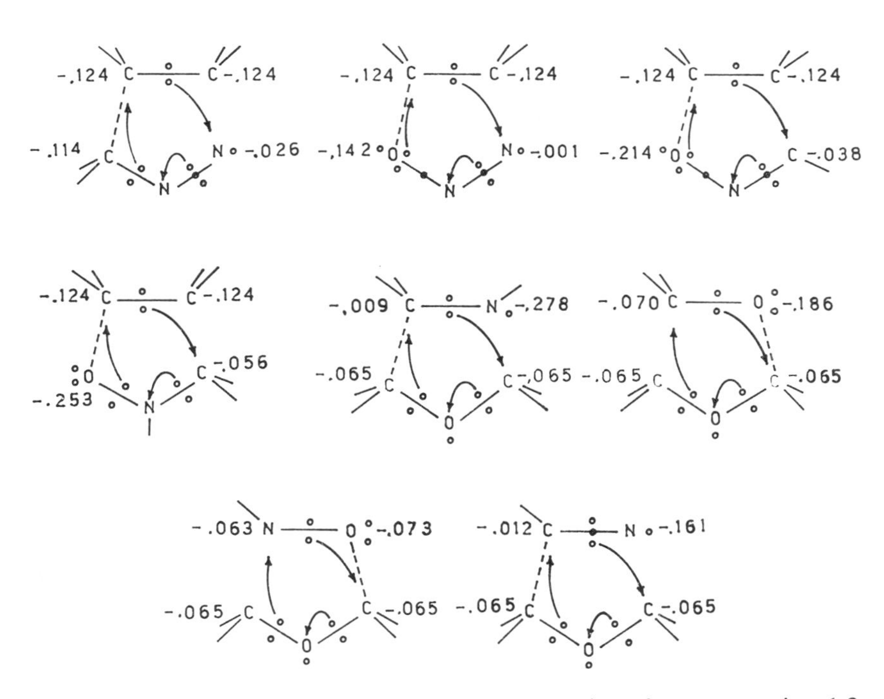

Figure 6.45 Movements of the centroids in a series of representative 1,3-dipolar cycloadditions

These results allow simple rules for anticipating the regiochemistry of 1,3-dipolar cycloadditions to be proposed and consequently a general definition of the concept of normal addition can be given:

(a) The most negative end of a dipole (dipolarophile) acts as a 'donor' of a σ bond; the other end of the molecule behaves as an 'acceptor' of the σ bond.
(b) In general, the dipole and the dipolarophile will react by placing their complementary centres face to face. By definition, this situation would correspond to the normal addition having the smallest activation energy.

Thus one may assume that the regiochemistry of a 1,3-dipolar cycloaddition essentially depends on the relative values of the net charges of the terminal atoms of the two molecules, which explains the various experimental results described in Figure 6.41.

Finally, these theoretical results may be used to propose original synthesis of heterocycles such as the ones described in Figure 6.46. It should be noted that, for the dipolarophiles, one may also use the π net charges whose qualitative values are easy to predict due to the theory of resonance.

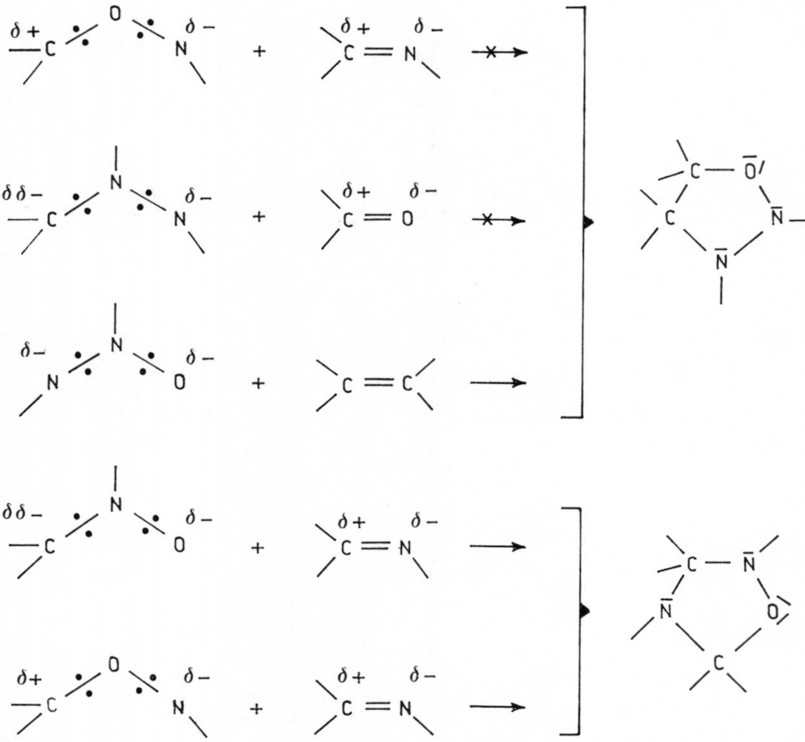

Figure 6.46 Tentative proposals of synthesis of some heterocycles

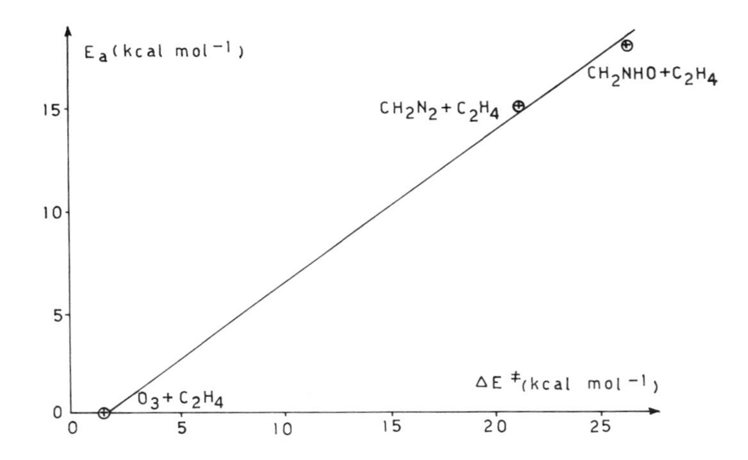

Figure 6.47 Activation energies versus activation barriers

(*c*) *Energetic results* The energetic properties calculated at the STO-3G level are very imprecise. Thus only their relative values in a homogeneous series of reactions may be considered as significant. Particularly, as pointed out above, the heats of reaction are largely overestimated. On the other hand, the theoretical activation barriers are very well correlated with the available experimental activation energies (see Figure 6.47). It should also be noted that the activation barrier for normal addition is generally lower than the barrier for the reverse reaction.

Recently, a statistical analysis of the theoretical results obtained on the concerted 1,3-dipolar cycloaddition has been performed in order to determine the leading parameters for the activation energies of these reactions.[81] We summarize below the main results obtained from this multivariate statistical analysis:

(a) The 1,3-dipolar cycloadditions are characterized more by the type of dipole than by the nature of the dipolarophile.
(b) The regression analysis between the variables characterizing the reactants and the dependent variable $\Delta E^{\neq}$ which characterizes the transition structure leads to the relation which best explains the activation barrier of the chosen reactions:

$$\begin{aligned}\Delta E^{\neq} = {} & 0.027\,\Delta E - 11.9\,q_{\mathrm{d}} - 14.1\,q_{\mathrm{e}} + 67.6\,q_{\mathrm{b}} \\ & - 396\,q_{\mathrm{a}} + 0.113\,E_{\mathrm{ed}} - 0.230\,E_{\mathrm{cb}} + 0.180\,E_{\mathrm{ab}} \\ & - 1.29\,\mu_{\mathrm{de}} + 3.59\,\mu_{\mathrm{abc}}\end{aligned}$$

where ΔE is the theoretical heat of reaction, q_i the net charge of atom i, E_{ij} the semiempirical bond energy of bond ij[82] and μ the theoretical dipole moment of a given molecule (abc or de).

The first term allows an estimation to be made of the relevance of the Evans–Polanyi relation. The regression coefficients associated with the atomic charges show that the feature for lowering activation energy is to make atom a of the dipole and atom e of the dipolarophile positive. This conclusion somewhat refines the previous one based on the negative charges of centres c and d. In terms of net charges the most favourable situation is described as:

$$\overset{\oplus}{\text{a}}{=}\text{b}{-}\text{c} \qquad \text{d}{=}\text{e}^{\oplus}$$

to compare with that previously proposed:

$$\text{a}{=}\text{b}{-}\text{c}^{\ominus} \qquad {}^{\ominus}\text{d}{=}\text{e}$$

We are now able to reexpress the normal/reverse rule as: 'If the net charges are such that q_e is more positive (or less negative) than q_d and q_a and more positive (or less negative) than q_c, the normal addition will lead to new ad and ce σ bonds and will have a lower activation energy than the reverse addition where the new σ bonds are ae and cd.' This rule is a very efficient tool for rationalizing the regiochemistry of the 1,3-dipolar cycloaddition. Note also that the regression coefficient associated with E_{de} explains why acetylenic dipolarophiles lead to larger activation barriers than ethylenic ones. Similarly, the regression coefficients corresponding to E_{bc} and E_{ab} explain why propargyl-type dipoles lead to higher activation barriers than allyl-type ones. We may conclude that this approach of 1,3-dipolar cycloadditions allows rationalization of a large amount of theoretical and experimental data and proposes a general mechanism to describe both the nuclear and the electronic motions responsible for these reactions. The results obtained are in good agreement with many experimental facts and also with other theoretical studies of the same level[(83)] or using a more sophisticated methodology.[(84)]

The ring closure of 1,5-dipoles

1,3-Dipoles substituted by an unsaturated group are often called 1,5-dipoles. Their thermal cyclization leads to a large variety of five-membered heterocycles, as shown in Figure 6.48. We shall describe here the mechanism of some typical ring closures determined using the theoretical approach. More precisely, we shall consider the thermal cyclization of propargyl-type species derived from HN_3, i.e. azidoazomethine and vinyl azide where d = e corresponds respectively to CH=NH and $CH{=}CH_2$.

(*a*) *The azidotetrazole isomerization*[(85)] The angular parameters used in the optimization procedure are defined in Figure 6.49. The distance between

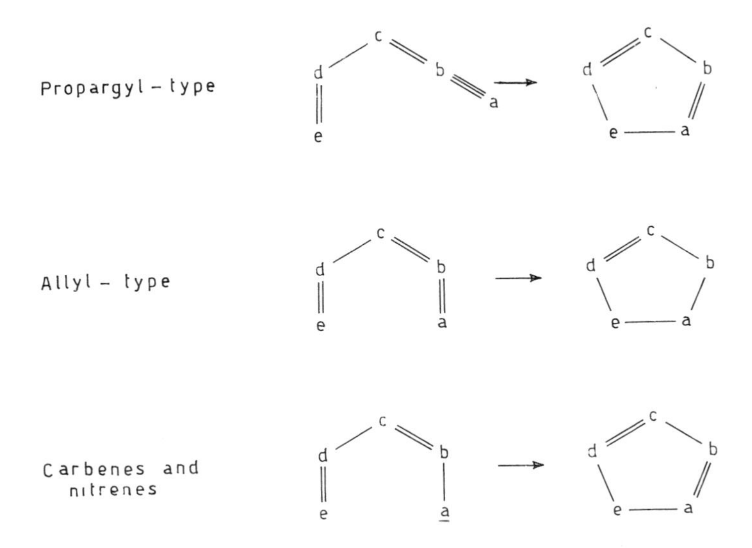

Figure 6.48 Ring closure of 1,5-dipoles

atoms N_1 and N_5 was chosen as the reaction coordinate for determining the reaction path.

It has been found that the cyclization process essentially involves three steps, as shown in Figure 6.50. The first step goes up to the transition state and is characterized by a shortening of the angle γ along with lengthening of the bonds N_3N_4 and N_4N_5. During this step, one of the N_4N_5 triple bond centroids moves to form a lone pair on N_4. The concomitant rise in energy is relatively small. This can be taken for the activation energy of the cyclization process.

The second step corresponds to shortening of the angle α to the value found in tetrazole. The angles γ and ν vary simultaneously but the only bond which undergoes an appreciable modification during this step is N_4N_5. Parallel to these geometric changes, one of the N_3N_4 centroids starts to migrate towards the bond C_2N_3, while stabilization of the system appreciably increases.

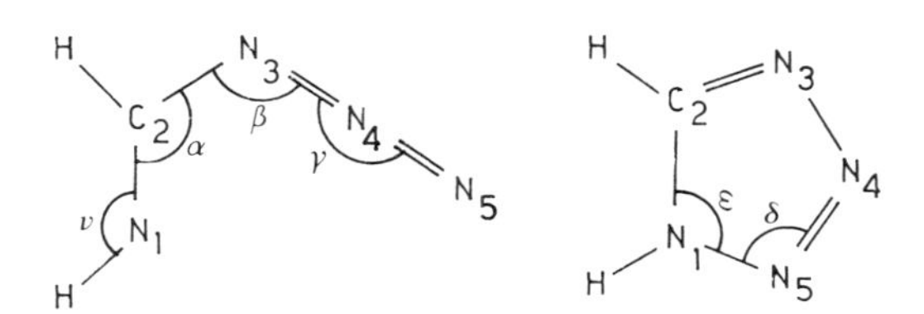

Figure 6 49 Angular parameters in the azidotetrazole isomerization. Reprinted with permission from Smith and Jones, copyright American Chemical Society 1982

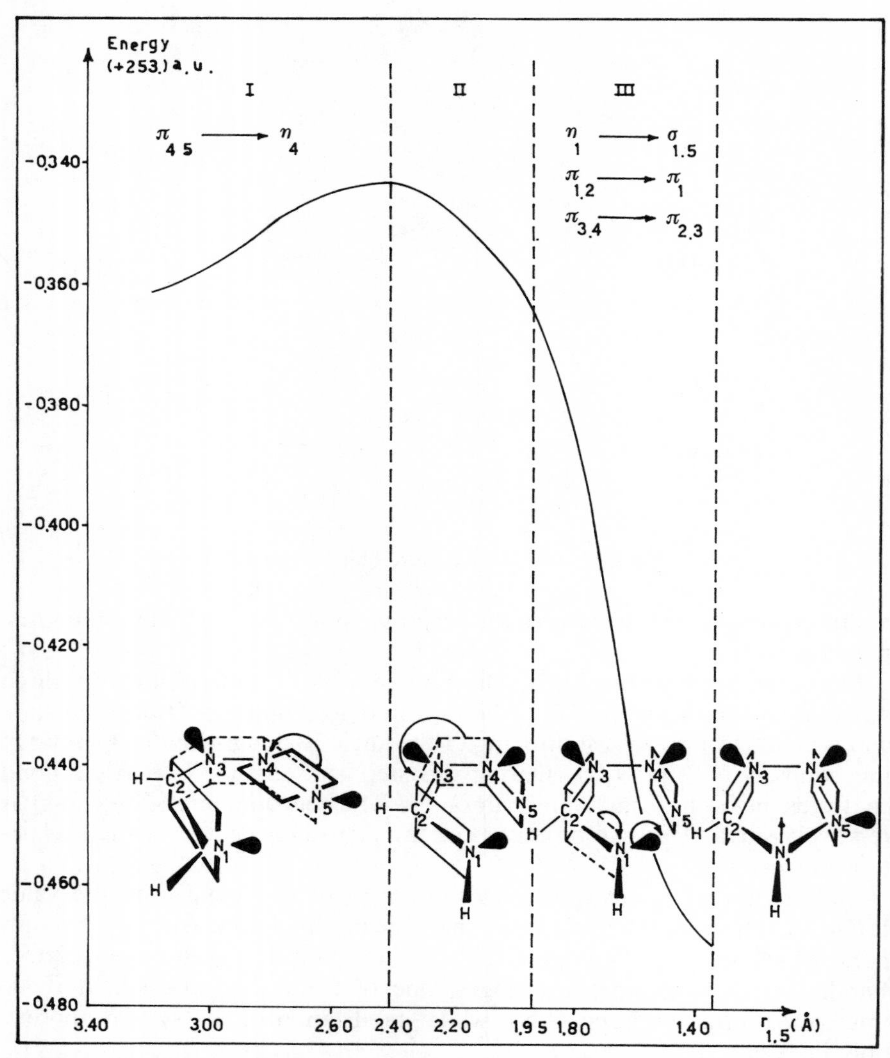

Figure 6.50 The reaction path of azidotetrazole isomerization. Reproduced by permission of the Société Scientifique de Bruxelles

It is during the third step that the most important modifications of the supersystem take place. All the structural parameters take up the values corresponding to the product; the σ bond between atoms N_1 and N_5 is formed at the expense of the lone pair on N_1 while the π electrons of N_3N_4 finally form the double bond on C_2N_3. In the last place, two π electrons of the C_2N_1 bond become a lone pair on atom N_1 in tetrazole.

A keynote of the whole process is the presence of a lone pair on N_1 in the open-chain system. This permits the formation of the new σ bond without having to rotate the NH group around the C_1N_1 double bond—a process which would be very disadvantaged from the energetic point of view. This is a quite general result which explains the relatively small activation energy of the ring closure of 1,5-dipoles, bringing such a lone pair on atom e. This activation energy is essentially due to the bending of the azide fragment, while that of the reverse process is due to the breaking of the N_1N_5 bond and to the loss of resonance energy when going from tetrazole to the activated complex.

From these results, it can be anticipated that the factors which increase the availability of the lone pair on atom e will favour the ring closure. Inversely, the protonation of this atom would prevent this reaction.

(*b*) *The vinyl azide–vinyl triazole isomerization*[86] It is interesting to compare the ring closure of vinyl azide to the azidotetrazole isomerization in neutral and in acidic media. The angular parameters for the optimization of vinyl azide closure are given in Figure 6.51. Here, again, it is found that the cyclization process involves three steps, as shown in Figure 6.52. The first step can be called the azide zone as the supermolecule resembles the reactant although the azide is bent at N_4. The second is a transition zone in which the CH_2 group begins to rotate. In the third, which may be called the triazole zone, the system resembles the product (vinyl triazole).

It is found that the azide zone is a process in which a lone pair is formed on N_4 while the π system of the molecule scarcely changes. In the transition zone, most of the major electronic reorganization takes place. Finally, in the third zone the bond between C_1 and N_5 is progressively formed. The

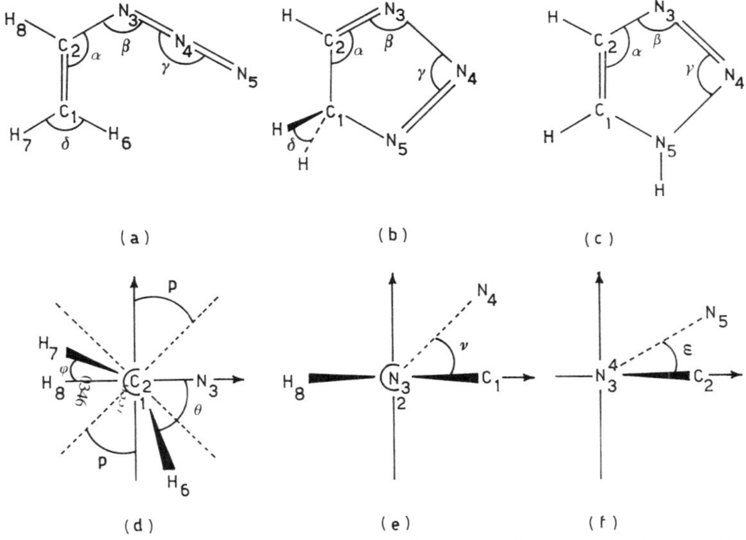

Fig 6.51 Angular parameters for the ring closure of vinyl azide. Reprinted with permission from Smith and Jones, copyright American Chemical Society 1982

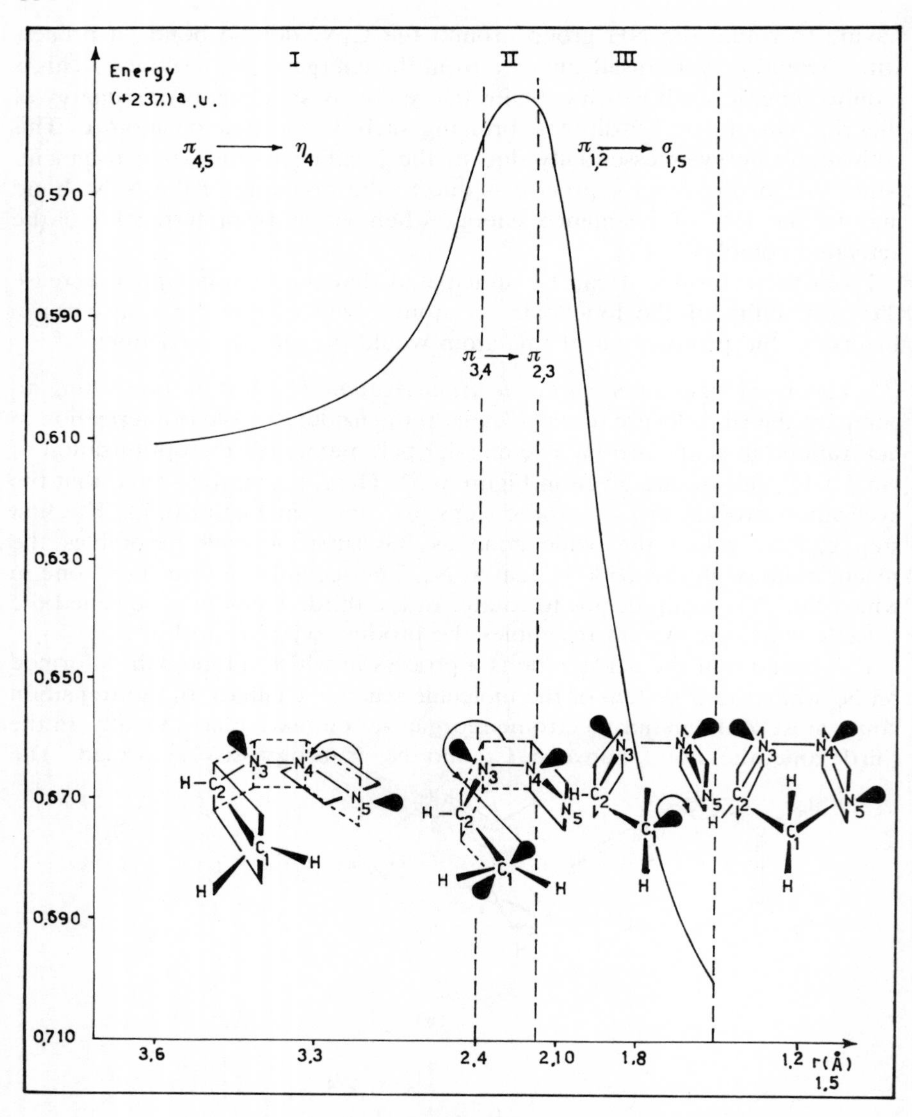

Figure 6.52 The reaction path of the ring closure of vinyl azide. Reproduced by permission of the Société Scientifique de Bruxelles

activation energy for the closure of vinyl azide is due to the bending of angle γ and to the rupture of the π system by turning of the terminal CH_2 group; the activation energy of the reverse process is merely due to the breaking of the C_1N_5 bond. This is also a general result which is valid for every 1,5-dipole which is not provided with a lone pair on atom e. This is precisely the case of protonated azidoazomethine for which the ring closure reaction

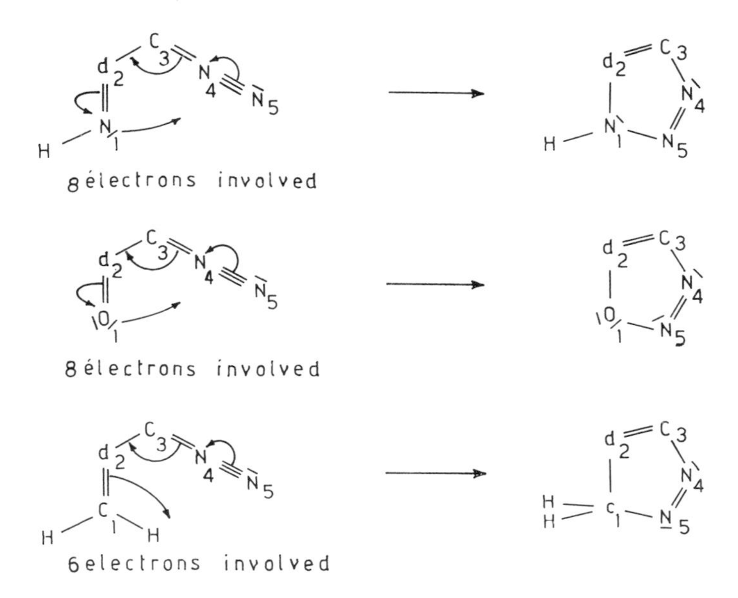

Figure 6.53 Electronic mechanism of the ring closure of propargyl-type 1,5-dipoles

path has been found very similar to that of vinyl azide, the two reactions being characterized by a large activation energy. Figure 6.53 summarizes the main features of the electronic mechanism of the ring closure of propargyl-type 1,5-dipoles. It can be seen that the reaction involves the migration of eight or six electrons according to whether the centre e has a lone pair or not.

These conclusions may certainly be generalized to other types of 1,5-dipoles without performing any additional computations.

The reaction of water with 1,3-dipoles[87]

The same methodology has been used to determine the mechanism of some addition reactions of water with 1,3-dipoles. We describe below the results obtained in the case of fulminic acid and acetonitrile oxide. Without going into much detail, we summarize the main features of the reaction of water with fulminic acid. It is found that this addition occurs in two phases. The first is the deformation of fulminic acid in the E (trans) mode as the oxygen of water approaches the carbon atom. This determines that the product will have the Z configuration as experimentally observed in all analogous reactions. Figure 6.54 shows the structure of the transition state and that of the product (hydroxyformaldoxime).

The second step starting after the transition state is the transfer of a hydrogen atom from the oxygen of water to that of the fulminic acid. From the electronic point of view, it is found that the deformation of the dipole

Figure 6.54 Structure of (a) activated complex and (b) product for the reaction $H_2O + HCNO$

leads to the formation of a lone pair on the central nitrogen from the original triple bond CN. The other electronic movements occur after the transition state. Then the hydrogen atom is progressively displaced from one oxygen to the other and the new σ(CO) bond is formed. This electronic reorganization is summarized in Figure 6.55 or, in more usual notation, by the following equation:

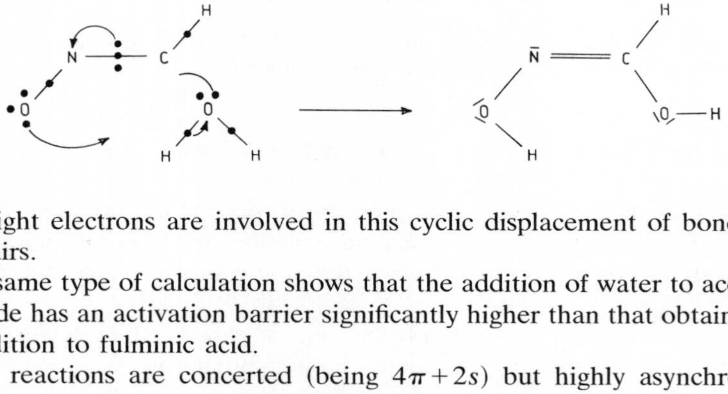

Thus eight electrons are involved in this cyclic displacement of bonds and lone pairs.

The same type of calculation shows that the addition of water to acetonitrile oxide has an activation barrier significantly higher than that obtained for the addition to fulminic acid.

Both reactions are concerted (being $4\pi + 2s$) but highly asynchronous. Furthermore, the proton slide at the transition state is found to occur without an energy barrier. The addition product is an oxime of configuration Z s-trans at the O_1 bond and s-cis about the N_3O_4 bond. This form of hydroxyformaldoxine is not the most stable: kinetically obtained, it may

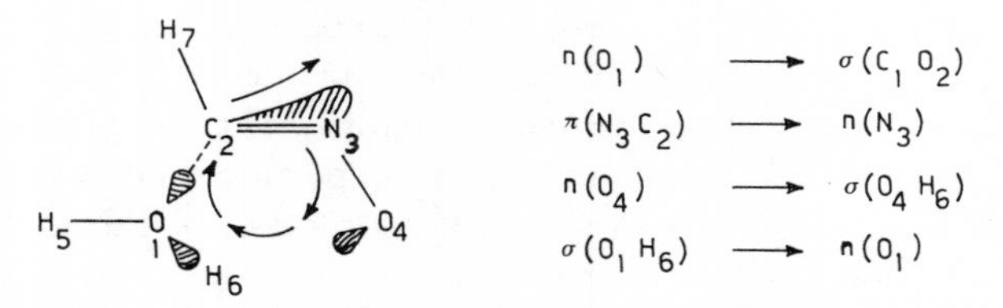

Figure 6.55 Electronic movements in the reaction of water with fulminic acid. Reprinted with permission from Smith and Jones, copyright American Chemical Society (1982)

eventually evolve to other structures.[88] However, such changes occur well after the transition state and do not affect the previous conclusions.

REFERENCES

1. The Chemical Society, *Tables of Interatomic Distances and Configuration in Molecules and Ions*, The Chemical Society Special Publication, no. 11, London, 1958.
2. J. A. Pople and M. S. Gordon, *J. Amer. Chem. Soc.*, **92,** 4796 (1970).
3. Quantum Chemistry Program Exchange, *Guide and Index to QCPE Catalog*, **13,** 44 (1981).
4. QCPE Programs 236, 368, 391, 406. *Q.C.P.E. Catalog*, 11 (1979).
5. P. W. Payne and L. C. Allen, Barriers to rotation and inversion, in *Modern Theoretical Chemistry* (Ed. H. F. Schaefer III), Vol. 4, Plenum Press, New York (1977).
6. R. Pauling, *The Nature of the Chemical Bond*, Cornell University Press, Ithaca, 1929.
7. J. W. Linnett, *The Electronic Structure of Molecules—A New Approach*, Methuen, London, 1966.
8. R. Daudel, *Quantum Theory of the Chemical Bond*, Reidel, Dordrecht, 1974.
9. G. W. Wheland, *The Theory of Resonance and its Application to Organic Chemistry*, Wiley, New York, 1953.
10. J. M. Foster and S. F. Boys, *Rev. Mod. Phys.*, **32,** 300 (1960); S. F. Boys, in *Quantum Theory of Atoms, Molecules and the Solid State*, (Ed. P. O. Löwdin), Academic Press, New York, 1966, p. 253.
11. J. A. Pople and K. Nesbet, *J. Chem. Phys.*, **22,** 571 (1954).
12. G. Leroy and D. Peeters, in *The Transferable Properties of Localized Orbitals* in *Localization and Delocalization in Quantum Chemistry*, (Eds. G. Chalvet *et al.*), Vol. 1, Reidel, Dordrecht, 1975, p. 207; G. Leroy, D. Peters, A. Deplus and M. Tihange, *Nouv. J. Chim.*, **3,** 213 (1979).
13. E. Kapuy, C. Kozmutza and M. E. Stephens, *Theoret. Chim. Acta*, **43,** 175 (1976); R. Daudel, M. E. Stephens, I. G. Csizmadia, C. Kozmutza, E. Kapuy and J. D. Goddard, *Internat. J. Quantum Chem.*, **11,** 665 (1977).
14. M. Tihange, Ph.D. Thesis, Catholic University of Louvain, 1983.
15. G. Herzberg, *Spectra of Diatomic Molecules*, Van Nostrand, New York, 1950 p. 343.
16. R. Daudel, A. Laforgue and C. Vroelant, *J. Chim. Phys.*, **5,** 44 (1952); M. Roux, S. Besnainou and R. Daudel, *J. Chim. Phys.*, **218,** 939 (1956).
17. G. Leroy, *Structure and properties of free-radicals—a theoretical contribution*, in *Computational Theoretical Organic Chemistry* (Eds. I. G. Czismadia and R. Daudel), Reidel, New York, 1981, p. 253.
18. W. C. Easley and W. Weltner, *J. Chem. Phys.*, **52,** 197 (1970).
19. H. M. McConnell, *J. Chem. Phys.*, **24,** 764 (1956); D. B. Chesnut, *J. Chem. Phys.*, **29,** 43 (1958).
20. R. W. Fessenden and R. H. Schuler, *J. Chem. Phys.*, **39,** 2147 (1963); D. Lazdins and M. Karplus, *J. Chem. Phys.*, **44,** 1600 (1966).
21. M. Sana and G. Leroy, *Theochem.*, **76,** 259 (1981).
22. J. B. Moffat and A. J. Knowles, *J. Chem. Eng. Data*, **14,** 215 (1966).
23. A. Langseth and C. Moller: *Acta Chem. Scand.*, **4,** 725 (1950); G. D. Craineat and K. H. Thompson, *Trans. Faraday Soc.*, **49,** 1273 (1953); T. Migazaiva, *J. Chem. Phys.*, **29,** 421 (1958).
24. G. Herzberg, *Molecular Spectra and Molecular Structure*, Vol. 2, Van Nostrand-Reinhold, New York, 1945.

25. W. J. Hehre, R. Ditchfield, H. Radom and J. A. Pople, *J. Amer. Chem. Soc.*, **92,** 4796 (1970).
26. W. J. Hehre, R. Ditchfield and J. A. Pople, *J. Chem. Phys.*, **56,** 2258 (1972).
27. S. W. Benson, F. R. Cruickshank, D. M. Golden, G. R. Haugen, H. E. O'Neal, A. S. Rodgers, R. Shaw and R. Walsh, *Chem. Rev.*, **69,** 279 (1969).
28. L. Radom, W. A. Lathan, W. J. Hehre and J. A. Pople, *J. Amer. Chem. Soc.*, **93,** 5339 (1971).
29. J. D. Cox and G. Pilcher, *Thermochemistry in Organic and Organometallic Compounds*, Academic Press, London, 1970.
30. M. Sana, *Internat. J. Quantum Chem.*, **19,** 139 (1981).
31. R. Stull and H. Prophet, *JANAF Thermochemical Tables*, 2nd ed., Department of Commerce, Washington D.C. (1971).
32. G. Leroy, C. Wilante, D. Peeters and M. Khalil, *Ann. Soc. Scient. Bruxelles*, **95,** III–IV, 157 and references therein (1981).
33. D. Griller and K. V. Ingold, *Acc. Chem. Res.*, **9,** 13 (1976).
34. M. J. S. Dewar and H. N. Schmeising, *Tetrahedron*, **5,** 166 (1959), **11,** 96 (1960).
35. K. J. Laidler, *Can. J. Chem.*, **34,** 626 (1956).
36. H. M. Rosenstock, K. Draxl, B. W. Steiner and J. T. Herron, Energetics of gaseous ions, *J. Phys. Chem. Ref. Data*, **6,** suppl. no. 1 (1977).
37. H. Hotop and W. C. Lineberger, *J. Phys. Chem. Ref. Data*, **6,** 539 (1975).
38. G. Leroy, D. Peeters and C. Wilante, *Theochem.*, **95,** III–IV, 157 and references therein (1981).
39. L. Stella, Z. Janousek, R. Merényi and H. G. Viehe, *Angew. Chem. Int. Ed. Engl.*, **9,** 691 (1978).
40. P. G. Gassman and J. J. Talley, *J. Amer, Chem. Soc.*, **102,** 1214 (1980); D. A. Dixon, P. A. Charlier and P. G. Gassman, *J. Amer. Chem. Soc.*, **102,** 3957 (1980); M. N. Paddon-Row, C. Santiago and K. N. Houk, *J. Amer. Chem. Soc.*, **102,** 6561 (1980); W. F. Reynolds and P. Dais, *Tetrahedron Lett.*, **22,** 1795 (1981).
41. S. G. Christov, *Lecture Notes in Chemistry. Collision Theory and Statistical Theory of Chemical Reactions*, Vol. 18, Springer-Verlag, 1980.
42. I. Glasstone, K. J. Laidler and H. Eyring, *The Theory of Rate Processes*, McGraw-Hill, 1941.
43. R. Fowler and E. Guggenheim, *Statistical Thermodynamics*, Cambridge University Press, 1939, p. 517; B. Swolinski and H. Eyring, *J. Amer. Chem. Soc.* **69,** 2702 (1947).
44. H. Eyring, *J. Chem. Phys.*, **3,** 107 (1935); *Chem. Rev.*, **17,** 65 (1935); *Trans. Far. Soc.*, **34,** 41 (1938).
45. R. Daudel, *Quantum Theory of Chemical Reactivity*, Reidel, Dordrecht, 1973.
46. S. W. Benson, *Thermochemical Kinetics*, 2nd ed., Wiley, 1976, Chap. I.
47. E. P. Wigner, *Z. Phys. Chem.*, **B19,** 203 (1933).
48. B. C. Garret and D. G. Thruhar, *J. Chem. Phys.*, **83,** 1079 (1979).
49. R. P. Bell, *Proc. Roy. Soc.*, **A139,** 466 (1933).
50. S. G. Christov, *Ber. Bunsenges. Phys. Chem.*, **76,** 507 (1972).
51. V. I. Goldanskii, *Chemica Scripta*, **13,** 1 (1978–79).
52. R. B. Bernstein, in *Atom–Molecule Collision Theory*, R. B. Bernstein (Ed.), Plenum Press, 1977, Chap. I, p. 1.
53. J. Ross, Molecular beams, *Adv. Chem. Phys.*, **10** (1966).
54. M. Karplus, R. N. Porter and R. D. Sharma, *J. Chem. Phys.*, **43,** 3259 (1965).
55. M. Karplus, in *The World of Quantum Chemistry*, Reidel, Dordrecht, 1974, p. 101.
56. D. A. Michz, *Adv. Chem. Phys.*, **30,** 7 (1975).
57. D. R. Herschbach, in *Potential Energy Surfaces in Chemistry*, W. A. Lester, Jr (Ed.) IBM Research Lab. Pub., 1971, p. 44.

58. H. Eyring, J. Walter and G. E. Kimball, *Quantum Chemistry*, Wiley, 1946.
59. B. J. Matkowsky and Z. Schuss, *SIAM J. Appl. Math.*, **33,** 365 (1977); Z. Schuss and B. J. Matkowsky, *SIAM J. Appl. Math.*, **35,** 604 (1979).
60. S. Chandrasekhar, in *Stochastic Problems in Physics and Astronomy—Selected Papers on Noise and Stochastic Processes*, N. Wax (Ed.), Dover, 1954.
61. H. A. Kramers, *Physica*, **7,** 284 (1940).
62. J-L. Villaveces, Ph. D. Thesis, Louvain la Neuve, 1981.
63. B. Huron, J-P. Malrieu and P. Rancurel, *J. Chem. Phys.*, **58,** 5745 (1973).
64. D. R. Stull and M. Prophet, *JANAF Thermochemical Tables*, Nat. Bur. Standards, Washington, D.C., 1971.
65. S. W. Benson, *Thermochemical Kinetics*, 2nd ed., Wiley, 1976, p. 86 and also p. 147.
66. M. G. Evans and M. Polanyi, *Trans. Far. Soc.*, **34,** 11 (1938).
67. G. S. Hammond, *J. Amer. Chem. Soc.*, **77,** 334 (1954).
68. R. R. Baldwin and R. W. Walker, *J. Chem. Soc.*, **Perkin II,** 361 (1973).
69. J. A. Kerr and M. J. Parsonage, *Evaluated Kinetic Data of Gas Phase Hydrogen Transfer Reactions of Methyl Radicals*, Butterworths, 1976.
70. M. Denissy and R. Lesclaux, *J. Amer. Chem. Soc.*, **102,** 2898 (1980).
71. N. R. Greiner, *J. Chem. Phys.*, **53,** 1070 (1970).
72. R. Foon and M. Kaufmann, *Prog. React. Kinetics*, **8,** 81 (1975).
73. R. Shaw, *J. Chem. Ref. Data*, **7,** 1179 (1978).
74. R. Zellner, *J. Phys. Chem.*, **83,** 18 (1979).
75. P. D. Pacey and J. H. Purnell, *J. Chem. Soc. Far. Trans.*, *I*, **68,** 1462 (1972).
76. D. D. Davis, S. Fischer and R. Shiff, *J. Chem. Phys.*, **61,** 2213 (1974).
77. W. A. Pryor, W. H. Davis and J. P. Staley, *J. Amer. Chem. Soc.*, **95,** 4754 (1973).
78. G. Leroy, M. Sana, L. A. Burke and M.-T. Nguyen, in *Quantum Theory of Chemical Reactions* (Eds. R. Daudel, A. Pullman, L. Salem and A. Veillard), Reidel, 1980, p. 91; M-T. Nguyen, Ph.D. Thesis, Louvain-la-Neuve, 1980.
79. R. Huisgen, *Angew. Chem. Int. Ed. Engl.*, **2,** 565 (1963).
80. *The Cycloaddition of Nitrone to Ethylene*, Film based on the work of G. Leroy and M. Sana, produced by G. Leroy and realized by J. Leyder (Louvain, 1978).
81. M. Sana, G. Leroy, G. Dive and M. T. Nguyen, *Theochem.*, **89,** 147 (1982).
82. G. Leroy and M. Sana, *Tetrahedron*, **32,** 709 (1976); G. Leroy, M. T. Nguyen and M. Sana, *Tetrahedron*, **34,** 2459 (1978).
83. D. Poppinger, *J. Amer. Chem. Soc.*, **98,** 486 (1976).
84. A. Kemernicki, J. D. Goddard and J. F. Schaefer III, *J. Amer. Chem. Soc.*, **102,** 1763 (1980).
85. L. A. Burke, G. Leroy, M. T. Nguyen and M. Sana, *J. Amer. Chem. Soc.*, **98,** 1685 (1976).
86. L. A. Burke, G. Leroy, M. T. Nguyen and M. Sana, *J. Amer. Chem. Soc.*, **100,** 3668 (1978).
87. M. T. Nguyen, M. Sana, G. Leroy, K. J. Dignam and F. Hegarty, *J. Amer. Chem. Soc.*, **102,** 573 (1980).
88. M. T. Nguyen, M. Sana and G. Leroy, *Bull. Soc. Chim.*, **90,** 681 (1981).

PART III

GOING FURTHER INTO THE NUCLEAR MOTIONS

CHAPTER 7

Basic Concepts

A. The potential energy hypersurfaces

7.1 A FUNCTION OF THE NUCLEAR COORDINATES

7.1.1 The Born–Oppenheimer Approximation

The time-independent Schrödinger equation for a system containing N nuclei and n electrons is:

$$\left[\left(-\frac{\hbar^2}{2}\sum_{j=1}^{N}\frac{1}{m_j}\Delta_j\right)+\left(-\frac{\hbar^2}{2m_e}\sum_{i=1}^{n}\Delta_i\right)+\left(\sum_{j<j'}\frac{Z_jZ_{j'}e^2}{r_{jj'}}\right)+\left(\sum_{i<i'}\frac{e^2}{r_{ii'}}\right)+\left(-\sum_{ij}\frac{Z_je^2}{r_{ij}}\right)\right]\Psi(\mathbf{R}_1\cdots\mathbf{R}_N,\mathbf{r}_1\cdots\mathbf{r}_n)$$
$$=W\Psi(\mathbf{R}_1\cdots\mathbf{R}_N,\mathbf{r}_1\cdots\mathbf{r}_n) \quad (7.1)$$

If we denote by T_N and T_E the kinetic operators relative respectively to the nuclei and the electrons and V the total potential energy, the preceding expression can be more simply written:

$$(T_N+T_E+V)\Psi=W\Psi \quad (7.2)$$

As T_N is proportional to the inverse of the nucleus mass and T_E to the inverse of the electron mass, T_N may be considered as a perturbation relative to T_E. Using perturbation theory we can write:

$$H^{(0)}=T_E+V$$
$$H^{(1)}=T_N \quad (7.3)$$

with:

$$H=T_N+T_E+V=H^{(0)}+\lambda H^{(1)}$$

λ only being used to indicate the lower order of magnitude of the perturbating operator $H^{(1)}$ relatively to $H^{(0)}$.

Similarly, for the wave function (Ψ) and for the energy (W) we can obtain the following developments:

$$\Psi=\Psi^{(0)}+\lambda\Psi^{(1)}+\lambda^2\Psi^{(2)}+\cdots$$
$$W=W^{(0)}+\lambda W^{(1)}+\lambda^2W^{(2)}+\cdots \quad (7.4)$$

We now insert such expansions in the Schrödinger equation:

$$(H^{(0)}+\lambda H^{(1)})(\Psi^{(0)}+\lambda\Psi^{(1)}+\cdots)=(W^{(0)}+\lambda W^{(1)}+\cdots)(\Psi^{(0)}+\lambda\Psi^{(1)}+\cdots) \tag{7.5}$$

Let us now separate this expression up to the first perturbational order:

$$\begin{aligned} H^{(0)}\Psi^{(0)} &= W^{(0)}\Psi^{(0)} \\ H^{(1)}\Psi^{(0)}+H^{(0)}\Psi^{(1)} &= W^{(1)}\Psi^{(0)}+W^{(0)}\Psi^{(1)} \\ \cdots\cdots \end{aligned} \tag{7.6}$$

The zeroth-order wave equation corresponds to the description of the electronic motion in the field of the fixed nuclei. The associated eigenvalues are only dependent on the nuclear coordinates ($R=\{\mathbf{R}_1\cdots\mathbf{R}_N\}$).

In the simplest approximation, we admit the separability of electronic and nuclear motions. Then the unperturbed wave function $\Psi^{(0)}$ will be of the form:

$$\Psi(\mathbf{R}_1\cdots\mathbf{R}_N,\mathbf{r}_1\cdots\mathbf{r}_n)\simeq\chi^{(0)}(\mathbf{R}_1\cdots\mathbf{R}_N)\Phi^{(0)}(\mathbf{R}_1\cdots\mathbf{R}_N,\mathbf{r}_1\cdots\mathbf{r}_n) \tag{7.7}$$

The Schrödinger equation becomes:

$$[T_{\mathrm{N}}(R)+T_{\mathrm{E}}(r)+V(Rr)]\chi^0(R)\Phi^0(Rr)=W\chi^0(R)\Phi^0(Rr) \tag{7.8}$$

Let us multiply this expression by $\Phi^*(Rr)$ and integrate over the electronic coordinates ($r=\{\mathbf{r}_1\cdots\mathbf{r}_n\}$):

$$\int\Phi^*(Rr)T_{\mathrm{N}}(R)\Phi(Rr)\,\mathrm{d}r\chi(R)+\int\Phi^*(Rr)\{T_{\mathrm{E}}(r)+V(Rr)\}\Phi(Rr)\,\mathrm{d}r\chi(R) = W(R)\int\Phi^*(Rr)\Phi(Rr)\,\mathrm{d}r\chi(R) \tag{7.9}$$

In the Born–Oppenheimer approximation, and under some particular conditions which are not discussed here, we admit that:

$$\int\Phi^*(Rr)T_{\mathrm{N}}(R)\Phi(Rr)\,\mathrm{d}r\chi(R)=\int\Phi^*(Rr)\Phi(Rr)\,\mathrm{d}rT_{\mathrm{N}}(R)\chi(R) \tag{7.10}$$

Let us choose for $\Phi(Rr)$ the normalized eigenfunctions of the operator $T_{\mathrm{E}}(r)+V(Rr)$. Then:

$$[T_{\mathrm{E}}(r)+V(Rr)]\Phi(Rr)=E(R)\Phi(Rr) \tag{7.11}$$

with:

$$\int\Phi^*(Rr)\Phi(Rr)\,\mathrm{d}r=1$$

where the eigenvalue $E(R)$ is the total energy of a molecular system for a particular set of nuclear coordinates:

$$E(R)=E_{\mathrm{e}}(R)+\sum_{j<j'}\frac{Z_jZ_{j'}e^2}{r_{jj'}} \tag{7.12}$$

where $E_e(R)$ stands for the electronic energy alone. It now remains that:

$$T_N(R)\chi(R)+E(R)\chi(R)=W\chi(R) \tag{7.13}$$

where $E(R)$ plays for the nuclei the same role as the potential V plays for the electrons. For such a reason we call $E(R)$ the potential energy surface associated with the nuclei motions.

7.1.2 The Internal Coordinates Frame

The potential energy $E(R)$ of a molecular system is a function of the nuclear positions (R). R is a column vector of $3N$ components containing the X, Y and Z cartesian positions for each atom. As the total energy of an isolated molecule does not depend on its relative position and its relative orientation in a fixed space coordinate system, we can remove six coordinates (except in the case of linear molecules). The coordinate system which contains a number of variables just large enough to express completely the total potential energy will be called the internal coordinate system and labelled $\mathbf{s}$:

$$\begin{aligned} R' &= \{\mathbf{R}_1 \cdots \mathbf{R}_N\} = \{X_1 Y_1 Z_1 \cdots X_N Y_N Z_N\} \\ \mathbf{s}' &= \{s_1 \cdots s_{3N-6}\} = \mathbf{s}'(R) \end{aligned} \tag{7.14}$$

We can choose the $3N-6$ components of the $\mathbf{s}$ vector in different ways; e.g.:

(a) Choose $3N-6$ bond lengths and bond angles (not linearly dependent).
(b) Eliminate from R the next components:

$$X_1 = Y_1 = Z_1 = X_2 = Y_2 = X_3 = 0$$

and use the remaining cartesian coordinates.
(c) The $3N-6$ symmetry coordinates may be chosen as a linear combination of bond lengths and bond angles.
etc.

In the particular case of linear molecules we have only $N-1$ degrees of freedom (i.e. the bond lengths) along the C_∞ axis. The $N-2$ other coordinates permit inclusion of all the molecular deformations in a plane. Those $2N-3$ degrees of freedom suffice to describe the molecular motion around the equilibrium position. The remaining $N-3$ coordinates which one needs in order to reach a total of $3N-6$ describe the out-of-plane nuclear motion. They may be the $N-3$ torsion angles. Nevertheless, it is sometimes more useful to replace the last $N-3$ set of coordinates by $N-2$ others which describe the nuclear motion in a perpendicular plane; so we are using $3N-5$ coordinates corresponding to the $3N-5$ molecular vibrators of the linear case (with $N-2$ degenerate motions). Further, we note the number of internal coordinates related to the potential energy surface (k).

In Figures 7.1 to 7.3, we illustrate our present purpose by giving the

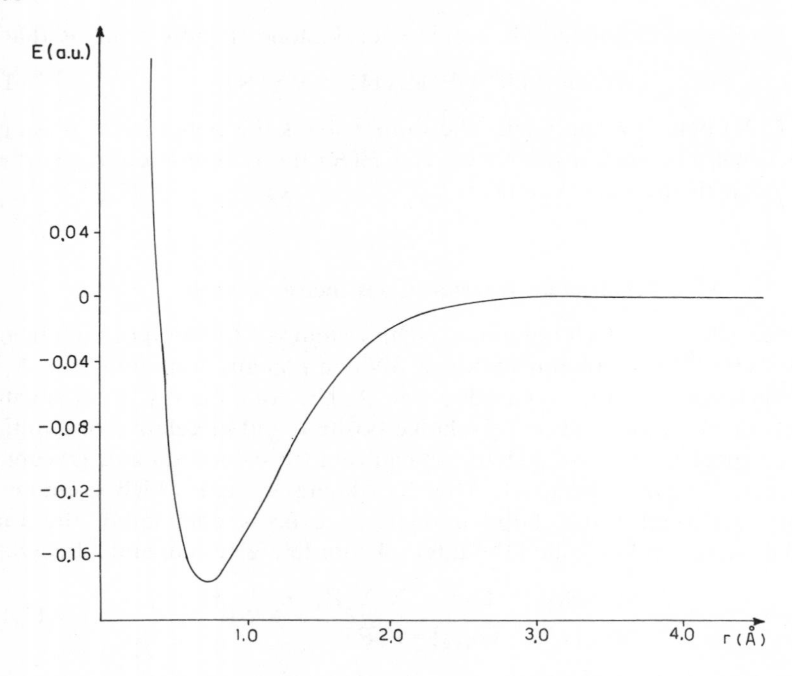

Figure 7.1 The dissociation curve of the H_2 molecule

potential energy surfaces for a few molecular systems. In the case of H_2 (see Figure 7.1),[1] as the potential energy depends only on one variable, the distance H—H, we talk about a potential energy curve. In other cases, as the number of variables is greater than two it is more exact to use the term the *potential energy hypersurface.* For a triatomic chemical system (e.g. DHF in Figure 7.2), we have three internal coordinates. Representation can be made by isoenergy surfaces in the three-dimensional internal coordinate space or by isoenergy curves, fixing one internal parameter at some value (see Figure 7.2).

In larger cases, as the number of variables is at least equal to six, the representation becomes more complicated. We often have to make some cuttings in the potential energy hypersurface (see Figure 7.3).[2]

7.2 THE CANONICAL POTENTIAL ENERGY SURFACE ANALYSIS

7.2.1 The Stationary Points

The potential energy surface (E) may be considered as a function of k internal coordinates ($\mathbf{s}' = \{s_1 \cdots s_k\}$). As long as the BO approximation is

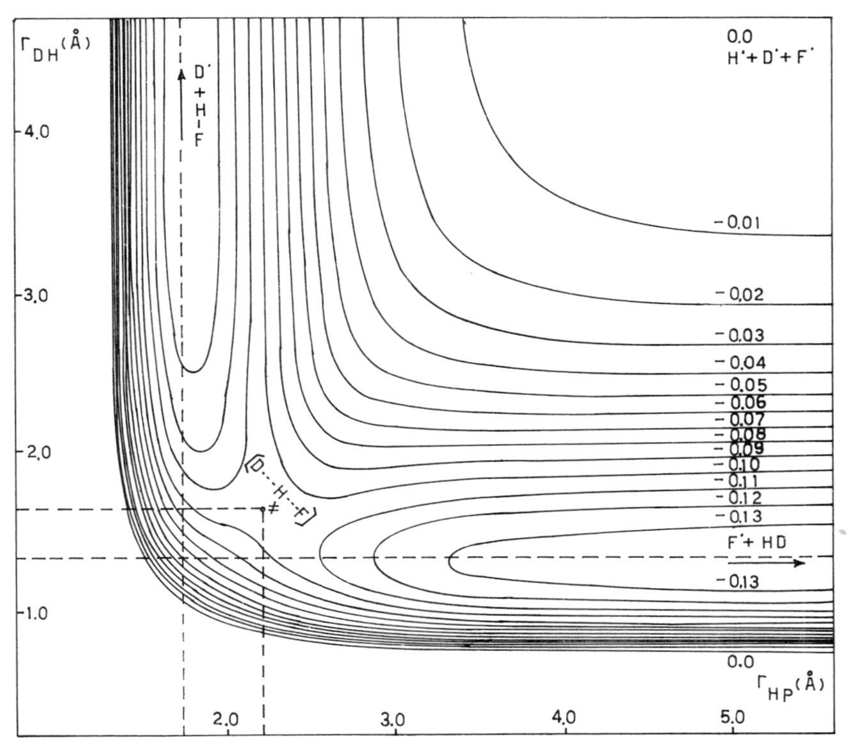

Figure 7.2 The potential energy surface of D + HF supersystem for the collinear abstraction reaction: $D + HF \rightarrow DH + F$

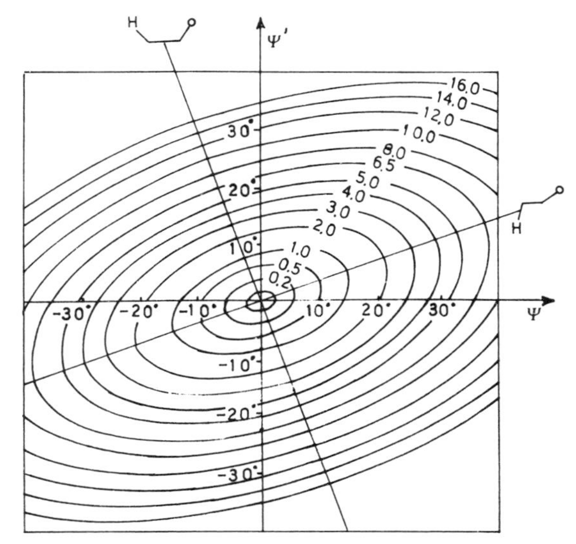

Figure 7.3 Two-dimensional view of the angular bending of fulminic acid (HCNO): $\Psi = H\hat{C}N$ and $\Psi' = C\hat{N}O$. Reprinted with permission of Smith and Jones, copyright American Chemical Society (1982)

valuable and the internal frame is continuous, the potential function must be continuous and continuously differentiable. At any point X of coordinates $\mathbf{s}_X$ we may write:

$$E_X = E(s_X) \tag{7.15a}$$

$$\mathbf{g}_X = (\boldsymbol{\nabla} E)_{s=s_X} = \left\{ \left(\frac{\partial E}{\partial s_i} \right)_{s=s_X} \right\} \quad \forall i = 1 \text{ to } k \tag{7.15b}$$

$$\mathbf{H}_X = (\boldsymbol{\nabla}\boldsymbol{\nabla}' E)_{s=s_X} = \left\{ \left(\frac{\partial^2 E}{\partial s_i\, \partial s_j} \right)_{s=s_X} \right\} \quad \forall i, j = 1 \text{ to } k \tag{7.15c}$$

where $\boldsymbol{\nabla}$ stands for the gradient operator and the prime superscript stands for the transposition operation, and $\mathbf{g}$ is the column vector of the first derivatives of E with respect to the internal coordinates. Then $-\mathbf{g}$ is nothing more than the forces acting on the nuclei. Similarly, $\mathbf{H}$ is a square symmetrical matrix (s_i and s_j being permutable in 7.15c) also called the force constant matrix.

Let us now develop the potential function around any point O of coordinates s_O. The displacement vector $\mathbf{S}_O$ will measure the distance from O to an other point X in the k-dimensional internal coordinates space:

$$\mathbf{S}_O = \mathbf{s}_X - \mathbf{s}_O \tag{7.16}$$

The Taylor expansion around O may be written as:

$$E(S_O) = E_O + \mathbf{g}_O' \mathbf{S}_O + \tfrac{1}{2} \mathbf{S}_O' \mathbf{H}_O \mathbf{S}_O + \cdots \tag{7.17}$$

We limit the development up to the second order. This approximation is fairly correct as long as the displacement from O remains small. This is expressed by the validity radius $\boldsymbol{r}$:

$$(\mathbf{SS})^{1/2} < \boldsymbol{r} \tag{7.18}$$

Some points (Σ) of the potential energy surface may present a gradient norm equal to zero:

$$\gamma_\Sigma = (\mathbf{g}_\Sigma' \mathbf{g}_\Sigma)^{1/2} = 0$$

This means that all first derivatives with respect to the internal coordinates are vanishing. Such points are named stationary points (Σ); they correspond to more or less stable molecular structures. Indeed, the energy gradient represents the internal force acting on the molecular structure; it disappears if $\mathbf{g} = \mathbf{0}$. Relative to the stationary point, the Taylor development (7.17) becomes:

$$\Delta E(\mathbf{S}_\Sigma) = E(\mathbf{S}_\Sigma) - E_\Sigma = \tfrac{1}{2} \mathbf{S}_\Sigma' \mathbf{H}_\Sigma \mathbf{S}_\Sigma + \cdots \tag{7.19}$$

where $\mathbf{S}_\Sigma$ stands for the displacement from the stationary point. The energy variations around such a point depend only on the second derivative matrix $\mathbf{H}$. This matrix is closely related to the nuclear vibration motions (see the next section). If we rotate our coordinate frame in such a way that $\mathbf{H}$

becomes diagonal (**h**), the energy expression (7.19) takes a simpler form. To do this we need a unitary matrix **U** such that:

$$\mathbf{HU} = \mathbf{Uh} \tag{7.20}$$

with:

$$\mathbf{U'U} = \mathbf{E} \qquad \text{(unitary condition)}$$

The last constraint guarantees preservation of the distances under rotation. The relation between the old and new coordinate systems (S and V) is given by:

$$\mathbf{S}_\Sigma = \mathbf{UV} \qquad \text{and} \qquad \mathbf{V} = \mathbf{U'S}_\Sigma \tag{7.21}$$

Relation (7.19) becomes:

$$\Delta E = \tfrac{1}{2}\mathbf{V'U'HUV} = \tfrac{1}{2}\mathbf{V'hV} \tag{7.22a}$$

With v_i, the ith component of the column vector $\mathbf{V}$, and h_i, the ith diagonal term of the matrix **h**:

$$\Delta E = \frac{1}{2}\sum_{i=1}^{k} v_i^2 h_i \tag{7.22b}$$

In the new coordinate system the sign of ΔE depends only on the signs of the eigenvalues of **H**.

7.2.2. The Equilibrium Structures

If the main diagonal of **h** contains only positive values ΔE must always become positive for any displacement from the stationary point. Such a structure is an absolute minimum and corresponds to an equilibrium geometry for the molecule of interest.

As an example we shall give the matrices computed for H_2O using a non-sophisticated *ab initio* calculation[3] with the 6-31G basis set of Pople. We shall use as the internal coordinates the three following parameters:

d_1: the OH_1 bond length

d_2: the OH_2 bond length

θ: the H_1OH_2 bond angle

If we choose for d_1 and d_2 a value of 0.9497 Å and for θ a value of 1.94605 rad, then the **g** vector vanishes (we shall see later how to arrive at such a conclusion). At this point the force constant matrix has the form:

$$\mathbf{H} = \begin{array}{c} d_1 \\ d_2 \\ \theta \end{array}\begin{pmatrix} 2.1155 & \text{Symmetric} & \\ -0.0436 & 2.1155 & \\ 0.0547 & 0.0547 & 0.1693 \end{pmatrix}$$
$$\qquad\qquad d_1 \qquad\quad d_2 \qquad\quad \theta$$

The diagonalization of **H** leads to the next eigenvectors:

$$\mathbf{U}=\begin{pmatrix} 0.707 & 0.706 & -0.030 \\ -0.707 & 0.706 & -0.030 \\ 0.000 & 0.043 & 0.999 \end{pmatrix}$$

the associated eigenvalue matrix being:

$$\mathbf{h}=\begin{pmatrix} 2.159 & & \\ 0.000 & 2.075 & \\ 0.000 & 0.000 & 0.166 \end{pmatrix}$$

All eigenvalues being positive, such a structure corresponds well to an equilibrium geometry for the water molecule.

In Figure 7.4 we show the mapping of the potential energy surface of water in two perpendicular planes. In a three-dimensional parameter space the isoenergy surface are ellipsoids centered on the equilibrium geometry (Figure 7.5).

If we now introduce the concept of supermolecule which appears when we study chemical systems containing more than one molecule, we must extend the conditions of equilibrium structure previously introduced. In such a case the requirement on **H** eigenvalues becomes:

$$h_i \geqslant 0 \qquad \forall i \mid 1 \leqslant i \leqslant k \tag{7.23}$$

This requirement, of course, completes the stationarity condition $g=0$ and means that in one direction in the parameter space the rotation or the translation of a part of the system relative to the rest of the supermolecule is completely free (it does not require any energy increase). For example, in the DHF supermolecule, when the separation distance between the deuterium atom and the HF molecule is infinite, two such directions exist in the parameter space and correspond to the free rotation of the HF molecule in the supermolecular plane (a) and the translation of HF relative to the deuterium atom (b).

(a) D - - - - R = ∞ - - - - H–F (rotation α)

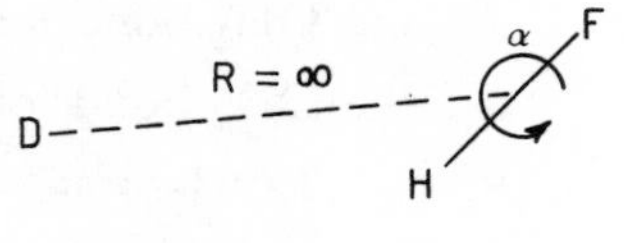

(b) D - - - - R = ∞ - - - - H–F (translation)

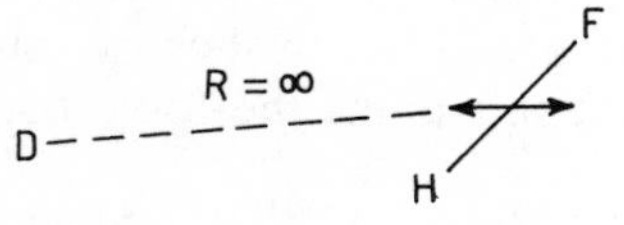

In Figure 7.6, we give the isoenergy plane for such a system in a three-dimensional mapping. If infinite distance between two subsystems does not exist such a situation has no physical meaning.

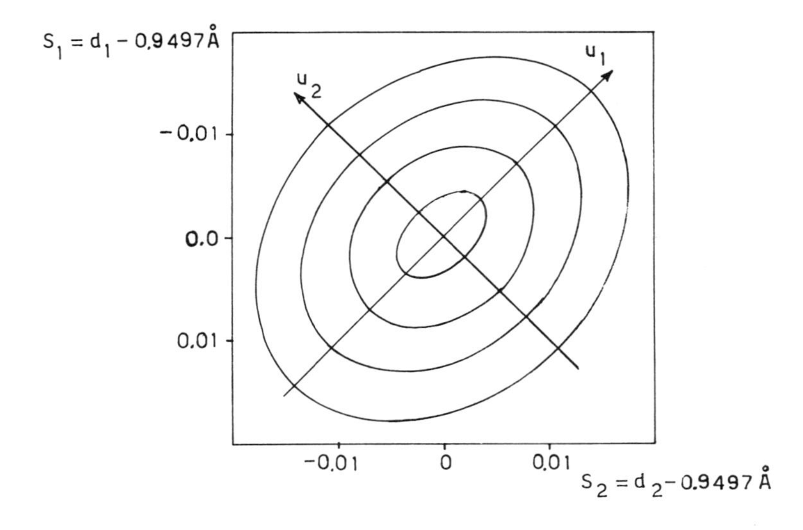

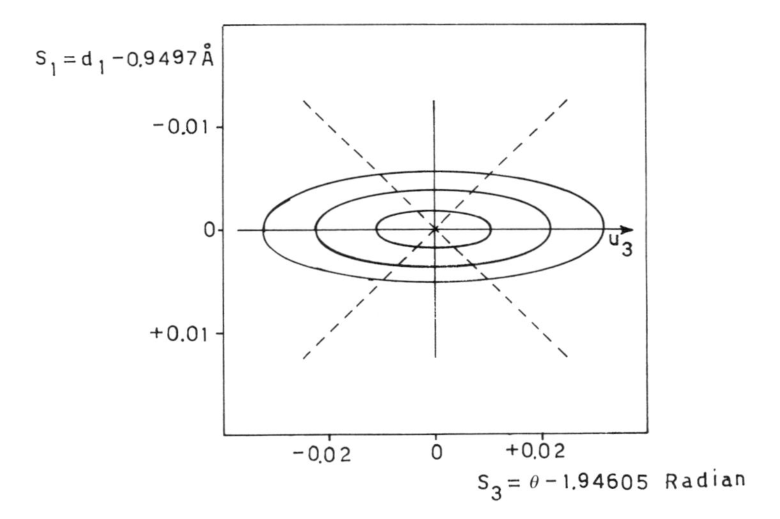

Figure 7.4 The potential energy hypersurface of the water molecule in a two-dimensional mapping

7.2.3 The Transition Structures

The last stationary point which is of particular interest for chemical systems is the *transition point* (or the transition structure). We define this as the highest energy point on the lowest energy pathway connecting two minima respectively called *reactant* and *product* structures.[4] The necessary condi-

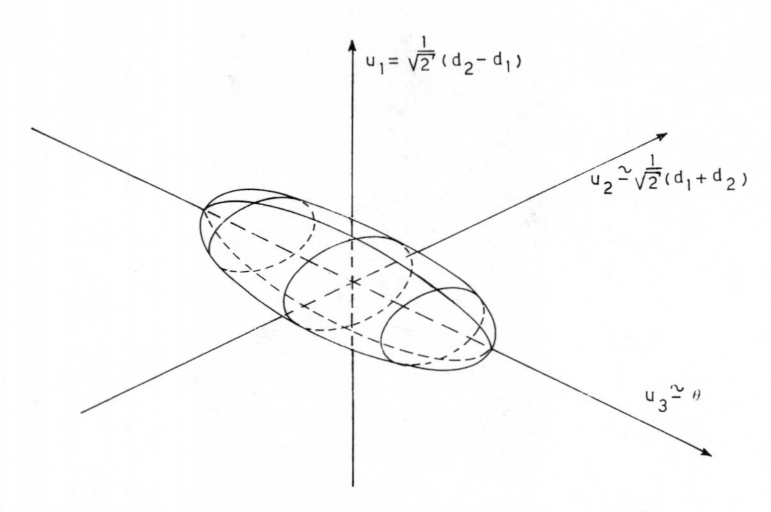

Figure 7.5 The potential energy hypersurface of the water molecule in a three-dimensional mapping

tions we want to satisfy are those of a first kind of minimax:

$$\begin{aligned} &\mathbf{g} = \mathbf{0} & & \\ &h_i > 0 & & \forall i \neq j \text{ such that } 1 \leqslant i \leqslant k \\ &h_j < 0 & & \text{in only one space direction} \end{aligned} \tag{7.24}$$

This means that the **H** matrix has only one negative eigenvalue, all others

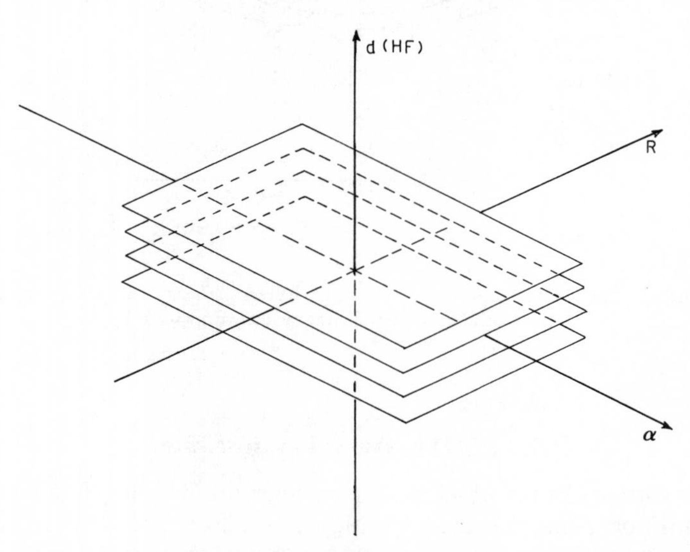

Figure 7.6 The potential energy hypersurface of the D+ HF system at infinite separation distance

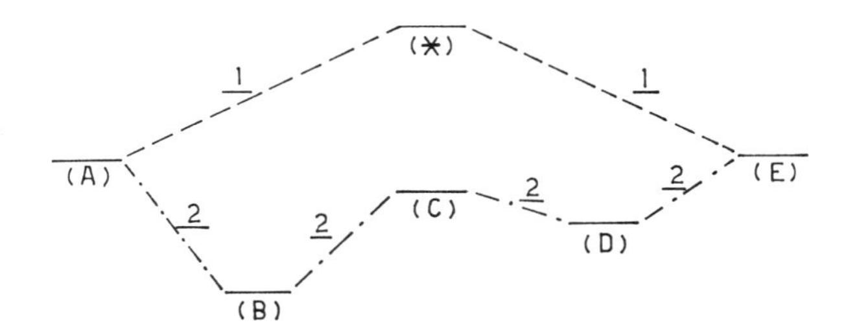

Figure 7.7 The transition point as a minimax of first order and not of higher order (at the point*)

being positive. This statement can easily be proved.[5] Let us suppose that the **h** matrix contains more than one negative eigenvalue for a particular point (* in Figure 7.7) on the energy hypersurface. We will now prove that there exists a pathway (A → B → C → D → E) of lower energy than those which pass through such a point (A → * → E). For this purpose, let us consider two nuclear motions. The first one connects the points of internal coordinates:

$$\mathbf{V}(\mathrm{A}) = (-d_1, 0, 0, \ldots, 0) \rightarrow \mathbf{V}(*) = (0, 0, 0, \ldots, 0) \rightarrow \mathbf{V}(\mathrm{E}) = (d_1, 0, 0, \ldots, 0)$$

and the second one:

$$\begin{aligned} \mathbf{V}(\mathrm{A}) = (-d_1, 0, 0, \ldots, 0) \rightarrow \mathbf{V}(B) = (-d_1, -d_2, 0, \ldots, 0) \rightarrow \mathbf{V}(\mathrm{c}) \\ = (0, -d_2, 0, \ldots, 0) \\ \rightarrow \mathbf{V}(\mathrm{D}) = (+d_1, -d_2, 0, \ldots, 0) \rightarrow \mathbf{V}(\mathrm{E}) \\ = (d_1, 0, 0, \ldots, 0) \end{aligned}$$

where d_1 and d_2 stand for infinitely small displacements. If the $\mathbf{h}_*$ matrix has the form:

$$\mathbf{h}_* = \begin{pmatrix} -\alpha^2 & & \\ 0 & -\beta^2 & \\ 0 & 0 & +\gamma^2 \end{pmatrix}$$

then the second way will require less energy to go from $V(\mathrm{A})$ to $V(\mathrm{E})$ than the first one and $\mathbf{V}(*)$ can never be the lowest possible pathway. In such a case the point labelled * cannot be considered as a transition state structure without contradicting our definition.

In Figure 7.8 we illustrate the local potential energy hypersurface at the transition point for the abstraction reaction:[6]

$$\mathrm{H} + \mathrm{HF} \rightarrow \mathrm{HH} + \mathrm{F}$$

We have chosen as internal coordinates the HH and HF bond lengths (respectively $d(\mathrm{HH})$ and $d(\mathrm{HF})$ in angstroms) and the HHF bond angle ($\theta(\mathrm{HHF})$ in radians):

$$\mathbf{s}' = \{d(\mathrm{HH})d(\mathrm{HF})\theta(\mathrm{HHF})\}$$

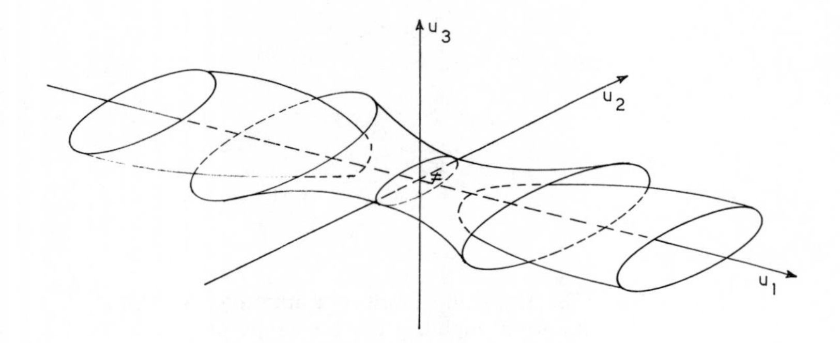

Figure 7.8 The potential energy surface at the transition structure for the reaction H+HF→HH+F. Each surface is an isoenergy surface

The transition structure (noted ≠) corresponds to:

$$\mathbf{s}'(\neq) = (0.912 \quad 1.144 \quad \pi)$$

The internal displacement vector **S** has the form:

$$\mathbf{S} = \begin{pmatrix} d(\mathrm{HH}) & -0.912 \\ d(\mathrm{HF}) & -1.144 \\ \theta(\mathrm{HHF}) & -\pi \end{pmatrix}$$

We obtain for the force constant matrix (in atomic units per angstrom squared, atomic units per angstrom-radian and atomic units per radian squared):

$$\mathbf{H} = \begin{pmatrix} 0.2269 & & \\ 0.6003 & -0.2234 & \\ 0.0003 & 0.0002 & 0.0112 \end{pmatrix}$$

Thus, the energy change for any small displacement from the stationary point ≠ becomes:

$$\Delta E(S) = \tfrac{1}{2}\mathbf{S}'\mathbf{H}\mathbf{S} \qquad \text{(in atomic units)}$$

The diagonalization of **H** leads to the following eigenvalues and eigenvectors:

$$h_1 = -0.6396; \quad u_1 = -0.570S_1 + 0.822S_2$$
$$h_2 = +0.0112; \quad u_2 = S_3$$
$$h_3 = +0.6428; \quad u_3 = 0.822S_1 + 0.570S_2$$

where h_1, h_2 and h_3 are the diagonal terms of **h** and u_1, u_2 and u_3 correspond to the column vectors of **U**.

In reactions containing more than three atoms, the representation of the potential energy hypersurface becomes more difficult; a two-dimensional map often remains the only possible way to illustrate the surface around a transition point. For internal hydrogen transfer between the two oxygen atoms in the anion:

we obtain the results reported in Figure 7.9.

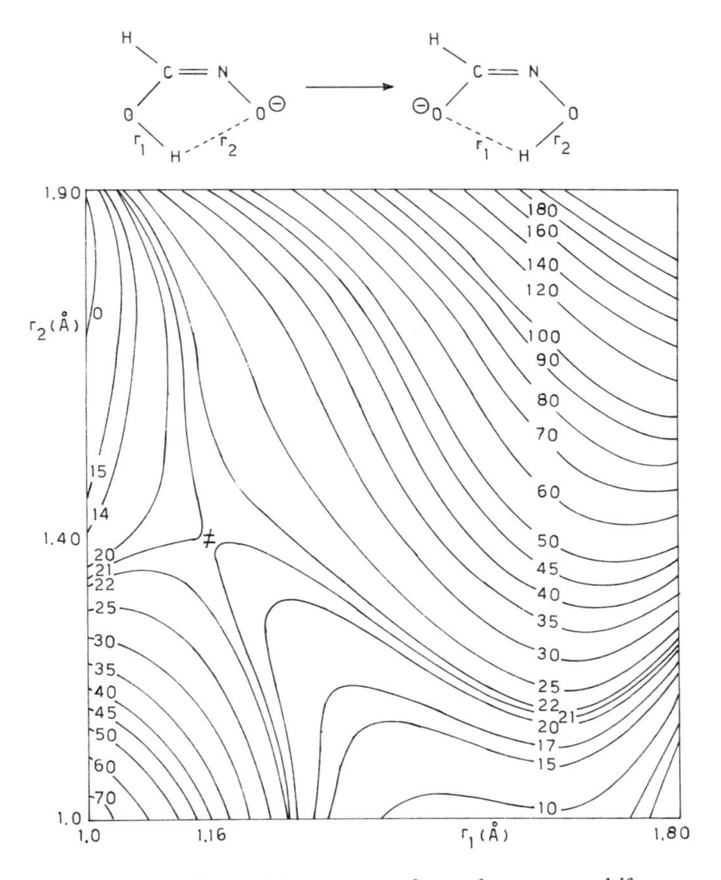

Figure 7.9 Potential energy surface of a proton shift reaction (the energy levels are in kilocalories per mole)

7.2.4 The Elementary Chemical Process

All nuclear motions can be represented on a potential energy surface by the displacement of one characteristic point. We shall distinguish different types of nuclear motions (in the internal coordinate frame):

(a) The first one is the *nuclear vibration*: in the subsystem of interest which contains two or more atoms at a finite distance from each other (normally between 0.7 and 1.9 Å for each bond) the nuclei are moving around their equilibrium position in the internal coordinate frame (one calls the subsystem a *molecule*).
(b) The second one is the *nuclear rotation*: if the system of interest contains two or more molecules or atoms at an infinite distance we find free rotations of any subsystem relative to the others. Such a motion does not change the potential energy.
(c) The third one is the *non-reactive translational process*: when the separation distance between two subsystems is appreciably decreasing from an initial infinite value, the process becomes non-reactive if after some time the system goes back to its initial position and does not fall into another absolute minimum of the potential energy surface.
(d) Lastly we find the *reactive process* which appears each time the characteristic point representing the supersystem on the potential energy surface goes from one minimum to another one. If only the initial and the final structures correspond to equilibrium geometries, such a reaction will be called an *elementary process* or *single-step reaction.*

In an elementary process, the initial point (R) represents the structure of the reactants and the final point (P) corresponds to the structure of the products. We call the *reaction energy* (ΔE^{R}) the energy difference between the products and reactants:

$$\Delta E^{R} = E^{P} - E^{R} \tag{7.25}$$

There must exist a lower energy pathway which connects the reactants and the products. The corresponding curve on the potential energy surface will be called the *reaction pathway*. As long as at least one point on such a pathway has an energy higher than that of the reactants (E^{R}) and that of the products (E^{P}), the reaction pathway must pass through one and only one transition point. The associated energy ($E^{\neq}$) is then the highest energy on the reaction pathway. We call the *transition barrier* ($\Delta E^{\neq}$) the energy difference:

$$\Delta E^{\neq} = E^{\neq} - E^{R} \tag{7.26}$$

Finally, let us note a possible reaction classification based on the potential energy properties:

(a) *The isomerization process*: two absolute minima (all $h_i > 0$) correspond to two different structures (equivalent or not), one for reactants and the other for products, e.g. the Z–E isomerization:

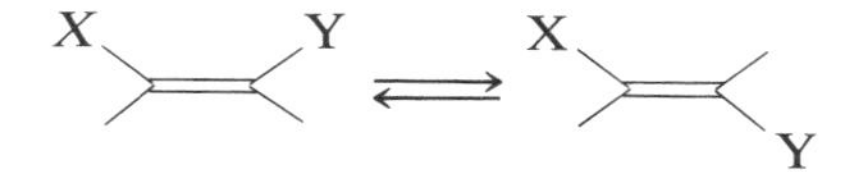

or the ammonia inversion:

$$\mathrm{N} \rightleftarrows \mathrm{N}$$

Thus we can have:

$$\mathrm{A} \rightleftarrows \mathrm{A} \qquad \text{or} \qquad \mathrm{A} \rightleftarrows \mathrm{B}$$

(b) *The dissociation and the addition processes*: one absolute minimum corresponds to the reactants in the case of dissociation or to the products in the case of addition; the other ones have at least one zero eigenvalue in the $\mathbf{h}$ matrix ($h_i \geqslant 0$), i.e.:

$$\mathrm{A} \rightarrow \mathrm{B} + \mathrm{C} \qquad \text{or} \qquad \mathrm{A} + \mathrm{B} \rightarrow \mathrm{C}$$

(c) *The rearrangement process*: in such a case the two minima have one or more zero eigenvalues in the $\mathbf{h}$ matrix ($h_i \geqslant 0$), i.e.:

$$\mathrm{A} + \mathrm{B} \rightleftarrows \mathrm{C} + \mathrm{D}$$

7.2.5 Reaction Pathway, Stationary Points and Coordinate Frame

The simplest way to find the reaction pathway is to select the steepest descent pathway starting from the transition point and going down respectively to the reactants and to the products (most often this way corresponds to a bottom valley but this is not an overall requirement). In practice, at the transition point one follows the eigendirection of the potential energy surface which corresponds to the negative curvature. When the potential gradient becomes non-vanishing one follows it down to the reactants and to the products. Nevertheless, such a procedure is of limited interest because the reaction pathway so generated depends on the internal coordinate frame in use.[7] Indeed, let us take two coordinate systems $\mathbf{s} = \{s_i\}$ and $\mathbf{v} = \{v_i\}$ related to each other by the transformation:

$$\mathbf{s} = \mathbf{P}\mathbf{v} \tag{7.27}$$

We can compute the first derivative vector with respect to the internal coordinates s or v anywhere:

$$\begin{aligned} \mathbf{g}_s &= \nabla_s E(s) = \left\{\frac{\partial E(s)}{\partial s_i}\right\} \\ \mathbf{g}_v &= \nabla_v E(v) = \left\{\frac{\partial E(v)}{\partial v_i}\right\} \end{aligned} \tag{7.28}$$

According to the chain rule the gradients $\mathbf{g}_s$ and $\mathbf{g}_v$ are related by:

$$\mathbf{g}_v = \mathbf{P}'\mathbf{g}_s \tag{7.29}$$

On the other hand, the gradient (i.e. $\mathbf{g}_s$) is a vector and must be transformed in $\mathbf{v}$ coordinates as a vector. This means the direction of $\mathbf{g}_s$ in v space (noted $\mathbf{d}_s^{(v)}$) is given by:

$$\mathbf{d}_s^{(v)} = \mathbf{P}^{-1}\mathbf{g}_s \tag{7.30}$$

Comparing the two last equations (7.29 and 7.30) we find that $\mathbf{d}_s^{(v)}$ and $\mathbf{g}_v$ coincide only in two cases:

(a) If $\mathbf{P}' = \mathbf{P}^{-1}$; this means if s and v spaces are related by an unitary transformation.
(b) If $\mathbf{g}_s = \mathbf{0}$, for any kind of $\mathbf{P}$ matrix, $\mathbf{g}_v = \mathbf{0}$ too; this means the stationary point location is independent of the coordinate frame in use.

Illustration of the non-intrinsic character of the reaction pathway defined as a steepest descent way is given in Figure 7.10, for the D+HF reaction. Let us use the two coordinates frames:

$$\mathbf{s}' = \{r_{\mathrm{DH}}, r_{\mathrm{HF}}, \mathrm{D\hat{H}F}\}$$
$$\mathbf{v}' = \{z, Z, \mathrm{D\hat{H}F}\}$$

with (see Kuntz[8]):

$$\mathbf{P} = \begin{pmatrix} \dfrac{-\cos\xi}{\sin\xi} & 1 & 0 \\ \dfrac{1}{\beta\sin\varepsilon} & 0 & 0 \\ 0 & 0 & 1 \end{pmatrix}$$

where:

$$\xi = \mathrm{arc}\ tg\left[\frac{m_{\mathrm{H}}(m_{\mathrm{H}} + m_{\mathrm{D}} + m_{\mathrm{F}})}{m_{\mathrm{D}}m_{\mathrm{F}}}\right]^{1/2}$$

and

$$\beta = \left\{\frac{(m_{\mathrm{D}} + m_{\mathrm{H}})m_{\mathrm{F}}}{(m_{\mathrm{F}} + m_{\mathrm{H}})m_{\mathrm{D}}}\right\}^{1/2}$$

We observe that in both cases we keep the stationary point location but elsewhere the pathway is different.

The second derivative matrix becomes also sensitive to the change of coordinate frame. The force constant matrix supports the next relation by applying the chain rule:

$$\mathbf{H}_v = \mathbf{P}'\mathbf{H}_s\mathbf{P} \tag{7.31}$$

According to the tensor transformation rules we can also write:

$$\mathbf{n}_s^{(v)} = \mathbf{P}^{-1}\mathbf{H}_s\mathbf{P} \tag{7.32}$$

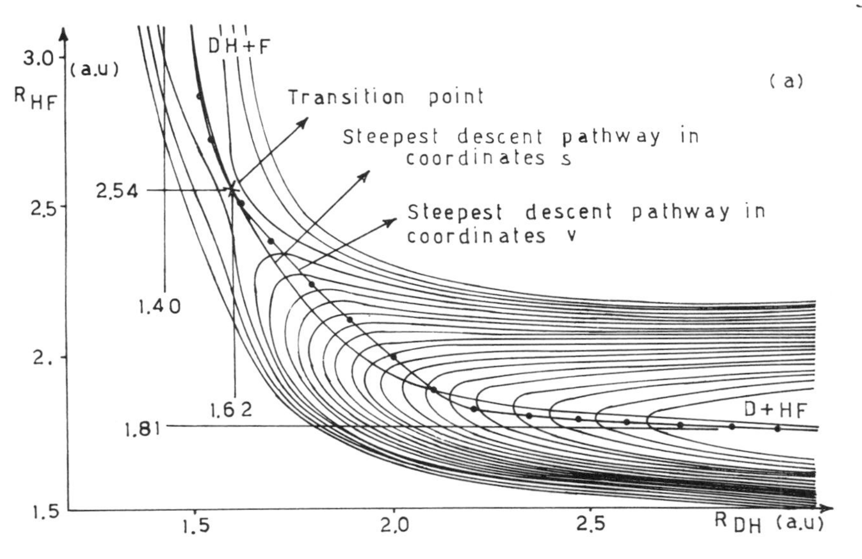

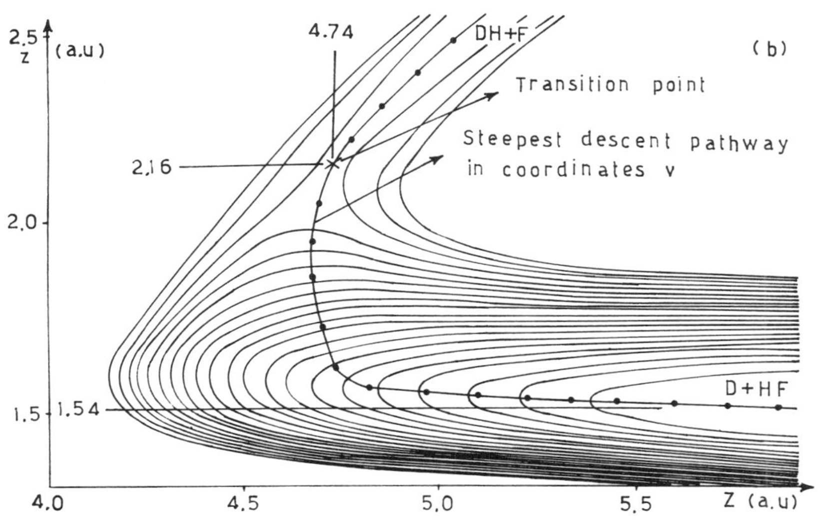

Figure 7.10 Reaction pathway in the D+HF reaction for two different reaction coordinate systems. Reproduced by permission of Springer-Verlag (Heidelberg)

This shows that $\mathbf{n}_s^{(v)}$ equals $\mathbf{H}_v$ only if $\mathbf{PP}' = \mathbf{E}$. Nevertheless, the sign of the $\mathbf{H}$ matrix determinant does not depend on the nature of the selected coordinate system (we assume that det $(\mathbf{P}) \neq 0$):[9]

$$\det(\mathbf{H}_v) = \det(\mathbf{H}_s)[\det(\mathbf{P})]^2 \tag{7.33}$$

As long as $\mathbf{H}_s$ does not have zero eigenvalues, the sign of the determinants of $\mathbf{H}_s$ and $\mathbf{H}_v$ must be identical. Moreover, we always regard $\mathbf{v}$ as having evolved continuously from $\mathbf{s}$. Then the number of negative or positive eigenvalues of $\mathbf{H}_s$ and $\mathbf{H}_v$ must remain constant; this means that the nature of a stationary point is invariant under coordinate transformation. The existence of zero eigenvalues of the $\mathbf{H}$ matrix has been excluded by some authors[9,10] for interacting atoms or molecules. In the asymptotic region of dissociation such an accident is nevertheless observed. In this case the $\mathbf{H}$ matrix can be partitioned and the last discussion remains valid for any part of $\mathbf{H}$ which does not correspond to zero eigenvalues.

In Section 8.4.3 we shall define a procedure leading to an intrinsic reaction pathway in the internal coordinate frame based on dynamical considerations.

7.3 FIRST AND SECOND DERIVATIVES

7.3.1 The Analytical Derivatives

7.3.1.1 The analytical first derivatives

Let us suppose we have an electronic wave function Ψ which depends on the nuclear cartesian coordinates $\mathbf{R}(\mathbf{R}=\{\mathbf{R}_1 \cdots \mathbf{R}_N\}=\{X_1 \cdots X_{3N}\})$ and on the SCF variational parameters $\mathbf{C}$ ($\mathbf{C}=\{\mathbf{C}_1 \cdots \mathbf{C}_p\}$). The variational parameters have been chosen in such a way as to minimize the total energy E:

$$\left(\frac{\partial E}{\partial \mathbf{C}_j}\right)=0 \qquad \forall j=1 \text{ to } p$$

This condition enables us to write:

$$\left(\frac{\mathrm{d}E}{\mathrm{d}X_i}\right)=\left(\frac{\partial E}{\partial X_i}\right) \qquad \forall i=1 \text{ to } 3N$$

even if $\mathbf{C}$ depends on the nuclear coordinates $\mathbf{X}$. We do not know how the variational parameters ($\mathbf{C}$) are related to the nuclear positions ($\mathbf{X}$) for computing the first-order derivatives of the potential energy surface. According to the hermiticity of the hamiltonian H, we write:[11]

$$f_i=-\frac{\mathrm{d}E}{\mathrm{d}X_i}=-\frac{\partial}{\partial X_i}\langle\Psi|\,H\,|\Psi\rangle=-\langle\Psi|\,\frac{\partial H}{\partial X_i}\,|\Psi\rangle-2\left\langle\frac{\partial \Psi}{\partial X_i}\right|H\left|\Psi\right\rangle \tag{7.34}$$

The first term describes the force applied to the nuclei by the electronic distribution (ρ). It is called the *Hellman–Feynman force* (f_i(HFF)). The differential operator $\partial H/\partial X_i$ only depends on the potential terms: nuclei–

nuclei ($V_{NN}(R)$) and nuclei–electrons ($V_{NE}(R, r)$):

$$\begin{aligned}\frac{\partial H}{\partial X_i} &= \frac{\partial}{\partial X_i}[T_E(r) + V_{NN}(R) + V_{NE}(R, r) + V_{EE}(r)] \\ &= \frac{\partial}{\partial X_i} V_{NN}(R) + \frac{\partial}{\partial X_i} V_{NE}(R, r)\end{aligned} \quad (7.35)$$

if $\mathbf{R} = \{X_1 \cdots X_i \cdots X_{3N}\}$ represents the nuclear coordinates and $\mathbf{r} = \{x_1 \cdots x_j \cdots x_{3n}\}$ represents the electron coordinates. As

$$V_{NN}(R) = \sum_{j<k=1}^{N} \sum^{N} \frac{Z_j Z_k}{|\mathbf{R}_j - \mathbf{R}_k|}$$

and

$$V_{NE}(R, r) = -\sum_{j=1}^{N} \sum_{k=1}^{n} \frac{Z_j}{|\mathbf{R}_j - \mathbf{r}_k|}$$

we find that:

$$\frac{\partial H}{\partial X_i} = \frac{\partial H}{\partial X_j(k)} = \sum_{\substack{l=1 \\ \neq k}}^{N} \frac{Z_l Z_k [X_j(l) - X_j(k)]}{|\mathbf{R}_l - \mathbf{R}_k|^3} - \sum_{l=1}^{n} \frac{Z_k [x_j(l) - X_j(k)]}{|\mathbf{r}_l - \mathbf{R}_k|^3} \quad (7.36)$$

where X_i stands for the jth cartesian coordinate ($1 \leqslant j \leqslant 3$) of the kth atom:

$$(1 \leqslant k \leqslant N) \text{ or } X_j(k) \quad \text{and} \quad |\mathbf{R}_l - \mathbf{R}_k| = \left\{\sum_{j=1}^{3} [X_j(l) - X_j(k)]^2\right\}^{1/2}$$

The Hellmann–Feynman force corresponds to the average value of a monoelectronic operator. It contains respectively one nuclei–nuclei and one nuclei–electrons term as shown in (7.36). The last part of f_i in (7.34) is called the *wave function force* (f_i(WFF)) and has a zero contribution for an exact function. Nevertheless, in practice we are using approximate wave functions and need to evaluate this wave function contribution as long as the total energy depends on the selected basis set (i.e. when we do not use exact Hartree–Fock wave functions). As the atomic basis set depends on the nuclear coordinates ($\chi = f(R)$) we need to compute a relaxation term; this means we consider that the atomic basis set may be moved at the same time as the nuclei and so modify the force applied to the nuclei. In order to find an explicit expression for f_i(WFF), let us differentiate the usual form of the electronic energy in the LCAO–SCF–MO frame[12] (in the closed-shell case):

$$E_e = 2 \sum_{jpq} C_{jp} C_{jq} \langle p | h^N | q \rangle + \sum_{jk} \sum_{pqrs} C_{jp} C_{jq} C_{kr} C_{ks} (2\langle pq \mid rs \rangle - \langle pr \mid qs \rangle) \quad (7.37)$$

where C_{jp} is the pth expansion coefficient of the jth molecular orbital ($\varphi_j = \sum_p C_{jp} \chi_p$), h^N stands for $T_E + V_{NE}$ and $\langle p | h^N | q \rangle$ and $\langle pq \mid rs \rangle$ are the mono- and bielectronic atomic integrals. If we note:

$$C_{jp}^{X_i} \quad \text{for} \quad \frac{\partial C_{jp}}{\partial X_i} \tag{7.38a}$$

$$\langle p|\, h^{\mathrm{N}}\, |q\rangle^{X_i'} \quad \text{or} \quad (h_{pq}^{\mathrm{N}})^{X_i'} \quad \text{for} \quad \left\langle \frac{\partial \chi_p}{\partial X_i} \right| h^{\mathrm{N}} \left| \chi_q \right\rangle + \left\langle \chi_p \right| h^{\mathrm{N}} \left| \frac{\partial \chi_q}{\partial X_i} \right\rangle \tag{7.38b}$$

and

$$\langle pq \mid rs\rangle^{X_i} \quad \text{for} \quad \left\langle \frac{\partial \chi_p}{\partial X_i}\chi_q \,\middle|\, \chi_r\chi_s \right\rangle + \left\langle \chi_p \frac{\partial \chi_q}{\partial X_i} \,\middle|\, \chi_r\chi_s \right\rangle + \left\langle \chi_p\chi_q \,\middle|\, \frac{\partial \chi_r}{\partial X_i}\chi_s \right\rangle + \left\langle \chi_p\chi_q \,\middle|\, \chi_r \frac{\partial \chi_s}{\partial X_i} \right\rangle \tag{7.38c}$$

Then the wave function force becomes:

$$\begin{aligned} f_i(\mathrm{WFF}) &= -2\left\langle \frac{\partial \Psi}{\partial X_i} \right| H \left| \Psi \right\rangle \\ &= -\Big[2 \sum_j \sum_{pq} C_{jp}C_{jq}\langle p|\, h^{\mathrm{N}}\, |q\rangle^{X_i'} \\ &\quad + \sum_{jk} \sum_{pqrs} C_{jp}C_{jq}C_{kr}C_{ks}(2\langle pq \mid rs\rangle^{X_i} - \langle pr \mid qs\rangle^{X_i}) \\ &\quad + 4 \sum_j \sum_{pq} C_{jp}^{X_i}C_{jq}\langle p|\, h^{\mathrm{N}}\, |q\rangle \\ &\quad + 4 \sum_{jk} \sum_{pqrs} C_{jp}^{X_i}C_{jq}C_{kr}C_{ks}(2\langle pq \mid rs\rangle - \langle pr \mid qs\rangle) \Big] \end{aligned} \tag{7.39}$$

or if:

$$\sum_j C_{jp}C_{jq} = D_{pq}$$

then:

$$\begin{aligned} &\sum_j \sum_{pq} C_{jp}C_{jq}\Big[\langle p|\, h^{\mathrm{N}}\, |q\rangle + \sum_k \sum_{rs} C_{kr}C_{ks}(2\langle pq \mid rs\rangle - \langle pr \mid qs\rangle)\Big] \\ &\quad = \sum_{pq} D_{pq}\langle p|\, h^{\mathrm{F}}\, |q\rangle = \sum_{pq} D_{pq}h_{pq}^{\mathrm{F}} \end{aligned}$$

and if:

$$(h_{pq}^{\mathrm{F}})^{X_i'} \quad \text{stands for} \quad \langle p|\, h^{\mathrm{N}}\, |q\rangle^{X_i'} + \sum_k \sum_{rs} C_{kr}C_{ks}(2\langle pq \mid rs\rangle^{X_i} - \langle pr \mid qs\rangle^{X_i}) \tag{7.40}$$

from (7.40) the wave function forces (7.39) may be expressed as:

$$f_i(\mathrm{WFF}) = -\Big\{ \sum_{pq} D_{pq}[(h_{pq}^{\mathrm{N}})^{X_i'} + (h_{pq}^{\mathrm{F}})^{X_i'}] + 4 \sum_j \sum_{pq} C_{jp}^{X_i}C_{jq}(h_{pq}^{\mathrm{F}}) \Big\} \tag{7.41}$$

As the molecular orbitals (φ_j) are eigenfunctions of the Fock operator (h^{F}) and as they are normalized:

$$h^{\mathrm{F}}\varphi_j = \varepsilon_j \varphi_j$$

with:

$$\langle \varphi_j \mid \varphi_j \rangle = 1 \qquad \text{or} \qquad \sum_{pq} C_{jp} C_{jq} S_{pq} = 1$$

Then:

$$\sum_j \sum_{pq} C_{jp}^{X_i} C_{jq} h_{pq}^{\mathrm{F}} = \sum_j \sum_{pq} \varepsilon_j C_{jp}^{X_i} C_{jq} S_{pq} \tag{7.42a}$$

and

$$2 \sum_{pq} C_{pq}^{X_i} C_{jq} S_{pq} + \sum_{pq} C_{jp} C_{jq} (S_{pq})^{X_i} = 0 \tag{7.42b}$$

where:

$$(S_{pq})^{X_i} = \left\langle \frac{\partial \chi_p}{\partial X_i} \,\middle|\, \chi_q \right\rangle + \left\langle \chi_p \,\middle|\, \frac{\partial \chi_q}{\partial X_i} \right\rangle$$

Thus:

$$\sum_j \sum_{pq} C_{jp}^{X_i} C_{jq} h_{pq}^{F} = -\frac{1}{2} \sum_j \sum_{pq} \varepsilon_j C_{jp} C_{jq} (S_{pq})^{X_i} \tag{7.43}$$

Introducing (7.43) in (7.41) we find that the terms $C_{jp}^{X_i}$ in the wave function force can disappear:

$$f_i(\mathrm{WFF}) = -\sum_{pq} D_{pq} [(h_{pq}^{\mathrm{N}})^{X_i'} + (h_{pq}^{\mathrm{F}})^{X_i'}] + 2 \sum_j \sum_{pq} \varepsilon_j C_{jp} C_{jq} (S_{pq})^{X_i} \tag{7.44}$$

The overall expression for the force becomes:

$$\begin{aligned} f_i = -\frac{\mathrm{d}E}{\mathrm{d}X_i} &= -2 \sum_{pq} [D_{pq} (h_{pq}^{\mathrm{N}})^{X_i} - \sum_j \varepsilon_j C_{jp} C_{jq} (S_{pq})^{X_i}] \\ &\quad - \sum_{pqrs} D_{pq} D_{rs} [2\langle pq \mid rs \rangle^{X_i} - \langle pr \mid qs \rangle^{X_i}] \\ &\quad + \sum_{l \neq k} \frac{Z_l Z_k [X_j(l) - X_j(k)]}{|\mathbf{R}_l - \mathbf{R}_k|^3} \end{aligned} \tag{7.45a}$$

where

$$(h_{pq}^{\mathrm{N}})^{X_i} = (h_{pq}^{\mathrm{N}})^{X_i'} + \langle p | \frac{\mathrm{d}h^{\mathrm{N}}}{\mathrm{d}X_i} | q \rangle$$

or

$$\begin{aligned} \left(\frac{\mathrm{d}E}{\mathrm{d}X_i}\right) = \Big\{ 2 \sum_{pq} [D_{pq} (h_{pq}^{\mathrm{N}})^{X_i} - \sum_j \varepsilon_j C_{jp} C_{jq} (S_{pq})^{X_i}] \\ + \sum_{pqrs} \langle pq \mid rs \rangle^{X_i} [D_{pr} D_{qs} - 2 D_{pq} D_{rs}] \Big\} \\ + \text{the purely nuclear term} \end{aligned} \tag{7.45b}$$

If N is the number of nuclei, we have to compute $3N$ such quantities $(\mathrm{d}E \mid \mathrm{d}X_i)$. This means we also need to compute the $3N$ corresponding first derivatives of the core (h^{N}), overlap (S) and bielectronic $(\langle pq \mid rs \rangle)$ integrals before any gradient evaluation. An analytical expression may also be found beyond the SCF level.[13]

7.3.1.2 The analytical second derivatives

The second-order analytical derivatives are available by differentiating once more the lastly defined equation for the first derivative vector (7.45b):

$$H_{ij} = \frac{\mathrm{d}}{\mathrm{d}X_j}\left(\frac{\mathrm{d}E}{\mathrm{d}X_i}\right)$$

Performing this derivation it follows that:[12,14]

$$\begin{aligned} H_{ij} = 2 \sum_{pq} \Big[& D_{pq}(h^{\mathrm{N}}_{pq})^{X_i X_j} + D^{X_j}_{pq}(h^{\mathrm{N}}_{pq})^{X_i} \\ & - (S_{pq})^{X_i X_j} \sum_k \varepsilon_k C_{kp} C_{kq} - (S_{pq})^{X_i} \sum_k C_{kq}(\varepsilon_k^{X_j} C_{kp} + 2\varepsilon_k C^{X_j}_{kq}) \Big] \\ & + \sum_{pqrs} [\langle pq \mid rs \rangle^{X_i X_j}(2D_{pr}D_{qs} - D_{pq}D_{rs}) \\ & + 2\langle pq \mid rs \rangle^{X_i}(2D^{X_j}_{pr}D_{qs} - D^{X_j}_{pq}D_{rs})] \\ & + \text{nuclei–nuclei contribution} \end{aligned} \tag{7.46}$$

where:

$$D^{X_i}_{pq} = \frac{\mathrm{d}}{\mathrm{d}X_i} \sum_k C_{kp} C_{kq} = \sum_k (C^{X_i}_{kp} C_{kq} + C_{kp} C^{X_i}_{kq})$$

The remaining problem is to find the derivatives of the expansion coefficients C_{kp} with respect to the nuclear displacements. In a general way, when a molecule is subjected to a small one-electron perturbation characterized by a parameter λ, we assume that for $\lambda = 0$ the solutions to the SCF equations are known:[15]

$$\mathbf{F}^{(0)}\mathbf{C}^{(0)} = \boldsymbol{\varepsilon}^{(0)}\mathbf{S}^{(0)}\mathbf{C}^{(0)} \tag{7.47}$$

the orthonormality condition being:

$$\mathbf{C}^{(0)\prime}\mathbf{S}^{(0)}\mathbf{C}^{(0)} = \mathbf{E}$$

For slightly different values of λ we obtain corrections to $\mathbf{C}^{(0)}$. The more general solution may be written as:

$$\mathbf{F}(\lambda)\mathbf{C}(\lambda) = \boldsymbol{\varepsilon}(\lambda)\mathbf{S}(\lambda)\mathbf{C}(\lambda) \tag{7.48}$$

with:

$$\mathbf{C}(\lambda)'\mathbf{S}(\lambda)\mathbf{C}(\lambda) = \mathbf{E}$$

Using the canonical molecular basis set instead of an atomic one we find for the preceding equation a simple form:

$$\mathbf{F}(\lambda)\mathbf{U}(\lambda)=\boldsymbol{\varepsilon}(\lambda)\mathbf{U}(\lambda) \tag{7.49}$$

with:

$$\mathbf{U}(\lambda)'\mathbf{O}(\lambda)\mathbf{U}(\lambda)=\mathbf{E}$$

where

$$\mathbf{F}(\lambda)=\mathbf{C}^{(0)\prime}\mathbf{F}(\lambda)\mathbf{C}^{(0)}$$
$$\mathbf{O}(\lambda)=\mathbf{C}^{(0)\prime}\mathbf{S}(\lambda)\mathbf{C}^{(0)}$$
$$\mathbf{U}(\lambda)=[\mathbf{C}^{(0)}]^{-1}\mathbf{C}(\lambda)$$

Let us now expand the matrices $\mathbf{F}(\lambda), \mathbf{O}(\lambda)$, (λ) and $\mathbf{U}(\lambda)$ as the Taylor series in λ:

$$\begin{aligned}\mathbf{F}(\lambda)&=\mathbf{F}^{(0)}+\lambda\mathbf{F}^{(1)}+\cdots\\ \mathbf{O}(\lambda)&=\mathbf{E}+\lambda\mathbf{O}^{(1)}+\cdots\\ \boldsymbol{\varepsilon}(\lambda)&=\boldsymbol{\varepsilon}^{(0)}+\lambda\boldsymbol{\varepsilon}^{(1)}+\cdots\\ \mathbf{U}(\lambda)&=\mathbf{U}^{(0)}+\lambda\mathbf{U}^{(1)}+\cdots\end{aligned} \tag{7.50}$$

Note that $\mathbf{U}^{(0)}$ is the identity matrix of the eigenvectors of $\mathbf{F}^{(0)}$ as this one is diagonal. Substituting these expansions and collecting together corresponding order in λ we have:

$$\mathbf{F}^{(0)}\mathbf{U}^{(0)}=\boldsymbol{\varepsilon}^{(0)}\mathbf{U}^{(0)} \tag{7.51a}$$
$$(\mathbf{F}^{(0)}-\boldsymbol{\varepsilon}^{(0)})\mathbf{U}^{(1)}=[\boldsymbol{\varepsilon}^{(0)}\mathbf{O}^{(1)}-(\mathbf{F}^{(1)}-\boldsymbol{\varepsilon}^{(1)}]\mathbf{U}^{(0)} \tag{7.51b}$$
$$\cdots$$

From the former equation, premultiplying it by $\mathbf{U}^{(0)\prime}$ we obtain, for any pair i, j:

$$\begin{aligned}(\varepsilon_i^{(0)}-\varepsilon_j^{(0)})\mathbf{U}_i^{(0)\prime}\mathbf{U}_j^{(1)}&=\varepsilon_j^{(0)}\mathbf{U}_i^{(0)\prime}\mathbf{O}^{(1)}\mathbf{U}_j^{(0)}-\mathbf{U}_i^{(0)\prime}\,\mathbf{F}^{(1)}\mathbf{U}_j^{(0)}+\varepsilon_j^{(1)}\,\delta_{ij}\\ (\varepsilon_i^{(0)}-\varepsilon_j^{(0)})U_{ij}^{(1)}&=\varepsilon_j^{(0)}O_{ij}^{(1)}-\mathscr{F}_{ij}^{(1)}+\varepsilon_j^{(1)}\,\delta_{ij}\end{aligned} \tag{7.52}$$

We find:

$$U_{ij}^{(1)}=\frac{\varepsilon_j^{(0)}O_{ij}^{(1)}-\mathscr{F}_{ij}^{(1)}}{\varepsilon_i^{(0)}-\varepsilon_j^{(0)}} \qquad \text{for } i\neq j \tag{7.53a}$$

$$\varepsilon_j^{(1)}=\mathscr{F}_{ij}^{(1)}-\varepsilon_j^{(0)}O_{ij}^{(1)} \qquad \text{for } i=j \tag{7.53b}$$

The elements $U_{jj}^{(1)}$ may be found from the orthonormality condition which gives for the first order in λ:

$$O_{ij}^{(1)}+U_i^{(1)\prime}\,U_j^{(0)}+U_i^{(0)\prime}\,U_j^{(1)}=0$$

or

$$O_{ij}^{(1)}+U_{ij}^{(1)\prime}+U_{ij}^{(1)}=0 \tag{7.54a}$$

Hence:

$$U_{jj}^{(1)}=-\tfrac{1}{2}O_{jj}^{(1)} \tag{7.54b}$$

In fact for perturbation calculations this is sufficient to determine those $U_{ij}^{(1)}$ elements between occupied and unoccupied orbitals. Lastly, the matrix elements such as $\mathscr{F}_{ij}^{(1)}$ may be obtained in the representation where $F^{(0)}$ is diagonal by differentiating $\mathcal{F}(\lambda)$ with respect to λ and computing it for $\lambda = 0$. Thus the problem may be solved iteratively. Finally, the terms $C_{ip}^{\mathbf{X}_i}$ are obtained from $C^{(0)}U^{(1)}$ since:

$$\mathbf{C}(\lambda) = \mathbf{C}^{(0)}\mathbf{U}(\lambda) \tag{7.55}$$

7.3.2 Numerical Derivatives, General Considerations

7.3.2.1 The fitting problem

Let us suppose we want to find the derivatives of a function (y) in the k-dimensional space of coordinates s ($\mathbf{s} = \{s_1 \cdots s_k\}$) up to the order q. This function may be the total energy or some of its derivatives. Let us also suppose we are unable to compute it for several points (m) of a domain of interest $\mathcal{D}$ so we dispose of a tabulated function. For our present purpose it is convenient to search for an analytical function $\hat{y}(s)$ which fits as well as possible the calculated values $y(s)$ within the domain $\mathcal{D}$. We choose an expression of the form:

$$y(s) = b_0 + \sum_{i=1}^{l} b_i f_i(s) \tag{7.56}$$

where $\{f_i(s)\}$ is a given set of expansion functions and $\{b_i\}$ stands for a set of unknown expansion coefficients. We would like to request that $\hat{y}(s)$ as well as its successive derivatives up to the order q are continuous over the domain $\mathcal{D}$. Moreover, $\hat{y}(s)$ and some of the functions $f_i(s)$ have to be q times differentiable.

The problem now becomes: 'If we dispose of m values of y, how do we find the l coefficients b_i such that $\hat{y}$ is everywhere in $\mathcal{D}$ as close as possible to y.' This problem of course can only be solved if m is greater than l. We would like also to check the fitting accuracy.

For numerical convenience we can normalize the quantities which are used. We replace y and f_i by Z and $\mathcal{F}_i$ such that:

$$Z_j = \frac{y(s_j) - \langle y \rangle}{\sigma_y} \tag{7.57a}$$

$$\mathcal{F}_i(j) = \frac{f_i(s_j) - \langle f_i \rangle}{\sigma_{f_i}} \quad \forall j = 1 \text{ to } m;\ i = 1 \text{ to } l \tag{7.57b}$$

with:

$$\langle y \rangle = \frac{1}{m} \sum_j y(s_j)$$

$$\sigma_y^2 = \frac{1}{m} \sum_j [y(s_j) - \langle y \rangle]^2$$

$$\langle f_i \rangle = \frac{1}{m} \sum_j f_i(s_j)$$

$$\sigma_{f_i}^2 = \frac{1}{m} \sum_j [f_i(s_j) - \langle f_i \rangle]^2$$

Note that Z and $\mathcal{F}_i$ have an average equal to zero and a standard deviation equal to one. Our fitting equation (7.56) becomes:

$$Z_j \simeq \hat{Z}_j = \sum_{i=1}^{l} \mathcal{B}_i \mathcal{F}_i(j) \tag{7.58}$$

There is no independent term to show the zero average on Z_j.

The connection between the unnormalized (7.56) and the normalized (7.58) expressions is given by the following equations:

$$b_i = \mathcal{B}_i \frac{\sigma_y}{\sigma_{f_i}} \tag{7.59a}$$

$$b_0 = \langle y \rangle - \sum_i \mathcal{B}_i \langle f_i \rangle \frac{\sigma_y}{\sigma_{f_i}} \tag{7.59b}$$

For convenience we prefer to use a matricial notation:

$$\mathbf{Z} \simeq \hat{\mathbf{Z}} = \mathbf{F}'\boldsymbol{B} \tag{7.60}$$

$$\mathbf{Z} = \begin{pmatrix} Z_1 \\ \vdots \\ Z_m \end{pmatrix}, \quad \hat{\mathbf{Z}} = \begin{pmatrix} \hat{Z}_1 \\ \vdots \\ \hat{Z}_m \end{pmatrix}, \quad \mathbf{B} = \begin{pmatrix} b_1 \\ \vdots \\ b_l \end{pmatrix}$$

and

$$\boldsymbol{F} = \begin{pmatrix} \mathcal{F}_1(1) & \cdots & \mathcal{F}_1(m) \\ \vdots & \ddots & \vdots \\ \mathcal{F}_1(1) & \cdots & \mathcal{F}_l(m) \end{pmatrix}$$

Even when we dispose of a correct expansion model, as the coefficients $\mathcal{B}$ are unknown, we can only find an estimated value ($\hat{\mathcal{B}}$) of such coefficients.

7.3.2.2 The least square fit

To estimate the $\hat{\mathcal{B}}_i$ terms as well as possible we use the least square fit method (also called the Gauss approximation). We shall choose the $\hat{\boldsymbol{B}}$ coefficients in such a way as to minimize the function:

$$\varepsilon^2 = (\mathbf{Z} - \hat{\mathbf{Z}})' \cdot \mathbf{W} \cdot (\mathbf{Z} - \hat{\mathbf{Z}}) \tag{7.61}$$

where $\mathbf{W}$ is a diagonal $m \times m$ matrix of the weighting factor which can be the unity matrix or not. We shall introduce it only to keep all the generality we need for our purpose. From (7.60) and (7.61) we write:

$$\begin{aligned} \varepsilon^2 &= \mathbf{Z}'\mathbf{W}\mathbf{Z} + \hat{\boldsymbol{B}}'\boldsymbol{F}\mathbf{W}\boldsymbol{F}'\hat{\boldsymbol{B}} - \hat{\boldsymbol{B}}'\boldsymbol{F}\mathbf{W}\mathbf{Z} - \mathbf{Z}'\mathbf{W}\boldsymbol{F}'\hat{\boldsymbol{B}} \\ &= \varepsilon_0^2 + \hat{\boldsymbol{B}}'\boldsymbol{A}\hat{\boldsymbol{B}} - \hat{\boldsymbol{B}}'\mathbf{R} - \mathbf{R}'\hat{\boldsymbol{B}} \end{aligned} \tag{7.62}$$

with:

$$\boldsymbol{A} = \boldsymbol{F}\mathbf{W}\boldsymbol{F}'$$

and

$$\boldsymbol{R} = \boldsymbol{F}\mathbf{W}\mathbf{Z}$$

The extremum requirement on ε^2 is met for

$$\left(\frac{\partial \varepsilon^2}{\partial \mathcal{B}_i}\right) = 0 \qquad \forall i = 1 \text{ to } l \tag{7.63}$$

Then:

$$\boldsymbol{R} - \boldsymbol{A}\hat{\boldsymbol{B}} = 0$$

or

$$\hat{\boldsymbol{B}} = \boldsymbol{A}^{-1}\boldsymbol{R} = (\boldsymbol{FWF'})^{-1}\boldsymbol{FWZ} \tag{7.64a}$$

Using the former defined relationship we can also write a similar expression in terms of unnormalized matrices. So the best estimator of $b(\hat{b})$ under Gauss approximation is:

$$\begin{aligned} \hat{\mathbf{b}} &= \mathbf{A}^{-1}\mathbf{R} \\ &= (\mathbf{FWF'})^{-1}(\mathbf{FWy}) \end{aligned} \tag{7.64b}$$

where $\mathbf{F}$, called the model matrix, has the form:

$$\mathbf{F} = \begin{pmatrix} 1 & 1 & & 1 \\ f_1(s_1) & f_1(s_2) & & f_1(s_m) \\ \vdots & \vdots & \ddots & \vdots \\ f_l(s_1) & f_l(s_2) & & f_l(s_m) \end{pmatrix} = [\mathbf{f}(s_1)\mathbf{f}(s_2) \cdots \mathbf{f}(s_m)]$$

The $\mathbf{A}^{-1}$ matrix is also called the variance covariance matrix. One important property of the $\hat{b}$ expression is to be an unbiased estimator of b:

$$\begin{aligned} \mathscr{E}[\hat{\mathbf{b}}] &= \mathscr{E}[(\mathbf{FWF'})^{-1}\mathbf{FWy}] \\ &= \mathscr{E}[(\mathbf{FWF'})^{-1}\mathbf{FWF'b}] \\ &= \mathbf{b} \end{aligned} \tag{7.65}$$

7.3.2.3 The model adequation: an *a posteriori* measurement

The regression being done, it may be interesting to check the quality of the model in use.

When m (the number of points) is greater than l (the number of coefficients) we could estimate the reproduction quality of our model by means of a Fischer test. This test enables us to compare two variances. If σ_y^2 is the variance of the function value and σ_{res}^2 is the residual variance:

$$\sigma_y^2 = \frac{1}{m}\sum_{j=1}^{m}(y_j - \langle y \rangle)^2$$

$$\sigma_{res}^2 = \frac{1}{m-1}\sum_{j=1}^{m}(y_j - \hat{y}_j)^2$$

then:

$$f = \frac{\sigma_y^2}{\sigma_{res}^2} \tag{7.66}$$

We accept the model at a confidence level p if:

$$f \geqslant F_p(m; m-1)$$

where F_p is the tabulated Fischer values with m and $m-1$ degrees of freedom. The multiple regression coefficient ρ^* is also used to evaluate the strength of the bond between the model and the reality:

$$\rho^* = \left(1 - \frac{1}{f}\right)^{1/2} \tag{7.67}$$

If the model reproduces as well as possible the correct $y(s)$ values then $\sigma^2_{\text{res}} = 0$ and ρ^* tends to be unity. Nevertheless, ρ^* can only be used as a discrete measurement of the fitting adequation, and never represents a criterion of interpolative accuracy of our analytical function anywhere in the experimental domain. *A posteriori*, the most guaranteed way of checking the representational model's accuracy is to use a set of m' testing points. Such points may be either randomly or suitably distributed around the point of interest. Defining the reproducibility variance ρ^2_{repr} as the variance between estimated and exact function values at the testing points, we use again the Fischer test:

$$f' = \frac{\sigma^2_y}{\sigma^2_{\text{repr}}} > F_p(m; m') \tag{7.68}$$

or a reproducibility coefficient (θ):

$$\theta = \sqrt{\left(1 - \frac{m'}{m}\frac{1}{f'}\right)} \tag{7.69}$$

which must tend to unity for an adequate model in the interpolative sense.

7.3.2.4 The model adequation: an *a priori* knowledge

The interpolative properties and the quality of the successive derivatives are closely bonded to the accuracy of the estimators $\hat{\mathbf{b}}$. The variance–covariance matrix provides such information:

$$\begin{aligned} \text{var}(\hat{b}) &= \mathscr{E}[(\hat{\mathbf{b}} - \mathbf{b})(\hat{\mathbf{b}} - \mathbf{b})'] \\ &= \mathscr{E}[(\mathbf{FWF}')^{-1}(\mathbf{FWF}')(\hat{\mathbf{b}} - \mathbf{b})(\hat{\mathbf{b}} - \mathbf{b})'(\mathbf{FWF}')(\mathbf{FWF}')^{-1}] \\ &= (\mathbf{FWF}')^{-1}\mathbf{FW}\mathscr{E}[(\hat{\mathbf{y}} - \mathbf{y})(\hat{\mathbf{y}} - \mathbf{y})']\mathbf{WF}'(\mathbf{FWF}')^{-1} \end{aligned} \tag{7.70}$$

Assuming the independence of errors we may write:

$$\mathscr{E}[(\hat{\mathbf{y}} - \mathbf{y})(\hat{\mathbf{y}} - \mathbf{y})'] = \mathbf{W}^{-1}\sigma^2_{\text{err}} \tag{7.71}$$

where σ^2_{err} stands for the error variance (on y) and $\mathbf{W}^{-1}$ stands for an $m \times m$ weighting matrix. Consequently, the variance–covariance matrix becomes:

$$\mathscr{E}[(\hat{\mathbf{b}} - \mathbf{b})(\hat{\mathbf{b}} - \mathbf{b})'] = (\mathbf{FWF}')^{-1}\sigma^2_{\text{err}} = \mathbf{A}^{-1}\sigma^2_{\text{err}} \tag{7.72}$$

This means that the precision of the regression coefficients depends only on the experimental design (which determines the model matrix) or that the point distribution in the k-dimensional space of the coordinates is not free at all but must be chosen in such a way as to improve the variance–covariance matrix. *Experimental plane* or *design* is used to describe the set of points which meet this requirement.[16]

Let us now summarize the most important requirements concerning the structure of the variance–covariance matrix.[17] As we should be able to estimate as closely as possible the regression coefficients, the **A** matrix determinant has to be maximized (the D-optimality condition). The relative measurement of such a property may be made by comparing the determinants of the moment matrix (**M**, which does not depend on the number of points (m)):

$$\mathbf{M} = \frac{1}{m}\mathbf{A} \tag{7.73}$$

or using the D-efficiency:

$$D = 100\left(\frac{\det(\mathbf{M}) \text{ for current plane}}{\det(\mathbf{M}) \text{ for best plane}}\right)^{1/r} \tag{7.74}$$

where r stands for the rank of matrix **M**. Furthermore, our design should not contain an excessively large number of experimental points; we measure this using the R-efficiency:

$$R = 100\,\frac{\text{number of points to be computed}}{\text{number of regression coefficients}} = 100\,\frac{m}{l} \tag{7.75}$$

or in a k-dimensional polynomial design of the qth order:

$$R = \frac{100m}{\binom{k+q}{q}} = \frac{100m\, q!\, k!}{(k+q)!} \tag{7.76}$$

It is also convenient to estimate independently all the coefficients. This condition is met every time the variance–covariance matrix is a diagonal one. The variance function at every point of the experimental domain must be constant at a constant distance from the origin of the design; such planes are called rotatable. If we write the function variance at any point **s**:

$$\operatorname{var}[\hat{\mathbf{y}}(\mathbf{s})] = \mathbf{f}(s)\cdot(\mathbf{FWF}')^{-1}\cdot\mathbf{f}'(s) \tag{7.77}$$

The plane is rotatable if this function depends only on the distance from the plane centre.

All these requirements cannot necessarily be met at the same time; one must find the best compromise according to the type of information required.[6]

7.3.2.5 The polynomial expansion

If $E(s)$ is the total energy, solution of the time-independent Schrödinger equation in the Born–Oppenheimer approximation for a molecular system is characterized by a set of internal coordinates $\mathbf{s}$ ($\mathbf{s}=\{s_1 \cdots s_k\}$). We assume that the region of immediate interest ($\mathscr{D}$), centered on a point of coordinates $\mathbf{s}_0$, may be accurately fitted by a polynomial expansion up to the order q. The former given regression equation (7.56):

$$y(s) \simeq \hat{y}(s) = b_0 + \sum_i b_i f_i(s)$$

becomes:

$$E(\mathbf{S}) \simeq \hat{E}(\mathbf{S}) = b_0 + \sum_{i=1}^{k} b_i S_i + \sum_{i \leqslant j}^{k} b_{ij} S_i S_j + \sum_{i \leqslant j \leqslant k}^{k} b_{ijk} S_i S_j S_k \cdots \tag{7.78}$$

where $\mathbf{S}$ stands for the internal displacement coordinate: $\mathbf{S}=\mathbf{s}-\mathbf{s}_0$ (see equation 7.16). Using matricial notation we have:

$$\hat{E} = \mathbf{S}^{[0,q]'} \hat{\mathbf{b}}^{[0,q]} \tag{7.79}$$

where $\mathbf{S}^{[0,q]}$ and $\hat{\mathbf{b}}^{[0,q]}$ are column vectors of terms in the order range $[0, q]$:

$$\mathbf{S}^{[0,q]'} = (1, S_1 \cdots S_k, S_1^2 \cdots S_k^2, S_2 S_1 \cdots S_k S_{k-1}, \ldots)$$
$$\hat{\mathbf{b}}^{[0,q]'} = (b_0, b_1 \cdots b_k, b_{11} \cdots b_{kk}, b_{21} \cdots b_{kk-1}, \ldots)$$

By differentiating this equation with respect to any internal coordinate we obtain:

$$\left(\frac{\mathrm{d}\hat{E}}{\mathrm{d}S_i}\right)_S = \hat{g}_i(S) = \left(\frac{\mathrm{d}\mathbf{S}^{[1,q]}}{\mathrm{d}S_i}\right)_S' \hat{\mathbf{b}}^{[1,q]} \tag{7.80}$$

$$\left(\frac{\mathrm{d}^2\hat{E}}{\mathrm{d}S_i\,\mathrm{d}S_j}\right)_S = \hat{H}_{ij}(S) = \left(\frac{\mathrm{d}^2\mathbf{S}^{[2,q]}}{\mathrm{d}S_i\,\mathrm{d}S_j}\right)_S' \hat{\mathbf{b}}^{[2,q]} \tag{7.81}$$

This shows that the regression coefficients measure the successive derivatives at the central point ($\mathbf{s}_0$) in the region of interest ($\mathscr{D}$):

$$\hat{E}(0) = \hat{b}_0 \tag{7.82a}$$

$$\hat{\mathbf{g}}(0) = \hat{\mathbf{b}}^{[1,1]} = \begin{pmatrix} \hat{b}_1 \\ \vdots \\ \hat{b}_k \end{pmatrix} \tag{7.82b}$$

$$\hat{\mathbf{H}}(0) = \begin{pmatrix} 2\hat{b}_{11} & \cdots & \hat{b}_{1k} \\ \vdots & \ddots & \vdots \\ \hat{b}_{k1} & \cdots & 2\hat{b}_{kk} \end{pmatrix} \tag{7.82c}$$

Estimation of the $\mathbf{b}$ vector only requires knowledge of the energy (E) or of its derivatives ($\mathbf{g}$ or $\mathbf{H}$) in m different points of the coordinate space around $\mathbf{s}_0$. The least square fit solution has the form (see equation 7.64):

For the energy:[6,17]

$$\mathbf{b}^{[0,q]} = (\mathbf{F}^{[0,q]}\mathbf{F}^{[0,q]\prime})^{-1}\mathbf{F}^{[0,q]}\mathbf{E} \tag{7.83}$$

For the gradient:[18]

$$\mathbf{b}^{[1,q]} = (\mathbf{F}^{[1,q]}\mathbf{F}^{[1,q]\prime})^{-1}\mathbf{F}^{[1,q]}\mathbf{g} \tag{7.84}$$

For the force constant matrix:

$$\mathbf{b}^{[2,q]} = (\mathbf{F}^{[2,q]}\mathbf{F}^{[2,q]\prime})^{-1}\mathbf{F}^{[2,q]}\mathbf{H} \tag{7.85}$$

where:

$$\mathbf{F}^{[i,q]} = [\mathbf{S}(1)^{[i,q]}, \mathbf{S}(2)^{[i,q]}, \ldots, \mathbf{S}(m)^{[i,q]}]$$

$\mathbf{S}(j)^{[i,q]}$ being the column vector $\mathbf{S}^{[i,q]}$ for the jth point.

The regressive formulas (7.83) to (7.85) show that every time a function ($p = 0$) or its derivatives at the order p is available it remains possible by a numerical approach to have at our disposal the next derivatives up to the order q ($q > p$). The next step is to find the point distribution which is to be chosen around the central point in order to find as good an estimation as possible for $\mathbf{b}^{[p,q]}$. The experience design technique meets this requirement for some q values greater than p by 1, 2 or 3 units.

7.3.3 The Experimental Planification

An experimental plane of order r is the design which is adequate for a regression of order q on a function p times differentiated if $r = q - p$.

The most classical experimental planes are known as equiradial planes. One regards such designs as being built up from a number of component sets of points, each set having all its points equidistant from the origin. For a first-order experimental plane one needs only one set of equiradial points and for a second-order plane one needs two sets of equiradial points at least, only one providing a singularity in matrix $\mathbf{A}$.

In order for our discussion to become dimension independent we define an adimensional experimental matrix $\mathbf{X}$. Each column of $\mathbf{X}$ corresponds to the adimensional coordinate vector $\mathbf{d}(j)$ of one point:

$$\mathbf{X} = \{\mathbf{d}(1) \cdots \mathbf{d}(j) \cdots \mathbf{d}(m)\} = \{d_i(j)\} \tag{7.86}$$

If the central point of our design is the point $\mathbf{s}_0$ of coordinate $\mathbf{S} = 0$ and if the step size in each space direction is given by the vector $\boldsymbol{\Delta}$ of k components, the ith coordinate of the jth point of our design is given by the relationship:

$$S_i(j) = 0 + \Delta_i d_i(j) \tag{7.87a}$$

or, from equation (7.16):

$$s_i(j) = s_i(0) + \Delta_i d_i(j) \tag{7.87b}$$

7.3.3.1 The first-order experimental designs

The simplex plane

In the k-dimensional coordinate space, the smallest design we can find is the simplex[19] which contains $k+1$ vertices on a sphere, all equidistant from each other:

For $k=2$ this is the equilateral triangle.
For $k=3$ this is the regular tetrahedron.
For $k \geqslant 4$ this is the regular polyhedron.

For practical convenience we prefer in general to add one central point at the original simplex plane. This plane may then have the form:

$$\mathbf{X} = \{\mathbf{d}(1) \cdots \mathbf{d}(k+2)\} = \{d_i(j)\} =$$

$$\begin{pmatrix} x_1 & -x_1 & 0 & \cdots & 0 & \cdots & 0 & 0 \\ x_2 & x_2 & -2x_2 & \cdots & 0 & \cdots & 0 & 0 \\ \vdots & \vdots & \vdots & & \vdots & & \vdots & \vdots \\ x_j & x_j & x_j & \cdots & -jx_j & \cdots & 0 & 0 \\ \vdots & \vdots & \vdots & & \vdots & & \vdots & \vdots \\ x_k & x_k & x_k & \cdots & x_k & \cdots & -kx_k & 0 \end{pmatrix} \tag{7.88}$$

with:

$$x_j = \{2j(j+1)\}^{-1/2} \qquad \text{for } 1 \leqslant j \leqslant k$$

It contains $k+2$ vertices ($m=k+2$). The distance between each vertex equals one in this pattern. In Figure 7.11 we illustrate our purpose for a three-dimensional centred simplex. The structure of the corresponding variance–covariance matrix has a diagonal form. Such a matrix is said to be orthogonal because each regression coefficient is estimated independently of the other ones. This structure maximizes the determinant of **A**. In such a case the R-efficiency is also optimum due to the fact that the simplex plane is the smallest that can be built.

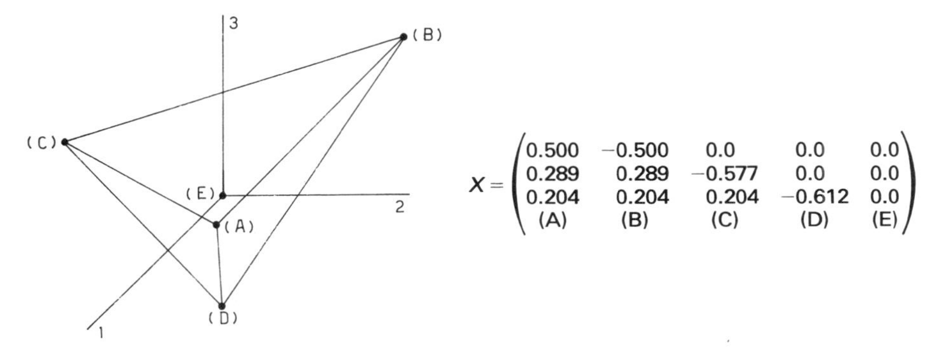

Figure 7.11 The three-dimensional simplex design

The regression coefficients will be:

For regression on the energy:

$$b_0 = \frac{1}{m} \sum_{j=1}^{m} E(s_j) \tag{7.89a}$$

$$b_i = \frac{2x_i}{\Delta_i} \left[\sum_{j=1}^{i} E(s_j) - iE(s_{i+1}) \right] \tag{7.89b}$$

For regression on the gradient:[18]

$$b_i = \frac{1}{m} \sum_{j=1}^{m} g_i(s_j) \tag{7.90a}$$

$$b_{ii} = \frac{x_i}{\Delta_i} \left[\sum_{j=1}^{i} g_i(s_j) - ig_i(s_{i+1}) \right] \tag{7.90b}$$

$$b_{il} = \frac{2}{\Delta_i^2 + \Delta_l^2} \left\{ x_l \Delta_l \left[\sum_{j=1}^{l} g_i(s_j) - lg_i(s_{l+1}) \right] + x_i \Delta_i \left[\sum_{j=1}^{i} g_l(s_j) - ig_l(s_{i+1}) \right] \right\} \tag{7.90c}$$

The same kind of equations may be deduced for regression on a second-order derivative matrix.

The cross plane

This plane is built up from a central point and a cross in the k-dimensional space of coordinates (Figure 7.12). All the cross vertices are equally spaced from the centre, the overall design contains $2k+1$ vertices ($m = 2k+1$) and it has the following adimensional form:

$$X = \begin{pmatrix} 1 & -1 & 0 & 0 & \cdots & 0 & 0 & 0 \\ 0 & 0 & 1 & -1 & \cdots & 0 & 0 & 0 \\ 0 & 0 & 0 & 0 & \cdots & 0 & 0 & 0 \\ \vdots & \vdots & \vdots & \vdots & & \vdots & \vdots & \vdots \\ 0 & 0 & 0 & 0 & \cdots & 1 & -1 & 0 \end{pmatrix} \tag{7.91}$$

The corresponding variance–covariance matrix keeps a diagonal form. Compared to the simplex plane the determinant of the moment matrix is reduced from $[2(k+2)]^k$ to $[2/(2k+1)]^k$ for an energy regression. The R-efficiency is also lowered

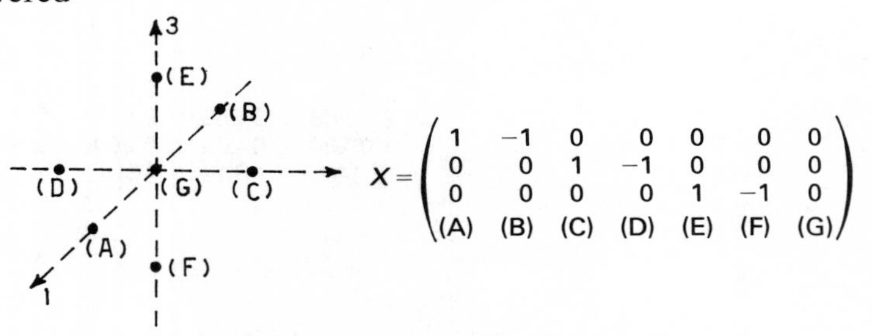

Figure 7.12 Three-dimensional cross-plane

The regression coefficients corresponding to the energy and gradient fitting are:

On the energy:

$$b_0 = \frac{1}{m} \sum_j E(s_j) \tag{7.92a}$$

$$b_i = \frac{1}{2\Delta_i^2} [E(s_{2i-1}) - E(s_{2i})] \tag{7.92b}$$

On the gradient:

$$b_i = \frac{1}{m} \sum_j g_i(s_j) \tag{7.93a}$$

$$b_{ii} = \frac{1}{4\Delta_i^2} g_i(s_{2i-1}) - g_i(s_{2i}) \tag{7.93b}$$

$$b_{il} = \frac{1}{2(\Delta_i^2 + \Delta_l^2)} \{\Delta_l[g_i(s_{2l-1}) - g_i(s_{2l})] + \Delta_i[g_l(s_{2i-1}) - g_l(s_{2i})]\} \tag{7.93c}$$

It seems from our experience[18] that such a cross plane provides more accurate estimates for gradient regressions than the simplex one.

7.3.3.2 The second-order experiental designs

The composite plane

Box and Hunter[16,17] have proposed the use of composite designs at order two. In a k-dimensional space we can build up such a design combining the following geometrical figures:

(a) m_0, central points ($m_0 \geq 1$).
(b) $m_1 = 2^k$ points disposed as the vertices of a square ($k = 2$), a cube ($k = 3$) or an hypercube ($k \geq 4$) or its fractional with $m_1 = 2^{k-p}$. The latter part of the design is also called factorial ($p = 0$) or fractional replicate ($p > 0$) of the factorial plane.
(c) $m_2 = 2k$ points as the vertices of a cross ($k = 2$) or a cross polytope ($k \geq 3$).

In Figure 7.13 we show the composite plane in a three-dimensional space. The total number of vertices is:

$$m = m_0 + m_1 + m_2 \tag{7.94}$$

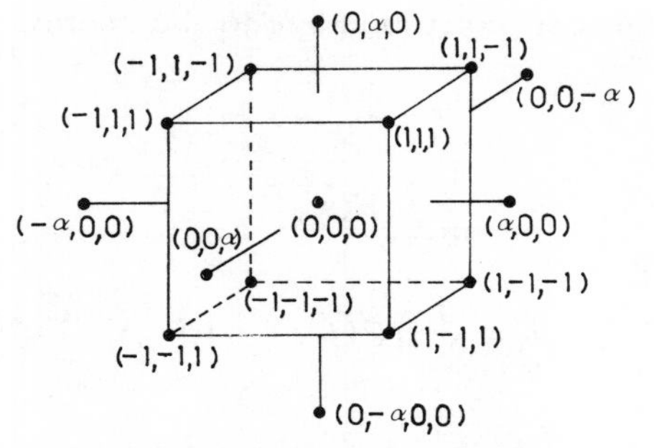

Figure 7.13 Three-dimensional composite design. Reproduced by permission of John Wiley and Sons, Inc., New York

$$X=\begin{pmatrix} 0 & 1 & 1 & 1 & 1 & -1 & -1 & -1 & -1 & \alpha & -\alpha & 0 & 0 & 0 & 0 \\ 0 & 1 & 1 & -1 & -1 & 1 & 1 & -1 & -1 & 0 & 0 & \alpha & -\alpha & 0 & 0 \\ 0 & 1 & -1 & 1 & -1 & 1 & -1 & 1 & -1 & 0 & 0 & 0 & 0 & \alpha & -\alpha \end{pmatrix}$$

Such a design meets well the D-efficiency requirement.[6] If we use complete designs ($p=0$), the R-efficiency goes down rapidly when k increases. However, fractional replicates $(1/2)^p$ of the 2^k factorial plane significantly reduce the number of experimental points.[6,20]

First of all we build up a 2^{k-p} factorial plane which provides values for the factors from 1 to $k-p$. For example, for $k=3$ and $p=1$:

$$\begin{bmatrix} +1 & +1 \\ -1 & +1 \\ +1 & -1 \\ -1 & -1 \end{bmatrix}$$

The missing part of this matrix (columns $p-k+1$ to k), which provides levels for the remaining factors, is obtained by multiplying term by term some of the first columns. In our example column 3 may be equal to the product of columns 1 and 2. Then the fractional replicate 1/2 of the factorial plane 2^3 is:

$$X=\begin{bmatrix} +1 & +1 & +1 \\ -1 & +1 & -1 \\ +1 & -1 & -1 \\ -1 & -1 & +1 \end{bmatrix}$$

and we write $3=1\circ 2$ (called contrast) or $I=1\circ 2\circ 3$ (called independent generator, I being the unit vector).

As we require the estimated regression coefficients to be independent from each other, some generators are not permissible. Analysis of the alias matrix shows that the generators must contain five terms at least.[6,21] That is why the first fractional replicate begins for $k=5$ with the contrast $5=1\circ 2\circ 3\circ 4$, and for $k=6$ we have only one contrast at our disposal: $6=1\circ 2\circ 3\circ 4\circ 5$. So the coefficients are only biased by terms of a higher order than two. In this kind of plane there remains one degree of freedom (α in Figure 7.13) to satisfy the orthogonality or the rotatable requirement.[6,17,22]

In practice, for energy regression[6] we use, for α[17]:

$$\alpha_R = 2^{(k-p)/4} \tag{7.95}$$

which is the rotatable value. If we then replicate the central point m_0 times, m_0 being the closest integer:

$$m_0 \simeq 4(1+2^{(k-p)/2}) - 2k \tag{7.96}$$

the plane becomes quasi-orthogonal. Then only the following covariance terms in the $\mathbf{A}^{-1}$ matrix remain:[23]

$$\text{cov}\,(b_0, b_{ii}) \neq 0$$

For gradient regressions, in order to select an α value, we know some way to build a quasi-orthogonal plane:[18]

The first one if the cross disappears:

$$\alpha = 0 \qquad \text{and} \qquad m_3 = 0$$

The second one if

$$\alpha = \left[\frac{\sqrt{(mm_1)} - m_1}{2}\right]^{1/2} \tag{7.97}$$

So we can progressively build the regression plane. We may start with the first-order cross plane and improve it up to the second order by adding the factorial plane. Or conversely we may start with some fractional replicate of the factorial plane and complete it by decreasing the p value and finally adding the cross with an appropriate α value.[18]

The Doehlert plane

Another type of design, the Doehlert plane,[24] may also be used. The basic figure is a simplex with $k+1$ vertices which generates by rotation a plane of $k(k+1)+1$ vertices. In Figure 7.14 we present a two-dimensional Doehlert plane. This plane is somewhat less efficient in the sense of D-optimality but provides a best R-efficiency.

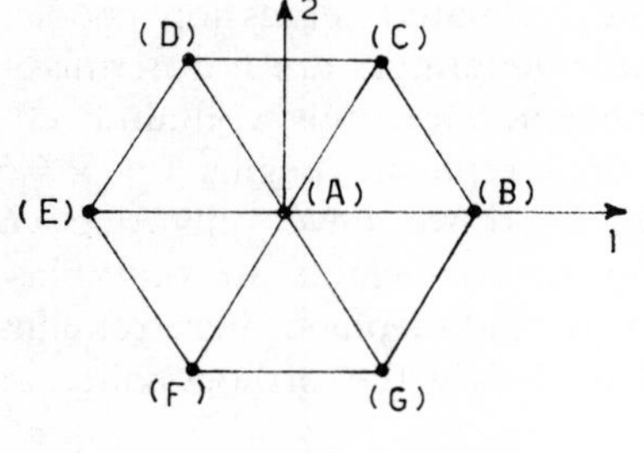

$$X = \begin{pmatrix} 0.0 & 1.0 & 0.5 & -0.5 & -1.0 & -0.5 & +0.5 \\ 0.0 & 0.0 & \sqrt{3}/2 & -\sqrt{3}/2 & 0.0 & -\sqrt{3}/2 & -\sqrt{3}/2 \\ (A) & (B) & (C) & (D) & (E) & (F) & (G) \end{pmatrix}$$

Figure 7.14 Two-dimensional plane of Doehlert

7.3.3.3 The third-order experimental designs

Third-order designs are also available but we do not discuss such experimental planes in this work. Information may be found elsewhere.[25]

7.3.3.4 An example of experience planification

Let us treat the water molecule as an example. Three internal parameters are considered: two OH distances (d_1 and d_2 in angstroms) and the angle HOH (θ in radians). The three-dimensional complete composite plane is built using the X matrix given in Figure 7.13. The energies come from an *ab initio* calculation at the 6-31G level. The centre of the composite plane is:

$$\mathbf{s}(0)' = (0.95 \text{ Å}, 0.95 \text{ Å}, 1.947 \text{ rad})$$

The interval on each variable is:

$$\mathbf{\Delta}' = (0.01 \text{ Å}, 0.01 \text{ Å}, 0.0087 \text{ rad})$$

Fifteen *ab initio* calculations have been performed according to equation (7.87b) and the regression coefficients found by equation (7.64) are:

$$b_0 = -73.59343 \text{ a.u.}$$

$$b_1 = b_2 = -2.0744 \text{ a.u. Å}^{-1}$$

$$b_3 = -0.4333 \text{ a.u. rad}^{-1}$$

$$b_{11} = b_{22} = 1.0578 \text{ a.u. Å}^{-2}$$

$$b_{33} = 0.0846 \text{ a.u. rad}^{-2}$$

$$b_{12} = -0.0436 \text{ a.u. Å}^{-2}$$

$$b_{13} = b_{23} = 0.0547 \text{ a.u. (Å rad)}^{-1}$$

with a residual variance of $(6.7 \times 10^{-6})^2$ a.u.; as the variance on the energy is

$(9.5\times10^{-5})^2$ a.u. the Fischer test gives 195.7 and the multiple regression coefficient equals 0.9974.

7.3.4 Other Kinds Second-order Numerical Derivatives

When we dispose only of the first derivative vector (**g**) by analytical computation it is always possible to construct progressively an estimation of the second derivative matrix (**H**) by the method described by Fletcher and Powell.[26] If ρ is the displacement from the current point (**O**) in the search direction σ of the k-dimensional space of coordinates to minimize the energy:

$$\boldsymbol{\rho}=\lambda\boldsymbol{\sigma} \tag{7.98}$$

If y is the gradient difference between the current point O and the point $\mathrm{O}+\rho$:

$$\mathbf{y}=\mathbf{g}(\mathrm{O}+\rho)-\mathbf{g}(\mathrm{O}) \tag{7.99}$$

then:

$$\mathbf{H}_{\mathrm{O}+\rho}^{-1}=\mathbf{H}_{\mathrm{O}}^{-1}+\mathbf{A}+\mathbf{B} \tag{7.100}$$

where:

$$\mathbf{A}=\frac{\boldsymbol{\rho}\boldsymbol{\rho}'}{\boldsymbol{\rho}'\mathbf{y}} \qquad \text{and} \qquad \mathbf{B}=\frac{(\mathbf{H}_{\mathrm{O}}^{-1}\mathbf{y})(\mathbf{H}_{\mathrm{O}}^{-1}\mathbf{y})'}{\mathbf{y}'\mathbf{H}^{-1}\mathbf{y}}$$

To start the procedure we can give a diagonal form to **H** (i.e. the unity matrix: $\mathbf{H}=\mathbf{E}$). This procedure may be considered as interesting for a stationary point search but nevertheless it does not provide an accurate force constant matrix. This is chiefly due to the fact that the good as well as the bad information are cumulated in the matrix **H** and can never be removed.

To improve the preceding algorithm one must start with a more accurate guess. The best way is to use the simplicial approach coupled to a gradient regression. More simply, if we dispose of the gradient vectors computed in k different and non-linearly related points, **g** and **H** being bonded by the following equation:

$$\mathbf{g}(S)=\mathbf{g}_0+\mathbf{H}_{\mathrm{O}}\mathbf{S} \tag{7.101}$$

we can deduce that:

$$\{[\mathbf{g}(S_1)\,|\,\mathbf{g}(S_2)\,|\cdots|\,\mathbf{g}(S_k)]-\mathbf{g}_0\}(\mathbf{S}_1\,|\,\mathbf{S}_2\,|\cdots|\,\mathbf{S}_k)^{-1}=\mathbf{H}_{\mathrm{O}} \tag{7.102}$$

Such an approach leads in practice to a non-symmetrical $\mathbf{H}_{\mathrm{O}}$ matrix. The initial guess then becomes:

$$\tfrac{1}{2}(\mathbf{H}_{\mathrm{O}}+\mathbf{H}_{\mathrm{O}}') \text{ instead of } \mathbf{H}_{\mathrm{O}} \tag{7.103}$$

Let us also note that equation (7.102) can only be solved if we use a set of k linearly independent points.

7.4 STATIONARY POINT SEARCH

7.4.1 General Considerations

The stationary point search on a response surface is an important problem of operational research. The search must be as fast (as economical) as possible and lead to the right point in a guaranteed way.

For a minimum, the function evaluation (E) provides, *a priori*, enough information and we look for the point $\mathbf{s}(m)$ such that:

$$E_m = \min_R \{E(s)\} \tag{7.104}$$

where $\mathbf{s}$ stands for the internal coordinate vector (equation 7.14). Nevertheless, we need information on derivatives in order to have a guided research method. For finding a minimax, the evaluation of the first-order derivatives vector ($\mathbf{g}$ in equation 7.15b) is requested (except where we know *a priori* the direction of the surface which corresponds to the negative curvature, but such a fact is highly improbable). With first derivatives we look for $\mathbf{s}^{\neq}$ in such a way that:[27]

$$\gamma_{\neq} = \{\mathbf{g}'_{\neq}\mathbf{g}_{\neq}\}^{1/2} = 0 \tag{7.105}$$

If we now require a guarantee concerning the kind of stationary point we have reached and the search quality, in all cases the second-order derivative matrix ($\mathbf{H}$ in equation 7.15c) is needed (as shown before in equations 7.23 and 7.24).

7.4.2 Random Search

The most simple fashion which could be imagined, but probably not the most efficient, is the random search. First of all let us define the research domain (R) in the k-dimensional space of the internal coordinates as:

$$[\mathbf{s}_i(\min), \mathbf{s}_i(\max)] \qquad \text{for } 1 \leqslant i \leqslant k \tag{7.106}$$

Such a domain corresponds to an hypercube. Using a series of uniformly distributed random numbers between 0 and 1 (see Figure 7.15) (ξ_α, for $\alpha = 1, 2, \ldots, \infty$) we can select any particular point in our search domain with an equal probability:

$$\mathbf{s}_i(\alpha) = \mathbf{s}_i(\min) + \xi_{\alpha,i}[\mathbf{s}_i(\max) - \mathbf{s}_i(\min)] \tag{7.107}$$

We keep as the stationary point the one which has the best energy or the gradient norm closest to zero:

$$\hat{E}(\min) = \min_\alpha \{E(\mathbf{s}_\alpha)\} \qquad \text{for minimum}$$

or

$$\hat{\gamma}(\mathbf{s}^{\neq}) = \min_\alpha \{\gamma(\mathbf{s}_\alpha)\} \qquad \text{for minimum or minimax}$$

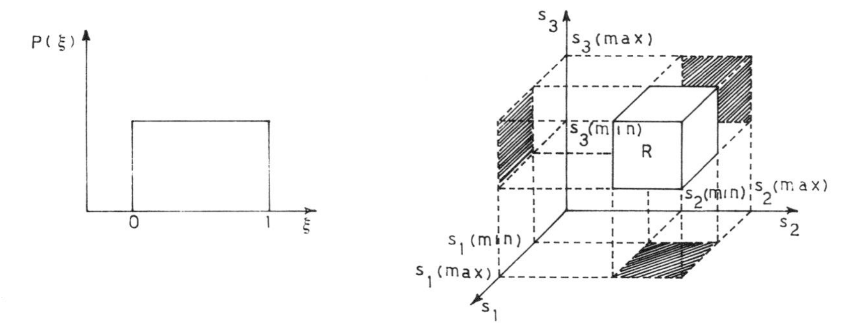

Figure 7.15 The ξ density function $(P(\xi) = dn(\xi)/d\xi)$ and three-dimensional research domain $\boldsymbol{R}$

More economical methods of course exist, but this one could give a very good idea of the position for faster algorithms.

The amount of work to perform is proportional to the search domain $\boldsymbol{R}$ $(V = \prod_i [s_i(\max) - s_i(\min)])$ and inversely proportional to the requested precision (v). If we divide the $\boldsymbol{R}$ domain into elementary cells of volume v, the probability of placing any point in one particular cell is given by:

$$p = \frac{v}{V} \tag{7.108}$$

This is true as long as we use a uniform distribution. The probability of finding at least one point after m trials in the cell labelled $\neq$ (centred on the stationary point of interest) is:

$$P(\neq) = \sum_{i=1}^{m} \binom{m}{i} p^i (1-p)^{m-i} = 1 - (1-p)^m \tag{7.109}$$

In Table 7.1 we report some values of $P(\neq)$ for different values of p and m. This table shows that to locate one stationary point with a probability of 95 per cent. and a precision v, it is necessary to use $3V/v$ points at least (or $3/p$). If we want to locate at the same time more than one stationary point

Table 7.1 The distribution law of m events of individual probability $p : P(\mathrm{A})$

p \ n	$m = \frac{1}{p}$	$m = \frac{2}{p}$	$m = \frac{3}{p}$	$m = \frac{4}{p}$	$m = \frac{5}{p}$
1/3	0.704	0.912	0.974	0.992	0.998
1/6	0.665	0.888	0.962	0.987	0.996
1/10	0.651	0.878	0.958	0.985	0.995
1/100	0.634	0.866	0.951	0.982	0.993
1/1000	0.632	0.865	0.950	0.981	0.993

we need to generalize the preceding formula. If A, B, . . . are the stationary points of interest, the probability of finding A and B simultaneously is given by:

$$P(\text{A and B}) = P(\text{A}) + P(\text{B}) + P(\bar{\text{A}} \text{ and } \bar{\text{B}}) - 1 \tag{7.110}$$

where $P(\bar{\text{A}} \text{ and } \bar{\text{B}})$ stands for the probability that there is no point around A as well as around B. This is equal to:

$$P(\bar{\text{A}} \text{ and } \bar{\text{B}}) = (1-2p)^m \tag{7.111}$$

Finally, we can write:

$$P(\text{A and B}) = 1 - 2(1-p)^m + (1-2p)^m \tag{7.112}$$

This equation may be generalized recursively for q stationary points as:

$$P(q \text{ events}) = 1 - q(1-p)^m + \prod_{i=2}^{q} (1-ip)^m \tag{7.113}$$

In Table 7.2 we give some values of this multinomial law. The table shows that we may reach three stationary points simultaneously with a probability of 95 per cent by computing $4/p$ points.

Table 7.2 Some values of the probability function $P_q(m, p, q > 1)$ for locating s stationary points simultaneously

p	q / m	2 / $m = \frac{3}{p}$	2 / $m = \frac{4}{p}$	2 / $m = \frac{5}{p}$	3 / $m = \frac{3}{p}$	3 / $m = \frac{4}{p}$	3 / $m = \frac{5}{p}$
$p = 1/10$		0.916	0.971	0.990	0.874	0.956	0.985
$p = 1/100$		0.904	0.964	0.987	0.855	0.946	0.980
$p = 1/1000$		0.903	0.963	0.987	0.853	0.945	0.980

We could use these results as a lower limit to estimate the efficiency of any other method. If m is the number of points which are requested, we write:

$$\text{For } q = 1: \quad m \simeq 3\frac{V}{v} = 3\left(\frac{c}{a}\right)^k \tag{7.114a}$$

$$\text{For } q = 3: \quad m \simeq 4\frac{V}{v} = 4\left(\frac{c}{a}\right)^k \tag{7.114b}$$

where c is the cube length equal to $\sqrt[k]{V}$ and a is $\sqrt[k]{v}$.

7.4.3 The Complete Search

We find that the exhaustive search method is the opposite of the random investigation. It consists of computing a set of points equally spaced in all

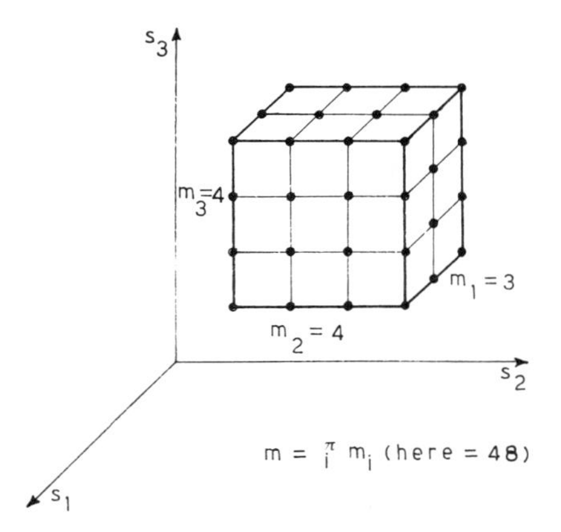

Figure 7.16 Complete research in a three-dimensional space of internal coordinates

the space directions (Figure 7.16). The resulting lattice contains m points with m equal to $\prod_{i=1}^{k} m_i$ when we dispose of m_i point on the ith space direction. The elementary cell has a volume of V/m. Then the probability of finding a point in a volume v around one stationary point is given by:

$$P_1(m) = \min\left(1; \frac{mv}{V}\right) \tag{7.115}$$

In order to reach the point of interest with 95 per cent. of success we need only to compute $m = 0.95V/v$ points. This method is then 3.16 times more efficient than the random search (which needs to compute $3V/v$ points); nevertheless it becomes rapidly prohibitive when the space dimension (k) is greater then 3. If we want to locate q stationary points at once we replace the last formula (7.115) by:

$$P_q(m) = \left[\min\left(1; \frac{mv}{V}\right)\right]^q \tag{7.116}$$

At a level of 95 per cent., m becomes:

$$m = (0.95)^{1/q} V/v \tag{7.117}$$

We note that:

$$\lim_{q \to \infty} m = V/v \tag{7.118}$$

So the complete search is always more efficient than the random search. We shall keep in mind the fact that the requested number of points is approximately equal to:

$$m \simeq \frac{V}{v} = \left(\frac{c}{a}\right)^k \tag{7.119}$$

7.4.4 Evolutive Experimental Designs

Let us suppose that we want to minimize the function $f(S)$ ($E(s)$ in equation 7.104 or $\mathbf{g}(s)$ in equation 7.105) in a k-dimensional space. Some techniques do not require any mathematical model to localize the stationary point of interest but may be controlled by a model of order one or two: we talk about the simplicial optimization or the Doehlert uniform shell.

7.4.4.1 The simplex search[28]

Note that this method is a general method for non-linear extremum search. We choose to start the optimization run at a point O of coordinate $\mathbf{s}_0$ and a step size $\boldsymbol{\Delta}$. Around the point O we build a simplex by using the adimensional simplicial matrix $\mathbf{X}$ (see the first-order experimental design of equation 7.88):

$$\mathbf{X}=\{\mathbf{d}(1), \mathbf{d}(2), \ldots, \mathbf{d}(k+1)\}=\{d_i(j)\}$$

where $\mathbf{d}(j)$ stands for a column vector containing the coordinates of point j. So the ith coordinate of the jth vertex is given by:

$$s_i(j)=s_i(\mathrm{O})+\Delta_i d_i(j)$$

Illustration of this in given in Figure 7.17.

By recomputing only one point we can translate the simplex. In order to improve our response as much as possible we replace in $\mathbf{X}$ the column which gives the worst value of $f(S)$. Labelling such a column w, the new point n

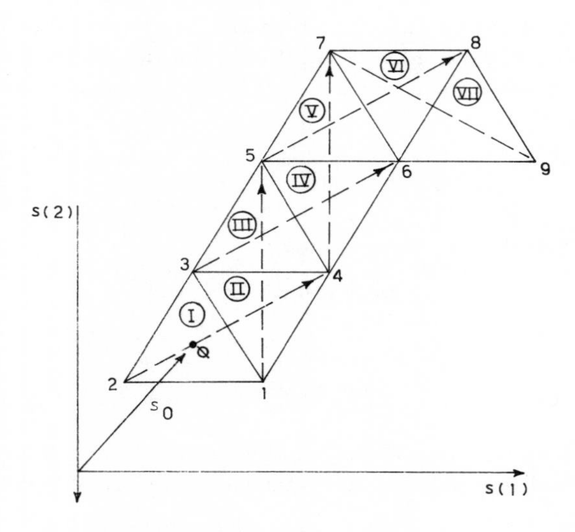

Figure 7.17 The simplicial evolution on a two-dimensional response surface

will be of the coordinates:

$$d_j(n)=\frac{1+\nu}{k}\sum_{i\neq w} d_j(i)-\nu d_j(w) \qquad \forall j=1 \text{ to } k \tag{7.120}$$

where $\nu=1$, to keep the simplex size, and $\nu>1$ to expand it and $\nu<1$ to contract it. Then $\mathbf{d}(n)$ is further used in place of $\mathbf{d}(w)$ and the next origin ($\mathbf{s}_0$) becomes the centre of the new simplex. Such an evolution operation is named reflexion (see Figure 7.17). If the computed response at the new point is worse than those at the point w, we keep it but we do not take it into account for the next reflexion. Finally, we stop the research when M successive simplices contain the same vertex; M in general is chosen as:

$$M=1.65k+0.05k^2 \tag{7.121}$$

It can easily be seen that the number of points (m) needed to travel the whole search domain (R) is only proportional to the rate c/a (where c is the size of the search domain and a the distance between two vertices in the simplex). After k reflexions the translation in the k-dimensional space is between a and $a/\sqrt{2}$. To scan a diagonal of R one needs the evaluation number:

$$m=\sqrt{\left(\frac{k^3}{2}\right)}\frac{c}{a}+k+1 \tag{7.122}$$

Compared to the random or to the complete search (where m is proportional to $(c/a)^k$; see equations 7.114) this method is much more efficient. Nevertheless, on a surface which is not smooth the evolution of the simplex may be much slower.

Each simplex can support a first-order mathematical model; we may then use it to control the progress of the search on the potential energy surface. So from total energy computation we can follow the gradient norm by numerical estimation along the optimization process; from analytical first derivatives we can numerically go up to the second-order analysis[18] and follow the stationary point prediction for each simplex (see Section 7.4.5 about the quadratically converging process). In this case the requested work becomes approximately proportional to k.

The simplex approach has also been adapted for transition point location without gradient evaluation.[29] The basic idea is to exclude the direction of the minimum energy search which must correspond to the reaction pathway; this constraint can then be used to find the minimum on the other space directions. This is performed as follows: given two points on the minimum energy path (they may correspond to the reactants and the products), a new point lying approximately on this path may be generated by minimizing the energy on a hypersphere centred on the highest of the two points with a radius defined as a fraction of the euclidian distance separating the two first points. The new point is further used in place of one of the former and the search restarted until convergence is reached.

7.4.4.2 The Doehlert shell[6,30]

As we have shown, the simplex method with a gradient facility must be considered as a pessimistic way to find an extremum (we neglect the worst point). A more optimistic procedure is given by Doehlert. Our basic pattern is now a Doehlert plane (instead of a simplex) which contains $k(k+1)+1$ vertices. By recomputing $k(k-1)+1$ points we may translate the plane in a k-dimensional space; thus we generate the uniform shell design of Doehlert (see Figure 7.18 for a two-dimensional example). As for simplex, another advantage is that we can add one variable and then increase the space dimension ($k \to k+1$) by only adding $2(k+1)$ new points.

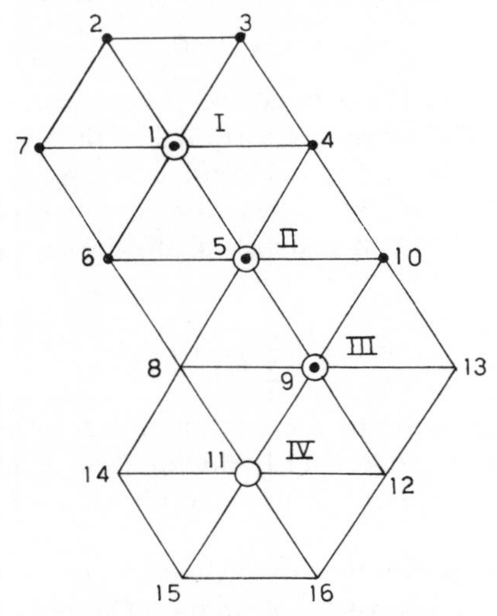

Figure 7.18 The Doehlert uniform shell in a two-dimensional space

The translation of the Doehlert plane is done in such a way that the best point already computed becomes the centre of the next plane. If this point corresponds to the column b then the new experimental matrix may be obtained by replacing each column by a new one of the coordinate:

$$\mathbf{d}'(i) = \mathbf{d}(i) + \mathbf{d}(b) \tag{7.123}$$

At each step we may follow the approach of the stationary point by using a second-order polynomial development. This is essential for minimax when gradients are not available. The requested work is now proportional to k^2.

7.4.5 The Quadratically Converging Process

Let us suppose that the surface may be developed in a Taylor series in a small domain around a point of coordinate $\mathbf{s}_0$:

$$E(S) = E(s_0) + \mathbf{S}'\mathbf{g}(s_0) + \tfrac{1}{2}\mathbf{S}'\mathbf{H}(s_0)\mathbf{S} + \cdots \tag{7.124a}$$

where $\mathbf{S}$ stands for $\mathbf{s}-\mathbf{s}_0$, and $E(s_0)$, $\mathbf{g}(s_0)$ and $\mathbf{H}(s_0)$ respectively stand for the total energy, the gradient vector and the force matrix at point $\mathbf{s}_0$. We can also write:

$$\mathbf{g}(s)=\nabla E(s)=\mathbf{g}(s_0)+\mathbf{H}(s_0)\mathbf{S}+\cdots \tag{7.124b}$$

and

$$\mathbf{H}(s)=\nabla\nabla' E(s)=\mathbf{H}(s_0)+\cdots \tag{7.124c}$$

At the stationary point ($\boldsymbol{\Sigma}$) the gradient vector will have zero norm; then:

$$\mathbf{g}(s_0)+\mathbf{H}(s_0)\boldsymbol{\Sigma}=0 \tag{7.125}$$

As $\mathbf{H}$ is a symmetrical $k\times k$ square matrix it can be inverted and:

$$\boldsymbol{\Sigma}=-\mathbf{H}(s_0)^{-1}\mathbf{g}(s_0) \tag{7.126}$$

As far as the second-order approximation is valid, $\mathbf{s}_0+\boldsymbol{\Sigma}$ gives the exact position of the stationary point of interest. In general, when $\boldsymbol{\Sigma}$ remains inside the investigated domain we can use an interpolative process to give a result which is quite acceptable. Nevertheless, when $\boldsymbol{\Sigma}$ increases and goes outside the experimental (or the validity) domain the method must then be called extrapolative and becomes dangerous for any large distance from the central point.

When we have enough information at our disposal the introduction of third-order corrections may provide more rapid convergence. To our preceding expression (7.124a) limited to the second order we add the third-order terms:

$$\tfrac{1}{6}\mathbf{S}'\mathbf{R}_3(S)\mathbf{S}$$

where $\mathbf{R}_3(S)$ stands for $\mathbf{S}^*\boldsymbol{T}(s_0)$ and $\mathbf{T}$ is the third-order derivative hypermatrix. At any point we have

$$\mathbf{g}(s)=\mathbf{g}(s_0)+[\mathbf{H}(s_0)+\tfrac{1}{2}\mathbf{R}_3(S)]\mathbf{S}=\mathbf{0} \tag{7.127a}$$

$$\mathbf{H}(s)=\mathbf{H}(s_0)+\mathbf{R}_3(S) \tag{7.127b}$$

Defining:

$$\langle\mathbf{H}(s_0\mid s)\rangle=\tfrac{1}{2}[\mathbf{H}(s_0)+\mathbf{H}(s)] \tag{7.128}$$

we can write:

$$\mathbf{g}(s)=\mathbf{g}(s_0)+\langle\mathbf{H}(s_0\mid s)\rangle\mathbf{S} \tag{7.129}$$

At a stationary point the gradient vector vanishes; then:

$$\boldsymbol{\Sigma}=-[\langle\mathbf{H}(s_0\mid s_0+\boldsymbol{\Sigma})\rangle]^{-1}\mathbf{g}(s_0) \tag{7.130}$$

which has formally the same form as the one previously found (equation 7.126).

The efficiency of such a process depends on the method which is used to evaluate the first and second derivatives. For many potential energy surfaces we often need only two or three times the derivative evaluations.[31]

7.4.6 Search Without Derivative Evaluation

If one does not dispose of any derivative of the potential energy surface the location of a minimum can be found using the Powell method of conjugate directions.[32] Directions $\mathbf{u}_i$ and $\mathbf{u}_j$ are said to be *conjugate* if:

$$\mathbf{u}_i \mathbf{H} \mathbf{u}_j = 0 \qquad \text{for } i \neq j \tag{7.131}$$

Let us suppose that we have at our disposal a set of k conjugate directions and we lead our optimization in one direction ($\mathbf{u}_i$) at a time. Starting at point $s(I)$, after k such extremizations, we reach the points $s(F)$:

$$\mathbf{s}_F = \mathbf{s}_I + \sum_{i=1}^{k} \lambda_i \mathbf{u}_i \tag{7.132}$$

where λ_i is a scalar which assumes we have an extremum in the direction $\mathbf{u}_i$.

The corresponding total energy for a quadratic local form is, from equations (7.131) and (7.132):

$$E(s_F) = E(s_I) + \sum_i \left[\lambda_i \mathbf{u}_i' \mathbf{g}(s_I) + \tfrac{1}{2}\lambda_i^2 \mathbf{u}_i' \mathbf{H}(s_I)\mathbf{u}_i\right] \tag{7.133}$$

This expression shows that the best λ_i values depend only on the direction $\mathbf{u}_i$. Consequently, the one direction at a time optimization leads to the stationary point of interest on a quadratic response surface every time the directions are conjugated. As an example we give in Figure 7.19 two optimization runs on the water molecule.

One question now remains: How do we find a set of conjugate directions? In fact we have at our disposal three types of information:

(a) As we can see in Figure 7.19 (in the H_2O example), it is powerful to replace any set of internal coordinates by equivalent symmetry coordinates. Such coordinates are symmetric or antisymmetric with respect to each symmetry operation. So in H_2O the C_{2v} point group consideration leads to three coordinates:

$$\left.\begin{matrix}\theta \\ d_1 + d_2\end{matrix}\right\} \text{symmetric with respect to the } C_{2v} \text{ axis}$$

$$d_1 - d_2 \quad \text{antisymmetric with respect to the } C_{2v} \text{ axis}$$

Using such variables the $\mathbf{H}$ matrix becomes quasi(block)-diagonal:

$$\mathbf{H} = \begin{pmatrix} 1.0796 & & \\ 0.0 & 1.0360 & \\ 0.0 & 0.0547 & 0.1692 \end{pmatrix} \begin{matrix} d_1 - d_2 \\ d_1 + d_2 \\ \theta \end{matrix}$$

$$\begin{matrix} d_1 - d_2 & d_1 + d_2 & \theta \end{matrix}$$

(b) Every time the force constant matrix is available from experimental data we can diagonalize it to find a set of conjugate directions which must be close enough to the theoretical results to represent an excellent choice

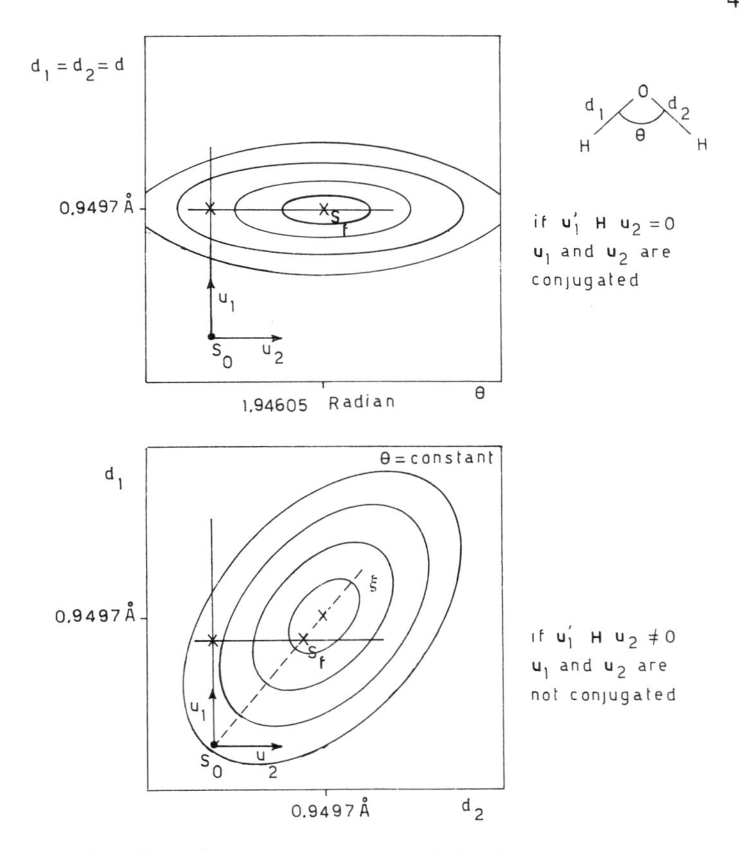

Figure 7.19 One direction at a time optimization: the use or not of the conjugate directions

(the **H** matrix can also be built using some tabulated values). We also note from experience that some interaction terms are weak; this is so for coupling terms between bond stretching and angular deformation (see supra, the structure of **H** in the water molecule).

(c) According to Powell,[33] we can prove that if $s(I)$ is an extremum on the direction $\mathbf{u}_k$ and if $s(F)$ (equation 7.132) is also an extremum on the same direction, then the direction $\mathbf{s}_F - \mathbf{s}_I$ is conjugated to $\mathbf{u}_k$ on a quadratic surface from equations (7.124a and b). Using the internal displacement vector **S** (see equation 7.16) we can write:

$$E(s+\lambda u_k) = E(s) + \lambda \mathbf{u}_k' \mathbf{g}(s) + \tfrac{1}{2}\lambda^2 \mathbf{u}_k' \mathbf{H}(s_0)\mathbf{u}_k \qquad (7.134)$$

For $\mathbf{s} = \mathbf{s}_I$ we require:

$$\frac{\partial}{\partial \lambda} E(s_I + \lambda u_k) = 0 \qquad \text{at } \lambda = 0$$

Therefore:

$$\mathbf{u}_k'[\mathbf{g}(s_0)+\mathbf{H}(s_0)(\mathbf{s}_I-\mathbf{s}_0)]+\lambda\mathbf{u}_k'\mathbf{H}(s_0)\mathbf{u}_k=0$$

Similarly, for $s=s_F$ we require:

$$\frac{\partial}{\partial\lambda}E(s_F+\lambda u_k)=0 \qquad \text{at } \lambda=0$$

Therefore:

$$\mathbf{u}_k'[\mathbf{g}(s_0)+\mathbf{H}(s_0)(\mathbf{s}_F-\mathbf{s}_0)]+\lambda\mathbf{u}_k'\mathbf{H}(s_0)\mathbf{u}_k=0$$

Hence:

$$\mathbf{u}_k'\mathbf{H}(s_0)(\mathbf{s}_F-\mathbf{s}_I)=\mathbf{u}_k'\mathbf{H}(s_0)\boldsymbol{\xi}=0 \tag{7.135}$$

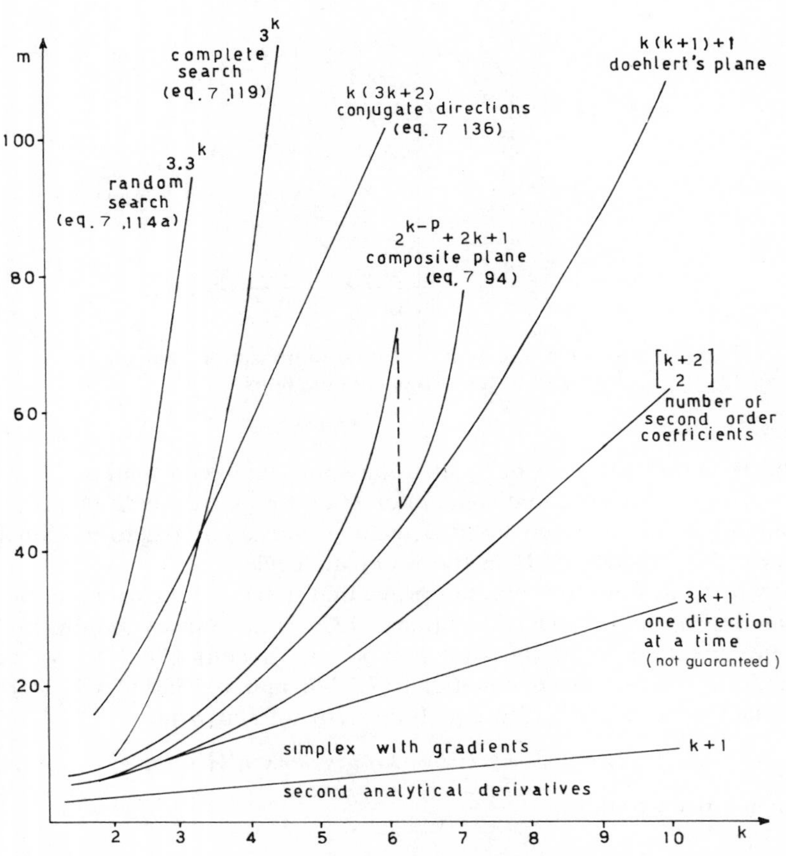

Figure 7.20 The optimization work for different methods (m = number of points and k = space dimension)

This suggests that we can use this theorem to introduce progressively conjugate directions in a one direction at a time search. The requested work to have a quadratically converging process is at least:

$$m = k(3k+2)+1 \tag{7.136}$$

In order to keep information about the surface curvature we must use a method which permits us to compute iteratively the second-order derivative matrix (see equation 7.100).

7.4.7 Some Efficiency Comparisons

To compare different optimization processes exactly is not easy. Nevertheless, in Figure 7.20 we try to accomplish this. The chart shows that the more sophisticated the method the more efficient it may be. Analytical derivatives always are most economical;(34) this is true as long as their programmation remains efficient. We must also keep in mind that experience planification enables us to obtain a polynomial development of one or two orders (at least) higher than the highest available derivative. This procedure can be used to compute the anharmonical effect due to the third and fourth orders.(11,35)

Before closing this section let us mention that in particular cases of the transition state location some methods (not very different from those explained here) have been developed. More details are available from the original paper.(36)

B. The molecular polyatomic motion for stable systems

7.5 THE CLASSICAL TREATMENT OF THE PURE HARMONIC POLYATOMIC VIBRATOR

7.5.1 The Differential Equations for the Nuclear Motion

We shall first use the classical approximation to describe the nuclear vibration motion in a polyatomic molecule. The nuclei are supposed to be moving on a potential energy hypersurface which may be built using the Born–Oppenheimer approximation method. Classical mechanics tells us that, without an external force field, the nuclear displacements are obtained by resolution of the Lagrange equations:

$$\frac{\mathrm{d}}{\mathrm{d}t}\frac{\partial T}{\partial \dot{s}_\mathrm{p}}+\frac{\partial V}{\partial s_\mathrm{p}}=0 \qquad \forall p \mid 1 \leqslant p \leqslant m \tag{7.137}$$

where T stands for the vibrational kinetic energy, V stands for the potential energy, s_p and $\dot{s}_p$ stand respectively for the pth internal coordinate and the related velocity.

This is a set of m differential equations; for non-linear molecules m equals $3N-6$, 6 being the three translational and the three rotational molecular degrees of freedom (N stands for the number of atoms). For linear molecules m equals $3N-5$ because one degree of rotation is lost. (The inertia component corresponding to the molecular axis is zero as well as the corresponding rotational energy; on the other hand, for each internal angle there are two ways to deform the molecule which correspond to bending in two perpendicular planes.)

The internal coordinates $\mathbf{s}_p$ *may be, for example, the bond lengths and the bond angles of the molecule. Thus in the water molecule we may choose* d_1, d_2 *and* θ:

Sometimes it is convenient to multiply the angular variables by a constant having length as a dimension so we will use $d_e\theta$ instead of θ where d_e may be taken to be the equilibrium bond length. If we collect the internal coordinates in a column vector $\mathbf{s}$ or $\mathbf{s}_e$ for the equilibrium position and if we call $\mathbf{S}$ the vector of internal displacements, we have:

$$\mathbf{S}=\mathbf{s}-\mathbf{s}_e \tag{7.138}$$

Then for water we may write:

$$\mathbf{S}=\begin{pmatrix} d_1 \\ d_2 \\ d_e\theta \end{pmatrix}-\begin{pmatrix} d_e \\ d_e \\ d_e\theta_e \end{pmatrix}=d_e\begin{pmatrix} d_1/d_e-1 \\ d_2/d_e-1 \\ \theta-\theta_e \end{pmatrix}$$

7.5.2 The Kinetic Energy Expression

The kinetic energy T is somewhat tedious to write in terms of internal coordinates. In fact we know its expression in cartesian coordinates. If $\mathbf{R}$ is a column vector of $3N$ components corresponding to the x, y, z cartesian coordinates of the N atoms forming the molecule of interest and if $\mathbf{R}_e$ is the same vector which corresponds to the equilibrium structure, then the cartesian displacement vector from the equilibrium $\boldsymbol{\xi}$ will be:

$$\boldsymbol{\xi}=\mathbf{R}-\mathbf{R}_e \tag{7.139}$$

Therefore for water:

$$\boldsymbol{\xi}_{H_2O} = \begin{pmatrix} x(H_1) - x_e(H_1) \\ y(H_1) - y_e(H_1) \\ z(H_1) - z_e(H_1) \\ x(H_2) - x_e(H_2) \\ y(H_2) - y_e(H_2) \\ z(H_2) - z_e(H_2) \\ x(O_3) - x_e(O_3) \\ y(O_3) - y_e(O_3) \\ z(O_3) - z_e(O_3) \end{pmatrix}$$

As the kinetic energy is:

$$T = \frac{1}{2} \sum_{i=1}^{3N} m_i \left(\frac{d\xi_i}{dt}\right)^2 \tag{7.140a}$$

using a matrical notation we write:

$$2T = \dot{\boldsymbol{\xi}}' \cdot \mathbf{M} \cdot \dot{\boldsymbol{\xi}} \tag{7.140b}$$

where $\dot{\boldsymbol{\xi}}$ is the column vector of the cartesian velocities $(\partial\xi_i/dt)$ and $\mathbf{M}$ is the $3N \times 3N$ diagonal square matrix of masses, each atom supposedly being mass isotrope. In Table 7.3 we give the isotopical masses of some current atoms.

Table 7.3 Isotope masses[37]

Atom	Mass (g mol^{-1})	Natural abundance (%)
H^1	1.007825	99.985
H^2	2.01410	0.015
C^{12}	12.00000	98.890
C^{13}	13.00335	1.110
N^{14}	14.00307	99.630
N^{15}	15.00011	0.370
O^{16}	15.99491	99.759
O^{17}	16.99914	0.037
O^{18}	17.99916	0.204
F^{19}	18.99840	100.000

We may also use an alternate coordinate system q defined as:

$$\mathbf{q} = \mathbf{M}^{1/2} \cdot \boldsymbol{\xi} \tag{7.141}$$

and a named mass weighted cartesian displacement coordinate. Then the kinetic energy expression becomes:

$$2T = \dot{\mathbf{q}}' \cdot \dot{\mathbf{q}} \tag{7.142}$$

We now need a linear transformation between the internal coordinates and the cartesian displacements:

$$\mathbf{S} = \mathbf{B} \cdot \xi \qquad (7.143)$$

or its inverse:

$$\xi = \mathbf{B}^{-1} \cdot \mathbf{S} = \mathbf{A} \cdot \mathbf{S} \qquad (7.144)$$

Such expressions are always valid as long as the displacement is infinitesimal, although they are usually not linear nor orthogonal. Notice that the **B** and **A** matrices have as dimensions $(m, 3N)$ and $(3N, m)$ respectively. Hence they are not simply the inverse of each other, but at least:

$$\mathbf{B} \cdot \mathbf{A} = \mathbf{E} \qquad (7.145)$$

where **E** stands for a unity matrix in an m-dimensional space. From equations (7.140) and (7.144) we find:

$$2T = \dot{\mathbf{S}}' \cdot \mathbf{A}' \cdot \mathbf{M} \cdot \mathbf{A} \cdot \dot{\mathbf{S}} \qquad (7.146)$$

Introduction of the **G** matrix of Wilson gives:[38]

$$\mathbf{G} = \mathbf{B} \cdot \mu \cdot \mathbf{B}' \qquad (7.147)$$

where μ stands for $\mathbf{M}^{-1}$. Then:

$$2T = \dot{\mathbf{S}}' \cdot \mathbf{G}^{-1} \cdot \dot{\mathbf{S}} \qquad (7.148)$$

Where **G** is an $m \times m$ symmetrical matrix which allows determination of the **A** matrix using the Crawford–Fletcher formula:[39]

$$\mathbf{A} = \mu \cdot \mathbf{B}' \cdot \mathbf{G}^{-1} \qquad (7.149)$$

The **B** matrix may be obtained using a set of m unity vectors per atom, properly distributed. More details may be found elsewhere[40] as this is not the purpose of this book. For the water molecule the structure of this matrix will be:

$$\mathbf{B} = \begin{pmatrix} \mathbf{e}_{31} & 0 & -\mathbf{e}_{31} \\ 0 & \mathbf{e}_{32} & -\mathbf{e}_{32} \\ \dfrac{\mathbf{e}_{31}}{\operatorname{tg}\theta} - \dfrac{\mathbf{e}_{32}}{\sin\theta} & \dfrac{\mathbf{e}_{32}}{\operatorname{tg}\theta} - \dfrac{\mathbf{e}_{31}}{\sin\theta} & (\mathbf{e}_{31} + \mathbf{e}_{32})\left(\dfrac{1-\cos\theta}{\sin\theta}\right) \end{pmatrix}$$

where $\mathbf{e}_{31}$ and $\mathbf{e}_{32}$ are unity vectors along OH_1 and OH_2 respectively:

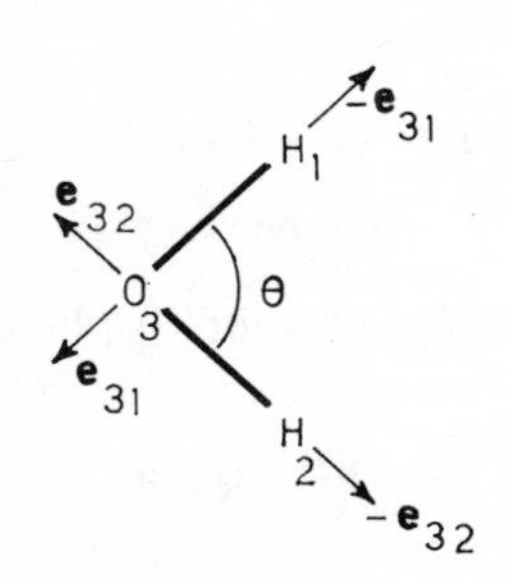

Numerically we find for the water molecule:

$$\mathbf{B}=\left(\begin{array}{ccc|ccc|ccc} 0 & 0.563 & 0.827 & 0 & 0.000 & 0.000 & 0 & -0.563 & -0.827 \\ 0 & 0.000 & 0.000 & 0 & 0.563 & -0.827 & 0 & -0.563 & 0.827 \\ 0 & -0.827 & 0.563 & 0 & -0.827 & -0.563 & 0 & 1.653 & 0.000 \end{array}\right)$$

7.5.3 The Potential Energy Expression

In order to obtain a solution as simple as possible we develop the potential energy (V) in a Taylor series limited to the second order with respect to the internal displacements:

$$V(S)=V_e+(\boldsymbol{\nabla}V)'_e\mathbf{S}+\tfrac{1}{2}\mathbf{S}'(\boldsymbol{\nabla}\boldsymbol{\nabla}'V)_e\mathbf{S}$$

where $\boldsymbol{\nabla}$ stands for the gradient operator. As the origin used to define the $\mathbf{S}$ coordinate frame is an equilibrium structure (or more generally a stationary point), we have:

$$(\boldsymbol{\nabla}V)_e=\mathbf{g}_e=\mathbf{0}$$

If we choose the value corresponding to the equilibrium structure as the zero potential energy then:

$$V_e=0$$

So $V(S)$ finds a simple form:

$$\begin{aligned} V(S)&=\tfrac{1}{2}\mathbf{S}'(\boldsymbol{\nabla}\boldsymbol{\nabla}'V)_e\mathbf{S} \\ &=\tfrac{1}{2}\mathbf{S}'\mathbf{H}\mathbf{S} \end{aligned} \tag{7.150}$$

where $\mathbf{H}$ stands for the second-order derivative matrix to the potential, in terms of internal displacement coordinates:†

$$H_{ij}=\left(\frac{\partial^2 V}{\partial S_i\,\partial S_j}\right)_e$$

Such an approximation on the form of the potential energy surface around the equilibrium position is found *a posteriori* to be efficient in computing the normal mode of vibration for polyatomic systems.[41]

† The H_{ij} components are generally expressed in millidynes per angstrom by spectroscopists, and in theoretical computation one uses the atomic unit system. The two systems are simply connected by the following relations:

$$\begin{aligned} 1\text{ a.u. of energy}&=\frac{m_e e^4}{(4\pi\varepsilon_0\hbar)^2} \\ &=4.359425081\times10^{-18}\text{ J} \\ &=4.359425081\text{ mdyn Å} \end{aligned}$$

and

$$1\text{ a.u. of distance}=\frac{4\pi\varepsilon_0\hbar^2}{m_e e^2}=0.5291671742\text{ Å}$$

Therefore: 1 a.u. of force constant = 15.56838012 mdyn Å^{-1}

In terms of cartesian displacements the potential energy becomes:

$$2V = \xi' \mathbf{B}' \mathbf{H} \mathbf{B} \xi = \xi' \mathbf{F}_\xi \xi \tag{7.151}$$

where $\mathbf{F}_\xi$ *is the corresponding force constant matrix related to* $\mathbf{H}$ *by:*

$$\mathbf{F}_\xi = \mathbf{B}' \mathbf{H} \mathbf{B} \tag{7.152}$$

Using a mass weighted cartesian displacement a set of equivalent expressions may be obtained:

$$\mathbf{S} = \mathbf{B}_q \cdot \mathbf{q} \quad \text{and} \quad \mathbf{q} = \mathbf{A}_q \cdot \mathbf{S} \tag{7.153}$$

with:

$$\mathbf{B}_q = \mathbf{B} \cdot \mathbf{M}^{-1/2} \quad \text{and} \quad \mathbf{A}_q = \mathbf{M}^{1/2} \cdot \mathbf{A} \tag{7.154}$$

Then for the kinetic energy:

$$2T = \dot{\mathbf{S}}' \cdot \mathbf{A}_q' \cdot \mathbf{A}_q \cdot \dot{\mathbf{S}} = \dot{\mathbf{S}}' \cdot \mathbf{G}^{-1} \cdot \dot{\mathbf{S}}$$

with:

$$\mathbf{G} = \mathbf{B}_q \cdot \boldsymbol{B}_q' \tag{7.155a}$$

and

$$\mathbf{A}_q = \mathbf{B}_q' \mathbf{G}^{-1} \tag{7.155b}$$

For the potential energy:

$$2V = \mathbf{q}' \cdot \mathbf{B}_q' \cdot \mathbf{H} \cdot \mathbf{B}_q \cdot \mathbf{q} = \mathbf{q}' \mathbf{F}_q \mathbf{q} \tag{7.156}$$

where:

$$\mathbf{F}_q = \mathbf{B}_q' \cdot \mathbf{H} \cdot \mathbf{B}_q = \mathbf{M}^{-1/2} \cdot \mathbf{F}_\xi \cdot \mathbf{M}^{-1/2} \tag{7.157}$$

7.5.4 The Normal Modes of Vibrations

The Lagrange equations (7.137), introducing the expressions of **S** (7.138), T (7.148) and V (7.150), will now be written as:

$$\mathbf{G}^{-1} \cdot \ddot{\mathbf{S}} + \mathbf{H} \cdot \mathbf{S} = \mathbf{0}$$

or, left multiplying by **G**:

$$\ddot{\mathbf{S}} + \mathbf{G} \cdot \mathbf{H} \cdot \mathbf{S} = \mathbf{0} \tag{7.158}$$

In order to get uncoupled differential equations we must find coordinates which diagonalize the $\mathbf{G} \cdot \mathbf{H}$ matrix. We call those coordinates the normal coordinates **Q** such that a coordinate exists per vibration mode. The **Q**'s are linearly related to **S** by:

$$\mathbf{S} = \mathbf{L} \cdot \mathbf{Q} \tag{7.159}$$

As the **L** matrix must be time independent:

$$\mathbf{L}\ddot{\mathbf{Q}} + \mathbf{GHLQ} = \mathbf{0}$$

or if $\mathbf{L}^{-1}(=\mathbf{K})$ is such that $\mathbf{L}^{-1}\mathbf{L} = \mathbf{E}$:

$$\ddot{\mathbf{Q}} + \mathbf{L}^{-1} \cdot \mathbf{GH} \cdot \mathbf{LQ} = \mathbf{0} \tag{7.160}$$

The matrix $\mathbf{L}$ we are looking for is the one that diagonalizes $\mathbf{GH}$. If:

$$\mathbf{L}^{-1}\mathbf{GHL} = \mathbf{\Lambda} \tag{7.161}$$

is a diagonal square matrix, then:

$$\ddot{\mathbf{Q}} + \mathbf{\Lambda Q} = \mathbf{0} \tag{7.162}$$

or

$$(\ddot{Q}_i + \lambda_i Q_i) = 0 \qquad \forall i \mid 1 \leqslant i \leqslant m$$

where λ_i stands for the ith diagonal term of $\mathbf{\Lambda}$.

The solution of these second-order differential equations (6.161) are of the form:

$$Q_i = a_i \cos(\omega_i t + \varphi_i) \tag{7.163}$$

where a, ω and φ respectively stand for the amplitude, the angular velocity ($\omega = 2\pi\nu$) and the phase of the vibrational motion. As $\ddot{Q}_i = -\omega_i^2 Q_i$ it appears that λ_i represents the square value of ω_i. The vibrational frequencies are given by the expression:

$$\nu_i = \frac{\omega_i}{2\pi} = \frac{\sqrt{\lambda_i}}{2\pi} \tag{7.164}$$

which is real as long as λ_i is positive. If $\mathbf{G}$ is in moles per gram and $\mathbf{H}$ in millidynes per angstrom we find that:

$$\nu_i = 3.905796227 \times 10^{13} \sqrt{\lambda_i}\ \text{Hz}$$

or

$$\tilde{\nu}_i = 1{,}302.833159 \sqrt{\lambda_i}\ \text{cm}^{-1}$$

where ν is the frequency of nuclear motion and $\tilde{\nu}$ is the wave number.

The $\mathbf{L}$ matrix still keeps one degree of freedom which is used to normalize the matrix (this does not change the λ_i values). We choose as the condition:

$$\mathbf{L}'\mathbf{G}^{-1}\mathbf{L} = \mathbf{E} \tag{7.165}$$

It appears then that:

$$\mathbf{LL}' = \mathbf{G} \quad \text{and} \quad \mathbf{K}'\mathbf{K} = \mathbf{G}^{-1} \tag{7.166}$$

Finally:

$$\mathbf{K} = \mathbf{L}'\mathbf{G}^{-1} \tag{7.167}$$

Multiplying (7.161) by (7.165) we find:

$$\mathbf{L}'\mathbf{HL} = \mathbf{\Lambda} \tag{7.168}$$

Note that $\mathbf{L}'$ is not the unitary transform which diagonalizes $\mathbf{H}$; in fact, $\mathbf{L}' \neq \mathbf{K}$. For dimensional convenience $\mathbf{L}$ has the same units as $\mathbf{G}^{1/2}$.

The kinetic and potential energies may now be written in terms of normal coordinates as:

$$2T = \dot{\mathbf{Q}}'\mathbf{E}\dot{\mathbf{Q}} \tag{7.169a}$$

$$2V = \mathbf{Q}'\mathbf{\Lambda Q} \tag{7.169b}$$

The inverse transformation between **Q** and **S** is:

$$\mathbf{Q} = \mathbf{K} \cdot \mathbf{S} \tag{7.170}$$

Each line of **K** defines the nuclear motion along the corresponding normal coordinate in terms of internal variables. In Table 7.4 we report the dimension which must be assigned to any vector and matrix.

Table 7.4 The vector and matrix dimensions (L stands for a length, M for a mass and T for a time

Vector or matrix	Dimension
$\mathbf{S}, \boldsymbol{\xi}$	L
$\mathbf{M}$	M
$\mathbf{H}$	MT^{-2}
$\mathbf{G}$	M^{-1}
$\mathbf{GH}, \boldsymbol{\Lambda}$	T^{-2}
$\mathbf{L}$	$M^{-1/2}$
$\mathbf{Q}, \mathbf{q}$	$M^{+1/2}L$
$\mathbf{B}$	—

The same deduction is available from the mass weighted coordinate frame where:

$$2T = \dot{\mathbf{q}}'\dot{\mathbf{q}}$$

$$2V = \mathbf{q}'\mathbf{F}_q\mathbf{q}$$

This leads to the following set of equations of motion:

$$\ddot{\mathbf{q}} + \mathbf{F}_q\mathbf{q} = \mathbf{0} \tag{7.171}$$

By diagonalization of $\mathbf{F}_q$ *we can uncouple the set of* $3N$ *second-order differential equations; as the potential has zero contribution for the three global translations and the three or two global rotations* (three for non-linear molecules and two for the linear), F_q *has 6 or 5 zero eigenvalues. Then we find*:

$$\begin{pmatrix} \mathbf{l}' \\ \mathbf{l}'_{\text{rt}} \end{pmatrix} \mathbf{F}_q (\mathbf{l}\mathbf{l}_{\text{rt}}) = \begin{pmatrix} \boldsymbol{\Lambda} & \mathbf{0} \\ \mathbf{0} & \mathbf{0} \end{pmatrix} \tag{7.172}$$

Notice that orthonormality requirements lead to the relations:

$$\left.\begin{matrix} \textit{from} & \begin{pmatrix} \mathbf{l}' \\ \mathbf{l}'_{\text{rt}} \end{pmatrix} (\mathbf{l}\mathbf{l}_{\text{rt}}) = \mathbf{E} \\ \textit{and} & \\ & (\mathbf{l}\mathbf{l}_{\text{rt}}) \begin{pmatrix} \mathbf{l}' \\ \mathbf{l}'_{\text{rt}} \end{pmatrix} = \mathbf{E} \end{matrix}\right\} \quad \textit{we find} \quad \begin{cases} \mathbf{l}'\mathbf{l} = \mathbf{E} \\ \mathbf{l}'_{\text{rt}}\mathbf{l} = \mathbf{0} \\ \mathbf{l}'_{\text{rt}}\mathbf{l}_{\text{rt}} = \mathbf{E} \\ \mathbf{l}\mathbf{l}' + \mathbf{l}_{\text{rt}}\mathbf{l}'_{\text{rt}} = \mathbf{E} \end{cases} \tag{7.173}$$

while the normal coordinates are now defined as:

$$\mathbf{Q} = \mathbf{l}'\mathbf{q} \tag{7.174a}$$

with as inverse equation:

$$\mathbf{q} = \mathbf{l}\mathbf{Q} \tag{7.174b}$$

Connection between this last expression for Q and the former in terms of internal displacement (7.159) *is available from the following equations:*

$$\mathbf{l} = \mathbf{M}^{1/2}\mathbf{A}\mathbf{L} = \mathbf{A}_q\mathbf{L} = \boldsymbol{\mu}^{1/2}\mathbf{B}'\mathbf{K}' = \mathbf{B}_q'\mathbf{K}' \tag{7.175a}$$

$$\mathbf{l}' = \mathbf{K}\mathbf{B}\boldsymbol{\mu}^{1/2} = \mathbf{K}\mathbf{B}_q = \mathbf{L}'\mathbf{A}'\mathbf{M}^{1/2} = \mathbf{L}'\mathbf{A}_q' \tag{7.175b}$$

From these we find convenient expressions for the symmetric product matrix $\mathbf{l}\mathbf{l}'$:

$$\mathbf{l}\mathbf{l}' = \mathbf{M}^{1/2}\mathbf{A}\mathbf{B}\boldsymbol{\mu}^{1/2} = \mathbf{A}_q\mathbf{B}_q \tag{7.176a}$$

or

$$\mathbf{l}\mathbf{l}' = \mathbf{M}^{1/2}\mathbf{A}\mathbf{G}\mathbf{A}'\mathbf{M}^{1/2} = \mathbf{A}_q\mathbf{G}\mathbf{A}_q' \tag{7.176b}$$

It follows that the **AB** *product has the form:*

$$\mathbf{A}\mathbf{B} = \mathbf{A}\mathbf{G}\mathbf{A}'\mathbf{M} \tag{7.177}$$

The **A** *matrix which has been defined as the right inverse of* **B** ($\mathbf{B}\mathbf{A} = \mathbf{E}$) *cannot be further considered as its left inverse too. Finally, we have for the product* $\mathbf{l}_{rt}\mathbf{l}_{rt}'$:

$$\mathbf{l}_{rt}\mathbf{l}_{rt}' = \mathbf{E} - \mathbf{l}\mathbf{l}' = \mathbf{E} - \mathbf{M}^{1/2}\mathbf{A}\mathbf{B}\boldsymbol{\mu}^{1/2} = \mathbf{E} - \mathbf{M}^{1/2}\mathbf{A}\mathbf{G}\mathbf{A}'\mathbf{M}^{1/2} \tag{7.178}$$

7.5.5 The Diagonalization Method

If **G** and **H** are symmetrical matrices separately, the product **GH** is in general unsymmetrical. So the classical diagonalization methods (Jacobi, Givens–Houscholder, etc.) cannot be used unless one finds a similarity transform. In order to find this, we first prove that **G** is a positive defined matrix. If **U** is an unitary matrix which diagonalizes the symmetrical **G** matrix then:

$$\mathbf{U}' \cdot \mathbf{G} \cdot \mathbf{U} = \boldsymbol{\gamma} \tag{7.179}$$

$$\mathbf{U}' = \mathbf{U}^{-1} \quad \text{and} \quad \mathbf{U}'\mathbf{U} = \mathbf{E}$$

U also diagonalizes $\mathbf{G}^{-1}$:

$$\mathbf{U}' \cdot \mathbf{G}^{-1} \cdot \mathbf{U} = \boldsymbol{\gamma}^{-1} \tag{7.180}$$

Let us define a new coordinate system $\mathbf{v}$ such that:

$$\mathbf{v} = \mathbf{U}'\mathbf{S} \tag{7.181}$$

The kinetic energy (7.148) then takes a diagonal form:

$$2T = \dot{\mathbf{S}}'\mathbf{G}^{-1}\dot{\mathbf{S}} = \dot{\mathbf{v}}'\mathbf{U}'\mathbf{G}^{-1}\mathbf{U}\dot{\mathbf{v}} = \dot{\mathbf{v}}'\boldsymbol{\gamma}^{-1}\dot{\mathbf{v}} \tag{7.182}$$

or

$$2T = \sum_i \gamma_i^{-1}\dot{v}_i^2$$

For all velocities different from zero, T must be strictly positive; thus all γ_i values must also be positive. In such a case the diagonalization of **GH** may simply be done by choosing a similarity transform matrix:

$$\mathbf{W} = \mathbf{G}^{1/2} \tag{7.183}$$

Then from (7.179) and (7.180) we have:

$$\mathbf{W} = \mathbf{U}\boldsymbol{\gamma}^{1/2} \quad \text{and} \quad \mathbf{W}^{-1} = \boldsymbol{\gamma}^{-1/2}\mathbf{U}' \tag{7.184}$$

Multiplying **GH** left by $\mathbf{W}^{-1}$ and right by **W** we get:

$$\mathbf{W}^{-1}(\mathbf{GH})\mathbf{W} = \mathbf{G}^{1/2\prime}\mathbf{H}\mathbf{G}^{1/2} \tag{7.185}$$

We have now a symmetrical matrix which is simply diagonalized by a **Y** transformation such that:

$$\mathbf{Y}^{-1}\mathbf{W}^{-1}(\mathbf{GH})\mathbf{WY} = \boldsymbol{\Lambda} \tag{7.186}$$

This expression shows that if $\boldsymbol{\Lambda}$ are the expected eigenvalues of **GH**, then **WY** are the corresponding eigenvectors:

$$\mathbf{L} = \mathbf{WY} \quad \text{or} \quad \mathbf{L}^{-1} = \mathbf{K} = \mathbf{Y}'\mathbf{W}^{-1} \tag{7.187}$$

Normalization of **L** is simply done by (see equation 7.166):

$$\mathbf{LL}' = \mathbf{WYY}'\mathbf{W}' = \mathbf{WW}' = \mathbf{G}$$

where:

$$\mathbf{Y}' = \mathbf{Y}^{-1}$$

7.6 CLASSICAL TREATMENT OF THE ROTATING–VIBRATING MOLECULES

7.6.1 Partitioning of the Kinetic Energy[42]

To express in a classical way the total kinetic energy we shall choose two cartesian coordinate systems. The first one is the space-fixed X, Y, Z system; the second one is a rotating body-fixed x, y, z system whose origin is placed at the centre of mass of the molecule (Figure 7.21).

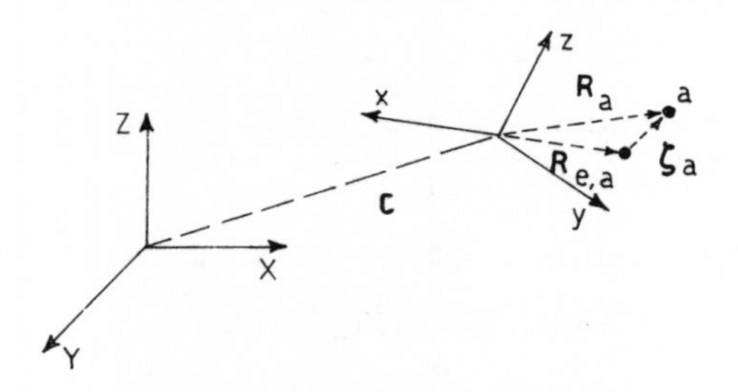

Figure 7.21 Position of coordinate systems and vectors of an atom 'a'

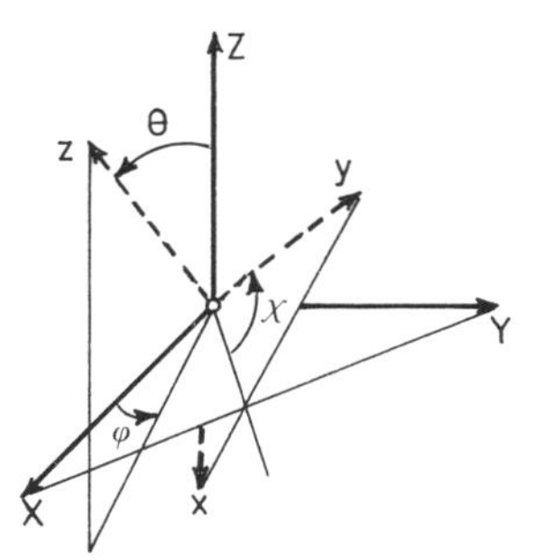

Figure 7.22 Definition of the eulerian angles

At any time the position of an atom a is given by the following vectors:

- **c**: The centre of mass position vector in the space-fixed frame with components c_X, c_Y and c_Z.
- **θ**: The eulerian angles (θ, φ, χ) which define the instantaneous orientation of the body-fixed frame relative to the space-fixed one (see Figure 7.22). The associated angular velocity will be given by the vector $\boldsymbol{\omega}$ of components ω_x, ω_y and ω_z.
- $\mathbf{R}_a$: The position vector of the atom a in the body-fixed frame, with components $R_{a,x}$, $R_{a,y}$ and $R_{a,z}$. All $\mathbf{R}_a$ vectors may be collected in a single vector $\mathbf{R} = \{\mathbf{R}_a, \forall a = 1, N\}$.

Let us also introduce the displacement vector $\boldsymbol{\xi}$ (in the body-fixed frame) from a reference atomic configuration defined by the position vector $\mathbf{R}_e$ (usually the equilibrium structure $\mathbf{R}_e$ vector):

$$\mathbf{R} = \mathbf{R}_e + \boldsymbol{\xi}$$

In addition, we assume that any displacement $(\boldsymbol{\xi}_a)$ relative to the moving cartesian coordinate system is subject to the Eckart conditions.

The Eckart conditions must define the rotating coordinate system completely. The condition that the origin of the moving frame (x, y, z) *be at the centre of mass of the whole molecule yields the equations*

$$\sum_a m_a x_a = \sum_a m_a y_a = \sum_a m_a z_a$$

or

$$\sum_a m_a \mathbf{R}_a = \mathbf{0} \tag{7.188}$$

It follows by time derivation that:

$$\begin{aligned}\sum_a m_a \dot{\mathbf{R}}_a &= \sum_a m_a \omega \times \mathbf{R}_a + \sum_a m_a \dot{\boldsymbol{\xi}}_a \\ &= \boldsymbol{\omega} \times \sum_a m_a \mathbf{R}_a + \sum_a m_a \dot{\boldsymbol{\xi}}_a \\ &= \sum_a m_a \dot{\boldsymbol{\xi}}_a = \mathbf{0}\end{aligned} \tag{7.189}$$

The last three conditions are chosen so that the axes will rotate with the molecule. The conditions used are:

$$\sum_a m_a \mathbf{R}_a^e \times \mathbf{R}_a = \mathbf{0} \tag{7.190}$$

This implies by time differentiation:

$$\begin{aligned}
\sum_a m_a \dot{\mathbf{R}}_a^e \times \mathbf{R}_a + \sum_a m_a \mathbf{R}_a^e \times \dot{\mathbf{R}}_a &= \mathbf{0} \\
\sum_a m_a (\boldsymbol{\omega} \times \mathbf{R}_a^e) \times \mathbf{R}_a + \sum_a m_a \mathbf{R}_a^e \times (\boldsymbol{\omega} \times \mathbf{R}_a) + \sum_a m_a \mathbf{R}_a^e \times \dot{\boldsymbol{\xi}}_a &= \mathbf{0} \\
\boldsymbol{\omega} \times \sum_a m_a (\mathbf{R}_a^e \times \mathbf{R}_a) + \sum_a m_a (\mathbf{R}_a^e \times \dot{\boldsymbol{\xi}}_a) &= \mathbf{0} \\
\sum_a m_a \mathbf{R}_a^e \times \dot{\boldsymbol{\xi}}_a &= \mathbf{0}
\end{aligned} \tag{7.191}$$

Our purpose is now to write the total (and not only the vibrational) kinetic energy by introducing the normal coordinates previously defined (7.170 and 7.143):

$$\mathbf{Q} = \mathbf{KS} = \mathbf{KB}\boldsymbol{\xi}$$

The total velocity vector of atom a can be expressed as the sum:

$$\mathbf{V}_a = \dot{\mathbf{c}} + \boldsymbol{\omega} \times \mathbf{R}_a + \dot{\boldsymbol{\xi}}_a \tag{7.192}$$

Then the kinetic energy becomes:

$$2T = \sum_a m_a \mathbf{V}_a \cdot \mathbf{V}_a \tag{7.193a}$$

$$\begin{aligned}
&= \dot{\mathbf{c}} \cdot \dot{\mathbf{c}} \sum_a m_a + \sum_a m_a (\boldsymbol{\omega} \times \mathbf{R}_a) \cdot (\boldsymbol{\omega} \times \mathbf{R}_a) + \sum_a m_a \dot{\boldsymbol{\xi}}_a \cdot \dot{\boldsymbol{\xi}}_a \\
&\quad + 2 \sum_a m_a \dot{\mathbf{c}} \cdot (\boldsymbol{\omega} \times \mathbf{R}_a) + 2 \sum_a m_a \dot{\mathbf{c}} \cdot \dot{\boldsymbol{\xi}}_a \\
&\quad + 2 \sum_a m_a (\boldsymbol{\omega} \times \mathbf{R}_a) \cdot \dot{\boldsymbol{\xi}}_a
\end{aligned} \tag{7.193b}$$

Remembering the vectorial property:

$$\mathbf{a} \cdot (\mathbf{b} \times \mathbf{c}) = (\mathbf{a} \times \mathbf{b}) \cdot \mathbf{c}$$

we find:

$$\begin{aligned}
2T &= \dot{\mathbf{c}} \cdot \dot{\mathbf{c}} \sum_a m_a + \sum_a m_a (\boldsymbol{\omega} \times \mathbf{R}_a) \cdot (\boldsymbol{\omega} \times \mathbf{R}_a) + \sum_a m_a \dot{\boldsymbol{\xi}}_a \cdot \dot{\boldsymbol{\xi}}_a \\
&\quad + 2(\dot{\mathbf{c}} \times \boldsymbol{\omega}) \cdot \sum_a m_a \mathbf{R}_a + 2\dot{\mathbf{c}} \cdot \left(\sum_a m_a \dot{\boldsymbol{\xi}}_a \right) \\
&\quad + 2\boldsymbol{\omega} \cdot \sum_a m_a (\mathbf{R}_a \times \dot{\boldsymbol{\xi}}_a)
\end{aligned} \tag{7.194}$$

As by the Eckart conditions (7.188 to 7.190):

$$\sum_a m_a \mathbf{R}_a = \sum_a m_a \dot{\boldsymbol{\xi}}_a = \sum_a m_a (\mathbf{R}_a^e \times \dot{\boldsymbol{\xi}}_a) = \mathbf{0}$$

the fourth and fifth terms vanish and the sixth term may be simplified. Only the three first terms remain, which correspond respectively to the translational, the rotational and the vibrational energies; the last one, which is the rotation–vibration coupling term, is also called the Coriolis term. The following partitioning of the total kinetic energy may be used:

For translational contribution:

$$2T_t = \sum_a m_a(\dot{\mathbf{c}} \cdot \dot{\mathbf{c}}) \tag{7.195a}$$

For rotational contribution:

$$2T_r = \sum_a m_a(\boldsymbol{\omega} \times \mathbf{R}_a) \cdot (\boldsymbol{\omega} \times \mathbf{R}_a) \tag{7.195b}$$

For vibrational contribution:

$$2T_v = \sum_a m_a \dot{\boldsymbol{\xi}}_a \cdot \dot{\boldsymbol{\xi}}_a \tag{7.195c}$$

For Coriolis contribution:

$$T_c = \boldsymbol{\omega} \sum_a m_a(\dot{\boldsymbol{\xi}}_a \times \dot{\boldsymbol{\xi}}_a) \tag{7.195d}$$

For total kinetic energy:

$$T = T_t + T_r + T_v + T_c \tag{7.195e}$$

Let us now expand each contribution to the total kinetic energy in terms of normal coordinates and introduce the matrix notation.

7.6.2 The Centre of Mass Translational Energy (T_t)

Defining a diagonal 3×3 matrix $\mathbf{m}$ with:

$$(m_{hh}) = \sum_a m_a \qquad \forall h = 1, 3 \qquad \text{or} \qquad h = x, y, z$$

the energy associated with the translation of the whole system has the form:

$$2T_t = \dot{\mathbf{c}}' \cdot \mathbf{m} \cdot \dot{\mathbf{c}} \tag{7.196}$$

7.6.3 The Rotational Kinetic Energy (T_r)

In order to express the vectorial product in terms of matrices let us introduce three partial $\mathbf{I}_a^x$, $\mathbf{I}_a^y$ and $\mathbf{I}_a^z$ matrices (matrices in Levi–Civita symbols) for each atom:

$$\mathbf{I}_a^x = \begin{pmatrix} 0 & 0 & 0 \\ 0 & 0 & 1 \\ 0 & -1 & 0 \end{pmatrix}; \quad \mathbf{I}_a^y = \begin{pmatrix} 0 & 0 & -1 \\ 0 & 0 & 0 \\ 1 & 0 & 0 \end{pmatrix}; \quad \mathbf{I}_a^z = \begin{pmatrix} 0 & 1 & 0 \\ -1 & 0 & 0 \\ 0 & 0 & 0 \end{pmatrix} \tag{7.197}$$

and three block diagonal matrices: $\mathbf{I}^x, \mathbf{I}^y$ and $\mathbf{I}^z$ ($3N \times 3N$ square matrices) where each block is the partial $\mathbf{I}_a^x$, $\mathbf{I}_a^y$ and $\mathbf{I}_a^z$ matrices:

$$\mathbf{I}^f = \begin{pmatrix} \mathbf{I}_1^f & & & 0 \\ & \mathbf{I}_2^f & & \\ & & \ddots & \\ 0 & & & \mathbf{I}_N^f \end{pmatrix} \tag{7.198}$$

If f, g or h are used instead of x, y or z components the rotational energy becomes:

$$2T_r = \sum_f \sum_a m_a(\boldsymbol{\omega} \times \mathbf{R}_a)_f^2 \tag{7.199}$$

where $(\boldsymbol{\omega} \times \mathbf{R}_a)_f$ stands for component f of $(\boldsymbol{\omega} \times \mathbf{R}_a)$ vector. Then:

$$\begin{aligned} 2T_r &= \sum_f \sum_a m_a(\boldsymbol{\omega}' \mathbf{I}_a^f \mathbf{R}_a)^2 \\ &= \sum_f \sum_g \boldsymbol{\omega}_f \mathbf{I}_{fg} \boldsymbol{\omega}_g = \boldsymbol{\omega}' \mathbf{I} \boldsymbol{\omega} \end{aligned} \tag{7.200}$$

with:

$$I_{fg} = \mathbf{R}'(\mathbf{I}^f)\mathbf{M}(\mathbf{I}^g)'\mathbf{R}$$

The 3×3 symmetric square matrix $\mathbf{I}$ is nothing more than the instantaneous inertia tensor. Let us now split the $\mathbf{R}$ vector into its two components $\mathbf{R}_e$ and $\boldsymbol{\xi}$. We then find:

$$\begin{aligned} I_{fg} &= (\mathbf{R}_e + \boldsymbol{\xi})'(\mathbf{I}^f)\mathbf{M}(\mathbf{I}^g)'(\mathbf{R}_e + \boldsymbol{\xi}) \\ &= J_{fg}^{(0)} + \boldsymbol{\xi}' \mathbf{J}_{fg,\boldsymbol{\xi}}^{(1)} + \boldsymbol{\xi}' \mathbf{J}_{fg,\boldsymbol{\xi}}^{(2)} \boldsymbol{\xi} \end{aligned} \tag{7.201}$$

If the next definitions are assumed:

(a)
$$J_{fg}^{(0)} = \mathbf{R}_e'(\mathbf{I}^f)\mathbf{M}(\mathbf{I}^g)'\mathbf{R}_e \tag{7.202a}$$

which is the inertia tensor of the reference structure. Furthermore, if it has been diagonalized by a proper choice of axis in the reference molecular configuration, then it becomes:

$$J_{ff}^{(0)} = \sum_a m_a(g_a^{e2} + h_a^{e2}) \qquad f \neq g \neq h \tag{7.202b}$$

$$J_{fg}^{(0)} = -\sum_a m_a f_a^e g_a^e = 0 \qquad f \neq g \tag{7.202c}$$

(b)
$$\mathbf{J}_{fg,\boldsymbol{\xi}}^{(1)} = 2(\mathbf{I}^f)\mathbf{M}(\mathbf{I}^g)'\mathbf{R}_e \tag{7.203}$$

(c)
$$\mathbf{J}_{fg,\boldsymbol{\xi}}^{(2)} = (\mathbf{I}^f)\mathbf{M}(\mathbf{I}^g)' \tag{7.204}$$

It must be mentioned that the same kind of expression is available in the normal coordinates Q previously defined:

$$I_{fg} = J_{fg}^{(0)} + \mathbf{Q}' \mathbf{J}_{fg,Q}^{(1)} + \mathbf{Q}' \mathbf{J}_{fg,Q}^{(2)} \mathbf{Q} \tag{7.205}$$

with:

$$\mathbf{J}^{(1)}_{fg,Q} = \mathbf{L'A'J}^{(1)}_{fg,\xi} \tag{7.206}$$

and

$$\mathbf{J}^{(2)}_{fg,Q} = \mathbf{L'A'J}^{(2)}_{fg,\xi}\mathbf{AL} \tag{7.207}$$

so we can write:

$$\mathbf{J}^{(1)}_{fg,Q} = \left\{ \left(\frac{\partial I_{fg}}{\partial Q_i} \right)_e \right\} \quad \text{and} \quad J^{(2)}_{fg,Q} = \frac{1}{2} \left\{ \left(\frac{\partial^2 I_{fg}}{\partial Q_i \, \partial Q_j} \right)_e \right\} \tag{7.208}$$

$\mathbf{J}^{(1)}_{fg,Q}$ and $\mathbf{J}^{(2)}_{fg,Q}$ are respectively the first- and second-order distortion terms of the rotational motion.

7.6.4 The Vibrational Kinetic Energy (T_v)

As previously found (7.169a), the kinetic energy for the vibrational motion may be written in terms of normal coordinates as:

$$2T_v = \dot{\mathbf{Q}}'\dot{\mathbf{Q}} \tag{7.209}$$

7.6.5 The Coriolis Coupling Energy (T_c)

The method presented above for the rotational term may be applied to this problem as well; then:

$$\begin{aligned} T_c &= \sum_f \omega_f \sum_a m_a(\boldsymbol{\xi}_a \times \dot{\boldsymbol{\xi}}_a)_f \\ &= \sum_f \omega_f \sum_a m_a(\boldsymbol{\xi}'_a \mathbf{I}^f_a \dot{\boldsymbol{\xi}}_a) \\ &= \sum_f \omega_f \boldsymbol{\xi}' \mathbf{M}(\mathbf{I}^f)\dot{\boldsymbol{\xi}} \end{aligned} \tag{7.210}$$

Going now through the normal coordinates it becomes:

$$T_c = \sum_f \omega_f \mathbf{Q}' \boldsymbol{\zeta}^f \dot{\mathbf{Q}} \tag{7.211}$$

where $\boldsymbol{\zeta}^f$ stands for the Meal–Polo coupling matrices:

$$\boldsymbol{\zeta}^f = \mathbf{L'A'M}(\mathbf{I}^f)\mathbf{AL} \tag{7.212}$$

Finally, it is convenient to introduce an auxiliary coupling matrix $\mathbf{Z}$ defined by:

$$\mathbf{Z} = \begin{pmatrix} \mathbf{Q}'\boldsymbol{\zeta}^x \\ \mathbf{Q}'\boldsymbol{\zeta}^y \\ \mathbf{Q}'\boldsymbol{\zeta}^z \end{pmatrix} \tag{7.213}$$

With this, the Coriolis term becomes:

$$T_c = \boldsymbol{\omega}'\mathbf{Z}\dot{\mathbf{Q}} \tag{7.214}$$

7.6.6 The Total Kinetic Energy (T)

Collecting together the above defined matrices in such a way that:

$$\boldsymbol{\alpha} = \left(\begin{array}{c|c|c} \mathbf{E} & \mathbf{Z}' & \mathbf{0} \\ \hline \mathbf{Z} & \mathbf{I} & \mathbf{0} \\ \hline \mathbf{0} & \mathbf{0} & \mathbf{m} \end{array}\right) \tag{7.215}$$

$$\boldsymbol{\eta} = \begin{pmatrix} \mathbf{Q} \\ \boldsymbol{\theta} \\ \mathbf{c} \end{pmatrix} \quad \text{and} \quad \dot{\boldsymbol{\eta}} = \begin{pmatrix} \dot{\mathbf{Q}} \\ \boldsymbol{\omega} \\ \dot{\mathbf{c}} \end{pmatrix} \tag{7.216}$$

of dimensions:

$\mathbf{E}$:	$(3N-6)\times(3N-6)$
$\mathbf{Z}$:	$(3)\times(3N-6)$
$\mathbf{I}$ and $\mathbf{m}$:	$(3)\times(3)$
$\mathbf{Q}$:	$(3N-6)\times(1)$
$\boldsymbol{\theta}$, $\boldsymbol{\omega}$ and $\mathbf{c}$:	$(3)\times(1)$

we obtain the kinetic energy expression (which joins equations 7.196, 7.200, 7.209 and 7.214):

$$2T = \dot{\boldsymbol{\eta}}'\boldsymbol{\alpha}\dot{\boldsymbol{\eta}} \tag{7.217}$$

We can express T as a function of the conjugate momenta to $Q_i(P_i)$, $c_f(p_f)$ and the components (m_f)of the total angular momentum:

$$\boldsymbol{\pi} = \frac{\partial T}{\partial \dot{\eta}} = \boldsymbol{\alpha}\dot{\boldsymbol{\eta}} = \left(\begin{array}{c} \mathbf{P} \\ \hline \boldsymbol{m} \\ \hline \mathbf{p} \end{array}\right) \tag{7.218}$$

then:

$$\dot{\boldsymbol{\eta}} = \boldsymbol{\alpha}^{-1} \cdot \boldsymbol{\pi} \tag{7.219}$$

and

$$2T = \boldsymbol{\pi}'\boldsymbol{\alpha}^{-1}\boldsymbol{\pi} \tag{7.220}$$

The relation between the angular velocities:

$$\boldsymbol{\omega} = \begin{pmatrix} \omega_x \\ \omega_y \\ \omega_z \end{pmatrix} \quad \text{and} \quad \dot{\boldsymbol{\theta}} = \begin{pmatrix} \dot{\theta} \\ \dot{\varphi} \\ \dot{\chi} \end{pmatrix}$$

has the form $\boldsymbol{\omega} = \boldsymbol{\rho}\dot{\boldsymbol{\theta}}$, *or*:[43]

$$\boldsymbol{\omega} = \begin{pmatrix} \sin\chi & -\sin\theta\cos\chi & 0 \\ \cos\chi & \sin\theta\sin\chi & 0 \\ 0 & \cos\theta & 1 \end{pmatrix}\dot{\boldsymbol{\theta}} \tag{7.221}$$

and the inverse relation: $\dot{\boldsymbol{\theta}}=\boldsymbol{\rho}^{-1}\boldsymbol{\omega}$ is:

$$\dot{\boldsymbol{\theta}}=\begin{pmatrix} \sin\chi & \cos\chi & 0 \\ \dfrac{\cos\chi}{\sin\theta} & \dfrac{\sin\chi}{\sin\theta} & 0 \\ \dfrac{\cos\chi}{\operatorname{tg}\theta} & -\dfrac{\sin\chi}{\operatorname{tg}\theta} & 1 \end{pmatrix}\boldsymbol{\omega} \tag{7.222}$$

So the components of the angular momentum **m** *are related to the conjugate momenta of the eulerian angles by the relation*:

$$\begin{aligned} m_f &= \frac{\partial T}{\partial \omega_f} = (\nabla_\omega \dot{\boldsymbol{\theta}}')(\nabla_{\dot\theta} T) \\ \boldsymbol{m} &= (\boldsymbol{\rho}^{-1})'\mathbf{p}_\theta \end{aligned} \tag{7.223}$$

Let us now write explicitly the momentum expression as:

$$\begin{aligned} 2T &= \dot{\boldsymbol{\eta}}'\boldsymbol{\alpha}\dot{\boldsymbol{\eta}} \\ &= \dot{\mathbf{Q}}'\dot{\mathbf{Q}}+\boldsymbol{\omega}'\mathbf{Z}\dot{\mathbf{Q}}+\dot{\mathbf{Q}}'\mathbf{Z}'\boldsymbol{\omega}+\boldsymbol{\omega}'\mathbf{I}\boldsymbol{\omega}+\dot{\mathbf{c}}'\mathbf{m}\dot{\mathbf{c}} \end{aligned} \tag{7.224}$$

We can obtain by differentiation:

$$\mathbf{P}=\frac{\partial T}{\partial \dot{Q}}=\dot{\mathbf{Q}}+\mathbf{Z}'\boldsymbol{\omega} \tag{7.225}$$

$$\boldsymbol{m}=\frac{\partial T}{\partial \omega}=\mathbf{Z}\dot{\mathbf{Q}}+\mathbf{I}\boldsymbol{\omega} \tag{7.226}$$

or, from (7.225), as:

$$\begin{aligned} \dot{\mathbf{Q}} &= \mathbf{P}-\mathbf{Z}'\boldsymbol{\omega} \\ \boldsymbol{m} &= \mathbf{Z}\mathbf{P}+(\mathbf{I}-\mathbf{Z}\mathbf{Z}')\boldsymbol{\omega} \end{aligned} \tag{7.227}$$

and defining:

$$\mathbf{I}^*=\mathbf{I}-\mathbf{Z}\mathbf{Z}' \tag{7.228}$$

$$\boldsymbol{m}=\mathbf{Z}\mathbf{P}+\mathbf{I}^*\boldsymbol{\omega} \tag{7.229}$$

$$\mathbf{p}=\frac{\partial T}{\partial \dot{\mathbf{c}}}=\mathbf{m}\dot{\mathbf{c}} \tag{7.230}$$

the total kinetic energy then becomes:

$$2T=\mathbf{P}'\mathbf{P}+\boldsymbol{\omega}'\mathbf{I}^*\boldsymbol{\omega}+\mathbf{p}'\mathbf{m}^{-1}\mathbf{p} \tag{7.231}$$

For convenience we can introduce the vector $\boldsymbol{m}^*$ defined by:

$$\boldsymbol{m}^*=\mathbf{I}^*\boldsymbol{\omega}=\boldsymbol{m}-\mathbf{Z}\mathbf{P} \tag{7.232}$$

The final expression of the total kinetic energy is of the form:

$$2T=\mathbf{P}'\mathbf{P}+\boldsymbol{m}^{*\prime}\mathbf{I}^{*-1}\boldsymbol{m}^*+\mathbf{p}'\mathbf{m}^{-1}\mathbf{p} \tag{7.233}$$

The modified inertia tensor $\mathbf{I}^*$ and its inverse can find simple forms. Let us separate the terms depending on Q to the zero, first and second order:

$$\mathbf{I}^* = \mathbf{I} - \mathbf{Z}\mathbf{Z}' = \mathbf{I}^{(0)} + \mathbf{I}^{(1)} + \mathbf{I}^{(2)} \tag{7.234}$$

where:

$$I_{fg}^{(0)} = J_{fg}^{(0)} \tag{7.235}$$

$$I_{fg}^{(1)} = \mathbf{Q}'\mathbf{J}_{fg,Q}^{(1)} \tag{7.236}$$

$$I_{fg}^{(2)} = \mathbf{Q}'(\mathbf{J}_{fg,Q}^{(2)} - \boldsymbol{\zeta}^f \boldsymbol{\zeta}^{g\prime})\mathbf{Q} \tag{7.237}$$

The last term may be reduced by using (from 7.207, 7.212, 7.166 and 7.178):

$$\begin{aligned} I_{fg}^{(2)} &= \mathbf{Q}'[\mathbf{L}'\mathbf{A}'(\mathbf{I}^f)\mathbf{M}(\mathbf{I}^g)'\mathbf{A}\mathbf{L} - \mathbf{L}'\mathbf{A}'\mathbf{M}(\mathbf{I}^f)\mathbf{A}\mathbf{L}\mathbf{L}'\mathbf{A}'(\mathbf{I}^g)'\mathbf{M}\mathbf{A}\mathbf{L}]\mathbf{Q} \\ &= \mathbf{Q}'[\mathbf{L}'\mathbf{A}'\mathbf{M}^{1/2}(\mathbf{I}^f)(\mathbf{E} - \mathbf{M}^{1/2}\mathbf{A}\mathbf{G}\mathbf{A}'\mathbf{M}^{1/2})(\mathbf{I}^g)'\mathbf{M}^{1/2}\mathbf{A}\mathbf{L}]\mathbf{Q} \\ &= \mathbf{Q}'[\mathbf{l}'(\mathbf{I}^f)\mathbf{l}_{rt}\mathbf{l}_{rt}'(\mathbf{I}^g)'\mathbf{l}]\mathbf{Q} \end{aligned} \tag{7.238}$$

From the Eckart conditions fixing the direction of axis, we find:

$$I_{fg}^{(2)} = \frac{1}{4}\sum_h \mathbf{Q}'\mathbf{J}_{fh,Q}^{(1)} \frac{1}{\mathbf{J}_{hh}^{(0)}} \mathbf{J}_{hg,Q}^{(1)\prime}\mathbf{Q} \tag{7.239}$$

or

$$\mathbf{I}^{(2)} = \tfrac{1}{4}\mathbf{I}^{(1)}(\mathbf{I}^{(0)})^{-1}\mathbf{I}^{(1)} \tag{7.240}$$

In the same way we shall express the inverse tensor $\mathbf{I}^{*-1}$ by separating it into terms that depend on Q to a degree up to two, neglecting higher orders:

$$\boldsymbol{\mu}^* = (\mathbf{I}^*)^{-1} = (\boldsymbol{\mu}^{(0)} + \boldsymbol{\mu}^{(1)} + \boldsymbol{\mu}^{(2)} + \cdots) \tag{7.241}$$

We can write:

$$(\boldsymbol{\mu}^{(0)} + \boldsymbol{\mu}^{(1)} + \boldsymbol{\mu}^{(2)} + \cdots)(\mathbf{I}^{(0)} + \mathbf{I}^{(1)} + \mathbf{I}^{(2)})$$

By collecting together terms of the same order all terms depending on Q must vanish; therefore the following relations arise:

(0) $$\boldsymbol{\mu}^{(0)}\mathbf{I}^{(0)} = \mathbf{E} \tag{7.242a}$$

(1) $$\boldsymbol{\mu}^{(0)}\mathbf{I}^{(1)} + \boldsymbol{\mu}^{(1)}\mathbf{I}^{(0)} = \mathbf{0} \tag{7.242b}$$

(2) $$\boldsymbol{\mu}^{(0)}\mathbf{I}^{(2)} + \boldsymbol{\mu}^{(1)}\mathbf{I}^{(1)} + \boldsymbol{\mu}^{(2)}\mathbf{I}^{(0)} = \mathbf{0} \tag{7.242c}$$

It follows that $\boldsymbol{\mu}^{(1)}$ and $\boldsymbol{\mu}^{(2)}$ may be expressed as:

$$\boldsymbol{\mu}^{(1)} = -\boldsymbol{\mu}^{(0)}\mathbf{I}^{(1)}\boldsymbol{\mu}^{(0)} \tag{7.243}$$

$$\boldsymbol{\mu}^{(2)} = \tfrac{3}{4}\boldsymbol{\mu}^{(1)}\mathbf{I}^{(0)}\boldsymbol{\mu}^{(1)} \tag{7.244}$$

In Table 7.5 we report the units corresponding to the matrices just defined, completing the previous Table 7.4.

Table 7.5 Dimension of matrices and vectors (continuation of Table 7.4)

Vector or matrix	Dimension
$\boldsymbol{\omega}$	T^{-1}
$\boldsymbol{\zeta}, \mathbf{J}_Q^{(2)}$	Adimensional
$\mathbf{J}^{(0)}, \mathbf{I}^*, \mathbf{I}^{(0)}, \mathbf{I}^{(1)}, \mathbf{I}^{(2)}$	ML^2
$\mathbf{Z}, \mathbf{J}_Q^{(1)}$	$M^{1/2}L$

7.7 CLASSICAL TREATMENT OF THE WATER MOLECULE

7.7.1 The Normal Vibrational Modes

Before closing this part concerning the classical nuclear motion in the neighbourhood of an equilibrium position, we wish to give a simple application of the theory. Using the 6-31G basis set of Pople, theoretical *ab initio* computations have been performed according to experimental planification, leading to an analytical potential energy surface up to the second order (see Sections 7.2.2 and 7.3.3.4). Results are summarized by the following formula:

$$\begin{aligned} V(d_1 d_2 \theta) = & -73.5934 - 1.0977(d_1 + d_2) - 0.4333\theta \\ & + 0.2962(d_1^2 + d_2^2) + 0.0846\theta^2 - 0.0122 d_1 d_2 \\ & + 0.0289(d_1 + d_2)\theta \end{aligned}$$

with d_1 and d_2 being the two bond lengths in atomic units, θ the HOH angle in radians and $V(d_1 d_2 \theta)$ in atomic units of energy.

Such an analytical form gives an absolute minimum (by equation 7.126) at $d_{1,\mathrm{e}} = d_{2,\mathrm{e}} = 1.7947$ a.u. and $\theta_\mathrm{e} = 1.9460$ rad. Replacing now θ by 1.7947θ and after units transformation, the force constant matrix $\mathbf{H}$ may be expressed in millidynes per angstrom by:

$$\mathbf{H} = \begin{array}{c} \\ d_1 \\ d_2 \\ d_\mathrm{e}\theta \end{array} \begin{array}{c} \begin{array}{ccc} d_1 & d_2 & d_\mathrm{e}\theta \end{array} \\ \begin{bmatrix} 9.2225 & & \\ -0.1901 & 9.2225 & \\ 0.2512 & 0.2512 & 0.8180 \end{bmatrix} \end{array}$$

The $\mathbf{G}$ matrix of Wilson which connects the cartesian and the internal displacements is of the form:

$$\mathbf{G}_{\mathrm{H_2O}} = \begin{pmatrix} \mu_{\mathrm{H}_1} + \mu_{\mathrm{O}_3} & & \text{Sym.} \\ \mu_{\mathrm{O}_3} \cos\theta_\mathrm{e} & \mu_{\mathrm{H}_2} + \mu_{\mathrm{O}_3} & \\ -\mu_{\mathrm{O}_3} \sin\theta_\mathrm{e} & -\mu_{\mathrm{O}_3} \sin\theta_\mathrm{e} & \mu_{\mathrm{H}_1} + \mu_{\mathrm{H}_2} + \mu_{\mathrm{O}_3}(1 - \cos\theta_\mathrm{e}) \end{pmatrix}$$

According to Table 7.3 and using the θ equilibrium value (θ_e), the next results may be found:

$$\mathbf{G}_{H_2O} = \begin{pmatrix} 1.0548 & & \text{Sym.} \\ -0.0229 & 1.0548 & \\ -0.0582 & -0.0582 & 2.1553 \end{pmatrix}$$

which is expressed in moles per gram. We have now to perform the **GH** product which has to be diagonalized in order to obtain the normal modes of vibration:

$$\mathbf{GH} = \begin{pmatrix} 0.7176 & -0.4263 & 0.2116 \\ -0.4263 & 9.7176 & 0.2116 \\ 0.0157 & 0.0157 & 1.7338 \end{pmatrix}$$

The diagonalization of the **G** matrix leads to the next eigenvalues (all positive) and eigenvectors:

$$\text{diag}(\boldsymbol{\gamma}) = (2.1613 \quad 1.0777 \quad 1.0259)$$

$$\mathbf{U} = \begin{pmatrix} -0.051 & -0.707 & 0.705 \\ -0.051 & 0.707 & 0.705 \\ 0.997 & 0.000 & 0.073 \end{pmatrix}$$

In order to diagonalize the **GH** product we are now able to construct the similarity transform **W**:

$$\mathbf{W} = \mathbf{U}\boldsymbol{\gamma}^{1/2} = \begin{pmatrix} -0.075 & -0.734 & 0.714 \\ -0.075 & 0.734 & 0.714 \\ 1.466 & 0.000 & 0.074 \end{pmatrix}$$

Consequently, we can find the eigenvectors and eigenvalues of the **GH** product:

$$\mathbf{L} = \begin{pmatrix} -0.734 & 0.717 & -0.041 \\ 0.734 & 0.717 & -0.041 \\ 0.000 & 0.003 & 1.468 \end{pmatrix}$$

$$\text{diag}(\boldsymbol{\Lambda}) = (10.1436 \quad 9.2918 \quad 1.7328)$$

The matrix **K** inverse of **L** is:

$$\mathbf{K} = \mathbf{L}^{-1} = \begin{pmatrix} -0.681 & 0.681 & 0.000 \\ 0.697 & 0.697 & 0.039 \\ -0.001 & -0.001 & 0.681 \end{pmatrix}$$

We can also transform the matrix $\boldsymbol{\Lambda}$ in reciprocal centimetre

$$\tilde{\nu}' = (4149 \quad 3971 \quad 1715)\ \text{cm}^{-1}$$

The **K** matrix shows the molecular motions corresponding to each of the three fundamental vibration frequencies:

Q = KS **Q = KBξ**

$\tilde{\nu}_1$: $Q_1 = 0.681(d_2 - d_1)$
the unsymmetrical stretching

$\tilde{\nu}_2$: $Q_2 \simeq 0.697(d_1 + d_2)$
the symmetrical stretching

$\tilde{\nu}_3$: $Q_3 \simeq 0.681 d_e \theta$
the bending deformation

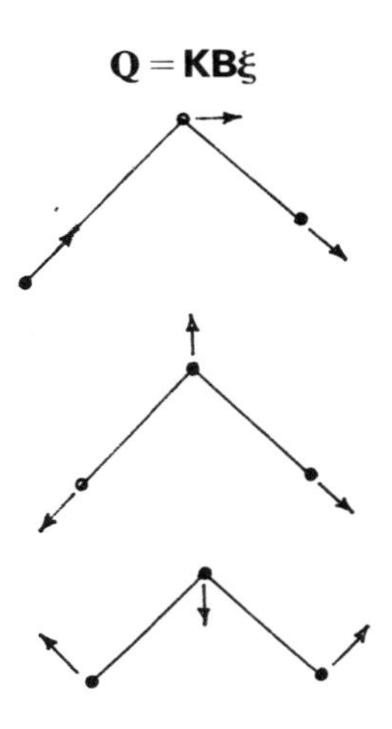

7.7.2 The Isotopic Shift

The same analysis has been performed for different isotopes. The results are summarized in Table 7.6. The correlation coefficient ρ^* which connects

Table 7.6 Experimental and theoretical values for the fundamental frequencies (cm^{-1}) of some isotopes of the water molecule†

	$\tilde{\nu}_1$		$\tilde{\nu}_2$		$\tilde{\nu}_3$	
	Theory(3)	Experiment(45)	Theory(3)	Experiment(45)	Theory(3)	Experiment(45)
$H_2{}^{16}O$	4,149	3,756	3,971	3,652	1,715	1,595
$D_2{}^{16}O$	3,049	2,788	2,855	2,671	1,259	1,178
$HD^{16}O$	4,067	3,707	2,945	2,727	1,504	1,402
$HT^{16}O$	4,065	3,720	2,472	—	1,428	1,324
$DT^{16}O$	2,972	2,735	2,456	—	1,164	—
$T_2{}^{16}O$	2,581	2,370	2,371	—	1,061	996
$H_2{}^{18}O$	4,131	3,743	3,964	3,647	1,707	1,586
$D_2{}^{18}O$	3,024	2,764	2,844	2,657	1,249	1,169

†

$$\sigma^2(\tilde{\nu}_{\text{exp}}) = \frac{1}{n} \sum (\tilde{\nu}_{\text{exp}} - \langle \tilde{\nu}_{\text{exp}} \rangle)^2 = 9.53162 \times 10^5 \text{ cm}^{-2}$$

$$\sigma^2(\tilde{\nu}_{\text{th}} - \tilde{\nu}_{\text{exp}}) = \frac{1}{n-1} \sum (\Delta\tilde{\nu} - \langle \Delta\tilde{\nu} \rangle)^2 = 0.59286 \times 10^5 \text{ cm}^{-2}$$

with $\langle \Delta\tilde{\nu} \rangle = 218 \text{ cm}^{-1}$:

$$\rho^* = \sqrt{\left[1 - \frac{\sigma^2(\tilde{\nu}_{\text{th}} - \tilde{\nu}_{\text{exp}})}{\sigma^2(\tilde{\nu}_{\text{exp}})}\right]} = 0.968$$

$$f = \frac{\sigma^2(\tilde{\nu}_{\text{th}} - \tilde{\nu}_{\text{exp}})}{\sigma^2(\tilde{\nu}_{\text{exp}})} = 16.08$$

experimental and theoretical frequencies presents a significantly good value ($\rho^* = 0.968$) and shows that the theoretical computations are not too bad. In Figure 7.23 we give a correlation diagram showing the experimental and the theoretical values; if the average error on the individual frequencies is about 218 cm^{-1} the correlation remains approximately linear. The error may be considerably decreased[(44)] by using theoretical computations beyond the SCF level.

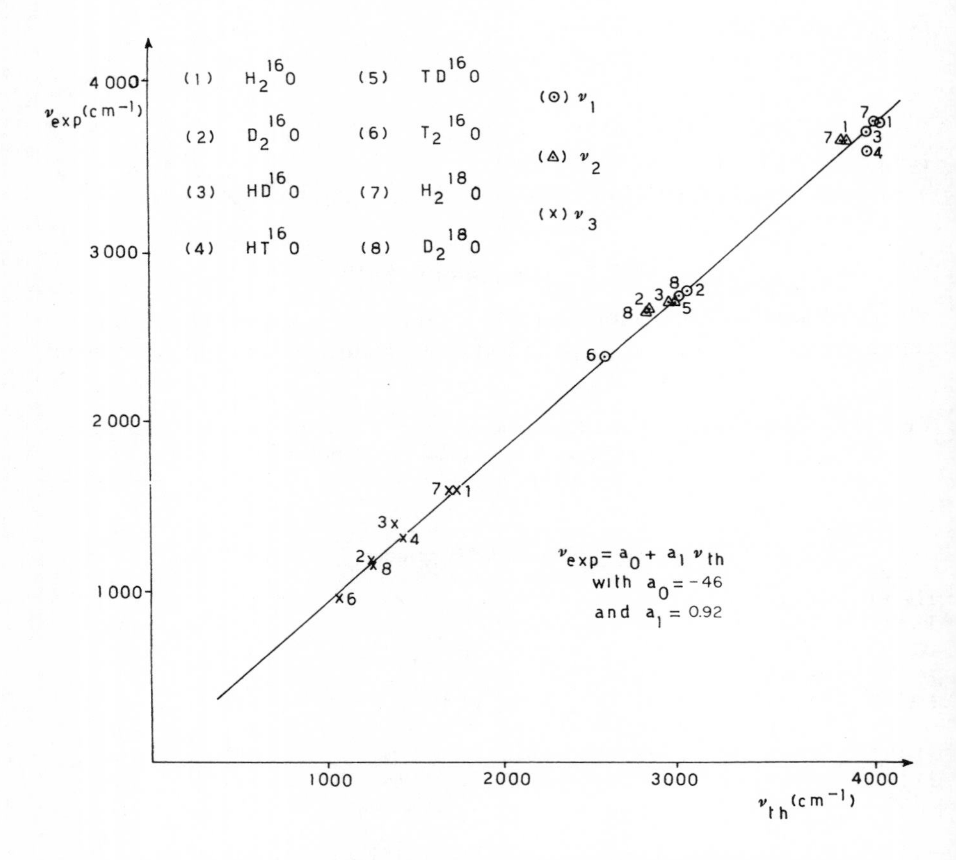

Figure 7.23 Correlation between theoretical and experimental fundamental frequencies for different isotopes of the water molecule

7.7.3 The Coriolis and the Inertia Matrices

Let us now compute the matrices which characterize the rotation and the coupling terms between rotation and vibration. The moving coordinate frame defined with respect to the equilibrium structure is the following:

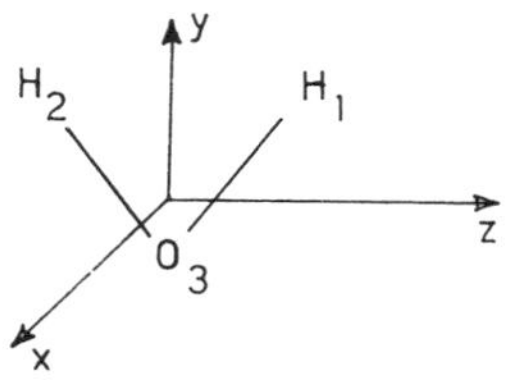

$$R_e = \begin{pmatrix} 0.000 \\ 0.475 \\ 0.785 \\ 0.000 \\ 0.475 \\ -0.785 \\ 0.000 \\ -0.060 \\ 0.000 \end{pmatrix}$$

The corresponding inertia tensor is diagonal and its main components are:

$$\text{diag}\,\mathbf{J}^{(0)} = (1.754 \quad 1.242 \quad 0.512)\ \text{g Å}^2\ \text{mol}^{-1}$$

The non-vanishing first- and second-order rotational distortions computed are:

$$\mathbf{J}^{(1)}_{xx} = \begin{pmatrix} 0.000 \\ 2.649 \\ -0.005 \end{pmatrix}; \quad \mathbf{J}^{(1)}_{yy} = \begin{pmatrix} 0.000 \\ 1.879 \\ 1.200 \end{pmatrix}; \quad \mathbf{J}^{(1)}_{zz} = \begin{pmatrix} 0.000 \\ 0.770 \\ -1.206 \end{pmatrix}; \quad \mathbf{J}^{(1)}_{yz} = \begin{pmatrix} 1.204 \\ 0.000 \\ 0.000 \end{pmatrix}$$

$$\mathbf{J}^{(2)}_{yy} = \begin{pmatrix} 0.708 & 0.000 & 0.000 \\ 0.000 & 0.710 & 0.454 \\ 0.000 & 0.454 & 0.290 \end{pmatrix}; \quad \mathbf{J}^{(2)}_{zz} = \begin{pmatrix} 0.292 & 0.000 & 0.000 \\ 0.000 & 0.290 & -0.454 \\ 0.000 & -0.454 & 0.710 \end{pmatrix}$$

$$\mathbf{J}^{(2)}_{yz} = \begin{pmatrix} 0.000 & 0.455 & 0.291 \\ 0.453 & 0.000 & 0.000 \\ -0.709 & 0.000 & 0.000 \end{pmatrix}; \quad \mathbf{J}^{(2)}_{xx} = \begin{pmatrix} 1.000 & 0.000 & 0.000 \\ 0.000 & 1.000 & 0.000 \\ 0.000 & 0.000 & 1.000 \end{pmatrix}$$

This shows that the terms of the current inertia tensor have the following form:

$$\mathbf{I}^{(1)} = \begin{pmatrix} 2.649Q_2 & & \\ 0.000 & 1.879Q_2 + 1.200Q_3 & \\ 0.000 & 1.204Q_1 & 0.770Q_2 - 1.206Q_3 \end{pmatrix}$$

$$\mathbf{I}^{(2)} = \begin{pmatrix} Q_1^2 + Q_2^2 + Q_3^2 & & \\ 0.000 & 0.708(Q_1^2 + Q_2^2) + 0.290Q_3^2 + 0.908Q_2Q_3 & \\ 0.000 & (0.908Q_2 - 0.418Q_3)Q_1 & 0.290(Q_1^2 + Q_2^2) + 0.710Q_3^2 - 0.90Q_2Q_3 \end{pmatrix}$$

$$\mathbf{ZZ}' = \begin{pmatrix} Q_1^2 + Q_3^2 & & \\ 0.000 & 0.000 & \\ 0.000 & 0.000 & 0.000 \end{pmatrix}$$

Similarly, the non-vanishing Coriolis term becomes:

$$\zeta_x = \begin{pmatrix} 0.000 & -0.002 & -1.000 \\ 0.002 & 0.000 & 0.000 \\ 1.000 & 0.000 & 0.000 \end{pmatrix}$$

So a coupling term between rotation and vibration arises chiefly for the out-of-plane motion; the corresponding Coriolis kinetic energy has the form:

$$T_c \simeq \omega_x (Q_3 - Q_1)\dot{Q}_3$$

The unsymmetrical vibrational motion (Q_2) has a weak importance as it corresponds to a zero average motion.

7.8 QUANTUM TREATMENT OF THE PURE POLYATOMIC VIBRATION

7.8.1 The General Nuclear Equation

In this section we extend the quantum treatment of the harmonic biatomic vibration to the polyatomic anharmonical case.[46] Due to the fact that the gap between the energy levels largely varies from one type to another (translational, rotational, vibrational or electronic), the total wave function for the atom motions can be written as a product of translational (ψ_T), rotational (ψ_R), vibrational (ψ_V) and electronic (ψ_E) wave functions:

$$\Psi = \psi_T \psi_R \psi_V \psi_E \tag{7.245}$$

In terms of normal coordinates Q, and using the previously developed expressions for T and V from equations (7.169), we can write the classical expression:

$$H = T + V = \tfrac{1}{2}\dot{Q}'\dot{Q} + \tfrac{1}{2}Q'\Lambda Q + f(Q) \tag{7.246}$$

where $f(Q)$ stands for the anharmonical contribution to the potential in terms of normal coordinates; i.e.:

$$\begin{aligned} f(Q) &= \frac{1}{6}\sum_{i,j,k} \frac{\partial^3 V}{\partial Q_i\, \partial Q_j\, \partial Q_k} Q_i Q_j Q_k \\ &\quad + \frac{1}{24}\sum_{i,j,k.l} \frac{\partial^4 V}{\partial Q_i\, \partial Q_j\, \partial Q_k\, \partial Q_l} Q_i Q_j Q_k Q_l + \cdots \\ &= \sum_{i,jk} \alpha^{(3)}_{ijk} Q_i Q_j Q_k + \sum_{i,j,k,l} \alpha^{(4)}_{ijkl} Q_i Q_j Q_k Q_l + \cdots \end{aligned} \tag{7.247}$$

Using the correspondence principle, equation (7.246) becomes:

$$H = -\frac{\hbar^2}{2}\Delta_Q^2 + \tfrac{1}{2}\mathbf{Q}'\boldsymbol{\Lambda}\mathbf{Q} + f(Q) \tag{7.248}$$

so the vibrational wave equation is of the form:

$$H\psi_V = \varepsilon_V \psi_V \tag{7.249}$$

Let us now use as a coordinate frame the adimensional following one:

$$\mathbf{z} = \sqrt{\left(\frac{\sqrt{\mathbf{\Lambda}}}{\hbar}\right)}\mathbf{Q} = \sqrt{\left[\frac{\sqrt{(2\mathbf{a}^{(2)})}}{\hbar}\right]}\mathbf{Q} = \sqrt{\left(\frac{\boldsymbol{\omega}}{\hbar}\right)}\mathbf{Q} \tag{7.250}$$

where $\mathbf{\Lambda}$ as $\mathbf{a}^{(2)}$ or $\boldsymbol{\omega}$ are square diagonal matrices. The matrix $\mathbf{\Lambda}$ is defined in equation (7.161) and ω in (7.164); by analogy to $\alpha^{(3)}$ and $\alpha^{(4)}$, $\alpha^{(2)}$ equals $\Lambda/2$.

Introducing now (7.247), (7.248) and (7.250) in equation (7.249) we find:

$$\left[\nabla_z^2 - z'z - \frac{2\omega^{-1}}{\hbar} f(\sqrt{(\hbar\omega^{-1})}\,\mathbf{z}) + \frac{2\omega^{-1}}{\hbar}\varepsilon_V\right]\psi_V = 0 \tag{7.251}$$

or if:

$$\varepsilon_0 = \frac{\hbar\omega}{2} \tag{7.252}$$

$$[\nabla_z^2 - z'z - \varepsilon_0^{-1} f(\sqrt{(\hbar\omega^{-1})}\,\mathbf{z}) + \varepsilon_V \cdot \varepsilon_0^{-1}]\psi_V = 0 \tag{7.253}$$

7.8.2 The Harmonic Approximation

As long as the anharmonic term $f(Q)$ does not exist in (7.246) the wave equation (7.253) is separable in m independent equations (m being the number of vibrators), each of them corresponding to one harmonic vibrational mode. Let us assume that:

$$\psi_V(z_1 \cdots z_m) = \psi(z_1)\cdot\psi(z_2)\cdots\psi(z_m) \tag{7.254a}$$

and

$$\varepsilon_V = \varepsilon(1) + \varepsilon(2) + \cdots + \varepsilon(m) \tag{7.254b}$$

Then (7.253) becomes:

$$\left[\nabla_{z_i}^2 - z_i^2 + \frac{\varepsilon(i)}{\varepsilon_0(i)}\right]\psi(z_i) = 0 \qquad \forall i \text{ from 1 to } m \tag{7.255}$$

Solution of this set of equations is known as the non-rotating harmonic vibrator wave function (for the diatomic case):

$$\psi_{v_i}(z_i) = N_{v_i}\exp\left(-\frac{z_i^2}{2}\right)H_{v_i}(z_i) \tag{7.256}$$

where v_i stands for a positive or null integer, N_{v_i} is a normalization factor and H_{v_i} is the v_ith Hermite orthogonal function.

Proof of this is simple. If we use as a trial solution the product of a gaussian function and an infinite polynomial function:

$$\psi(z) = \exp\left(-\frac{z^2}{2}\right)H(z) = \exp\left(-\frac{z^2}{2}\right)\sum_{i=0}^{\infty} C_i z^i \tag{7.257}$$

Differentiating twice and inserting into equation (7.255) leads to:

$$\exp\left(-\frac{z^2}{2}\right)\left[H''(z)-2zH'(z)+\left(\frac{\varepsilon}{\varepsilon_0}-1\right)H(z)\right]=0$$

or

$$\exp\left(-\frac{z^2}{2}\right)\left\{\sum_{i=2} C_i\left[i(i-1)z^{i-2}+\left(\frac{\varepsilon}{\varepsilon_0}-1-2i\right)z^i\right]\right. \qquad (7.258)$$

$$\left.+C_1\left(\frac{\varepsilon}{\varepsilon_0}-3\right)z+C_0\left(\frac{\varepsilon}{\varepsilon_0}-1\right)\right\}=0$$

We satisfy this expression for any value of z if the coefficients corresponding to each power of z cancel. In this way we find the next relation between all the undefined values of C_i.

$$C_{i+2}=C_i\frac{2i+1-\varepsilon/\varepsilon_0}{(i+1)(i+2)} \qquad (7.259)$$

Two degrees of freedom remain: C_0 and C_1; we choose two kinds of solution corresponding respectively to an odd or to an even function. Lastly, as we want to dispose of a converging wave function for large z values, it appears for (7.259) that the infinite expansion introduced in (7.257) may be limited if:

$$\frac{\varepsilon}{\varepsilon_0}=2v+1 \qquad (7.260)$$

where v stands for a positive or null integer.

The function $H(z)$ so defined is known as Hermite's polynome. In Table 7.7 we report the first Hermite functions. The even as well as the odd

Table 7.7 The first Hermite function

v	$H_v(z)$	v	$H_v(z)$
0	1	4	$16z^4-48z^2+12$
1	$2z$	5	$32z^5-160z^3+120z$
2	$4z^2-2$	6	$64z^6-480z^4+720z^2-120$
3	$8z^3-12z$	7	$128z^7-1344z^5+3360z^3-1680z$

functions are related by the following recurrence relationship (deduced from 7.259 and 7.260):

$$H_{v+1}(z)-2zH_v(z)+2vH_{v-1}(z)=0 \qquad (7.261)$$

As we expect to dispose of a normalized function we insert in the general expression a normalizing coefficient N_v. The final expression of $\psi(z)$ is then:

$$\psi_{v_i}(z_i)=N_{v_i}\exp\left(-\frac{z_i^2}{2}\right)H_{v_i}(z_i) \qquad \forall i=1 \text{ to } m$$

where:

$$N_{v_i} = \left[\frac{\sqrt{(1/\pi)}}{2^{v_i} v_i!}\right]^{1/2} \qquad \text{such that} \int \psi^*_{v_i} \psi_{v_i} \, \mathrm{d}z_i = 1 \tag{7.262}$$

The vibration energy associated to any of those functions is:

$$\varepsilon_i = (2v_i + 1)\varepsilon_0 = (v_i + \tfrac{1}{2})\hbar\omega \tag{7.263}$$

The only difference with respect to the classical treatment concerns the discontinuity of the vibrational levels.

The molecular vibrational levels are now characterized by m integers: $V = \{v_1, \ldots, v_m\}$. If all these quantum numbers are zero we call this level the *ground level*, if only one quantum number differs from zero the resulting level is called the *fundamental level* or if the non-zero quantum number respectively is equal to or greater than one it is called the *overtone level*. Other vibrational levels are called *combination levels*. Let us also notice that at the harmonical approximation the classical and the quantum treatments infer the same frequencies $\{\omega_i\}$.

7.8.3 The Anharmonic Perturbation

To continue, we can consider the anharmonical term $f(Q)$ as a perturbation. By applying the Rayleigh–Schroedinger method it is evident that the two first corrections are:

The first-order correction:

$$\Delta\varepsilon_{\mathrm{V}}^{(1)} = \langle\psi_{\mathrm{V}}^{0}|\, f(Q)\, |\psi_{\mathrm{V}}^{0}\rangle \tag{7.264}$$

The second-order correction:

$$\Delta\varepsilon_{\mathrm{V}}^{(2)} = \sum_{\mathrm{V}'} \frac{\langle\psi_{\mathrm{V}'}^{0}|\, f(Q)\, |\psi_{\mathrm{V}}^{0}\rangle\langle\psi_{\mathrm{V}}^{0}|\, f(Q)\, |\psi_{\mathrm{V}'}^{0}\rangle}{\varepsilon_{\mathrm{V}}^{0} - \varepsilon_{\mathrm{V}'}^{0}} \tag{7.265}$$

The matrix elements between the unperturbed (harmonic) vibrational wave functions which appear in (7.264) and (7.265) may be analytically or numerically obtained. For polynomial expansions of $f(Q)$ some useful integral solutions are reported in Table 7.8.

Let us now introduce the explicit Taylor expansion of the potential up to the quartic terms (7.247). The first-order perturbation energy is then:

$$\Delta\varepsilon_{\mathrm{V}}^{(1)} = \sum_{ijk}^{m} \alpha_{ijk}^{(3)} \left\langle \prod_{n}^{m} \psi_{v_n}^{(0)}(Q_n) \right| Q_i Q_j Q_k \left| \prod_{n}^{m} \psi_{v_n}^{(0)}(Q_n) \right\rangle$$

$$+ \sum_{ijkl}^{m} \alpha_{ijkl}^{(4)} \left\langle \prod_{n}^{m} \psi_{v_n}^{(0)}(Q_n) \right| Q_i Q_j Q_k Q_l \left| \prod_{n}^{m} \psi_{v_n}^{(0)}(Q_n) \right\rangle + \cdots$$

Table 7.8 Analytical integral $\int_{-\infty}^{+\infty} \psi_V^{0*} Q^n \psi_{V'}^0 \, dQ$†

$\langle\psi_V^0\| Q_n \|\psi_{V'}^0\rangle$	Resulting expression‡
$\langle v\| Q \|v\rangle$	a_0
$\langle v\| Q \|v+1\rangle$	$a_1\sqrt{(v+1)}$
$\langle v\| Q^2 \|v+2\rangle$	$a_2\sqrt{[(v+2)!/v!]}$
$\langle v\| Q^2 \|v\rangle$	$a_2\times 2(v+\frac{1}{2})$
$\langle v\| Q^3 \|v+3\rangle$	$a_3\sqrt{[(v+3)!/v!]}$
$\langle v\| Q^3 \|v+1\rangle$	$a_3\times 3(v+1)\sqrt{(v+1)}$
$\langle v\| Q^4 \|v+4\rangle$	$a_4\sqrt{[(v+4)!/v!]}$
$\langle v\| Q^4 \|v+2\rangle$	$a_4\times 4(v+\frac{3}{2})\sqrt{[(v+2)!/v!]}$
$\langle v\| Q^4 \|v\rangle$	$a_4\times 6[(v+\frac{1}{2})^2+\frac{1}{4}]$
$\langle v\| Q^5 \|v+5\rangle$	$a_5\sqrt{[(v+5)!/v!]}$
$\langle v\| Q^5 \|v+3\rangle$	$a_5\times 5(v+2)\sqrt{[(v+3)!/v!]}$
$\langle v\| Q^5 \|v+1\rangle$	$a_5\times 10[(v+1)^2+\frac{1}{2}]\sqrt{(v+1)}$
$\langle v\| Q^6 \|v+6\rangle$	$a_6\sqrt{[(v+6)!/v!]}$
$\langle v\| Q^6 \|v+4\rangle$	$a_6\times 6(v+\frac{5}{2})\sqrt{[(v+4)!/v!]}$
$\langle v\| Q^6 \|v+2\rangle$	$a_6\times 15[(v+\frac{3}{2})^2+\frac{3}{4}]\sqrt{[(v+2)!/v!]}$
$\langle v\| Q^6 \|v\rangle$	$a_6\times 5(v+\frac{1}{2})[4(v+\frac{1}{2})^2+5]$
$\langle v\| Q^7 \|v+7\rangle$	$a_7\sqrt{[(v+7)!/v!]}$
$\langle v\| Q^7 \|v+5\rangle$	$a_7\times 7(v+3)\sqrt{[(v+5)!/v!]}$
$\langle v\| Q^7 \|v+3\rangle$	$a_7\times 21[(v+2)^2+1]\sqrt{[(v+3)!/v!]}$
$\langle v\| Q^7 \|v+1\rangle$	$a_7\times 35(v+1)[(v+1)^2+2]\sqrt{(v+1)}$
$\langle v\| Q^8 \|v+8\rangle$	$a_8\sqrt{[(v+8)!/v!]}$
$\langle v\| Q^8 \|v+6\rangle$	$a_8\times 8(v+\frac{7}{2})\sqrt{[(v+6)!/v!]}$
$\langle v\| Q^8 \|v+4\rangle$	$a_8\times 28[(v+\frac{5}{2})^2+\frac{5}{4}]\sqrt{[(v+4)!/v!]}$
$\langle v\| Q^8 \|v+2\rangle$	$a_8\times 56(v+\frac{3}{2})[(v+\frac{3}{2})^2+\frac{11}{4}]\sqrt{[(v+2)!/v!]}$
$\langle v\| Q^8 \|v\rangle$	$a_8\times\frac{35}{8}[16(v+\frac{1}{2})^4+56(v+\frac{1}{2})^2+9]$

† The matrix element not given $\langle v| Q^n |v'\rangle$ for $v \leq v'$ and $0 \leq n \leq 8$ vanishes.
‡ $a_n = (\hbar/2\omega)^{n/2}$.

As shown in Table 7.8 the odd integrals disappear. There remains only:

$$\Delta\varepsilon_V^{(1)} = \sum_i \alpha_{iiii}^{(4)}\langle v_i| Q_i^4 |v_i\rangle + \sum_{i\neq j} \alpha_{iijj}^{(4)}\langle v_i| Q_i^2 |v_i\rangle\langle v_j| Q_j^2 |v_j\rangle + \cdots$$

or

$$\Delta\varepsilon_V^{(1)} = \sum_i \left(\frac{2\omega_i}{\hbar}\right)^{-2} 6\alpha_{iiii}^{(4)}[(v_i+\tfrac{1}{2})^2+\tfrac{1}{4}] + \sum_{i\neq j}\left(\frac{4\omega_i\omega_j}{\hbar^2}\right)^{-1} 4\alpha_{iijj}^{(4)}(v_i+\tfrac{1}{2})(v_j+\tfrac{1}{2}) \tag{7.266}$$

The cubic terms contribute only to the second-order perturbation. For each V state ($V = \{\cdots v_i \cdots v_j \cdots v_k \cdots\}$) we find, from Table 7.8, 18 interactions term summarized as:

$$\begin{aligned} &\alpha_{ijk}^{(3)} \text{ for } v_i' = v_i \pm 1,\ v_j' = v_j \pm 1,\ v_k' = v_k \pm 1 \text{ (8 terms)} \\ &\alpha_{iij}^{(3)} \text{ for } v_i' = v_i \pm 2 \text{ or } v_i,\ v_j' = v_j \pm 1 \text{ (6 terms)} \\ &\alpha_{iii}^{(3)} \text{ for } v_i' = v_i \pm 3 \text{ or } v_i \pm 1 \text{ (4 terms)} \end{aligned} \tag{7.267}$$

7.8.4 The Interaction Model

Another way to go further than the harmonic approximation is to develop the vibrational wave function as a linear combination of the harmonic solutions:

$$\psi_{\mathrm{V}}(Q)=\sum_{\mu} C_{\mu}\psi^0_{\mathrm{V}_\mu}(Q) \tag{7.268}$$

where V_μ stands for the vector $\{v_{1\mu}\cdots v_{p\mu}\cdots v_{m\mu}\}$ which characterizes the μth harmonic eigenstate and $\psi^0_{\mathrm{V}_\mu}$ is the product of the one-dimensional harmonic wave function (equation 7.254a):

$$\psi^0_{\mathrm{V}_\mu}(Q)=\prod_{p=1}^{m}\psi^0_{\mathrm{V}_{p\mu}}(Q_p) \tag{7.269}$$

By integration of the vibrational equation $H\psi_{\mathrm{V}}=\varepsilon_{\mathrm{V}}\psi_{\mathrm{V}}$ over any function $\psi^0_{\mathrm{V}_\nu}$ we find the set of equations:

$$\sum_{\mu} C_\mu(\langle\psi^0_{\mathrm{V}_\nu}|\,H\,|\psi^0_{\mathrm{V}_\mu}\rangle-\varepsilon_{\mathrm{V}}\langle\psi^0_{\mathrm{V}_\nu}\mid\psi^0_{\mathrm{V}_\mu}\rangle)=0 \tag{7.270}$$

where:

$$\langle\psi^0_{\mathrm{V}_\nu}\mid\psi^0_{\mathrm{V}_\mu}\rangle=\prod_p\langle\psi^0_{v_{p\nu}}\mid\psi^0_{v_{p\mu}}\rangle=\prod_p\delta(v_{p\nu},v_{p\mu})$$

and

$$\langle\psi^0_{\mathrm{V}_\nu}|\,H\,|\psi^0_{\mathrm{V}_\mu}\rangle=\left\langle\prod_p\psi^0_{v_{p\nu}}\right|H\left|\prod_p\psi^0_{v_{p\mu}}\right\rangle$$

Let us now develop the former equation, by remembering equation (7.246):

$$H=\tfrac{1}{2}P^2+\tfrac{1}{2}Q'\Lambda Q+f(Q)$$

(a)
$$\begin{aligned}\langle\psi_{\mathrm{V}_\nu}|\,P^2\,|\psi_{\mathrm{V}_\mu}\rangle&=\sum_p\langle\psi_{\mathrm{V}_\nu}|\,P^2_p\,|\psi_{\mathrm{V}_\mu}\rangle\\&=\sum_p\langle\psi^0_{v_{p\nu}}(Q_p)|\,P^2_p\,|\psi^0_{v_{p\mu}}(Q_p)\rangle\prod_{q\neq p}\langle\psi^0_{v_{q\nu}}(Q_q)\mid\psi^0_{v_{q\mu}}(Q_q)\rangle\\&=\sum_p\langle v_{p\nu}|\,P^2_p\,|v_{p\mu}\rangle\prod_{q\neq p}\langle v_{q\nu}\mid v_{q\mu}\rangle\\&=\sum_p\langle v_{p\nu}|\,P^2_p\,|v_{p\mu}\rangle\prod_{q\neq p}\delta(v_{q\nu},v_{q\mu})\end{aligned} \tag{7.271}$$

(b)
$$\begin{aligned}\langle\psi^0_{\mathrm{V}_\nu}|\,Q'\Lambda Q\,|\psi^0_{\mathrm{V}_\mu}\rangle&=\sum_p\lambda_p\langle\psi^0_{\mathrm{V}_\nu}|\,Q^2_p\,|\psi^0_{\mathrm{V}_\mu}\rangle\\&=\sum_p\lambda_p\langle v_{p\nu}|\,Q^2_p\,|v_{q\mu}\rangle\prod_{q\neq p}\delta(v_{q\nu},v_{q\mu})\end{aligned} \tag{7.272}$$

From Tables 7.8 and 7.9, the only remaining contribution is:

$$\tfrac{1}{2}\langle\psi^0_{\mathrm{V}_\nu}|\,P^2+Q'\Lambda Q\,|\psi^0_{\mathrm{V}_\mu}\rangle=\left[\sum_p\hbar\omega_p(v_p+\tfrac{1}{2})\right]\left[\prod_p\delta(v_{p\nu},v_{p\mu})\right] \tag{7.273}$$

where:

$$\omega_p=\sqrt{\lambda_p}$$

Table 7.9 Analytical integral $\int_{-\infty}^{+\infty} \psi_V^{0*} P^n \psi_V^0 \, dQ$ where P stands for $(\hbar/i)\,\partial/\partial Q$†

Integral	Result‡
$\langle v \vert P \vert v+1\rangle$	$b_1 i\sqrt{(v+1)}$
$\langle v \vert P^2 \vert v+2\rangle$	$-b_2\sqrt{[(v+2)!/v!]}$
$\langle v \vert P^2 \vert v\rangle$	$b_2 2(v+\frac{1}{2})$

† The matrix elements not given $\langle v| P^n |v\rangle$ with $v' \geqslant v$ and $n = 1$ or 2 vanish.
‡ $b_n = (\hbar\omega/2)^{n/2}$.

The off-diagonal matrix elements $\langle v| P^2 |v+2\rangle$ and $\lambda\langle v| Q^2 |v+2\rangle$ have equal absolute values but opposite sign and the diagonal terms $\langle v| P^2 |v\rangle$ and $\lambda\langle v| Q^2 |v\rangle$ are equal.

(c)
$$\langle \psi^0_{V_\nu}| f(Q) |\psi^0_{V_\mu}\rangle = \left\langle \prod_p \psi^0_{v_{p\nu}}(Q_p) \right| \sum_{ijk} \alpha^{(3)}_{ijk} Q_i Q_j Q_k + \sum_{ijkl} \alpha^{(4)}_{ijkl} Q_i Q_j Q_k Q_l + \cdots \left| \prod_q \psi^0_{v_{q\mu}}(Q_q) \right\rangle \tag{7.274}$$

The unknown coefficients C_μ may be found by diagonalizing the so-generated matrix:

$$\|\langle \psi^0_{V_\nu}| H - \varepsilon |\psi^0_{V_\mu}\rangle\| = 0 \tag{7.275}$$

The only problem is to limit the expansion of the vibrational wave function in terms of harmonic contributions. Table 7.8 shows that, if the highest power of Q in the potential expansion is q, all the integrals $\langle v| Q^q |v'\rangle$ disappear, except when v' belongs to the interval $[v-q, v+q]$. This interval may be considered as defining the first neighbourhood of the state v. In practice, the stability of the eigenvalues corresponding to the state v of interest is reached by considering two or three neighbourhoods (between $[v-2q, v+2q]$ and $[v-3q, v+3q]$).

7.8.5 Some Characteristic Values of the Vibrational Motion

The classical turning point corresponds to the point where the speed is vanishing ($\dot{Q} = 0$). Two such points exist: the first ($Q^{(-)}$) for a negative value of Q and the second ($Q^{(+)}$) for a positive value of Q. In the case of a purely harmonic vibrator we must have at any time:

$$\hbar\omega_i(v_i + \tfrac{1}{2}) = \tfrac{1}{2}\dot{Q}_i^2 + \tfrac{1}{2}\omega_i^2 Q_i^2 \qquad \forall i \mid 1 \leqslant i \leqslant m \tag{7.276}$$

Then for $\dot{Q}_i^{(\pm)} = 0$ it is evident that:

$$Q_i^{(\pm)} = \pm\sqrt{\left[\frac{\hbar(2v_i+1)}{\omega_i}\right]} \tag{7.277}$$

For an anharmonic potential we have to find the roots of the following equation:

$$\varepsilon(v_i) - V(Q^{(\pm)}) = 0 \tag{7.278}$$

This may be done by the Newton–Raphson method. Starting from a current point as close as possible to the value of $Q^{\pm}$, we can write in terms of Taylor's expansion:

$$\varepsilon(v_i) - V(Q^{(\pm)} + h) \simeq -\left(\frac{dV}{dQ}\right)_{Q^{(\pm)}} h$$

where the displacement (h) from the current point ($Q^{\pm} + h$) is approximated by:

$$h \simeq \frac{V(Q^{(\pm)} + h) - \varepsilon(v_i)}{(dV/dQ)_{Q^{(\pm)}+h}} \tag{7.279}$$

We iterate until h vanishes.

Another characteristic value is the mean normal coordinate value for any vibrator. This is simply obtained by solving the integral (which corresponds to a first moment):

$$\langle Q_l \rangle = \langle \psi_V | \, Q_l \, | \psi_V \rangle \qquad \forall l = 1, n \tag{7.280}$$

In the case of the interacting model it turns out that, if:

$$\psi_V = \sum_i C_i \prod_p \psi^0_{pi}(Q_p)$$

and according to Table 7.8, then:

$$\begin{aligned}
\langle Q_l \rangle &= \sum_{ij} C_i C_j \langle v_{l_i} | \, Q_l \, | v_{l_j} \rangle \prod_{p \neq l} \delta(v_{p_i}, v_{p_j}) \\
&= \sqrt{\left(\frac{\hbar}{2\omega_l}\right)} \sum_{ij} C_i C_j \sqrt{(v_{l_i} + 1)} \, \delta(v_{l_i}, v_{l_j} \pm 1) \prod_{p \neq l} \delta(v_{p_i}, v_{p_j})
\end{aligned} \tag{7.281}$$

For the pure harmonic vibrator this set of means always remains zero; this is quite different in the anharmonic case.

The second moment of vibration gives the dispersion of the normal coordinates around their mean values. This is a square symmetrical matrix of the form:

$$\langle \mathbf{QQ}' \rangle = \langle \psi_V | \, \mathbf{QQ}' \, | \psi_V \rangle \tag{7.282}$$

In the case of purely harmonic motion this matrix becomes diagonal. The values corresponding to any vibrator are:

$$\langle Q_l^2 \rangle = \frac{\hbar}{\omega_l} (v + \tfrac{1}{2}) \tag{7.283}$$

From the standpoint of current statistical thermodynamics this former quantity is replaced by the Δ matrix bonded to the spectral density, where Δ is called the mean square amplitude matrix with the diagonal elements:[47]

$$\Delta_k = Q_k^2 = \frac{\hbar}{2\omega_k} \coth\left(\frac{\hbar\omega_k}{2kT}\right) \tag{7.284}$$

In terms of internal displacements we find:

$$\boldsymbol{\Sigma} = \mathbf{SS'} = \mathbf{L\Delta L'} \tag{7.285}$$

In other coordinates frames connected to S by the transformation $\mathbf{Q} = \mathbf{RS}$ this matrix becomes:

$$\mathbf{P} = \mathbf{qq'} = \mathbf{R\Sigma R'}$$

In practice the mean square amplitude quantities for the bonded or non-bonded interatomic distances are the most interesting ones.

7.9 QUANTUM TREATMENT OF THE H_2 MOLECULE

7.9.1 The Potential Energy Curve

First of all we have to find an analytical form which is able to reproduce the tabulated potential energy curve (Figure 7.24). We choose the next polynomial expansion:[1,48]

$$V(R) = \sum_{i=0}^{n} C_i (R - R_e)^i$$

In Figure 7.25 we show that C_0 to C_4 converge to a fixed value when n increases to 8. The analytical function thus generated reproduces accurately the potential curve between $R = 0.75$ a.u. and $R = 2.9$ a.u.; the highest deviate is smaller than 0.1 kcal mol^{-1}. For the equilibrium distance and the corresponding derivatives up to order four we use the following values:

$$R_e = 1.406 \text{ a.u.}$$

$$\left(\frac{d^2 V}{dR^2}\right)_{R=R_e} = 0.36208 \text{ a.u.}$$

$$\left(\frac{d^3 V}{dR^3}\right)_{R=R_e} = -1.28704 \text{ a.u.}$$

$$\left(\frac{d^4 V}{dR^4}\right)_{R=R_e} = 3.38294 \text{ a.u.}$$

7.9.2 The Harmonic Treatment

Let us suppose the H_2 molecule is oriented along the z axis. The $\mathbf{B}$ matrix of Wilson is then:

$$\mathbf{B} = (0 \quad 0 \quad -1 \quad 0 \quad 0 \quad +1)$$

The mass of the hydrogen atom is 1.007825 g mol^{-1} or in atomic units (where the electron has a mass of 1), 1,836.5396 a.u.

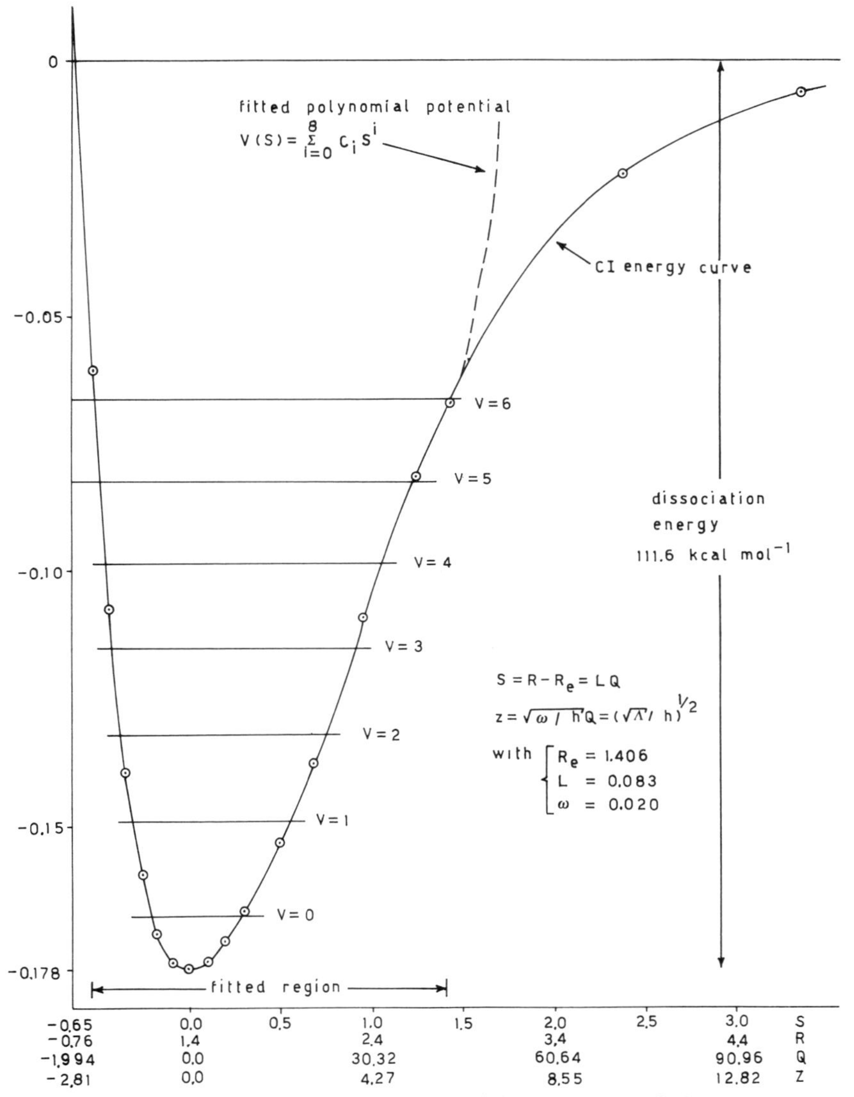

Figure 7.24 The H_2 potential energy curve fitting

The relations previously defined are now given in atomic units:†

$$\mathbf{G} = \mathbf{B}\mathbf{M}^{-1}\mathbf{B}' \qquad = 1.0890 \times 10^{-3} \text{ a.u.}$$

$$\omega = \sqrt{\Lambda} = \sqrt{(\mathbf{GH})} = 1.9857 \times 10^{-2} \text{ a.u.}$$

$$\mathbf{L} = \mathbf{G}^{1/2} \qquad = 3.3000 \times 10^{-2} \text{ a.u.}$$

* 1 a.u. mass = 1,822.2803 atom mass, 1 a.u. time = 2.418887×10^{-17} s, Speed of light = c = 2.9979251×10^{10} cm s^{-1}, ω in atomic units may be converted to cm^{-1} by multiplying by 219,474.3531.

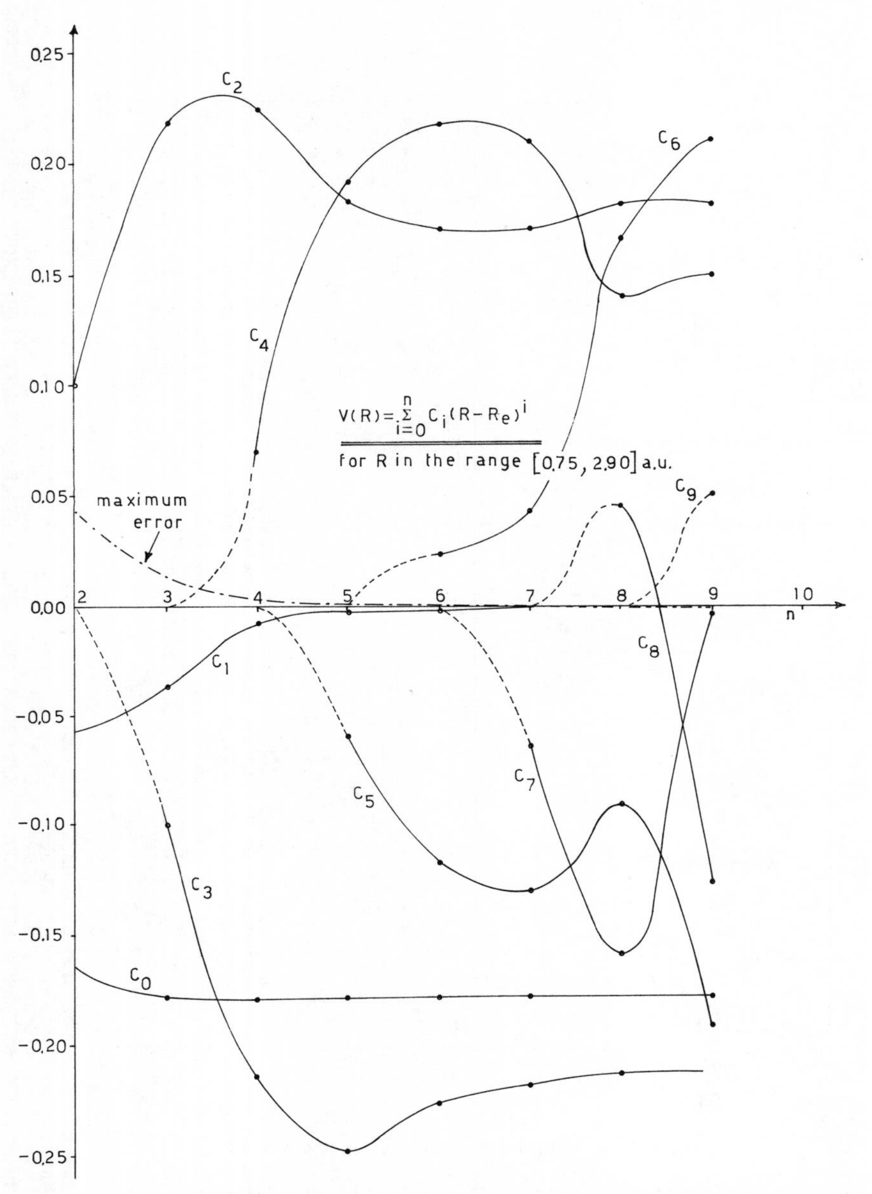

Figure 7.25 Polynomial expansion of the H_2 potential energy curve (in atomic units)

The associated wave length in reciprocal centimetre is:

$$\bar{\nu} = \frac{\omega}{2\pi c} = 4{,}355.6 \text{ cm}^{-1}$$

We are now able to express the higher order derivatives of the potential energy with respect to the internal coordinate in terms of normal coordinates. If:

$$Q = L^{-1}S = L^{-1}(R - R_e)$$

then:

$$\left(\frac{d^3V}{dQ^3}\right)_e = L^3\left(\frac{d^3V}{dR^3}\right)_e = 4.6253 \times 10^{-5} \text{ a.u.}$$

$$\left(\frac{d^4V}{dQ^4}\right)_e = L^4\left(\frac{d^4V}{dR^4}\right)_e = 4.0119 \times 10^{-6} \text{ a.u.}$$

7.9.3 The Perturbational Treatment

The first-order correction to the vibrational energy depends on the fourth-order derivatives by the expression:

$$\Delta\varepsilon_v^{(1)} = \frac{1}{24}\left(\frac{d^4V}{dQ^4}\right)_e\left(\frac{3\hbar^2}{2\omega^2}\right)[(v+\tfrac{1}{2})^2+\tfrac{1}{4}] = 6.3592 \times 10^{-4}[(v+\tfrac{1}{2})^2+\tfrac{1}{4}]$$

In terms of wave length we obtain:

$$\Delta\tilde{\nu}_v^{(1)} = \frac{\Delta\varepsilon_v^{(1)}}{2\pi\hbar c} = 139.6 \text{ cm}^{-1}[(v+\tfrac{1}{2})^2+\tfrac{1}{4}]$$

At the second order we want to consider four terms at most:

The first two are: $\frac{1}{6}\left(\frac{d^3V}{dQ^3}\right)_e \frac{(\langle v|\, Q^3\, |v \pm 1\rangle)^2}{\pm\hbar\omega}$

The last two are: $\frac{1}{6}\left(\frac{d^3V}{dQ^3}\right)_e \frac{(\langle v|\, Q^3\, |v \pm 3\rangle)^2}{\mp 3\hbar\omega}$

The total second-order correction is of the form:

$$\Delta\varepsilon_v^{(2)} = \left[\frac{1}{24}\left(\frac{d^3V}{dQ^3}\right)_e\right]^2 \frac{\hbar^2}{8\omega^4} f(v) = 4.7776 \times 10^{-5} f(v)$$

where $f(v)$ is a function of the quantum number v. According to Table 7.8 we find:

$$f(v=0) = -11$$
$$f(v=1) = -71$$
$$f(v=2) = -191$$

All the $f(v)$ values may be found from the general expression:

$$f(v) = -30[(v+\tfrac{1}{2})^2+\tfrac{7}{60}]$$

In the case of H_2 the second-order perturbation gives:

$$\Delta\tilde{\nu}_v^{(2)} = -314.6[(v+\tfrac{1}{2})^2+\tfrac{7}{60}] \text{ cm}^{-1}$$

The wave number associated with any vibrational level becomes:

$$\begin{aligned}\tilde{\nu}_v &= \tilde{\nu}_v^{(0)} + \Delta\tilde{\nu}_\nu^{(1)} + \Delta\tilde{\nu}_v^{(2)} \\ &= -1.8 + 4{,}355.6(v+\tfrac{1}{2}) - 175.0(v+\tfrac{1}{2})^2 \text{ cm}^{-1}\end{aligned}$$

7.9.4 The Interaction Model

In this case we have only to build the matrix elements of the vibrational hamiltonian between the harmonic functions (ψ_V^0). As the expansion previously given for the potential energy, limited to the fourth order, reproduces it accurately enough around the minimum, we shall keep it in the future. The computed matrix $H_{ij} = \langle\psi_i^0|\, H\, |\psi_j^0\rangle$ is a symmetrical square matrix ($n \times n$) and is reported below for the first terms (in cm^{-1}):

$$(H_{ij}) = \left(\begin{array}{r|rrrr|rr}
2{,}247.5 & & & & & & \\
\hline
-640.7 & 6{,}882.0 & & & & & \\
197.1 & -1{,}812.3 & 11{,}795.2 & & & & \\
-523.2 & 569.1 & -3{,}329.4 & 16{,}987.3 & & & \\
113.8 & -1{,}046.3 & 1{,}126.8 & -5{,}126.0 & 22{,}458.2 & & \\
\hline
0.0 & 254.5 & -1{,}654.4 & 1{,}870.3 & -7{,}163.8 & 28{,}207.9 & \\
0.0 & 0.0 & 440.8 & -2{,}339.7 & 2{,}799.7 & -9{,}417.1 & 34{,}236.4 \\
0.0 & 0.0 & 0.0 & 673.4 & -3{,}095.1 & 3{,}914.9 & -11{,}866.9 \\
0.0 & 0.0 & 0.0 & 0.0 & 952.3 & -3{,}915.1 & 5{,}216.1
\end{array}\right)$$

We diagonalize this matrix to find the eigenvectors and the eigenvalues. Convergence of the eigenvalues to a fixed value is obtained by increasing the matrix size (n) as shown in Figure 7.26. We find:

$$\tilde{\nu}_0 = 2{,}143.0 \text{ cm}^{-1} \quad \text{with } \psi_0 = 0.99\psi_0^{(0)} + 0.15\psi_1^{(0)} + 0.02\psi_2^{(0)}$$

$$\tilde{\nu}_1 = 6{,}257.9 \text{ cm}^{-1} \quad \text{with } \psi_1 = -0.15\psi_0^{(0)} + 0.91\psi_1^{(0)} + 0.37\psi_2^{(0)} + 0.10\psi_3^{(0)} + 0.07\psi_4^{(0)}$$

This corresponds to the following expression:

$$\tilde{\nu}_v = -4.7 + 4{,}355.6(v+\tfrac{1}{2}) - 120.4(v+\tfrac{1}{2})^2$$

A similar expression is obtained from experimental results:

$$\tilde{\nu}_v = \omega_e(v+\tfrac{1}{2}) - \omega_e x_e(v+\tfrac{1}{2})^2$$

For H_2 we have $\omega_e = 4{,}395 \text{ cm}^{-1}$ and $\omega_e x_e = 117.9 \text{ cm}^{-1}$.

Finally, the anharmonic contamination is illustrated in Figure 7.27 by comparing the computed wave function (ψ_v) to the harmonic dominant contribution (ψ_v). From the former treatment we are enable to compute

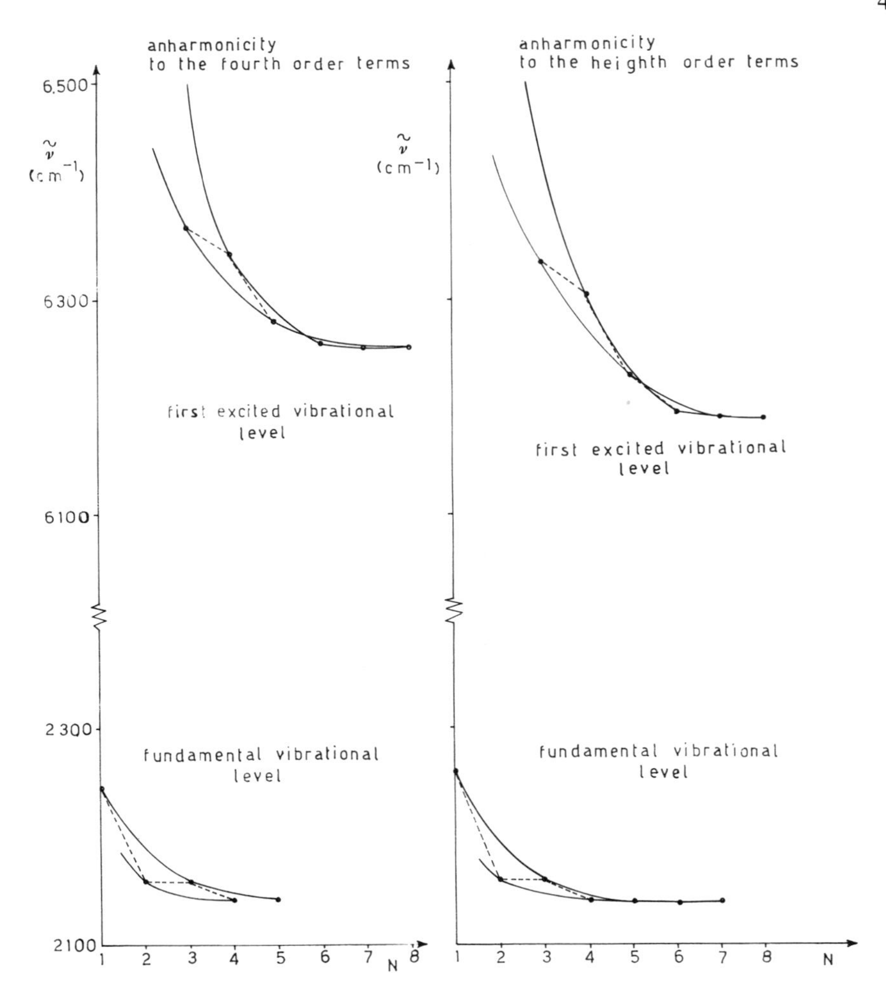

Figure 7.26 Eigenvalue convergence as a function of the interaction matrix size

some characteristic values of the vibrator. Table 7.10 summarizes these results.

Figure 7.27 as well as Table 7.10 show that the anharmonic vibrator is shifted (by ± 0.05 a.u. for $v = 0$ and by ± 0.10 a.u. for $v = 1$) with respect to the harmonic description. On the other hand, the mean square amplitude displacement (**SS'**) has approximately the same value as the mean square displacement ($\langle S^2 \rangle$) due to the fact that at 298.16 K the fundamental vibrational level is the only occupied level.

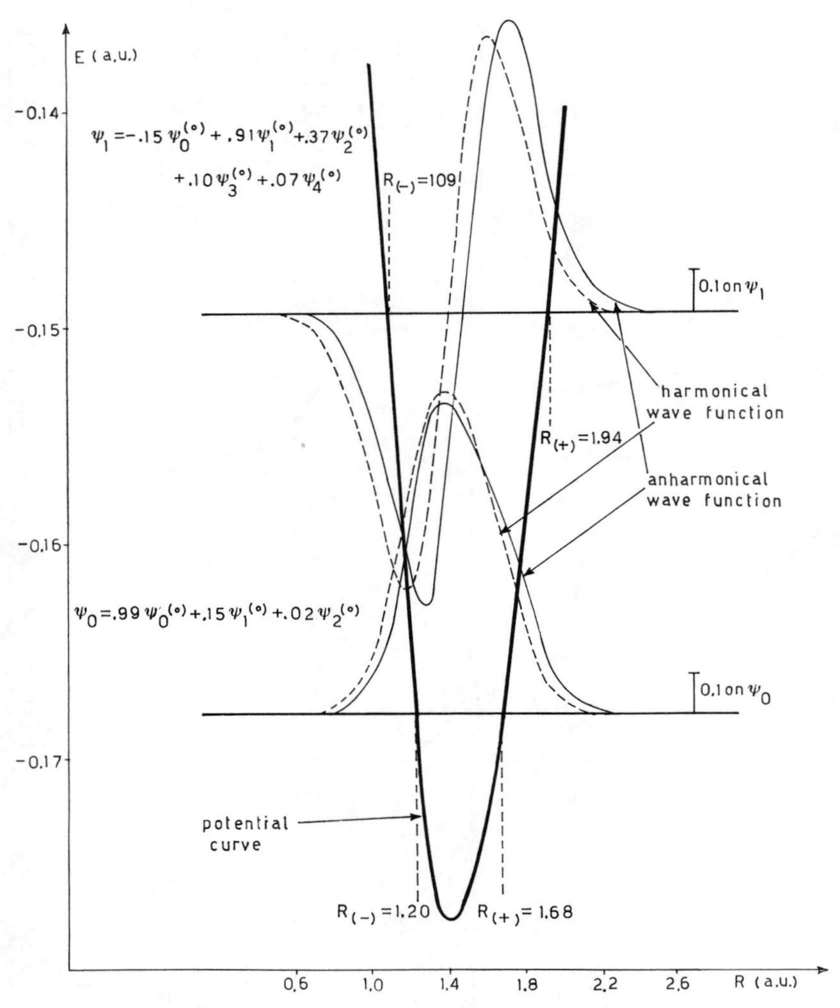

Figure 7.27 The vibrational wave function in the H_2 molecule

Table 7.10 Some characteristics of the H_2 vibrator (in atomic units) at the harmonic as well as at the anharmonic level

	$v = 0$		$v = 1$	
	Harmonic	Anharmonic	Harmonic	Anharmonic
$R_{(-)}$	1.17	1.20	0.99	1.09
$R_{(+)}$	1.63	1.68	1.81	1.94
$\langle R \rangle$	1.41	1.46	1.41	1.54
$\langle S^2 \rangle$	0.027	0.031	0.082	0.103
SS'	0.031 (harmonic case at 298.16 K)			

7.10 THE ELECTRICAL DIPOLAR TRANSITION IN THE VIBRONIC SPECTRUM

The most satisfactory treatment of systems containing radiations as well as matter is due to Dirac.[49] Because of the complexity of this theory we shall prefer a simpler one in which only absorption and induced emission are treated; the connection between those two phenomena and spontaneous emission are given by the Einstein coefficients.[50]

7.10.1 The Einstein Coefficient

Let us suppose that a molecular system has two non-degenerate vibrational levels W_m (for the upper one) and W_n (for the lower one). The emission or absorption radiation frequency will be:

$$\nu_{mn} = \frac{W_m - W_n}{h} \tag{7.286}$$

Let us suppose also that the molecule is in a bath of radiations for which the probability of finding radiations between ν_{mn} and $\nu_{mn} + d\nu$ per unit of volume is given by the density function $\rho(\nu_{mn})$. According to Planck's law the density of radiant emission at temperature T is:

$$\rho(\nu) = \frac{8\pi h\nu^3}{c^3} (e^{h\nu/kT} - 1)^{-1} \tag{7.287}$$

Now we shall assume that the probability of radiation absorption per unit of time is:

$$B_{nm}\rho(\nu_{mn}) \tag{7.288a}$$

Similarly, the probability of emission is:

$$A_{mn} + B_{mn}\rho(\nu_{mn}) \tag{7.288b}$$

The coefficients A_{mn}, B_{mn} and B_{nm} are respectively known as the Einstein coefficients for spontaneous emission, for induced emission and for absorption. From statistical distribution we know that the number of systems in the energy states $m(N_m)$ and $n(N_n)$ are related by the expression:

$$\frac{N_n}{N_m} = e^{-(W_n - W_m)/kT} = e^{h\nu_{mn}/kT} \tag{7.289}$$

If the total system containing both radiations and molecules is at equilibrium we can write from (7.288):

$$N_n B_{nm}\rho(\nu_{mn}) = N_m[A_{mn} + B_{mn}\rho(\nu_{mn})] \tag{7.290}$$

Then:

$$\rho(\nu_{mn}) = A_{mn}(B_{nm} e^{h\nu_{mn}/kT} - B_{mn})^{-1} \tag{7.291}$$

As (7.291) has to be identical to equation (7.287), we must assume that the Einstein coefficients are related by:

$$B_{mn} = B_{nm} \tag{7.292a}$$

$$A_{mn} = \frac{8\pi h\nu_{mn}^3}{c^3} B_{mn} \tag{7.292b}$$

7.10.2 The Infrared Absorption Coefficient

In infrared spectra a cell of length l which contains an absorbing gas is submitted to a parallel beam of radiation with intensity I. Decreasing intensity due to the absorption is taken to be proportional to the incident intensity and to the layer which is crossed. So the experimental absorption law is of the form:

$$-\mathrm{d}I = KI\,\mathrm{d}l \tag{7.293}$$

where K is the absorption coefficient.

On the other hand, according to Einstein's coefficients, we can write for the net rate of transitions in a unitary volume:

$$(B_{nm}N_n - B_{mn}N_m)\rho(\nu_{mn}) \tag{7.294}$$

As $B_{nm} = B_{nm}$, for a unit cross-section area and a transition of energy $h\nu_{mn}$ we find:

$$-\mathrm{d}I = h\nu_{mn}B_{mn}(N_n - N_m)\rho(\nu_{mn})\,\mathrm{d}l \tag{7.295}$$

The radiation flux intensity and density are related by:

$$I = c\rho \tag{7.296}$$

Then:

$$K = \frac{h\nu_{mn}}{c} B_{mn}(N_n - N_m) \tag{7.297}$$

By statistical mechanics the populations of the states n and m (N_n and N_m) are of the form:

$$N_n = \frac{N}{\Omega} \mathrm{e}^{-W_n/kT} \tag{7.298}$$

with:

$$\Omega = \sum_i \mathrm{e}^{-W_i/kT}$$

Ω being the partition function and N the total number of molecules per unit of volume.

7.10.3 The Transition Probabilities

Let us consider a molecule in a particular initial state $|A_0\rangle$ at the time t_0. According to the superposition principle $|A_0\rangle$ may in general be considered as a linear combination of all possible eigenstates of the system ($|i\rangle$):

$$|A_0\rangle = \mathbf{a}_0 \, |i\rangle \tag{7.299}$$

If it is submitted to an external electromagnetic field, the total hamiltonian of the molecule–field system is now time dependent ($H(t)$). Nevertheless, it may be decomposed into two parts: the first (E) does not involve the time explicitly and the second is the perturbating potential $V(t)$ which is time dependent and governs the interaction:

$$H(t) = E + V(t) \tag{7.300}$$

By the superposition principle which must hold at any time we can write:

$$|A_t\rangle = \mathbf{T} \, |A_0\rangle \tag{7.301}$$

where T is a linear operator depending only on t, which has the unitary property:

$$\mathbf{T}^*\mathbf{T} = \mathbf{E} \tag{7.302}$$

(T^* being the conjugate complex of T). The operator T is then closely related to $V(t)$ and describes what happens when an interaction occurs between the molecule and the external field. The principle of spectral decomposition tells us that the probability of finding the molecule in a particular eigenstate $|m\rangle$ is found to be the square of its expansion coefficient. At any time we shall find:

$$P(|m\rangle) = |\langle m \mid A_t\rangle|^2 = |\langle m| \, T \, |A_0\rangle|^2 \tag{7.303}$$

Then the transition probability between states $|n\rangle$ and $|m\rangle$ has the form:

$$P(n, m) = |\langle m| \, T \, |n\rangle|^2 \tag{7.304}$$

It remains to express the relation between T and $H(t)$. If we now pass to the infinitesimal case by making $t \rightarrow t_0$:

$$\left(\frac{\mathrm{d}\,|A_t\rangle}{\mathrm{d}t}\right)_{t=t_0} = \lim_{t\rightarrow t_0} \frac{|A_t\rangle - |A_0\rangle}{t - t_0} = \left[\lim_{t\rightarrow t_0}\left(\frac{T - \mathbf{E}}{t - t_0}\right)\right] |A_0\rangle \tag{7.305}$$

If we put this limit operator multiplied by $i\hbar$ equal to $H(t)$ we find:

$$i\hbar \frac{\mathrm{d}\,|A_t\rangle}{\mathrm{d}t} = H(t)\,|A_t\rangle$$

and

$$i\hbar\left(\frac{\mathrm{d}T}{\mathrm{d}t}\right)|A_0\rangle = H(t)T\,|A_0\rangle$$

valid for any ket $|A_0\rangle$, then:

$$i\hbar\left(\frac{\mathrm{d}T}{\mathrm{d}t}\right) = H(t)T \tag{7.306}$$

As $H(t)$ contains only one part which is time dependent ($V(t)$), it would be more useful to relate T and V directly. Defining τ as:

$$\boldsymbol{\tau} = e^{iE(t-t_0)/\hbar}\mathbf{T} \tag{7.307}$$

which gives by time differentiation:

$$i\hbar\left(\frac{d\tau}{dt}\right) = e^{iE(t-t_0)/\hbar}\left(i\hbar\frac{dT}{dt} - ET\right)$$

and remembering equations (7.306) and (7.300), we may write:

$$i\hbar\left(\frac{d\tau}{dt}\right) = e^{iE(t-t_0)/\hbar}V(t)T \tag{7.308}$$

Definining now $\mathscr{V}(t)$ as:

$$\mathscr{V}(t) = e^{iE(t-t_0)/\hbar}V(t)e^{-iE(t-t_0)/\hbar} \tag{7.309}$$

equation (7.308) becomes:

$$i\hbar\left(\frac{d\tau}{dt}\right) = \mathscr{V}(t)\tau \tag{7.310}$$

Note that τ and T are equally good for determining the transition probabilities, while:

$$\langle m|\,\tau\,|n\rangle = e^{iE_m(t-t_0)/\hbar}\langle m|\,T\,|n\rangle$$

so that:

$$P(n, m) = |\langle m|\,\tau\,|n\rangle|^2 \tag{7.311}$$

is equivalent to the equation (7.304).

Let us now assume that $\mathscr{V}(t)$ is a small quantity compared to E and express τ in the form:

$$\boldsymbol{\tau} = \mathbf{E} + \boldsymbol{\tau}_1 + \cdots \tag{7.312}$$

(because for $\mathscr{V} = 0$ it would make $\tau = 1$). By replacing such an expansion into the expression for $d\tau/dt$ in (7.310) and equating the terms of equal order:

$$i\hbar\frac{d\tau_1}{dt} = \mathscr{V}(t) \tag{7.313}$$

then:

$$\tau_1 = \int_{t_0}^{t} d\tau_1 = -\frac{i}{\hbar}\int_{t_0}^{t} \mathscr{V}(t)\,dt \tag{7.314}$$

For many practical problems this first-order expression is sufficiently accurate; the probability transition (7.311) takes the form (if $m \neq n$):

$$\begin{aligned} P(n, m) &= \frac{1}{\hbar^2}\left|\langle m|\int_{t_0}^{t} \mathscr{V}(t)\,dt\,|n\rangle\right|^2 \\ &= \frac{1}{\hbar^2}\left|\langle m|\int_{t_0}^{t} e^{iE_m(t-t_0)/\hbar}\mathscr{V}(t)e^{-iE_n(t-t_0)/\hbar}\,dt\,|n\rangle\right|^2 \end{aligned} \tag{7.315}$$

We also discard the magnetic field of the incident radiation and assume its wave length is large compared to the molecular dimension (1 Å equals 10^{-8} cm^{-1}). The perturbating potential simply becomes the scalar product between the electric displacement $\boldsymbol{\mu}$ and the electric field $\mathscr{E}(t)$.

$$V = \boldsymbol{\mu} \cdot \mathscr{E} = \sum_{\alpha} \mu_\alpha \cdot \mathscr{E}_\alpha \qquad \forall \alpha = X, Y \text{ or } Z \tag{7.316}$$

Hence the expression for the transition probability becomes for any cartesian direction (α):

$$P_\alpha(n, m) = \frac{1}{\hbar^2} \left| \langle m | \, \mu_\alpha \, | n \rangle \int_{t_0}^{t} e^{i(E_m - E_n)(t - t_0)/\hbar} \mathscr{E}_\alpha(t) \, dt \right|^2 \tag{7.317}$$

According to classical electrodynamics the density of the energy of radiation of frequency $\nu = (E_m - E_n)/h$ in space is:

$$\rho(\nu) = \frac{3}{2\pi} \left| \left| \int_{t_0}^{t} e^{2\pi i \nu (t - t_0)} \mathscr{E}_\alpha(t) \, dt \right|^2 \tag{7.318}$$

Then:

$$P_\alpha(n, m) = \frac{8\pi^3}{3h^2} |\langle m | \, \mu_\alpha \, | n \rangle|^2 \, \rho(\nu)$$

Taking into account all space directions at a time we have the following expression to be compared with the former equation (7.288a):

$$\begin{aligned} P(n, m) &= \frac{8\pi^3}{3h^2} (|\langle m | \, \mu_X \, | n \rangle|^2 + |\langle m | \, \mu_Y \, | n \rangle|^2 \\ &\quad + |\langle m | \, \mu_Z \, | n \rangle|^2) \rho(\nu_{n,m}) \\ &= B_{n,m} \rho(\nu_{n,m}) \end{aligned} \tag{7.319}$$

At this point, by (7.319) we are able to compute the absorption coefficient B as long as we can calculate the dipolar electrical transition integrals $\langle m | \, \mu_\alpha \, | n \rangle$. This is the purpose of the following paragraph. The absorption coefficients are related to the infrared measurement by equations (7.293) and (7.297) and we can write:[(51)]

$$K = \frac{8\pi^3}{3ch} \nu_{mn} \sum_{\alpha} |\langle m | \, \mu_\alpha \, | n \rangle|^2 \, (N_n - N_m) \tag{7.320a}$$

Nevertheless, it is more realistic to treat the Einstein coefficient as a constant for a given infrared line. So we write for the total line absorption coefficient:[(51)]

$$\int_{\text{line } n \to m} K(\nu) \, d\nu = \frac{8\pi^3}{3ch} \nu_{mn} (N_n - N_m) \sum_{\alpha} |\langle m | \, \mu_\alpha \, | n \rangle|^2 \tag{7.320b}$$

7.10.4 The Vibrational Selection Rules

The dipolar electric transition intensities between the states n and m depend on the value of the integrals:

$$\langle n|\ \mu_\alpha\ |m\rangle = \int \Psi_n^* \mu_\alpha \Psi_m \, \mathrm{d}T \qquad \forall \alpha = X, Y \text{ or } Z$$

It can be stated that the total wave function may be approximated by a simple product of electronic, vibrational, rotational and translational functions:

$$\Psi_n = \psi_\mathrm{E} \cdot \psi_\mathrm{V} \cdot \psi_\mathrm{R} \cdot \psi_\mathrm{T} \tag{7.321}$$

According to our actual preoccupation the electronic function may be ignored. As the dipole moment is not origin dependent, the translational function may be factorized. For a rotating molecule the components of the dipole moments ($\mu_X \mu_Y \mu_Z$) in terms of the space-fixed (SF) coordinate system (XYZ) can be expressed in the body-fixed (BF) coordinate system (xyz) using the direction cosines (ϕ) connecting the various pair of axis:

$$\begin{aligned} \mu_X &= \phi_{Xx}\mu_x + \phi_{Xy}\mu_y + \phi_{Xz}\mu_z \\ \mu_Y &= \phi_{Yx}\mu_x + \phi_{Yy}\mu_y + \phi_{Yz}\mu_z \\ \mu_Z &= \phi_{Zx}\mu_x + \phi_{Zy}\mu_y + \phi_{Zz}\mu_z \end{aligned}$$

or with matrical notation:

$$\boldsymbol{\mu}_\mathrm{SF} = \boldsymbol{\phi}\boldsymbol{\mu}_\mathrm{BF} \tag{7.322}$$

Then the integral $\langle n|\ \mu_\alpha\ |m\rangle$ becomes:

$$\langle n|\ \mu_\mathrm{SF}\ |m\rangle = \int \psi_{\mathrm{T}_n}^* \psi_{\mathrm{T}_m} \, \mathrm{d}\tau_\mathrm{T} \int \psi_{\mathrm{R}_n}^* \boldsymbol{\phi} \psi_{\mathrm{R}_m} \, \mathrm{d}\tau_\mathrm{R} \int \psi_{\mathrm{V}_n}^* \mu_\mathrm{BF} \psi_{\mathrm{V}_m} \, \mathrm{d}\tau_\mathrm{V} \tag{7.323}$$

As the set of translational functions $\{\psi_\mathrm{T}\}$ is orthonormal, the transition vanishes unless the initial and final translational quantum numbers are identical. The remaining part of the preceding equation provides rotational and vibrational selection rules. The electrical dipole in the molecular or body-fixed coordinate system $\boldsymbol{\mu}_\mathrm{BF}$ is not constant during the vibrational motion.

In order to take this into account we develop $\boldsymbol{\mu}$ in the Taylor series:

$$\boldsymbol{\mu}_\mathrm{BF} = \overset{0}{\boldsymbol{\mu}}_\mathrm{BF} + (\boldsymbol{\nabla}\mu_\mathrm{BF})_0' \mathbf{S} + \tfrac{1}{2}\mathbf{S}'(\boldsymbol{\nabla\nabla}'\mu_\mathrm{BF})_0 \mathbf{S} + \cdots \tag{7.324}$$

where $(\boldsymbol{\nabla}\mu_\mathrm{BF})_0$ and $(\nabla\nabla'\mu)_0$ are respectively the first derivative vector and second derivative matrix of μ at the equilibrium geometry ($\mathbf{S} = \mathbf{0}$), relative to the internal displacement coordinates $\mathbf{S}$ (equation 7.138). It is also possible to express $\boldsymbol{\mu}$ in terms of normal coordinates $\mathbf{Q}$ using the transformation (7.159), $\mathbf{S} = \mathbf{LQ}$:

$$\begin{aligned} \boldsymbol{\mu}_Q &= \overset{0}{\boldsymbol{\mu}}_\mathrm{BF} + (\nabla\mu_\mathrm{BF})'\mathbf{LQ} + \tfrac{1}{2}\mathbf{Q}'\mathbf{L}'(\nabla\nabla'\mu_\mathrm{BF})\mathbf{LQ} + \cdots \\ \text{or} \quad \boldsymbol{\mu}_Q &= \boldsymbol{\mu}^{(0)} + \boldsymbol{\mu}^{(1)}\mathbf{Q} + \tfrac{1}{2}\mathbf{Q}'\boldsymbol{\mu}^{(2)}\mathbf{Q} + \cdots \end{aligned} \tag{7.325}$$

if

$$\boldsymbol{\mu}^{(1)} = (\nabla \mu_{\text{BF}})' \mathbf{L}$$

and

$$\boldsymbol{\mu}^{(2)} = \mathbf{L}'(\nabla\nabla' \mu)\mathbf{L}$$

As we have seen before, the vibrational wave function ψ_V for an harmonic polyatomic oscillator is given by (equation 7.254a):

$$\psi_V(Q_1 \cdots Q_i \cdots) = \Pi_i \psi^0_{V_i}(Q_i)$$

The vibrational part $\langle n|\,\boldsymbol{\mu}\,|m\rangle_V$ of the integral is:

$$\begin{aligned}\langle n|\,\boldsymbol{\mu}\,|m\rangle_V = {} & \boldsymbol{\mu}^{(0)} \Pi_i \langle V_{i,n} \mid V_{i,m}\rangle \\ & + \sum_k \boldsymbol{\mu}_k^{(1)} \Big(\langle V_{k,n}|\, Q_k \,|V_{k,m}\rangle \prod_{i \neq k} \langle V_{i,n} \mid V_{i,m}\rangle \Big) \\ & + \frac{1}{2} \sum_k \boldsymbol{\mu}_{kk}^{(2)} \Big(\langle V_{k,n}|\, Q_k^2 \,|V_{k,m}\rangle \prod_{i \neq k} \langle V_{i,n} \mid V_{i,m}\rangle \Big) \\ & + \sum_{\substack{k,l \\ k \neq l}} \boldsymbol{\mu}_{kl}^{(2)} \Big(\langle V_{k,n}|\, Q_k \,|V_{k,m}\rangle \\ & \qquad \times \langle V_{l,n}|\, Q_l \,|V_{l,m}\rangle \prod_{i \neq k,l} \langle V_{i,n} \mid V_{i,m}\rangle \Big) + \cdots \end{aligned} \tag{7.326}$$

If the ith vibrator of the states n and m are respectively characterized by the quantum number $V_{i,n} = v_i'$ and $V_{i,m} = v_i''$ (with $v_i' < v_i''$), then the resolution of the integrals over the vibrational functions gives (see Table 7.8):

$$\langle v_i' \mid v_i''\rangle = \delta_{v_i'', v_i'}$$

$$\langle v_i'|\, Q \,|v_i''\rangle = \left[\frac{\hbar(v_i'+1)}{2\omega_i}\right]^{1/2} \delta_{v_i'', v_i'+1}$$

$$\langle v_i'|\, Q^2 \,|v_i''\rangle = \left[\frac{\hbar(v_i'+2)(v'+1)}{2\omega_i}\right]^{1/2} \delta_{v_i'', v_i'+2}$$

According to these results transitions occur only in the following cases:

(a) If all the vibrational quantum numbers in the initial and final states are identical except one which corresponds to the ith normal mode and such that $v_i'' = v_i' + 1$, then:

$$\langle n|\,\boldsymbol{\mu}\,|m\rangle_V = \boldsymbol{\mu}_i^{(1)} \sqrt{\left[\frac{(v'+1)\hbar}{2\omega_i}\right]} \tag{7.327a}$$

The associated frequency will be given at the harmonical level by:

$$\nu_{nm} = \frac{W_{v_i''} - W_{v_i'}}{h} = \nu_i \tag{7.327b}$$

This frequency is called fundamental.

(b) If all vibrational quantum numbers in the initial and final states are identical except one which corresponds to the ith normal mode and such that $v''_i = v'_i + 2$, then:

$$\langle n|\, \boldsymbol{\mu}\, |m\rangle_V = \boldsymbol{\mu}_{ii}^{(2)} \sqrt{[(v'_i+2)(v'_i+1)]} \frac{\hbar}{2\omega_i} \tag{7.328a}$$

The associated absorption frequency will be given at the harmonical level by:

$$\nu_{nm} = 2\nu_i \tag{7.328b}$$

Such a frequency is called overtone.

(c) If all vibrational quantum numbers in the initial and final states are equal except two which correspond to the ith and jth vibrational mode and such that $v''_i = v'_i + 1$ and $v''_j = v'_j + 1$, then:

$$\langle n|\, \boldsymbol{\mu}\, |m\rangle = \boldsymbol{\mu}_{ij}^{(2)} \sqrt{\left[\frac{(v'_i+1)(v'_j+1)}{\omega_i \omega_j}\right]} \frac{\hbar}{2} \tag{7.329a}$$

The associated absorption frequency will be given at the harmonic level by:

$$\nu_{nm} = \nu_i + \nu_j \tag{7.329b}$$

Such a frequency is called a combination frequency.

Generalization to the case of anharmonic vibrator may easily be done. If we develop the vibrational function ψ_V as a combination of the harmonic function $(\psi^0_{V_i})$:

$$\psi_V(Q) = \sum_p C_{pV} \left[\prod_i \psi^0_{V_i}(Q_i)\right]$$

the integral $\langle n|\, \boldsymbol{\mu}\, |m\rangle_V$ may now be written as:

$$\langle n|\, \boldsymbol{\mu}\, |m\rangle = \sum_p \sum_q C_{pn} C_{qm} \left(\left\langle \prod_i \psi^0_{V_{i,n}} \middle|\, \boldsymbol{\mu}\, \middle| \prod_i \psi^0_{V_{i,m}} \right\rangle\right) \tag{7.330}$$

which corresponds to a double sum over integrals previously defined for the harmonic case.

The equations (7.327) to (7.330) show that the electrical dipolar transitions which are infrared active may be classified in two groups:

(a) Those which occur when the dipole moment derivatives are non-vanishing.
(b) Those where we find both a non-vanishing derivative and an anharmonic contribution.

The more intense infrared lines are normally those of the first type due to the term $\boldsymbol{\mu}^{(1)}$ (equation 7.327). They may be called first-order harmonic lines; the associated selection rule is well known:

$$\Delta v = m - n = 1 \tag{7.331}$$

7.10.5 The Dipole Moment Derivatives

It only remains to compute an infrared spectrum to be able to calculate the dipole moment derivatives ($\boldsymbol{\mu}^{(1)}$ and $\boldsymbol{\mu}^{(2)}$) which appear in (7.325) or in (7.324). It is usually preferable to evaluate these derivatives with respect to the internal displacement **S** (7.324) rather than with respect to the normal coordinates **Q** (7.325). Such a derivation may be done numerically.[53] In this case experimental planification, previously explained, gives reliable results.[54] The first derivatives of the dipolar moment may also be computed in an analytical way (usually in terms of cartesian displacement). They may be defined[55] (as previously done for the force constant matrix) as:

$$\frac{\partial \mu}{\partial X_i} = \frac{\partial \mu}{\partial X_h(a)} = -\left[\frac{\partial^2 E(\mathscr{E}X)}{\partial \mathscr{E}_h \, \partial X_h(a)}\right]_{\mathscr{E}=0;\, X=e} \tag{7.332}$$

where X_i stands for the ith cartesian coordinate of the chemical system of interest or the cartesian coordinate $h(x$, y or $z)$ of the atom a, and E is the corresponding total energy of the molecule in an electric field $\mathscr{E}$ of components $\mathscr{E}_x \mathscr{E}_y \mathscr{E}_z$.

The general expression for the hamiltonian in a constant electric field (we exclude at this level the contribution which arise from the electric field gradient) may be written as:[56]

$$H = H^0 + \sum_n \mathbf{r}_n \mathscr{E} + \sum_N Z_N \mathbf{R}_N \mathscr{E} \tag{7.333}$$

where H^0 stands for the unperturbed usual hamiltonian ($H = H^0$ if $\mathscr{E} = 0$). The hamiltonian part which contains the electric field may be considered as a perturbation relative to H^0 and contains two parts:

(a) The first one describes the interaction between the electric field and the electronic distribution (**r** stands for the electron cartesian coordinates).
(b) The second one describes the interaction betwen the electric field and the nuclear charges (**R** stands for the nucleus cartesian coordinates).

By using the same coupled perturbational Hartree–Fock method as the one described in Section 7.3.1.2 for the force constant matrix we can easily find the expected values of $\partial \mu_h / \partial X_h(a)$:[55,57]

$$\frac{\partial \mu}{\partial X_i} = \frac{\partial \mu_h}{\partial X_f(a)} = -2 \sum_i^{\text{occ}} \langle \varphi_i | \, X_f \, | \varphi_i \rangle^{X_h} - 4 \sum_i^{\text{occ}} \sum_j^{\text{occ+virt}} \langle \varphi_i | \, X_f \, | \varphi_j \rangle U_{ji}(\mathscr{E}_h) + Z_a \tag{7.334}$$

From this, and using equations (7.144) and (7.159) we can deduce:

$$\mu^{(1)} = \left(\frac{\partial \mu}{\partial Q}\right) = (\mathbf{AL})' \left(\frac{\partial \mu}{\partial X}\right) \tag{7.335}$$

REFERENCES

1. G. Reckinger, Ph.D. Thesis, Université Catholique de Louvain (1983).
2. M. T. Nguyen, M. Sana, G. Leroy, K. J. Dignam and A. F. Hegarty, *J. Amer. Chem. Soc.*, **102,** 573 (1980).
3. J. L. Villaveces, Ph.D. Thesis, Université Catholique de Louvain, 1981.
4. J. W. McIver Jr. and A. Komornicki, *J. Amer. Chem. Soc.*, **94,** 2625 (1972).
5. J. N. Murrel and K. J. Laidler, *Trans. Faraday Soc.*, **64,** 371 (1968).
6. M. Sana, *Internat. J. Quantum Chem.*, **19,** 139 (1981).
7. K. Fukui, S. Kato and H. Fujimoto, *J. Amer. Chem. Soc.*, **97,** 1 (1975); A. Tachibana and K. Fukui, *Theoret. Chim. Acta*, **49,** 821 (1978), **51,** 189, 275 (1979); M. Sana, G. Reckinger and G. Leroy, *Theoret. Chim. Acta*, **58,** 145 (1981).
8. P. J. Kuntz, in *Molecular Collisions* (Ed. H. Miller), Part B, Plenum Press, 1976, Chap. 2, p. 74.
9. R. E. Stanton and J. W. McIver Jr., *J. Amer. Chem. Soc.*, **97,** 3632 (1975).
10. J. N. Murrel and K. J. Laidler, *Trans. Faraday Soc.*, **64,** 371 (1968).
11. P. Pulay, *Mol. Phys.*, **17,** 197 (1969).
12. S. Bratoz, *Colloque Internationaux du CNRS*, **82,** 287 (1958).
13. H. B. Schlegel, in *Computational Theoretical Organic Chemistry* (Eds. I. G. Csizmadia and R. Daudel), Vol. 67, Reidel, 1981, p. 128; J. D. Goddard, in *Computational Theoretical Organic Chemistry* (Eds. I. G. Csizmadia and R. Daudel), Vol. 67, Reidel, 1981, p. 161.
14. P. Swanstrom, K. Thomsen and P. B. Yde, *Mol. Phys.*, **20,** 1135 (1971).
15. J. Gerratt and I. M. Mills, *J. Chem. Phys.*, **49,** 1719 (1968); see also P. Swanstrom, K. Thomsen and P. B. Yde, *Mol. Phys.*, **20,** 1135 (1971); K. Thomsen and P. Swanstrom, *Mol. Phys.*, **26,** 735 (1973).
16. G. E. P. Box, *Biometrics*, **10,** 16 (1954).
17. G. E. P. Box and J. S. Hunter, *Ann. of Math. Stat.*, **28,** 195 (1957).
18. M. Sana, *Theoret. Chim. Acta*, **60,** 543 (1982).
19. W. Spendley, G. R. Hext and F. R. Himsworth, *Technometrics*, **4,** 41 (1962); W. Spendley in *Symposium of the Institute of Mathematics and its Applications* (Ed. R. Fletcher), Hoswell, 1968, Chap. 16, p. 259; V. Kafarov, *Méthodes Cybernétiques et Technologie Chimique*, MIR, 1974, p. 226.
20. H. O. Hartley, *Biometrics*, **15,** 611 (1959).
21. V. Kafarov, *Méthodes Cybernétiques et Technologie Chimique*, MIR, 1974, p. 211.
22. J. M. Lucas, *Technometrics*, **16,** 561 (1974).
23. R. J. Hader and S. H. Park, *Technometrics*, **20,** 413 (1978).
24. D. H. Doehlert and V. L. Klee, *Discrete Math.*, **2,** 309 (1972).
25. D. A. Gardiner, A. H. E. Grandage and R. J. Hader, *Ann. Math. Stat.*, **30,** 1082 (1959); N. R. Draper, *Technometrics*, **31,** 865, 875 (1960), **32,** 910 (1961); P. J. Thaker and M. N. Das, *J. Ind. Soc. Agric. Stat.*, **13,** 218 (1961); N. R. Draper, *Technometrics*, **4,** 219 (1962).
26. R. Fletcher and M. J. D. Powell, *Computer J.*, **6,** 163 (1963); D. Poppinger, *Chem. Phys. Lett.*, **34,** 332 (1975); J. Pacir, *Czechoslovak Chem. Comm.*, **40,** 2726 (1975).
27. K. Müller, *Angew. Chem. Int. Ed.*, **19,** 1 (1980).
28. J. A. Nelder and R. Mead, *Computer J.*, **7,** 308 (1965); S. L. Morgan and S. N. Deming, *Anal. Chem.*, **46,** 1170 (1974).
29. K. Müller and L. D. Brown, *Theoret. Chim. Acta*, **53,** 75 (1979).
30. D. H. Doehlert, *Appl. Stat.*, **19,** 231 (1970).
31. H. B. Schlegel in *Computational Theoretical Organic Chemistry* (ed. I. G. Csizmadia and R. Daudel), Reidel, Dordrecht, 1981, p. 129.

32. M. J. D. Powell, *Comp. J.*, **7,** 155 (1964) and **7,** 303 (1965).
33. M. J. D. Powell, *Computer J.*, **7,** 155 (1964).
34. D. M. Bishop and M. Randic, *J. Chem. Phys.*, **44,** 2480 (1966).
35. H. B. Schlegel, S. Wolfe and F. Bernardi, *J. Chem. Phys.*, **63,** 3632 and references 1–10 (1975); P. Pulay and G. Fogarasi, *J. Chem. Phys.*, **74,** 3999 (1981).
36. P. G. Mezey, M. R. Pieterson and I. G. Csizmadia, *Can. J. Chem.*, **55,** 2941 (1977).
37. R. C. Weast and S. M. Selby, in *Handbook of Chemistry and Physics*, 47th ed., The Chemical Rubber Co. (1966–67).
38. E. B. Wilson Jr., *J. Chem. Phys.*, **7,** 1047 (1939).
39. B. L. Crawford Jr. and W. H. Fletcher, *J. Chem. Phys.*, **19,** 141 (1951).
40. E. B. Wilson Jr., J. C. Decius and P. C. Cross, *Molecular Vibrations*, McGraw-Hill, 1955, Chap. 4, p. 54.
41. S. J. Cyvin, *Molecular Structures and Vibrations*, Elsevier, 1972, Chap. 1, p. 29.
42. E. B. Wilson Jr., J. C. Decius and P. C. Cross, *Molecular Vibrations*, McGraw-Hill, 1955, Chap. 11, 273; G. O. Sorensen, in *Molecular Structures and Vibrations*, (Ed. S. J. Cyvin), Elsevier, 1972, Chap. 2, p. 32.
43. E. B. Wilson Jr., J. C. Decius and P. C. Cross, *Molecular Vibrations*, McGraw-Hill, 1955, Chap. 11, p. 282.
44. H. B. Schlegel and also J. D. Goddard, *Computational Theoretical Organic Chemistry* (Eds. I. G. Csizmadia and R. Daudel), Reidel, 1981, pp. 155 and 168 respectively.
45. S. D. Ross, *Inorganic Infrared and Raman Spectra*, McGraw-Hill, 1972.
46. D. M. Dennison, *Rev. Mod. Phys.*, **3,** 280 (1931); H. H. Nielsen, *Phys. Rev.*, **38,** 1432 (1931); L. Pauling and E. B. Wilson Jr. in *Introduction to Quantum Mechanics*, Int. Stud. Ed., 1935; D. M. Dennison, *Rev. Mod. Phys.*, **12,** 175 (1940); G. Herzberg, *Molecular Spectra and Molecular Structure*, Vol. I, Van Nostrand-Reinhold, 1950, Chap. 2, p. 61.
47. F. Block, *Z. Phys.*, **74,** 295 (1932); R. W. James, *Physik Z.*, **33,** 737 (1932); S. J. Cyvin, *Spect. Chim. Acta*, **10,** 828 (1959).
48. M. Sana, *Rapport d'Activité Scientifique du CECAM*, CECAM, 1978, p. 176.
49. P. A. M. Dirac, *The Principles of Quantum Mechanics*, Oxford Press, 1958, Chap. 8, p. 168.
50. L. Pauling and E. B. Wilson Jr., *Introduction to Quantum Mechanics*, Int. Stud. Ed., 1935, Chap. XI.
51. R. Daudel and S. Bratoz, *Cah. Phys.*, **75,** 1 (1956).
52. J. H. Van Vleck and V. Weiskopf, *Rev. Mod. Phys.*, **17,** 227 (1945); W. Heitler, *Quantum Theory of Radiation*, Oxford, 1954; E. B. Wilson Jr., J. C. Decius and P. C. Cross, *Molecular Vibrations*, McGraw-Hill, 1955, Chap. 7.
53. P. Pulay, G. Fogarasi and J. E. Boggs, *J. Chem. Phys.*, **74,** 3999 (1981).
54. M. Sana and G. Leroy, *Theochem*, **76,** 259 (1981).
55. K. Thomsen and P. Swanstrom, *Mol. Phys.*, **26,** 735 (1976).
56. T. P. Das and R. Bersohn, *Phys. Rev.*, **115,** 897 (1959).
57. J. Gerratt and I. M. Mills, *J. Chem. Phys.*, **49,** 1719 (1968).

CHAPTER 8

The Collisional Chemical Process

8.1 THE DYNAMIC ORIENTED POTENTIAL ENERGY SURFACES

The potential energy surface is a function of $3N-6$ internal coordinates, where N is the number of atoms. The internal coordinates (s) chosen to describe the potential energy surface may for convenience be different from those used in the dynamical process (q). As we want to dispose of the potential energy ($V(q)$) and of its first-order derivatives ($\partial V/\partial q$) for any nuclear configuration, the simplest selection is to build an analytical potential energy surface $V(s)$ which is everywhere continuous and continuously differentiable. Moreover, we admit that we have at our disposal the relations connecting the coordinate frames s and q:

$$\mathbf{s}=\mathbf{s}(q) \tag{8.1}$$

So for each value of q we can compute s and $V(s)$. Using the chain rule, the first derivatives of V are related by:

$$\frac{\partial V}{\partial q_i}=\sum_j \frac{\partial V}{\partial s_j}\frac{\partial s_j}{\partial q_i}$$

or matricially:

$$(\boldsymbol{\nabla}_q V)=(\boldsymbol{\nabla}_q \mathbf{s}')[\boldsymbol{\nabla}_s V(s)] \tag{8.2}$$

Whatever may be the coordinate system s, it is obvious that the obtention of an analytical surface is a very tedious job. It is our intention not to do a general survey of the analytical potentials but rather to present some general features of this problem.

8.2 THE DIATOMIC POTENTIAL ENERGY CURVES

The solution of the time-independent Schroedinger equation in the Born–Oppenheimer approximation leads to a tabulated energy function. In the case of a diatomic molecule we have only one internal coordinate, the interatomic distance r. Some interpolating functions have been proposed to fit the theoretical curve. Hereafter we give the most frequently used:

(a)

$$\hat{V}(r)=\sum_{j=0}^{l} C_j \mathrm{e}^{-j\alpha(r-r_e)} \tag{8.3}$$

where r_e is the equilibrium distance, C_0 is the energy of the dissociated form $V(\infty)$ and $\sum_i C_i$ is the equilibrium energy $V(r_e)$.

Such an exponential basis set is only a generalization of the well-known Morse function ($k=+1$) or the anti-Morse function ($k=-1$):[1]

$$\begin{aligned}\hat{V}(r) &= D_e(1-k\mathrm{e}^{-\alpha(r-r_e)})^2 \\ &= D_e - k\cdot 2D_e{}^{-\alpha(r-r_e)} + D_e\mathrm{e}^{-2\alpha(r-r_e)}\end{aligned} \tag{8.4}$$

Such a fit is not too bad except for intermediate interatomic distances (see Figure 8.1). An alternate and more accurate solution has been proposed by Mohammad[2] using the hyperbolical functions:

$$\hat{V}(r) = A\tanh\left(\frac{r^m}{d}\right) - kC\,\mathrm{sech}^2\left(\frac{r^m}{d}\right) \tag{8.5}$$

with:

$$C = \frac{V(r_e)}{[1-\tanh(r_e^m/d)]^2}$$

$$A = -2C\tanh\left(\frac{r_e^m}{d}\right)$$

$$m = \alpha + \beta\Delta^{1/2} + \gamma\Delta$$

$$\Delta = r_e\left[\frac{(\partial^2 V/\partial r^2)_{r_e}}{2V(r_e)}\right]$$

$k=\pm 1$ according to the fact that $V(r)$ is a bonding or a non-bonding function

(b)

$$\hat{V}(r) = \sum_{j=0}^{l} C_j \frac{1}{r^j} \tag{8.6}$$

Such a fitting is particularly interesting as it leads to a linear expression whose coefficients C_j may be obtained using a regression method. Moreover, a basis set which contains something like 13 terms ($l=12$) enables us to reach a precision better than 0.1 kcal mol^{-1} everywhere (see Figure 8.1). Contrary to the exponential expansion the ($1/r$) one goes to infinity when r becomes zero. This type of function is not adapted to reproduce the potential energy for small interatomic distances.

(c) The third-order splines.[3] The space domain to be fitted by an analytical function is divided into $k-1$ intervals ($[r_i, r_{i+1}[, \forall i = 1, k-1$) for k computed energies ($V_i, \forall i = 1$ to k). In each interval we use the following third-order polynome:

$$S_i(x) = a_i x^3 + b_i x^2 + c_i x + d_i \qquad 1 \leqslant i < k \tag{8.7}$$

with:

$$x = \frac{r - r_i}{r_{i+1} - r_i}$$

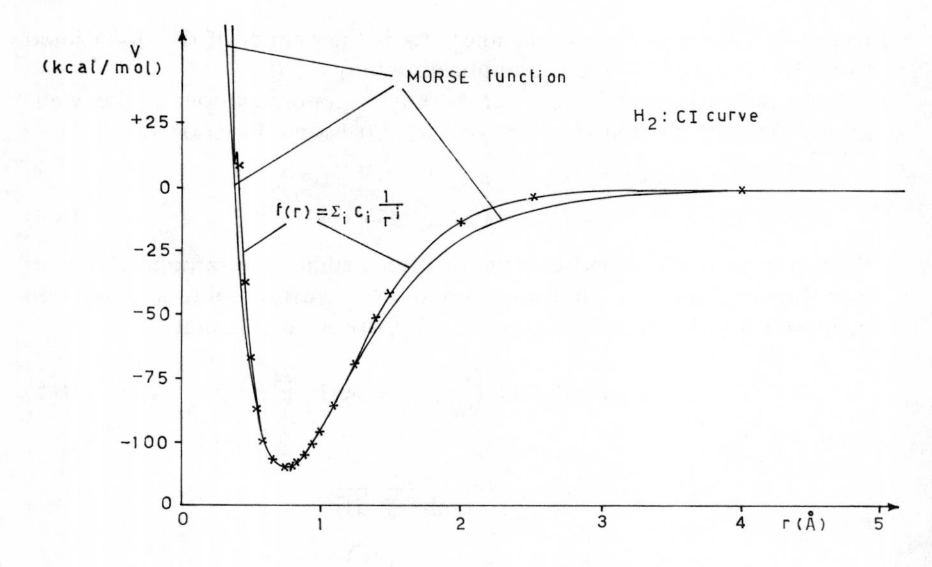

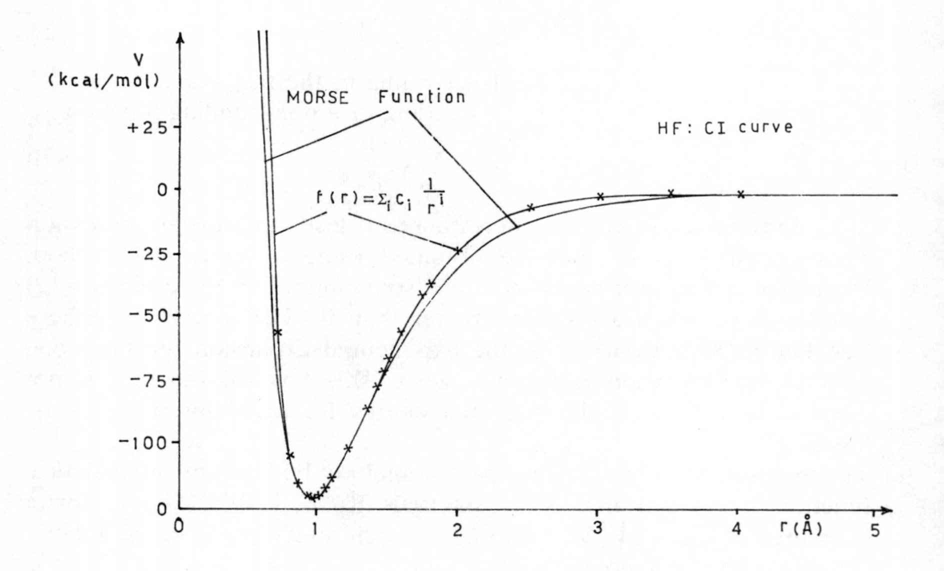

Figure 8.1 Analytical fitting for the dissociation curve of H_2 and HF molecules at the CI level

where:

$$r_i \leqslant r < r_{i+1} \qquad \text{and} \qquad 0 \leqslant x < 1$$

The associated derivatives with respect to r are:

$$S_i'(x) = \frac{3a_i x^2 + 2b_i x + c_i}{r_{i+1} - r_i} \tag{8.8a}$$

$$S_i''(x) = \frac{6a_i x_i + 2b_i}{(r_{i+1} - r_i)^2} \tag{8.8b}$$

$$S_i'''(x) = \frac{6a_i}{(r_{i+1} - r_i)^3} \tag{8.8c}$$

We dispose of $4(k-1)$ independent coefficients $(a_1 b_1 c_1 d_1 \cdots a_{k-1} b_{k-1} c_{k-1} d_{k-1})$ which are used in such a way that energy and derivatives are continuously defined when we go from one interval to the next one. Spline expansion will be univocally defined if we know the potential values for the k points between the $k-1$ intervals and if we know the first derivative of the potential with respect to the interatomic distance (or an approximate value) at the two limits of the domain to fit (r_1 and r_k). Illustrations of spline functions are given in Figure 8.2.

This interpolation method is certainly the most accurate, but it requires between 25 and 30 points in order to have a non-oscillating interpolative function (especially for the large distances). It has been shown that this method is able to reproduce locally the behaviour of a polynomial expansion of order seven around the equilibrium position.[4]

(d)

$$\hat{V}(r) = \sum_{i=0}^{l} C_i r^i \tag{8.9}$$

This expansion gives accurate results without including terms of too high an order if we limit the region of interest[5] (see Section 7.9). The fitting problem appears to be linear and reaches a finite potential value for r. Nevertheless, its convergence becomes a difficult working problem in the

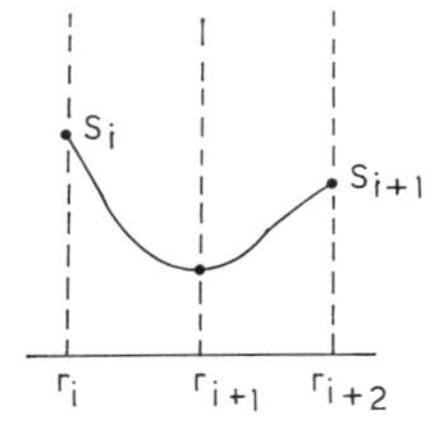

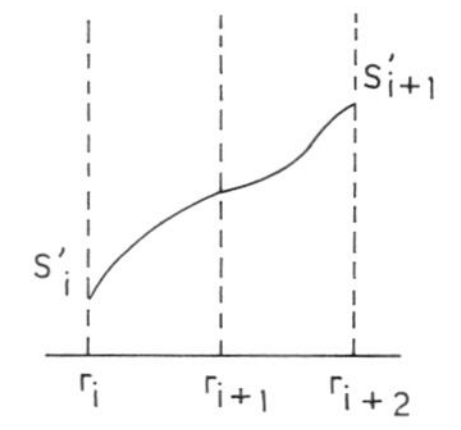

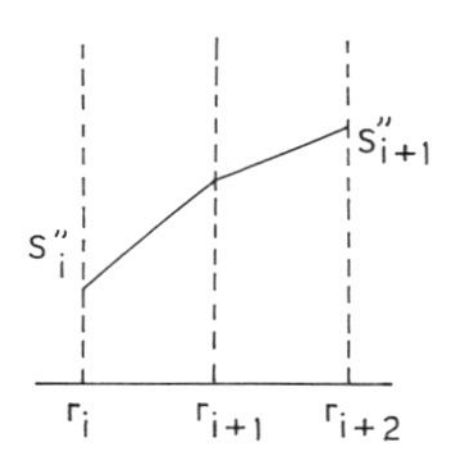

Figure 8.2 The cubic spline function and its derivatives for two consecutive intervals

dissociative region where the potential becomes flat. In fact, the Dunham series cannot converge beyond $2r_e$, as shown by Beckel and Engelke.[6] Improvement to expansion (8.9) has been proposed by using Euler transformation[7] or Padé approximants.[8]

(e) Other approaches using Padé approximants[9] or continued fractions[10] are useful methods to use in the extrapolation of data and treatment of avoided crossing problems.

8.3 THE TRIATOMIC POTENTIAL SURFACES

8.3.1 The Diatom-In-Molecule (DIM) Surfaces and Related LEPS Procedures

The obtention of an accurate analytical form for the triatomic potential energy surface is somewhat harder. The DIM method is based on the London equation.[11] For a triatomic system, when the states of the three atoms are all ^2S, we can write:

$$\hat{V}(r_1r_2r_3) = Q_1 + Q_2 + Q_3 - (J_1^2 + J_2^2 + J_3^2 - J_1J_2 - J_1J_3 - J_2J_3)^{1/2} \quad (8.10)$$

with:

$$Q_i = \frac{^1V_i(r_i) + {}^3V_i(r_i)}{2}$$

$$J_i = \frac{^1V_i(r_i) - {}^3V_i(r_i)}{2}$$

where $^1V_i(r_i)$ is a Morse potential function and $^3V_i(r_i)$ is an anti-Morse potential function. For the ith couple of atoms:

$$^1V_i(r_i) = {}^1D_{e,i}[(1 - e^{-{}^1\alpha_i(r_i - {}^1r_{e,i})})^2 - 1]$$

$$^3V_i(r_i) = {}^3D_{e,i}[(1 + e^{-{}^3\alpha_i(r_i - {}^3r_{e,i})})^2 - 1]$$

where the interatomic distances r_1, r_2, r_3 are given by:

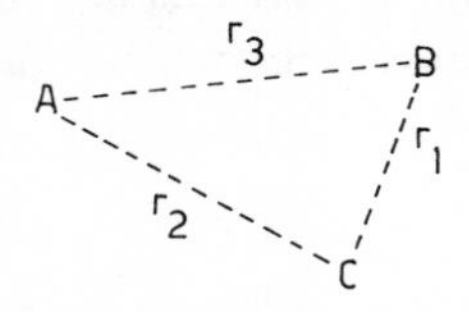

Such a function is useful to build a semi-empirical potential surface. If only the ground state potential surface is required the $^1V_i(r_i)$ diatomic functions are usually chosen to represent the corresponding dissociation curve, especially in the asymptotic region, while the $^3V_i(r_i)$ functions are adjusted to provide a good representation of the surface in the interaction region. Such a procedure is called the LEPS procedure (LEPS for London–Eyring–Polanyi–Sato). Many calculations have been performed using this approach.[12] Nevertheless, when one fits an *ab initio* or CI surface the standard error often reaches a value up to 10 kcal mol^{-1}. As we dispose of a limited set of non-linear adjustable parameters (at most 18 independent parameters when the three atoms are different), this suggests the use of modified LEPS functions.[13] It is easy to improve first the diatomic parts or Morse functions ($[{}^1V_i(r_i)]$) by better and more flexible diatomic curves (equations 7.3 and 7.9). In this way the asymptotic behaviour of the potential energy surface may be guaranteed. The triatomic interaction potential may be improved in two different ways. First of all we can use more flexible analytical functions for the anti-Morse curve ($[{}^3V(r_i)]$) and lastly it always remains possible to augment the LEPS basis function by a correction function:

$$\hat{V}(r) = \mathrm{LEPS}(r_1r_2r_3) + f(r_1r_2r_3) \tag{8.11}$$

It should be remembered that the anti-Morse as well as the correction functions must tend quickly to zero with $\{r_i\}$ increasing in order to preserve the asymptotic behaviour. The residual $f(r_1r_2r_3)$ function may be chosen as an expansion of the form:

$$f(r_1r_2r_3) = V(r_1r_2r_3) - \mathrm{LEPS}(r_1r_2r_3) = \sum_i C_i\varphi_i(r_1r_2r_3) \tag{8.12}$$

The advantage of this type of expansion is the linearity in C which allows the use of the least square fit method (see Section 7.3.2). The basis set ($\{\varphi_i(r_1r_2r_3)\}$) must be chosen in order to take into account the symmetry properties of the surface to be fitted. This problem occurs if two or three of the atoms are identical. In the following scheme we define the coordinates used to analyse the symmetry problem:

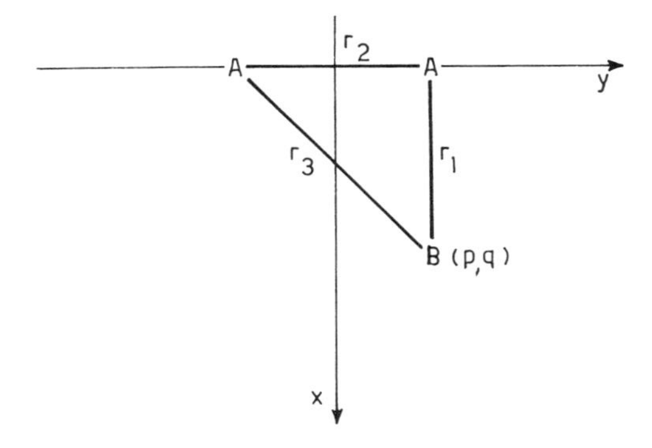

p and q are the cartesian coordinates of B:

$$r_1^2 = p^2 + \left(q - \frac{r_2}{2}\right)^2$$

$$r_3^2 = p^2 + \left(q + \frac{r_2}{2}\right)^2$$

The two symmetry elements (one on x and one on y) reduce the computation by four.

We have then to account for the following requests:

(a) The potential is symmetric under a change of sign in the x direction:

$$V(pqr_2) = V(-pqr_2) \tag{8.13a}$$

which becomes on first derivatives:

$$\left(\frac{\partial V}{\partial x}\right)_{x=0} = 0 \tag{8.13b}$$

and

$$\left(\frac{\partial V}{\partial x}\right)_{x=p} = -\left(\frac{\partial V}{\partial x}\right)_{x=-p} \tag{8.13c}$$

By the chain rule we have:

$$\frac{\partial}{\partial x} = \frac{x}{\sqrt{[x^2+(y-r_2/2)^2]}}\frac{\partial}{\partial r_1} + \frac{x}{\sqrt{[x^2+(y+r_2/2)^2]}}\frac{\partial}{\partial r_3} \tag{8.14}$$

This shows that the preceding requirement (8.13) is always met.

(b) The potential is symmetric under a change of sign in the y direction:

$$V(pqr_2) = V(p-qr_2) \tag{8.15a}$$

and, on first derivatives:

$$\left(\frac{\partial V}{\partial y}\right)_{y=0} = 0 \tag{8.15b}$$

and

$$\left(\frac{\partial V}{\partial y}\right)_{y=q} = -\left(\frac{\partial V}{\partial y}\right)_{y=-q} \tag{8.15c}$$

By the chain rule we find:

$$\frac{\partial}{\partial y} = \frac{r_2/2+y}{\sqrt{[x^2+(y+r_2/2)^2]}}\frac{\partial}{\partial r_3} - \frac{r_2/2-y}{\sqrt{[x^2+(y-r_2/2)^2]}}\frac{\partial}{\partial r_1} \tag{8.16}$$

This shows that we must use symmetrized basis set functions under r_1r_3 exchange, i.e.:

$$\varphi_{i*}(r_1r_2r_3) = \varphi_i(r_1r_2r_3) + \varphi_i(r_3r_2r_1) \tag{8.17}$$

(c) The third requirement comes from the residual function of $(r_1r_2r_3)$ and its zero asymptotic behaviour. We have then:

$$\lim_{r_j\to\infty} f(r_1r_2r_3) = \lim_{r_j\to\infty}\left[\frac{\partial f(r_1r_2r_3)}{\partial r_j}\right] = 0 \qquad \forall j = 1, 2 \text{ or } 3 \tag{8.18}$$

We can build more than one functional basis set which meets these requirements. Let us mention the two following expansions:

$$f(r_1r_2r_3)=\sum_{l,m,n=0}^{LMN} C_{lmn}\frac{1}{r_2^m}\left(\frac{1}{r_1^l r_3^n}+\frac{1}{r_1^n r_3^l}\right)\dagger \tag{8.19}$$

or

$$f(r_1r_2r_3)=\sum_{l,m,n=0}^{LMN} C_{lmn}\frac{r_2^{2m-2n-1}(r_3-r_1)^{2n}}{(r_3+r_1)^{2m}} \tag{8.20}$$

Such correction decreases the standard error below 1 kcal mol^{-1}.

8.3.2 The Spline-type Function

In each space direction (α, β, γ) we take a functional basis set which contains two cubic cardinal spline functions B (SB_1 and SB_k) and $k-1$ cubic cardinal spline functions A (SA_i, $\forall i=1$ to $k-1$), where k stands for the number of points we have at our disposal in the space directions α, β or γ. The form of any of those cubic splines is the same as the one previously defined in the one-dimensional problem (8.7). We impose that the cardinal spline functions satisfy the following conditions:

$$SB_1(\alpha_i)=SB_k(\alpha_i)=0 \qquad \forall i \text{ from 1 to } k \tag{8.21a}$$

$$SB_1'(\alpha_1)=SB_k'(\alpha_k)=1 \tag{8.21b}$$

$$SB_1'(\alpha_k)=SB_k'(\alpha_1)=0 \tag{8.21c}$$

$$SA_i(\alpha_j)=\delta_{ij} \qquad \forall i \text{ and } j \text{ from 1 to } k \tag{8.21d}$$

$$SA_i'(\alpha_1)=SA_i'(\alpha_k)=0 \qquad \forall i \text{ from 1 to } k \tag{8.21e}$$

† This is the symmetry adapted function basis set of Conroy.[14]

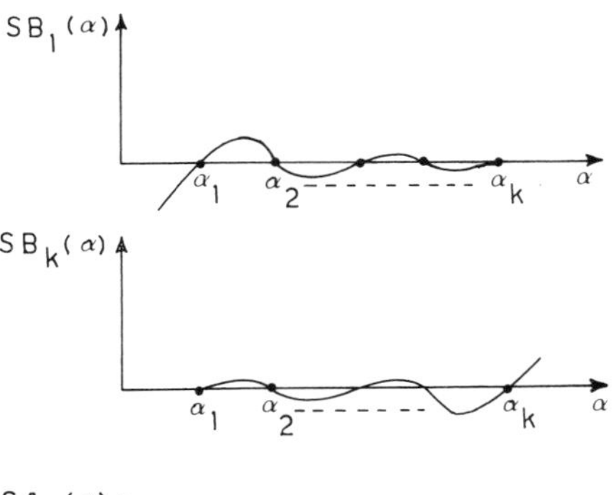

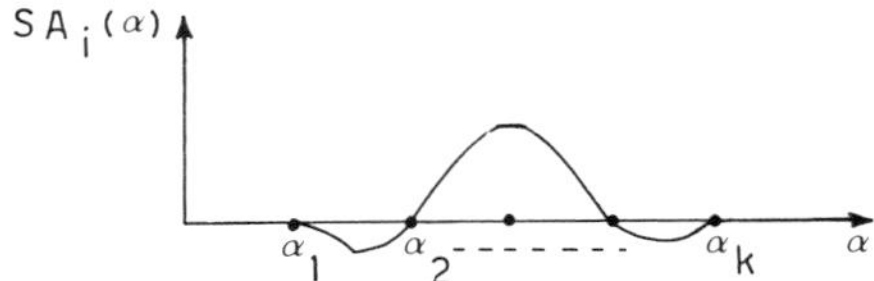

Figure 8.3 The cardinal A and B spline functions

We illustrate the form of the A and B type functions in Figure 8.3. In this functional basis set the potential energy is developed below with N_α, N_β and N_γ, the number of points in each space direction (α, β and γ):

$$
\begin{aligned}
V(\alpha,\beta,\gamma) = & \sum_{i=1}^{N_\alpha-1}\sum_{j=1}^{N_\beta-1}\sum_{k=1}^{N_\gamma-1} c_{ijk} SA_i(\alpha) SA_j(\beta) SA_k(\gamma) \\
& + \sum_{i}^{N_\alpha-1}\sum_{j}^{N_\beta-1} SA_i(\alpha) SA_j(\beta)[d_{ij1} SB_1(\gamma) + d_{ijN} SB_N(\gamma)] \\
& + \sum_{i}^{N_\alpha-1}\sum_{k}^{N_\gamma-1} SA_i(\alpha) SA_k(\gamma)[e_{i1k} SB_1(\beta) + e_{iNk} SB_N(\beta)] \\
& + \sum_{j}^{N_\beta-1}\sum_{k}^{N_\gamma-1} SA_j(\beta) SA_k(\gamma)[f_{1jk} SB_1(\alpha) + f_{Njk} SB_N(\alpha)] \\
& + \sum_{i}^{N_\alpha-1} SA_i(\alpha)[g_{i11} SB_1(\beta) SB_1(\gamma) \\
& \qquad + g_{i1N} SB_1(\beta) SB_N(\gamma) + g_{iN1} \cdots + g_{iNN} \cdots] \\
& + \sum_{j}^{N_\beta-1} SA_j(\beta)[h_{i11} SB_1(\alpha) SB_1(\gamma) + h_{i1N} SB_1(\alpha) SB_N(\gamma) \\
& \qquad + h_{iN1} \cdots + h_{iNN} \cdots] \\
& + \sum_{k}^{N_\gamma-1} SA_k(\gamma)[i_{i11} SB_1(\alpha) SB_1(\beta) \\
& \qquad + i_{i1N} SB_1(\alpha) SB_N(\beta) + i_{iN1} \cdots + i_{iNN} \cdots] \\
& + p_{111} SB_1(\alpha) SB_1(\beta) SB_1(\gamma) + p_{11N} SB_1(\alpha) SB_1(\beta) SB_N(\gamma) + \cdots \\
& \qquad + p_{NNN} SB_N(\alpha) SB_N(\beta) SB_N(\gamma)
\end{aligned}
\tag{8.22}
$$

According to the basic set properties (8.21) we find the following relations between the expansion coefficients and the potential energy or its successive derivatives:

$$V(\alpha_i \beta_j \gamma_l) = c_{ijl} \tag{8.23a}$$

$$\left(\frac{\partial V}{\partial \gamma}\right)_{\alpha_i \beta_j \gamma_1} = d_{ij1} \tag{8.23b}$$

$$\left(\frac{\partial^2 V}{\partial \alpha\, \partial \beta}\right)_{\alpha_i \beta_1 \gamma_1} = g_{i11} \tag{8.23c}$$

$$\left(\frac{\partial^3 V}{\partial \alpha\, \partial \beta\, \partial \gamma}\right)_{\alpha_1 \beta_1 \gamma_1} = p_{111} \tag{8.23d}$$

$$\left(\frac{\partial^3 V}{\partial \alpha\, \partial \beta\, \partial \gamma}\right)_{\alpha_N \beta_N \gamma_N} = p_{NNN} \tag{8.23e}$$

As the B spline functions break down rapidly, one often discards the terms including a product of two or more B splines ($g\ldots, h\ldots, i\ldots, p\ldots$). Finally, an appropriate choice of coordinate frame (α, β, γ) leads to some vanishing first-order derivatives.

The multidimensional interpolation process using the spline functions is accurate,[15] but often requires many points to give non-oscillating surfaces. In the case of three-dimensional surfaces (the triatomic problem) one needs between 1,000 and 10,000 points.

8.3.3 The Non-reactive Triatomic Surfaces

When studying non-reactive atom–diatom long-range potential energy hypersurfaces more simple analytical forms are available. In such cases the Van der Waals forces are the leading terms in the potential and we can write:

$$V(R, \gamma) = v_0(R) + v_2(R)P_2(\cos \gamma) \tag{8.24}$$

where:

$$v_0(R) = \tfrac{1}{3}[V(R, \gamma = 0) + 2V(R, \gamma = 90)]$$

and

$$v_2(R) = \tfrac{2}{3}[V(R, \gamma = 0) - 2V(R, \gamma = 90)]$$

according to the following definition for R and γ:

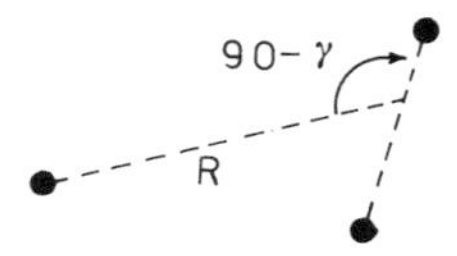

If $P_2(\cos \gamma)$ represents the second Legendre polynome this becomes:

For $\gamma = 0$: $P_2(\cos \gamma) = 1$

then: $V(R, \gamma) = V(R, \gamma = 0)$

For $\gamma = 90$: $P_2(\cos \gamma) = -\frac{1}{2}$

then: $V(R, \gamma) = V(R, \gamma = 90)$

This shows that v_2 stands for the anisotropy measurement. The radial $V(R)$ behaviour is written as:

$$V(R) = A \exp(-bR) + \sum_{n \geq 3} \frac{C_{2n}}{R^{2n}} + (M_1 + M_2)\left(b^2 - \frac{2b}{R}\right) A \exp(-bR) \tag{8.25}$$

Repulsion term (SCF term) — Dispersion term (correlation term) — Coupling term

An example is given in Figure 8.4.[16]

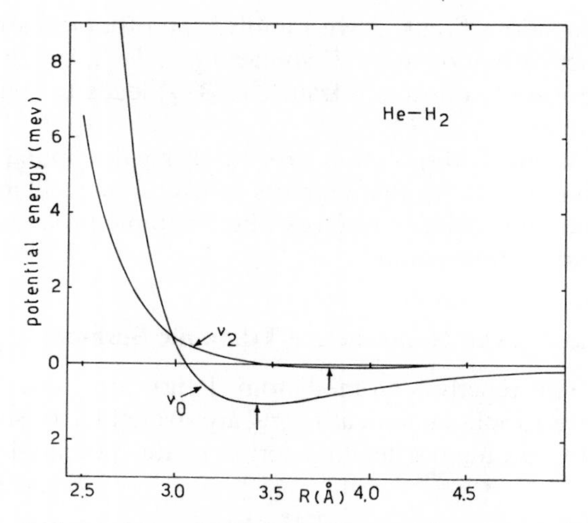

Figure 8.4 The spherically symmetric potential component v_0 and the anisotropic component v_2 (both in 10^{-3} eV ± 1 meV) predicted by the present model potential are plotted as a function of the centre of mass distance $\mathbf{R}$ (in angstroms) for He—H_2. The arrows indicate the locations of the respective minima. Reproduced by permission of The American Institute of Physics

For large anisotropic characters or large vibrational motions the coefficients of the $V(\mathbf{R})$ potential shape may be developed in series as a function of the atom–molecule distance ($\mathbf{R}$) and the atom–atom distance in the molecule ($\mathbf{r}$).

8.4 THE POLYATOMIC POTENTIAL ENERGY HYPERSURFACES

As the number of variables increases as $3N-6$ where N is the number of atoms, the computational effort becomes increasingly harder for a large number of atoms. It follows that the spline approach is too time-consuming in practice. Generalization of LEPS-type functions remains in use for the diatoms to diatoms potential[17] as it does not contain too many degrees of freedom (at least six parameters per atom couple). This approach presents the same advantages and disadvantages as in the the triatomic case.

For many problems we are more interested in an analytical form which fits the potential accurately up to a given energy limit than in a global analytical form including the atomization energy:

$$\hat{V} \simeq V \qquad (\text{for } V \leqslant V_{\text{max}}) \tag{8.26}$$

where $\hat{V}$ stands for an analytical estimator of the tabulated V exact function. The next proposals use this assertion and separate the non-reactive (elastic or inelastic) scattering problem from the reactive case. Thus one needs to make the local approach feasible in order to have a reference curve available in the domain of interest—also named the intrinsic reaction pathway.

8.4.1 The Non-reactive Surface

In the case of molecular interactions the inverse power of the interatomic distance expansion is often accurate enough. We then find:

$$V(A+B)=V(A)+V(B)+\sum_{i\in A}\sum_{j\in B} V(i,j) \tag{8.27}$$

where

$$V(i,j)=\sum_k A_k(i,j)\frac{1}{R_{ij}^k}$$

The i and j sums are expanded over all atomic pairs and sometimes include additional dummy atom pairs.

Examples of this kind of approach may be found elsewhere:

$$H_2O+CH_4$$

with k values of 1, 3 and 12;[18]

$$HF+HF$$

with k values of 1, 3, 6 and 12;[19]

$$H_2O+H_2O$$

with one additional exponential in $V(i,j)$;[20]

$$\exp(-b_{ij}R_{ij})$$

with k values of 1, 6, 8 and 10. Such a potential seems to be useful in a statistical approach of non-perfect gas or liquid properties.[21]

8.4.2 The Internal Intrinsic Reaction Pathway

In Section 7.2.5 we have shown that knowledge of the potential energy surface does not allow us to define the notion of the intrinsic reaction pathway. In fact, the gradient as well as the force constant matrix are quantities which are not invariant with the choice of the internal coordinate system. They only allow characterization of a set of coordinate frames related by unitary transforms. Nevertheless, the stationary points location remains invariant with respect to any set of coordinates.

In order to define a more interesting reaction pathway than the one based on the steepest descent pathway from the transition structure to the reactants or the products, we want to find a curve which is invariant under the coordinate transformation and passes through the three stationary points characterizing the reactants, the transition structure and the products.[22] To do this we express the lagrangian equations in terms of internal displacement coordinates (**S** in equation 7.16):

$$\frac{\mathrm{d}}{\mathrm{d}t}\frac{\partial T}{\partial \dot{\mathbf{S}}}+\frac{\partial V}{\partial \mathbf{S}}=\mathbf{0} \tag{8.28}$$

where:

$$2T=\dot{\mathbf{S}}'\mathbf{G}^{-1}\dot{\mathbf{S}} \qquad \text{(see equation 7.148)}$$

and

$$V(S)=V(0)+(\boldsymbol{\nabla}V)_0'\mathbf{S}+\tfrac{1}{2}\mathbf{S}'(\boldsymbol{\nabla}\boldsymbol{\nabla}'V)_0\mathbf{S}$$

$$=V(0)+\mathbf{g}_0'\mathbf{S}+\tfrac{1}{2}\mathbf{S}'\mathbf{H}_0\mathbf{S} \qquad \text{(see equation 7.17)}$$

Then:

$$\boldsymbol{\nabla}V(\mathbf{S})=\mathbf{g}_0+\mathbf{H}_0\mathbf{S}$$

These equations lead to:

$$\mathbf{G}^{-1}\ddot{\mathbf{S}}+\mathbf{g}_0+\mathbf{H}_0\mathbf{S}=\mathbf{0}$$

or

$$\ddot{\mathbf{S}}+\mathbf{G}\mathbf{g}_0+\mathbf{G}\mathbf{H}_0\mathbf{S}=\mathbf{0} \tag{8.29}$$

We can now prove that **Gg** as well as **GH** are two invariants. Let us take another coordinate system **v** related to **S** by:

$$\mathbf{S}=\mathbf{P}\mathbf{v} \tag{8.30}$$

as:

$$\mathbf{S}=\mathbf{B}\boldsymbol{\xi} \qquad \text{(equation 7.143)}$$

and

$$\mathbf{G}_S=\mathbf{B}\boldsymbol{\mu}\mathbf{B}' \qquad \text{(equation 7.147)}$$

Then **G** transforms as:

$$\mathbf{G}_v=\mathbf{P}^{-1}\mathbf{G}_S\mathbf{P}^{-1'} \tag{8.31}$$

having seen that **g** and **H** are transformed as:

For **g**: $\mathbf{g}_v=\mathbf{P}'\mathbf{g}_S$ (equation 7.29)

For **H**: $\mathbf{H}_v=\mathbf{P}'\mathbf{H}_S\mathbf{P}$ (equation 7.30)

Then:

$$\mathbf{G}_v\mathbf{g}_v=\mathbf{P}^{-1}\mathbf{G}_S\mathbf{g}_S \tag{8.32}$$

$$\mathbf{G}_v\mathbf{H}_v=\mathbf{P}^{-1}\mathbf{G}_S\mathbf{H}_S\mathbf{P} \tag{8.33}$$

These expressions show that **Gg** must be transformed as a vector and that **GH** must be transformed as a tensor. Thus the equations of motion may be considered as invariant and used to build an intrinsic reaction pathway. At the transition state the gradient to the potential energy surface is zero; then:

$$(\neq)\ \ddot{\mathbf{S}}+\mathbf{GHS}=\mathbf{0} \tag{8.34}$$

In terms of normal coordinates **Q** which diagonalize the **GH** product ($\mathbf{S}=\mathbf{LQ}$ in equation 7.159 with $\mathbf{L}^{-1}\mathbf{GHL}=\mathbf{\Lambda}$) we obtain:

$$\ddot{\mathbf{Q}}=-\mathbf{\Lambda Q} \tag{8.35}$$

As we have a transition structure, one and only one negative eigenvalue (λ^*) exists in the matrix $\mathbf{\Lambda}$ (Section 7.2.3). This corresponds to an imaginary vibrational frequency. The associated eigenvector ($\mathbf{L}^{-1*}$) may be chosen locally as a particular way to bypass the transition structure.

Elsewhere on the potential surface where the gradient with respect to the energy is significantly different from zero we limit the Taylor expansion of the surface up to the first order. Then we obtain for the equation of motion:

$$\ddot{\mathbf{S}}+\mathbf{Gg}=\mathbf{0} \tag{8.36}$$

We can now define an invariant reaction pathway by applying the next iterative algorithm:[23]

(a) Starting from the transition structure we do a small step alternately with the reactants and with the products in the $\mathbf{L}^{-1*}$ direction:

$$\mathbf{s}_{\pm}^{(0)}=\mathbf{s}_{\neq}\pm\frac{\varepsilon^2\mathbf{L}^{-1*}}{(\mathbf{L}^{-1*'}\cdot\mathbf{L}^{-1*})^{1/2}} \tag{8.37a}$$

where ε^2 stands for an infinitely small quantity.

(b) From the new point ($s_{\pm}^{(0)}$) we follow the direction defined by the vector **Gg**:

$$\mathbf{s}_{\pm}^{(i+1)}=\mathbf{s}_{\pm}^{(i)}-\frac{\varepsilon^2(\mathbf{Gg})_i}{(\mathbf{g}'\mathbf{G}'\mathbf{Gg})_i^{1/2}} \tag{8.37b}$$

(c) We stop the process when the norm of vector **g** becomes smaller than a given threshold η^2:

$$\mathbf{g}'\mathbf{g}\leqslant\eta^2 \tag{8.37c}$$

This definition of the reaction pathway will not be confused with any possible trajectory because we exclude considerations on the rotation of the overall system which may disturb the evolution of the internal coordinate by the presence of the Coriolis terms. Moreover, at any step of our process we keep (artificially) the instantaneous velocity to zero; in fact, for dynamical reasons we should have anywhere.

$$\mathbf{s}^{(i+1)}=\mathbf{s}^{(i)}+\dot{\mathbf{s}}^{(i)}t+(\mathbf{Gg})_i\frac{t^2}{2} \tag{8.38a}$$

and

$$\dot{\mathbf{s}}^{(i+1)} = \dot{\mathbf{s}}^{(i)} + (\mathbf{Gg})_i t \tag{8.38b}$$

As an example, in Figure 8.5 the reaction pathway is shown according to the previous definition for the reaction D+HF (the curve in the **v** coordinate frame).

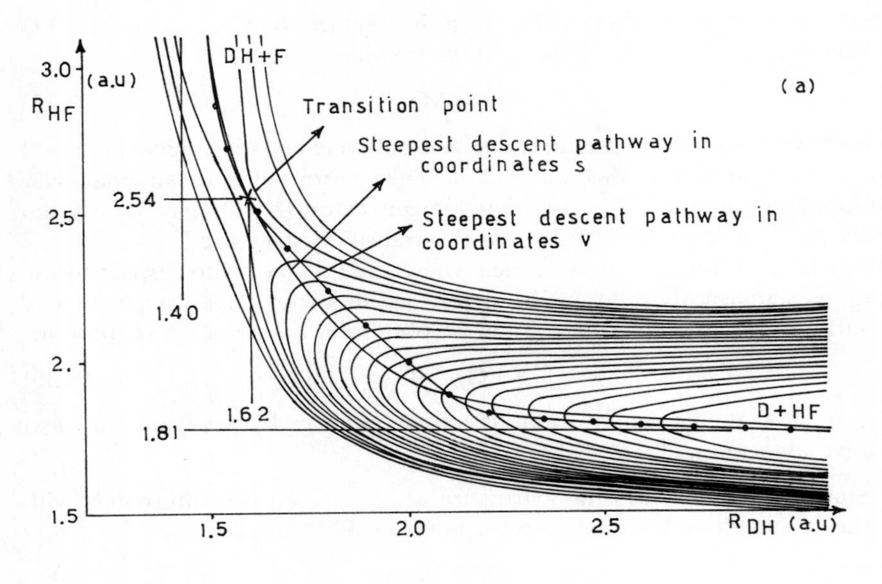

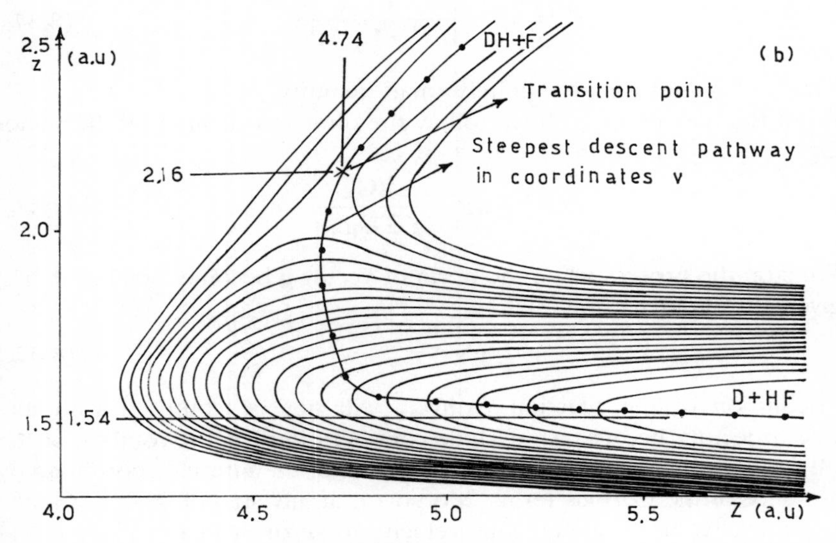

Figure 8.5 The coordinate invariant reaction pathway for the D+HF system compared to the steepest descent way. Reproduced by permission of Spinger-Verlag (Heidelberg)

The present reaction pathway definition is similar to the one of Fukui, Kato and Fujimoto who use the $3N$ weighted cartesian coordinates (**q** in equation 7.141) as coordinates instead of the $3N-6$ ones (**S**). In such a case $\mathbf{G}_q$ is the unity matrix $\mathbf{E}$ and $\ddot{\mathbf{q}} = -\mathbf{g}_q$. This means that the usual steepest descent procedure becomes available.

8.4.3 Many-dimensional Surfaces Around the Reaction Pathway

The problem is now to describe the potential along the intrinsic internal reaction pathway. We call the reaction coordinate (σ) the coordinate which follows the reaction pathway everywhere. In addition, we want to find a set of the $3N-7$ remaining space directions in order to dispose of a complete set of coordinates. Let us write the equations of motion previously defined in terms of the internal coordinates (**S** in equation 8.29) in a new frame **R** related to **S** by:

$$\mathbf{R} = \mathbf{G}^{-1/2}\mathbf{S} \tag{8.39}$$

This frame is in fact the one used to define the reaction pathway. Then we find:

$$\ddot{\mathbf{R}} + \mathbf{g}_R + \mathbf{H}_R\mathbf{R} = \mathbf{0} \tag{8.40}$$

where:

$$\mathbf{g}_R = \mathbf{G}^{1/2'}\mathbf{g}_s$$

and

$$\mathbf{H}_R = \mathbf{G}^{1/2'}\mathbf{H}_s\mathbf{G}^{1/2}$$

As $\mathbf{H}_R$ is a symmetrical matrix there exists a unitary transformation $\mathbf{Y}$ which diagonalizes H_R ($\mathbf{Y}'\mathbf{H}_R\mathbf{Y} = \mathbf{\Lambda}$). Replacing $\mathbf{H}_R$ by $\mathbf{Y\Lambda Y}'$ and premultiplying the former equation of motion by $\mathbf{Y}'$ we find that:

$$\mathbf{Y}'\ddot{\mathbf{R}} + \mathbf{Y}'\mathbf{g}_R + \mathbf{\Lambda Y}'\mathbf{R} = \mathbf{0}$$

or if:

$$\mathbf{Q} = \mathbf{Y}'R \tag{8.41}$$

then:

$$\ddot{\mathbf{Q}} + \mathbf{g}_Q + \mathbf{\Lambda Q} = \mathbf{0} \tag{8.42}$$

This corresponds to a set of $3N-6$ uncoupled differential equations:

$$\ddot{Q}_i + g_{Q_i} + \lambda_i\mathbf{Q}_i = 0 \qquad \forall i \text{ from } 1 \text{ to } 3N-6$$

The Q's define a normal coordinate frame which, at any stationary point, becomes the normal vibrational coordinates.

In addition, as a unitary transformation (**Y**) connects the Q and R frames, the directions of $\mathbf{g}_Q$ and $\mathbf{g}_R$ are identical in the two Q and R representations.

Close to the transition point, if we follow the eigendirection of G_sH_s corresponding to its negative curvature (λ^*), all the g_Q components are 0 except one which corresponds to the reaction coordinate previously defined. For reasons of continuity, this constraint remains valid anywhere on the reaction pathway as long as it follows the bottom of a valley. If this statement cannot be satisfied, before diagonalizing the $\mathbf{H}_R$ matrix we want to throw out of $\mathbf{H}_R$ the direction corresponding to the reaction coordinates:[25]

$$\mathbf{H}_R^P = (\mathbf{E}-\mathbf{P})\mathbf{H}_R(\mathbf{E}-\mathbf{P}) \tag{8.43}$$

where $\mathbf{P}$ is the square projection matrix corresponding to the reaction pathway direction. Diagonalization of $\mathbf{H}_R^P$ gives $3N-7$ eigenvectors orthogonal to the reaction pathway. The last eigendirection has a zero eigenvalue due to the projection operation. The S and Q coordinates frames are related by:

$$\mathbf{Q} = \mathbf{Y}'\mathbf{G}^{-1/2}\mathbf{S} \quad \text{and} \quad \mathbf{S} = \mathbf{G}^{1/2}\mathbf{Y}\mathbf{Q}\dagger$$

or

$$\mathbf{Q} = \mathbf{K}\mathbf{S} \quad \text{and} \quad \mathbf{S} = \mathbf{L}\mathbf{Q}$$

with:

$$\mathbf{K} = \mathbf{L}^{-1} \quad \text{and} \quad \mathbf{L}\mathbf{L}' = \mathbf{G}$$

This set of normal reaction coordinates may be used to describe the potential anywhere in a not too large region along the reaction pathway:

$$\hat{V}(\sigma, Q_1, \ldots, Q_{3N-7}) = V_0 + V(\sigma) + \sum_{i=1}^{3N-7} \lambda_i(\sigma)Q_i + \cdots \tag{8.44}$$

with:

$$\mathbf{Q} = \mathbf{K}(\sigma)\mathbf{S}$$

In practice the following procedure can be used to build such a type of analytical function:

1. Locate the stationary point of interest.
2. Search the intrinsic internal reaction pathway.
3. For some values of the reaction coordinates compute the $\mathbf{H}_R$ force constant matrices.
4. Diagonalizing $\mathbf{H}_R^P$, define the normal reaction coordinates and the associated curvature (λ_i).
5. Replace the $V(\sigma)$ and $\lambda(\sigma)$ tabulated value by a one-dimensional interpolation function, continuous in value and in derivatives.

† To be compared with equations (7.187) and (7.183).

8.5 THE HAMILTON EQUATIONS OF MOTION

The total energy of a polyatomic system is given by:

$$H = T + V \tag{8.45a}$$

where V, the potential energy, is only a function of the $3N-6$ internal coordinates, while T also depends on the coordinates associated with the rotation and the translation of the whole molecular system (see Section 7.6.1):

$$V = V(s) \tag{8.45b}$$

$$T = T(s, \theta, c, \dot{s}, \dot{\theta}, \dot{c}) = T(\eta, \dot{\eta}) = T(\eta, p) \tag{8.45c}$$

where $\dot{\eta}$ stands for $d\eta/dt$ and $p = \partial T/\partial \dot{\eta}$ stands for the conjugated moment to η. The hamiltonian formalism compared to the lagrangian one has the advantage of leading to a set of $6N$ first-order differential equations instead of $3N$ second-order equations:

$$\dot{\boldsymbol{\eta}} = \frac{\partial H(\eta, p)}{\partial \mathbf{p}} \tag{8.46a}$$

$$\mathbf{p} = -\frac{\partial H(\eta, p)}{\partial \boldsymbol{\eta}} \tag{8.46b}$$

In order to write these expressions in more detail we shall present the hamiltonian introduced for the classical description of the nuclear rotation–vibration motion in a more general form.[26,25] Before continuing this section it is advisable to reconsider Section 7.6 as deductions made there are now generalized.

8.5.1 The Kinetic Energy Expression

As for the deduction of the rotation–vibration hamiltonian, let us use two cartesian coordinate frames: the first is a space-fixed frame XYZ(SF) and the second a body-fixed frame xyz(BF) (see Figures 7.21 and 7.22). The last one is defined with respect to the former by the six Eckart constraints (equations 7.188 and 7.190). According to the notation previously introduced let us write:

$\mathbf{r}_a$ for the cartesian vector position of atom a
$\mathbf{r}_a^0$ for the same vector in the reference configuration†
$\mathbf{p}_a$ for the cartesian displacement vector $\mathbf{r}_a - \mathbf{r}_a^0$
$\tilde{m}_a$ for the mass of the atom a

† For the present purpose r_a^0 replaces the equilibrium vector r_a^e in Section 7.6; it corresponds to the point of the intrinsic reaction pathway (σ) which is closest to the current point r_a.

We write for the BF origin and orientation:

$$\sum_a m_a \mathbf{r}_a(\text{BF}) = \mathbf{0} \tag{8.47a}$$

$$\sum_a m_a \mathbf{r}_a^0(\text{BF}) = \mathbf{0} \tag{8.47b}$$

Then:

$$\sum_a m_a \boldsymbol{\rho}_a(\text{BF}) = \mathbf{0} \tag{8.47c}$$

$$\sum_a m_a \mathbf{r}_a^0(\text{BF}) \times \mathbf{r}_a(\text{BF}) = \mathbf{0} \tag{8.48a}$$

Then

$$\sum_a m_a \mathbf{r}_a^0(\text{BF}) \times \boldsymbol{\rho}_a(\text{BF}) = \mathbf{0} \tag{8.48b}$$

and

$$\sum_a m_a \mathbf{r}_a(\text{BF}) \times \boldsymbol{\rho}_a(\text{BF}) = \mathbf{0} \tag{8.48c}$$

As our purpose is to express the hamiltonian in terms of $3N-7$ (for non-linear cases) vibrational coordinates (S or Q), one reaction coordinate (σ), three eulerian angles (θ, φ, χ) and three centre of mass translations ($\mathbf{c}$), we may introduce one more constraint in order to show that cartesian displacements are orthogonal to the reaction path:

$$\sum_a m_a \left[\frac{\mathrm{d}\mathbf{r}_a^0(\text{BF})}{\mathrm{d}\sigma}\right] \boldsymbol{\rho}_a(\text{BF}) = 0. \tag{8.49a}$$

Then:

$$\sum_a m_a \mathbf{r}_a^0(\text{BF}) \times \left[\frac{\mathrm{d}\mathbf{r}_a^0(\text{BF})}{\mathrm{d}\sigma}\right] = \mathbf{0} \tag{8.49b}$$

if:

$$\frac{\mathrm{d}r_a^0}{\mathrm{d}\sigma} = -N\left(\frac{\partial V}{\partial r_a}\right)_{r_a = r_a^0} \tag{8.50}$$

where N stands for the normalization constant. The total velocity of any atom a (equation 7.192) may now be expressed as:

$$\mathbf{V}_a = \dot{\mathbf{c}}(\text{SF}) + \boldsymbol{\omega} \times \mathbf{r}_a(\text{BF}) + \mathbf{v}_a(\text{BF}) \tag{8.51}$$

where $\boldsymbol{\omega}$ is the angular velocity vector of the moving coordinate system. This relation can be introduced in the kinetic energy relation (7.193a) (dropping the BF subscript):

$$2T = \sum_a m_a \mathbf{V}_a \cdot \mathbf{V}_a$$

$$= \sum_a m_a \dot{\mathbf{c}} \cdot \dot{\mathbf{c}} + \sum_a m_a (\boldsymbol{\omega} \times \mathbf{r}_a) \cdot (\boldsymbol{\omega} \times \mathbf{r}_a) + \sum_a m_a \mathbf{v}_a \cdot \mathbf{v}_a$$

$$+ 2 \sum_a m_a (\boldsymbol{\omega} \times \mathbf{r}_a) \cdot \mathbf{v}_a + 2 \sum_a m_a \dot{\mathbf{c}} \cdot (\boldsymbol{\omega} \times \mathbf{r}_a) + 2 \sum_a m_a \dot{\mathbf{c}} \cdot \mathbf{v}_a \quad (8.52)$$

Let us now come back to the seven constraints previously given, as well as their time derivatives:

$$\sum_a m_a \mathbf{v}_a = \sum_a m_a \dot{\boldsymbol{\rho}}_a = \sum_a m_a \mathbf{v}_a^0 = \mathbf{0} \quad (8.53a)$$

$$\sum_a m_a (\mathbf{v}_a \times \mathbf{r}_a^0 + \mathbf{r}_a \times \mathbf{v}_a^0) = \sum_a m_a (\dot{\boldsymbol{\rho}}_a \times \mathbf{r}_a^0 + \boldsymbol{\rho}_a \times \mathbf{v}_a^0) = \sum_a m_a (\mathbf{v}_a \times \boldsymbol{\rho}_a + \mathbf{r}_a \times \dot{\boldsymbol{\rho}}_a) = \mathbf{0} \quad (8.53b)$$

$$\sum_a m_a \mathbf{v}_a^0 \cdot \boldsymbol{\rho}_a = 0; \qquad \sum_a m_a \mathbf{r}_a^0 \times \mathbf{v}_a^0 = 0; \qquad \sum_a m_a (\boldsymbol{\gamma}_a^0 \cdot \dot{\boldsymbol{\rho}}_a + \dot{\sigma} \boldsymbol{\Gamma}_a^0 \cdot \boldsymbol{\rho}_a) = 0 \quad (8.53c)$$

$$\mathbf{v}_a^0 = \frac{d\mathbf{r}_a^0}{dt} = \frac{d\mathbf{r}_a^0}{d\sigma} \frac{d\sigma}{dt} = \boldsymbol{\gamma}_a^0 \dot{\sigma} \quad (8.54a)$$

and

$$\dot{\boldsymbol{\gamma}}_a^0 = \frac{d\boldsymbol{\gamma}_a^0}{dt} = \frac{d\boldsymbol{\gamma}_a^0}{d\sigma} \frac{d\sigma}{dt} = \frac{d^2\mathbf{r}_a^0}{d\sigma^2} \dot{\sigma} = \boldsymbol{\Gamma}_a^0 \dot{\sigma} \quad (8.54b)$$

We find then for the kinetic energy (8.52):

$$2T = \left(\sum_a m_a\right) \dot{\mathbf{c}} \cdot \dot{\mathbf{c}} + \sum_a m_a (\boldsymbol{\omega} \times \mathbf{r}_a) \cdot (\boldsymbol{\omega} \times \mathbf{r}_a)$$

$$+ \dot{\sigma}^2 \sum_a m_a \boldsymbol{\gamma}_a^0 \cdot \boldsymbol{\gamma}_a^0 - 2\dot{\sigma}^2 \sum_a m_a \boldsymbol{\Gamma}_a^0 \cdot \boldsymbol{\rho}_a + \sum_a m_a \dot{\boldsymbol{\rho}}_a \cdot \dot{\boldsymbol{\rho}}_a$$

$$+ 2\boldsymbol{\omega} \cdot \sum_a m_a \boldsymbol{\rho}_a \times \dot{\boldsymbol{\rho}}_a - 4\dot{\sigma} \boldsymbol{\omega} \cdot \sum_a m_a \boldsymbol{\gamma}_a^0 \times \boldsymbol{\rho}_a \quad (8.55)$$

Collecting now all the vectors in matrix form:

$$\mathbf{c} = (c_g); \qquad \boldsymbol{\omega} = (\omega_g); \qquad \mathbf{R} = \{\mathbf{r}_a\}; \qquad \mathbf{R}^0 = \{\mathbf{r}_a^0\}; \qquad \boldsymbol{\xi} = \{\boldsymbol{\rho}_a\} = \mathbf{R} - \mathbf{R}^0$$

$$\boldsymbol{\gamma}^0 = \{\boldsymbol{\gamma}_a^0\}; \qquad \boldsymbol{\Gamma}^0 = \{\boldsymbol{\Gamma}_a^0\}; \qquad \text{diag}\,\mathbf{m} = \{\sum_a m_a\} \text{ and diag}\,\mathbf{M} = \{m_a\} \quad (8.56)$$

we write:

$$2T = \dot{\mathbf{c}}'\mathbf{m}\dot{\mathbf{c}} + \boldsymbol{\omega}'\mathbf{I}\boldsymbol{\omega} + \dot{\sigma}^2(\boldsymbol{\gamma}^{0\prime}\mathbf{M}\boldsymbol{\gamma}^0 - 2\boldsymbol{\Gamma}^{0\prime}\mathbf{M}\boldsymbol{\xi})$$

$$+\dot{\boldsymbol{\xi}}'\mathbf{M}\dot{\boldsymbol{\xi}} + 2\sum_g \omega_g \boldsymbol{\xi}'\mathbf{M}\mathbf{I}^g\dot{\boldsymbol{\xi}} - 4\dot{\sigma}\sum_g \omega_g \gamma^{0\prime}\mathbf{M}\mathbf{I}^g\boldsymbol{\xi} \qquad (8.57)$$

where g stands for the x, y or z cartesian components, $\mathbf{I}^g$ is the $3N \times 3N$ Meal–Polo matrix (equation 7.197) and $\mathbf{I}$ is the instantaneous inertia tensor:

$$\begin{aligned} \mathbf{I}_{gh} &= \mathbf{R}'\mathbf{I}^g\mathbf{M}\mathbf{I}^{h\prime}\mathbf{R} \\ &= \mathbf{R}^{0\prime}\mathbf{I}^g\mathbf{M}\mathbf{I}^{h\prime}\mathbf{R}^0 + 2\mathbf{R}^{0\prime}\mathbf{I}^g\mathbf{M}\mathbf{I}^{h\prime}\boldsymbol{\xi} + \boldsymbol{\xi}'\mathbf{I}^g\mathbf{M}\mathbf{I}^{h\prime}\boldsymbol{\xi} \end{aligned} \qquad (8.58)$$

Let us note that for our present purpose the reference configuration ($\mathbf{r}_a^0$) varies in each point of the reaction pathway and the associated velocity ($\mathbf{v}_a^0$) is not vanishing as it usually does in the treatment of molecular vibration (see Section 7.6). We note also that $\boldsymbol{\gamma}^0$ is a $3N$-dimensional unit vector which points along the reaction path and $\boldsymbol{\Gamma}_a^0$ (its derivative with respect to σ) is associated with the reaction pathway curvature.

We are now able to express T in terms of any internal displacements $\mathbf{S}$ (equation 7.16) by using the traditional **GH** formalism of Wilson (see Chapter 7 about the molecular vibrations). We introduce the following linear transformations between $\mathbf{S}$ and $\boldsymbol{\xi}$:

$$\mathbf{S} = \mathbf{B}\boldsymbol{\xi} \qquad (8.59)$$

The inverse relationship is:

$$\boldsymbol{\xi} = \mathbf{A}\mathbf{S} \qquad (8.60)$$

The matrices $\mathbf{B}$ and $\mathbf{A}$ are formally those defined in (7.143) and (7.144) but do not necessarily coincide with them numerically. $\mathbf{A}$ will exist as long as the $3N-7$ vibrational coordinates and the seven constraint coordinates are all linearly independent. The kinetic energy expression also requests the knowledge of the time derivative of ξ. We shall write:

$$\dot{\boldsymbol{\xi}} = \mathbf{A}\dot{\mathbf{S}} + \dot{\sigma}\boldsymbol{a}\mathbf{S} \qquad (8.61)$$

where:

$$\boldsymbol{a} = \frac{\mathrm{d}\mathbf{A}}{\mathrm{d}\sigma}$$

We now introduce the following definitions:

$$\mathbf{Z}^g_{\sigma,S} = (\mathbf{S}'\mathbf{A}'\mathbf{M}\mathbf{I}^g\boldsymbol{a} - 2\boldsymbol{\gamma}^{0\prime}\mathbf{M}\mathbf{I}^g\mathbf{A})\mathbf{S} \qquad (8.62a)$$

$$N_{\sigma\sigma,S} = \boldsymbol{\gamma}^{0\prime}\mathbf{M}\boldsymbol{\gamma}^0 + (\mathbf{S}'\boldsymbol{a}'\mathbf{M}\boldsymbol{a} - 2\boldsymbol{\Gamma}^{0\prime}\mathbf{M}\mathbf{A})\mathbf{S} \qquad (8.62b)$$

$$\mathbf{Z}^g_S = \mathbf{S}'\mathbf{A}'\mathbf{M}\mathbf{I}^g\mathbf{A} \qquad (8.62c)$$

$$\mathbf{N}_{S,\sigma} = \mathbf{S}'\boldsymbol{a}'\mathbf{M}\mathbf{A} \qquad (8.62d)$$

$$\mathbf{N}_{SS} = \mathbf{A}'\mathbf{M}\mathbf{A} (= \mathbf{G}^{-1} \text{ in Wilson notation}) \qquad (8.62e)$$

$$\mathbf{Z}_S = \begin{pmatrix} \mathbf{Z}_S^x \\ \mathbf{Z}_S^y \\ \mathbf{Z}_S^z \end{pmatrix} \quad \text{and} \quad \mathbf{Z}_{\sigma,S} = \begin{pmatrix} Z_{\sigma,S}^x \\ Z_{\sigma,S}^y \\ Z_{\sigma,S}^z \end{pmatrix} \tag{8.62f}$$

The kinetic energy becomes in matrix form:

$$2T = \begin{pmatrix} \dot{\mathbf{S}} \\ \dot{\sigma} \\ \boldsymbol{\omega} \\ \dot{\mathbf{c}} \end{pmatrix} \begin{pmatrix} \mathbf{N}_{SS} & & \text{Sym.} & \\ \mathbf{N}_{S,\sigma} & N_{\sigma\sigma,S} & & \\ \mathbf{Z}_S & \mathbf{Z}_{\sigma,S} & \mathbf{I} & \\ \mathbf{0} & \mathbf{0} & \mathbf{0} & \mathbf{m} \end{pmatrix} \begin{pmatrix} \dot{\mathbf{S}} \\ \dot{\sigma} \\ \boldsymbol{\omega} \\ \dot{\mathbf{c}} \end{pmatrix} \tag{8.63}$$

We may now go through the $3N-7$ normal reaction coordinates Q of Section 8.4.3:

$$\mathbf{S} = \mathbf{L}\mathbf{Q} \tag{8.64}$$

Differentiating this last relation with respect to time yields:

$$\dot{\mathbf{S}} = \dot{\sigma}\mathfrak{L}\mathbf{Q} + \mathbf{L}\dot{\mathbf{Q}} \tag{8.65}$$

where $\mathfrak{L}$ stands for $\mathrm{d}\mathbf{L}/\mathrm{d}\sigma$. We now rewrite the kinetic energy as:

$$2T = \begin{pmatrix} \dot{\mathbf{Q}} \\ \dot{\sigma} \\ \boldsymbol{\omega} \\ \dot{\mathbf{c}} \end{pmatrix} \begin{pmatrix} \mathbf{N}_{QQ} & & \text{Sym.} & \\ \mathbf{N}_{Q,\sigma} & N_{\sigma\sigma,Q} & & \\ \mathbf{Z}_Q & \mathbf{Z}_{\sigma,Q} & \mathbf{I} & \\ \mathbf{0} & \mathbf{0} & \mathbf{0} & \mathbf{m} \end{pmatrix} \begin{pmatrix} \dot{\mathbf{Q}} \\ \dot{\sigma} \\ \boldsymbol{\omega} \\ \dot{\mathbf{c}} \end{pmatrix} \tag{8.66}$$

where the matrix components correspond to the following expressions:

$$\mathbf{l} = \mathbf{M}^{1/2}\mathbf{A}\mathbf{L} \tag{8.67a}$$

$$\boldsymbol{A} = \mathbf{M}^{1/2}\boldsymbol{aL} \tag{8.67b}$$

$$\boldsymbol{B} = \mathbf{M}^{1/2}\mathbf{A}\mathfrak{L} \tag{8.67c}$$

and

$$\mathfrak{L} = \frac{\mathrm{d}\mathbf{l}}{\mathrm{d}\sigma} = \boldsymbol{A} + \boldsymbol{B} \tag{8.67d}$$

Then:

$$\mathbf{N}_{QQ} = \mathbf{l}'\mathbf{l} \tag{8.68a}$$

$$N_{\sigma\sigma,Q} = \boldsymbol{\gamma}^{0\prime}\mathbf{M}\boldsymbol{\gamma}^0 - 2\boldsymbol{\Gamma}^{0\prime}\mathbf{M}^{1/2}\mathbf{l}\mathbf{Q} + \mathbf{Q}'\mathfrak{L}'\mathfrak{L}\mathbf{Q} \tag{8.68b}$$

$$\mathbf{N}_{Q,\sigma} = \mathbf{Q}'\mathfrak{L}'\mathbf{l} = \mathbf{Q}'\boldsymbol{\xi}_\sigma \tag{8.68c}$$

$$\mathbf{Z}_Q^g = \mathbf{Q}'\mathbf{l}'\mathbf{I}^g\mathbf{l} = \mathbf{Q}'\boldsymbol{\xi}_Q^g \tag{8.68d}$$

$$Z_{\sigma,Q}^g = -2\boldsymbol{\gamma}^{0\prime}\mathbf{M}^{1/2}\mathbf{I}^g\mathbf{l}\mathbf{Q} + \mathbf{Q}'(\mathbf{l}'\mathbf{I}^g\mathfrak{L})\mathbf{Q} \tag{8.68e}$$

$$I^{gh} = \mathbf{R}^{0\prime}\mathbf{I}^g\mathbf{M}\mathbf{I}^{h\prime}\mathbf{R}^0 + 2\mathbf{R}^{0\prime}\mathbf{M}^{1/2}\mathbf{I}^g\mathbf{I}^{h\prime}\mathbf{l}\mathbf{Q} + \mathbf{Q}'\mathbf{l}'\mathbf{I}^g\mathbf{I}^{h\prime}\mathbf{l}\mathbf{Q} \tag{8.68f}$$

The matrix $\mathbf{l}$ is simply the linear transformation between the normal coordinates and the mass weighted cartesian displacements (q), as previously introduced by equations (7.174) and (7.175):

$$\mathbf{q}=\mathbf{l}\mathbf{Q} \qquad \text{and} \qquad \mathbf{Q}=\mathbf{l}'\mathbf{q} \qquad \text{with } \mathbf{l}'\mathbf{l}=\mathbf{E} \tag{8.69}$$

The above kinetic energy can be expressed in terms of conjugate momenta. In order to have a more simple expression for T let us first use the following definitions:

$$\mathbf{i}=\begin{pmatrix} N_{\sigma\sigma,Q} & \mathbf{Z}'_{\sigma,Q} \\ \mathbf{Z}_{\sigma,Q} & \mathbf{I} \end{pmatrix} \tag{8.70a}$$

$$\mathbf{Z}=\begin{pmatrix} \mathbf{N}_{Q,\sigma} \\ \mathbf{Z}_Q \end{pmatrix}=\begin{pmatrix} \mathbf{Q}'\boldsymbol{\xi}_\sigma \\ \mathbf{Q}'\boldsymbol{\xi}_Q \end{pmatrix} \tag{8.70b}$$

$$\dot{\boldsymbol{\Omega}}=\begin{pmatrix} \dot{\sigma} \\ \boldsymbol{\omega} \end{pmatrix} \tag{8.70c}$$

As the centre of mass global translation is completely separated from the other motions we will further discard this movement assuming $\dot{c}=0$. The kinetic energy becomes:

$$2T=\begin{pmatrix} \dot{\mathbf{Q}} \\ \dot{\boldsymbol{\Omega}} \end{pmatrix}'\begin{pmatrix} \mathbf{E} & \mathbf{Z}' \\ \mathbf{Z} & \mathbf{i} \end{pmatrix}\begin{pmatrix} \dot{\mathbf{Q}} \\ \dot{\boldsymbol{\Omega}} \end{pmatrix} \tag{8.71}$$

Then we have:

$$\mathbf{P}=\frac{\partial T}{\partial \dot{\mathbf{Q}}}=\dot{\mathbf{Q}}+\mathbf{Z}'\dot{\boldsymbol{\Omega}} \tag{8.72}$$

$$\begin{aligned} \boldsymbol{\Pi} &= \frac{\partial T}{\partial \dot{\boldsymbol{\Omega}}}=\mathbf{i}\dot{\boldsymbol{\Omega}}+\mathbf{Z}\dot{\mathbf{Q}} \\ &= (\mathbf{i}-\mathbf{Z}\mathbf{Z}')\dot{\boldsymbol{\Omega}}+\mathbf{Z}\mathbf{P} \\ &= \mathbf{i}^*\dot{\boldsymbol{\Omega}}+\boldsymbol{\pi} \end{aligned} \tag{8.73}$$

We find from this:

$$2T=\mathbf{P}'\mathbf{P}+(\boldsymbol{\Pi}-\boldsymbol{\pi})'\mathbf{i}^{*-1}(\boldsymbol{\Pi}-\boldsymbol{\pi}) \tag{8.74}$$

8.5.2 The Equations of Motion

The derivatives of T with respect to $\eta(\{Q\sigma\theta\}$ where θ stands for the Eulerian angles θ, φ, χ) and p (the conjugate moments: $p=\partial T/\partial\dot{\eta}$) needed to define the set of differential equations are of the form:

$$\frac{\partial T}{\partial \dot{\boldsymbol{\eta}}}=\begin{pmatrix} \partial T/\partial \dot{\mathbf{Q}} \\ \partial T/\partial \dot{\sigma} \\ \partial T/\partial \dot{\boldsymbol{\theta}} \end{pmatrix}=\begin{pmatrix} \partial T/\partial \dot{\mathbf{Q}} \\ \partial T/\partial \dot{\sigma} \\ \boldsymbol{\rho}'\partial T/\partial \boldsymbol{\omega} \end{pmatrix}=\begin{pmatrix} \mathbf{P} \\ P_\sigma \\ \boldsymbol{\rho}'\mathbf{M} \end{pmatrix} \tag{8.75}$$

where $\boldsymbol{\rho}$ is the matrix which connects $\boldsymbol{\omega}$ and $\dot{\boldsymbol{\theta}}$ by:

$$\boldsymbol{\omega} = \boldsymbol{\rho}\dot{\boldsymbol{\theta}} \qquad \text{(equation 7.221)}$$

with:

$$\boldsymbol{\rho} = \begin{pmatrix} \sin\chi & -\sin\theta\cos\chi & 0 \\ \cos\chi & \sin\theta\sin\chi & 0 \\ 0 & \cos\theta & 1 \end{pmatrix}$$

From the previously established relations (8.71), (8.72) and (8.73), we have:

$$2T = \begin{pmatrix} \dot{\mathbf{Q}} \\ \dot{\boldsymbol{\Omega}} \end{pmatrix}' \begin{pmatrix} \mathbf{E} & \mathbf{Z}' \\ \mathbf{Z} & \mathbf{i} \end{pmatrix} \begin{pmatrix} \dot{\mathbf{Q}} \\ \dot{\boldsymbol{\Omega}} \end{pmatrix} = \begin{pmatrix} \mathbf{P} \\ \boldsymbol{\Pi} \end{pmatrix}' \begin{pmatrix} \mathbf{E} & \mathbf{Z}' \\ \mathbf{Z} & \mathbf{i} \end{pmatrix}^{-1} \begin{pmatrix} \mathbf{P} \\ \boldsymbol{\Pi} \end{pmatrix} \tag{8.76}$$

Then:

$$\begin{pmatrix} \partial T/\partial\mathbf{P} \\ \partial T/\partial\boldsymbol{\Pi} \end{pmatrix} = \begin{pmatrix} \mathbf{E} & \mathbf{Z}' \\ \mathbf{Z} & \mathbf{i} \end{pmatrix}^{-1} \begin{pmatrix} \mathbf{P} \\ \boldsymbol{\Pi} \end{pmatrix} \tag{8.77}$$

where a numerical inversion of the square matrix is not hard to obtain even though an analytical inverse is not provided. On the other hand:

$$\frac{\partial T}{\partial\boldsymbol{\eta}} = \begin{pmatrix} \partial T/\partial\mathbf{Q} \\ \partial T/\partial\sigma \\ \partial T/\partial\boldsymbol{\theta} \end{pmatrix} \tag{8.78}$$

$$\frac{\partial T}{\partial\boldsymbol{\eta}} = \frac{1}{2}(\boldsymbol{\Pi}-\boldsymbol{\pi})' \frac{\partial\mathbf{i}^{*-1}}{\partial\boldsymbol{\eta}}(\boldsymbol{\Pi}-\boldsymbol{\pi}) - \mathbf{P}'\frac{\partial\mathbf{Z}'}{\partial\boldsymbol{\eta}}\dot{\boldsymbol{\Omega}} \tag{8.79}$$

from:

$$\frac{\partial(\boldsymbol{\Pi}-\boldsymbol{\pi})}{\partial\boldsymbol{\eta}} = -\frac{\partial\boldsymbol{\pi}}{\partial\boldsymbol{\eta}} = -\frac{\partial\mathbf{Z}}{\partial\boldsymbol{\eta}}\mathbf{P}$$

and

$$\boldsymbol{\Pi}-\boldsymbol{\pi} = \mathbf{i}^*\dot{\boldsymbol{\Omega}}$$

As:

$$\mathbf{i}^{*-1}\mathbf{i}^* = \mathbf{E}$$

we find, in order to avoid analytical derivatives of $\mathbf{i}^*$ inverse:

$$\frac{\partial\mathbf{i}^{*-1}}{\partial\boldsymbol{\eta}} = -\mathbf{i}^{*-1}\frac{\partial\mathbf{i}^*}{\partial\boldsymbol{\eta}}i^{*-1} \tag{8.80}$$

where:

$$I^* = \begin{pmatrix} N_{\sigma\sigma,Q} - \mathbf{N}_{Q,\sigma}\mathbf{N}'_{Q,\sigma} & \mathbf{Z}'_{\sigma,Q} - \mathbf{N}_{Q,\sigma}\mathbf{Z}'_Q \\ \mathbf{Z}_{\sigma,Q} - \mathbf{Z}_Q\mathbf{N}'_{Q,\sigma} & \mathbf{I} - \mathbf{Z}_Q\mathbf{Z}'_Q \end{pmatrix} \tag{8.81}$$

Then

$$\frac{\partial \mathbf{i}^*}{\partial \boldsymbol{\eta}} = \begin{pmatrix} \dfrac{\partial N_{\sigma\sigma,Q}}{\partial \boldsymbol{\eta}} - 2\dfrac{\partial \mathbf{N}_{Q,\sigma}}{\partial \boldsymbol{\eta}}\mathbf{N}'_{Q,\sigma} & \\ \dfrac{\partial \mathbf{Z}_{\sigma,Q}}{\partial \boldsymbol{\eta}} - \dfrac{\partial \mathbf{Z}_Q}{\partial \boldsymbol{\eta}}\mathbf{N}'_{Q,\sigma} - \mathbf{Z}_Q\dfrac{\partial \mathbf{N}_{Q,\sigma}}{\partial \boldsymbol{\eta}} & \dfrac{\partial \mathbf{I}}{\partial \boldsymbol{\eta}} - 2\dfrac{\partial \mathbf{Z}_Q}{\partial \boldsymbol{\eta}}\mathbf{Z}'_Q \end{pmatrix} \tag{8.82}$$

and

$$\frac{\partial \mathbf{Z}}{\partial \boldsymbol{\eta}} = \begin{pmatrix} \partial \mathbf{N}_{\sigma,Q}/\partial \boldsymbol{\eta} \\ \partial \mathbf{Z}_Q/\partial \boldsymbol{\eta} \end{pmatrix} \tag{8.83}$$

For normal coordinate derivatives it appears that:

$$\frac{\partial N_{\sigma\sigma,Q}}{\partial \mathbf{Q}} = 2\mathfrak{L}'\mathfrak{L}\mathbf{Q} - 2\mathbf{l}'\mathbf{M}^{1/2}\boldsymbol{\Gamma}^0 \tag{8.84a}$$

$$\frac{\partial \mathbf{N}_{Q,\sigma}}{\partial \mathbf{Q}} = \mathfrak{L}'\mathbf{l} = \boldsymbol{\xi}_\sigma \tag{8.84b}$$

$$\frac{\partial \mathbf{Z}_Q^g}{\partial \mathbf{Q}} = \mathbf{l}'\mathbf{I}^g\mathbf{l} = \boldsymbol{\xi}_Q^g \tag{8.84c}$$

$$\frac{\partial \mathbf{Z}_{\sigma,Q}^g}{\partial \mathbf{Q}} = 2\mathbf{l}'\mathbf{I}^g\mathfrak{L}\mathbf{Q} - 2\mathbf{l}'\mathbf{I}^g\mathbf{M}^{1/2}\boldsymbol{\gamma}^0 \tag{8.84d}$$

$$\frac{\partial I^{gh}}{\partial \mathbf{Q}} = 2\mathbf{l}'\mathbf{I}^g\mathbf{I}^{h'}\mathbf{M}^{1/2}\mathbf{R}^0 + 2\mathbf{l}'\mathbf{I}^g\mathbf{I}^{h'}\mathbf{l}\mathbf{Q} \tag{8.84e}$$

For derivatives with respect to the reaction coordinate we have:

$$\begin{aligned} \frac{\partial N_{\sigma\sigma,Q}}{\partial \sigma} = {} & 2\boldsymbol{\Gamma}^{0'}\mathbf{M}\boldsymbol{\gamma}^0 - 2\frac{\partial \boldsymbol{\Gamma}^{0'}}{\partial \sigma}\mathbf{M}^{1/2}\mathbf{l}\mathbf{Q} \\ & - 2\boldsymbol{\Gamma}^{0'}\mathbf{M}^{1/2}\mathfrak{L}\mathbf{Q} + 2\mathbf{Q}'\frac{\partial \mathfrak{L}'}{\partial \sigma}\mathfrak{L}\mathbf{Q} \end{aligned} \tag{8.85a}$$

$$\frac{\partial \mathbf{N}_{Q,\sigma}}{\partial \sigma} = \mathbf{Q}'\frac{\partial \mathfrak{L}'}{\partial \sigma}\mathbf{l} + \mathbf{Q}'\mathfrak{L}'\mathfrak{L} \tag{8.85b}$$

$$\begin{aligned} \frac{\partial \mathbf{Z}_{\sigma Q}^g}{\partial \sigma} = {} & -2(\boldsymbol{\Gamma}^{0'}\mathbf{M}^{1/2}\mathbf{I}^g\mathbf{l} + \boldsymbol{\gamma}^{0'}\mathbf{M}^{1/2}\mathbf{I}^g\mathfrak{L})\mathbf{Q} \\ & + \mathbf{Q}'\left(\mathfrak{L}'\mathbf{I}^g\mathfrak{L} + \mathbf{l}'\mathbf{I}^g\frac{\partial \mathfrak{L}}{\partial \sigma}\right)\mathbf{Q} \end{aligned} \tag{8.85c}$$

$$\frac{\partial \mathbf{Z}_Q^g}{\partial \sigma} = 2\mathbf{Q}'\mathfrak{L}'\mathbf{I}^g\mathbf{l} \tag{8.85d}$$

$$\begin{aligned} \frac{\partial I^{gh}}{\partial \sigma} = {} & 2\boldsymbol{\gamma}^{0'}\mathbf{I}^g\mathbf{M}\mathbf{I}^{h'}\mathbf{R}^0 + 2(\boldsymbol{\gamma}^{0'}\mathbf{M}^{1/2}\mathbf{I}^g\mathbf{I}^{h'}\mathbf{l} + \mathbf{R}^{0'}\mathbf{M}^{1/2}\mathbf{I}^g\mathbf{I}^{h'}\mathfrak{L})\mathbf{Q} \\ & + 2\mathbf{Q}'(\mathfrak{L}'\mathbf{I}^g\mathbf{I}^{h'}\mathfrak{L})\mathbf{Q} \end{aligned} \tag{8.85e}$$

In the particular cases of $\partial T/\partial \boldsymbol{\theta}$ there only remain:

$$\frac{\partial T}{\partial \boldsymbol{\theta}} = \boldsymbol{\omega}' \frac{\partial \boldsymbol{m}}{\partial \boldsymbol{\theta}} = \boldsymbol{\omega}' \frac{\partial \rho^{-1'}}{\partial \boldsymbol{\theta}} \mathbf{p}_\theta \tag{8.86}$$

with:

$$\frac{\partial (\boldsymbol{\rho}^{-1})'}{\partial \theta} = \begin{pmatrix} 0 & \dfrac{\cos\chi \cos\theta}{\sin^2\theta} & -\dfrac{\cos\chi}{\sin^2\theta} \\ 0 & -\dfrac{\sin\chi\cos\theta}{\sin^2\theta} & \dfrac{\sin\chi}{\sin^2\theta} \\ 0 & 0 & 0 \end{pmatrix} \tag{8.87a}$$

$$\frac{\partial (\boldsymbol{\rho}^{-1})'}{\partial \varphi} = \begin{pmatrix} 0 & 0 & 0 \\ 0 & 0 & 0 \\ 0 & 0 & 0 \end{pmatrix} \tag{8.87b}$$

$$\frac{\partial (\boldsymbol{\rho}^{-1})'}{\partial \chi} = \begin{pmatrix} \cos\chi & \dfrac{\sin\chi}{\sin\theta} & \dfrac{\sin\chi}{\mathrm{tg}\,\theta} \\ -\sin\chi & \dfrac{\cos\chi}{\sin\theta} & \dfrac{\cos\chi}{\mathrm{tg}\,\theta} \\ 0 & 0 & 0 \end{pmatrix} \tag{8.87c}$$

8.5.3 The Trajectories in Phase Space

Given a starting point on the reactant side which can be expressed in terms of generalized coordinates ($\mathbf{q}$) and their associated kinetic moment ($\mathbf{p}$) the hamiltonian equations enable us to know the future evolution of this point when the time increases. So a trajectory in phase space for a system of N atoms may be obtained by integrating the $6N$ first-order coupled differential equations of Hamilton (8.46) from a set of initial conditions chosen in the reactant region. We know that:

$$\dot{\mathbf{q}} = \frac{\partial H(p, q)}{\partial \mathbf{p}} \quad \text{and} \quad \dot{\mathbf{p}} = -\frac{\partial H(p, q)}{\partial \mathbf{q}}$$

or more generally we write, if y is set for $\mathbf{q}$ as well as for $\mathbf{p}$:

$$\dot{y} = f[y(t)] \tag{8.88}$$

We may integrate that set of equations:

$$y(t_\nu + h) = y(t_\nu) + \int_{t_\nu}^{t_\nu + h} \dot{y}(t_\nu)\, dt \tag{8.89}$$

To do that we replace the last integral by the finite quadrature formula:

$$y(t_\nu + h) = y(t_\nu) + h \sum_{i=0}^{r-1} w_i \dot{y}(t_\nu + \rho_i h) \tag{8.90}$$

with:

$$0 \leq \rho_i \leq 1$$

and r being the degree of the quadrature. The r, w_i parameters are chosen in such a way as to improve the efficiency of the numerical method. We shall give the general philosophy of two algorithms later.[27] In each case we use at most quadratic formulas (for simplicity), even if available programs employ higher order expansions. This level of detail is beyond our immediate intention; we shall confine ourselves to describing how to obtain a typical estimate of $y(t_\nu + h)$. The most usual algorithm is the predictor–corrector one. Let us suppose that we have four functions (y) and their derivatives ($\dot{y}$) already computed for different times, each connected to the closest one by a constant value (h). The first prediction may be performed

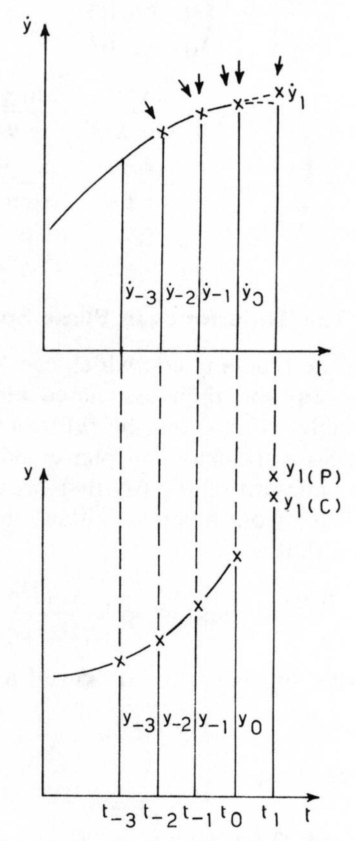

Figure 8.6 The predictor–corrector propagation algorithm. Reproduced by permission of Harper & Row Publishers Inc.

according to (see Figure 8.6):

$$y_1(P) = y_{-3} + \frac{4h}{3}(2\dot{y}_{-2} - y_{-1} + 2\dot{y}_0) \tag{8.91}$$

Calculating the time derivative at the prediction ($\dot{y}_1(P)$), correction becomes available from:

$$y_1(c) = y_{-1} + \frac{h}{3}(\dot{y}_{-1} + 4\dot{y}_0 + \dot{y}_1) \tag{8.92}$$

In order to avoid too long or too short a step size (h) on the time scale we impose an absolute difference between $y_1(P)$ and $y_1(c)$ in a defined range:

$$\alpha < |y_1(P) - y_1(c)| < \beta \tag{8.93}$$

For too small differences the step size is doubled and for too large values the step size is divided by two.

Another algorithm which seems to be really efficient is the Bulirsch–Stoer propagator. The h step is divided into an even number (n) of subdivisions (four in Figure 8.7). The basic substep is a linear extrapolation of y from t to

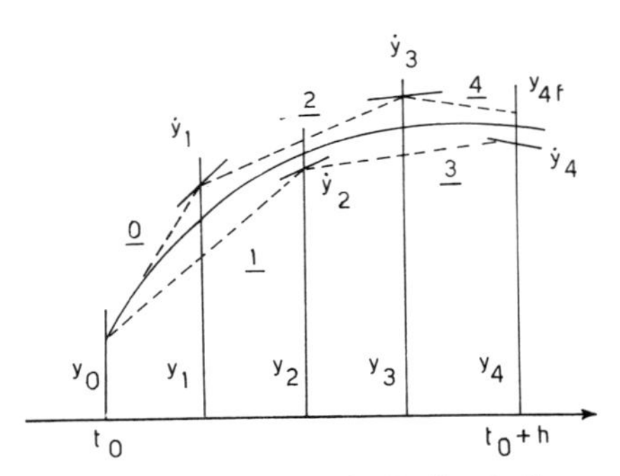

Figure 8.7 The basic Bulirsch–Stoer integration step. Reproduced by permission of Harper & Row Publishers Inc.

$t+2h/n$ using the slope at the midpoint $t+h/n$; thus:

$$y\left(t+2\frac{h}{n}\right) = y(t) + 2\frac{h}{n}\dot{y}\left(t+\frac{h}{n}\right) \tag{8.94}$$

To start the process, the first step must be taken as:

$$y\left(t+\frac{h}{n}\right) = y(t) + \frac{h}{n}\dot{y}(t) \tag{8.95}$$

Likewise, the final step is one of length h/n. This algorithm produces an overestimated prediction for the odd extrapolations and an underestimated value for the even extrapolations; the upper and lower limits so generated converge for $t+h$. A complete Bulirsch–Stoer step for four subintervals is given in Figure 8.7 and corresponds to the following formulas:

The starting step:

$$y_1 = y_0 + \frac{h}{n}\,\dot{y}_0$$

The propagation step:

$$y_2 = y_0 + \frac{2h}{n}\,\dot{y}_1$$

$$y_3 = y_1 + \frac{2h}{n}\,\dot{y}_2$$

$$y_4 = y_2 + \frac{2h}{n}\,\dot{y}_3$$

The final step:

$$y_{4f} = y_3 + \frac{h}{n}\,\dot{y}_4 \tag{8.96}$$

We keep for the final ordinate $(y(t_0+h))$ an average value between the two estimates y_4 and y_{4f}:

$$y(t_0+h) = \frac{y_4 + y_{4f}}{2} \tag{8.97}$$

This shows that we are able from a given starting point (a set of $\mathbf{q}$ and $\mathbf{p}$) to know what happens with the chemical system of interest after some time. The evolution of the representative point for the supersystem in the phase space is named trajectory. Such a trajectory may be reactive or not according to whether we cross the transition point to find nuclear configurations which are product-like. In the three next figures we illustrate our actual purpose for the F+HD collision.

We use the internal interatomic distances coordinate frame ($S = \{r_{HF}, r_{DH}, r_{DF}\}$) to show the nuclear movements. At the beginning of those charts we find the vibration of the HD molecule (sinusoidal comportment of the $r_{HD}(t)$ curve) as well as the relative approach of the fluorine atom. After at least 10^{-13} s we reach the region of strong triatomic interactions. Lastly, the reaction process may occur (Figures 8.8 and 8.9) or not (Figure 8.10), depending on the initial conditions which have been assumed. There exist $6N-6$ (here 12) initial available conditions which are summarized here by only three, corresponding to translation, rotation and vibration states.

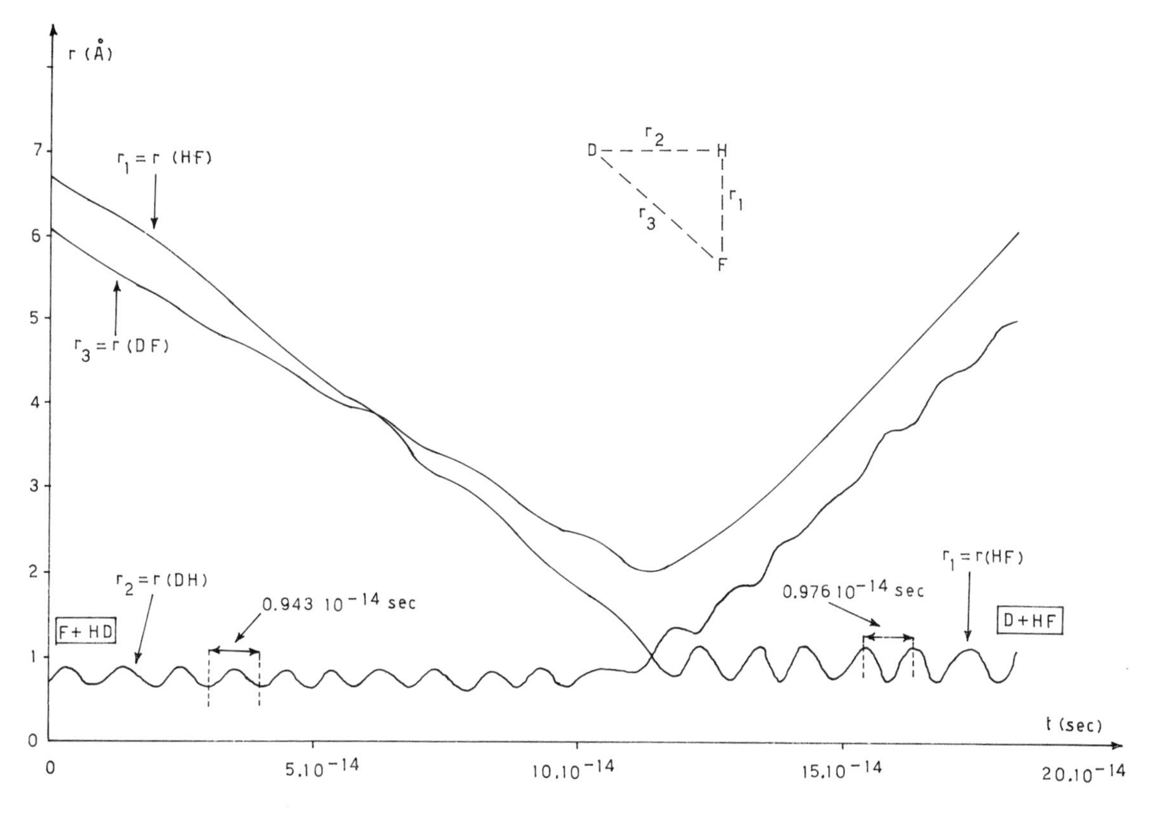

Figure 8.8 The F + HD collision process with initially $V = 0$, $J = 1$ and $v_r = 3.9\ \text{cm s}^{-1}$

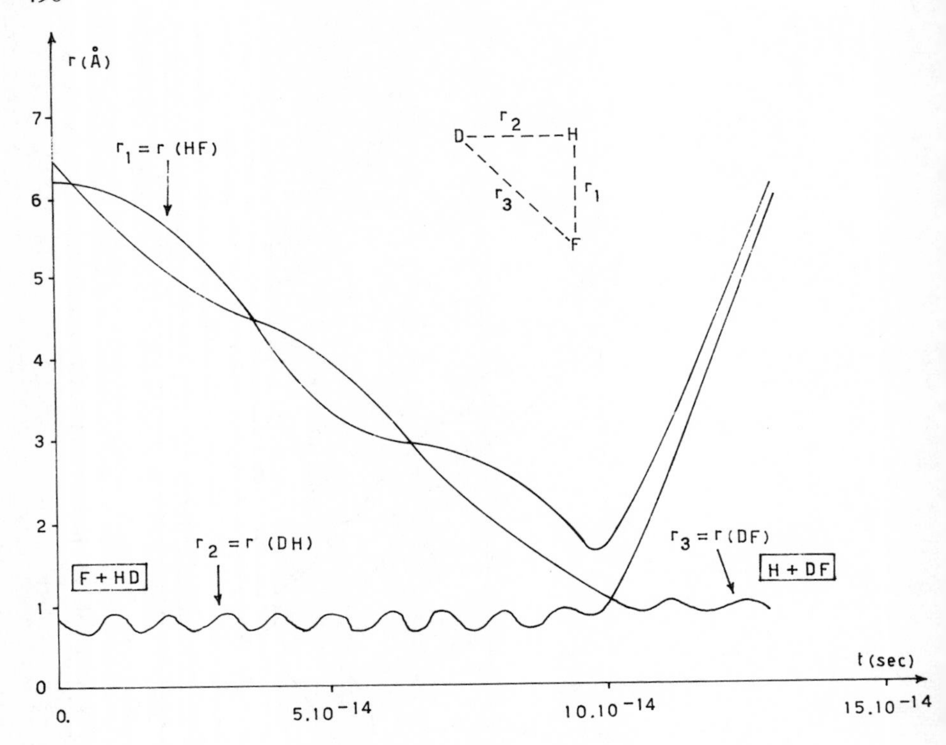

Figure 8.9 The F+HD collision process with initially $V=0$, $J=6$ and $v_r = 5.0\ \text{cm s}^{-1}$

8.6 THE CLASSICAL REACTION RATE CONSTANT EXPRESSION

The most simple reaction process we can imagine is the collision of two point masses A and B which are moving in a potential V_{AB}. We shall divide the rate constant calculation in two parts:[28]

(a) First, the volumic collision frequency Z_{AB} which counts the number of collision per unit of time and volume;
(b) Second, the efficiency of each collision P_{AB} which depends on the number of reactive collisions.

8.6.1 The Volumic Collision Frequency

The volumic collision frequency is only a function of the initial conditions we choose for the system:

(a) The relative velocity ($\mathbf{v}_r$) is defined as the difference between the laboratory velocities of each collision partner ($\mathbf{v}_A$ and $\mathbf{v}_B$): $\mathbf{v}_r = \mathbf{v}_A - \mathbf{v}_B$.

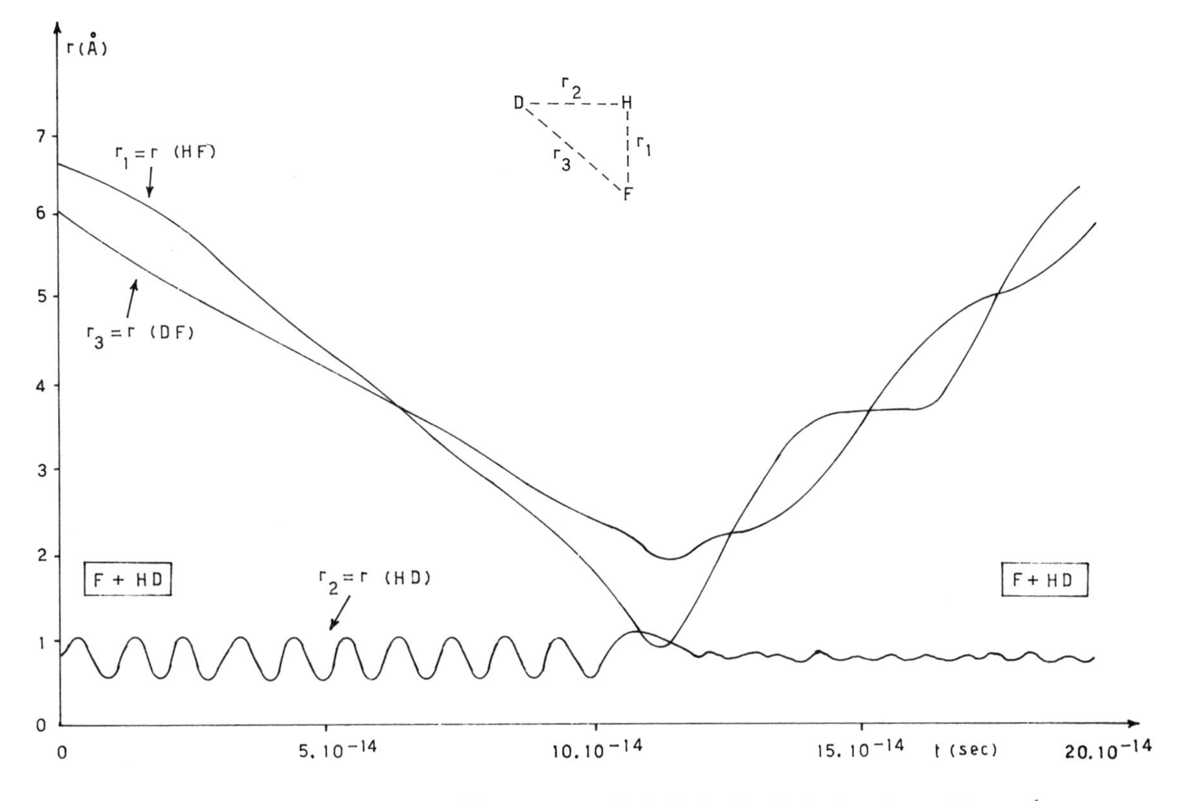

Figure 8.10 The F + HD collision process with initially $V = 1$, $J = 1$ and $v_r = 4.1\ \mathrm{cm\,s^{-1}}$

(b) The impact parameter (b) stands for the distance between the relative velocity direction applied to the centre of mass of A and the centre of mass of B at the starting point.
(c) The number of A and B systems per unit of volume (N_A and N_B).
(d) If A and/or B are not single atoms but molecules we want to include the vibrational–rotational states, the internal molecular coordinates, the orientations of the reacting molecules and the orientation of their angular momentum. We will come back later to such parameters.

The initial separation distance (ρ_0) between the centres of mass of A and B is not to be taken into consideration but must be such that at this distance the two systems do not know each other ($(\mathrm{d}V_{AB}/\mathrm{d}\rho)_{\rho \geqslant \rho_0} = 0$). On the other hand, the centre of mass velocity $\mathbf{c}$ of course does not influence the volumic collision frequency. Figure 8.11 shows the parameters previously defined.

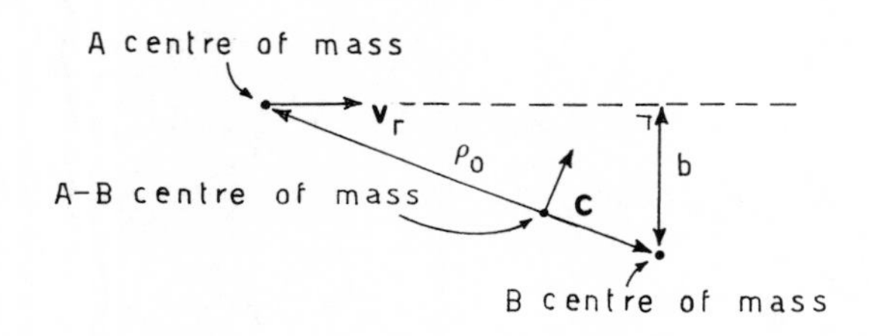

Figure 8.11 Two-particle collision

8.6.1.1 The reaction volume

The reaction volume τ swept out per unit time is the product of the relative velocity magnitude $|v_r|$ and the surface in the range b and $b + db$, as illustrated in Figure 8.12:

$$\tau = |v_r| \cdot 2\pi b \, \mathrm{d}b \tag{8.98}$$

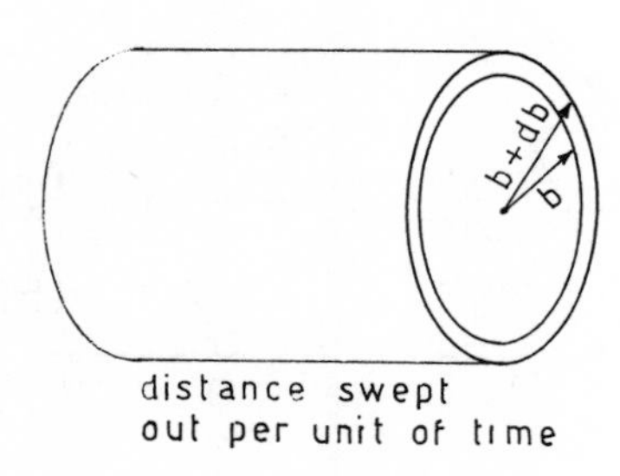

Figure 8.12 Reaction volume swept out per unit of time

8.6.1.2 The normalized velocity density function

The fraction of AB pairs (dp) having a relative velocity $\mathbf{v}_r$ is the fraction of AB pairs which are found to be simultaneously moving with velocities in the range $[\mathbf{v}_A, \mathbf{v}_A + \mathrm{d}\mathbf{v}_A]$ for A and $[\mathbf{v}_B, \mathbf{v}_B + \mathrm{d}\mathbf{v}_B]$ for B. If we note $f(\mathbf{v}_A)$ and $f(\mathbf{v}_B)$, the normalized density functions, then:

$$\mathrm{d}p(\mathbf{v}_A\mathbf{v}_B) = f(\mathbf{v}_A) \cdot f(\mathbf{v}_B) \cdot \mathrm{d}\mathbf{v}_A \cdot \mathrm{d}\mathbf{v}_B \tag{8.99}$$

where:

$$f(\mathbf{v}) = \frac{\mathrm{d}p(\mathbf{v})}{\mathrm{d}\mathbf{v}} \tag{8.100}$$

with:

$$\iiint_{-\infty}^{+\infty} f(\mathbf{v})\,\mathrm{d}\mathbf{v} = \iiint_{-\infty}^{+\infty} \mathrm{d}p(\mathbf{v}) = 1$$

The explicit form of the functions $f(\mathbf{v})$ follows from the experimental situation we have to reproduce. We can find two opposite kinds of experiments:

(a) *The thermal equilibrium.* In this case one usually chooses the Boltzmann distribution for $f(v)$:

$$f(v) = \left(\frac{m}{2\pi kT}\right)^{3/2} \exp\left(-\frac{mv^2}{2kT}\right) \tag{8.101}$$

where m stands for the mass of the moving system and T for the absolute equilibrium temperature.

(b) *The selected speed.* By using an ideal apparatus the density function becomes a Dirac's delta function:

$$f(v) = \delta(\mathbf{v} - \mathbf{v}^0) \tag{8.102}$$

This expression may be replaced by a gaussian or a lorentzian function in order to take into account a more realistic situation.

Different experiments are reported in Figure 8.13. *In cross-beam reactions* we can choose the individual speed of the two partners; a focalizer and a speed selector are placed on the two beams in order to reduce the velocity range in direction as well as in magnitude. This kind of apparatus is shown in Figure 8.14. The corresponding $p(v_A v_B)$ function becomes for the ideal case:

$$\mathrm{d}p(\mathbf{v}_A\mathbf{v}_B) = \delta(\mathbf{v}_A - \mathbf{v}_A^0) - \delta(\mathbf{v}_B - \mathbf{v}_B^0)\,\mathrm{d}\mathbf{v}_A\,\mathrm{d}\mathbf{v}_B \tag{8.103}$$

and the relative velocity (v_r) norm:

$$v_r = (v_A^{02} - 2v_A^0 v_B^0 \cos\gamma + v_B^{02})^{1/2} \tag{8.104}$$

where γ stands for the beams crossing angle.

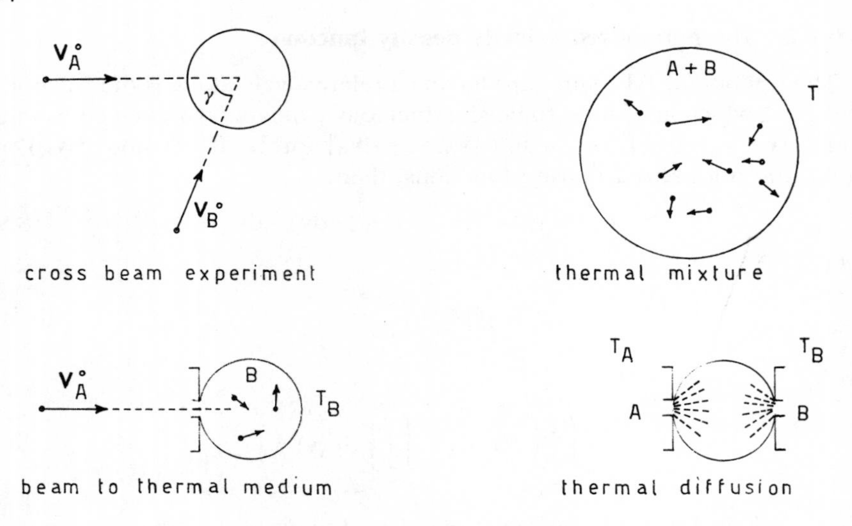

Figure 8.13 Some experimental designs for the collision process

Conversely, we find collisions between A and B systems in translational thermal equilibrium. Let us first suppose (for generality) that the two equilibrium temperatures are respectively T_A and T_B. Using the Boltzmann distribution function (8.101), we write:

$$\mathrm{d}p(\mathbf{v}_A\mathbf{v}_B) = \left(\frac{m_A m_B}{4\pi^2 k^2 T_A T_B}\right)^{3/2} \exp\left(-\frac{m_A v_A^2}{2kT_A} - \frac{m_B v_B^2}{2kT_B}\right) \mathrm{d}\mathbf{v}_A\, \mathrm{d}\mathbf{v}_B \tag{8.105}$$

As the collision frequency is independent on the centre of mass motion it is convenient to transform this equation in new variables $\mathbf{v}_r$, the relative velocity, and $\mathbf{c}$, the centre of mass velocity. From Newton's diagram we find:

$$\mathbf{v}_A = \mathbf{c} + \frac{m_B}{m_A + m_B}\mathbf{v}_r \tag{8.106a}$$

$$\mathbf{v}_B = \mathbf{c} + \frac{m_A}{m_A + m_B}\mathbf{v}_r \tag{8.106b}$$

If we define M and μ as:

$$M = m_A + m_B \qquad \text{(the total mass)} \tag{8.107a}$$

$$\mu = \frac{m_A m_B}{m_A + m_B} \qquad \text{(the reduced mass)} \tag{8.107b}$$

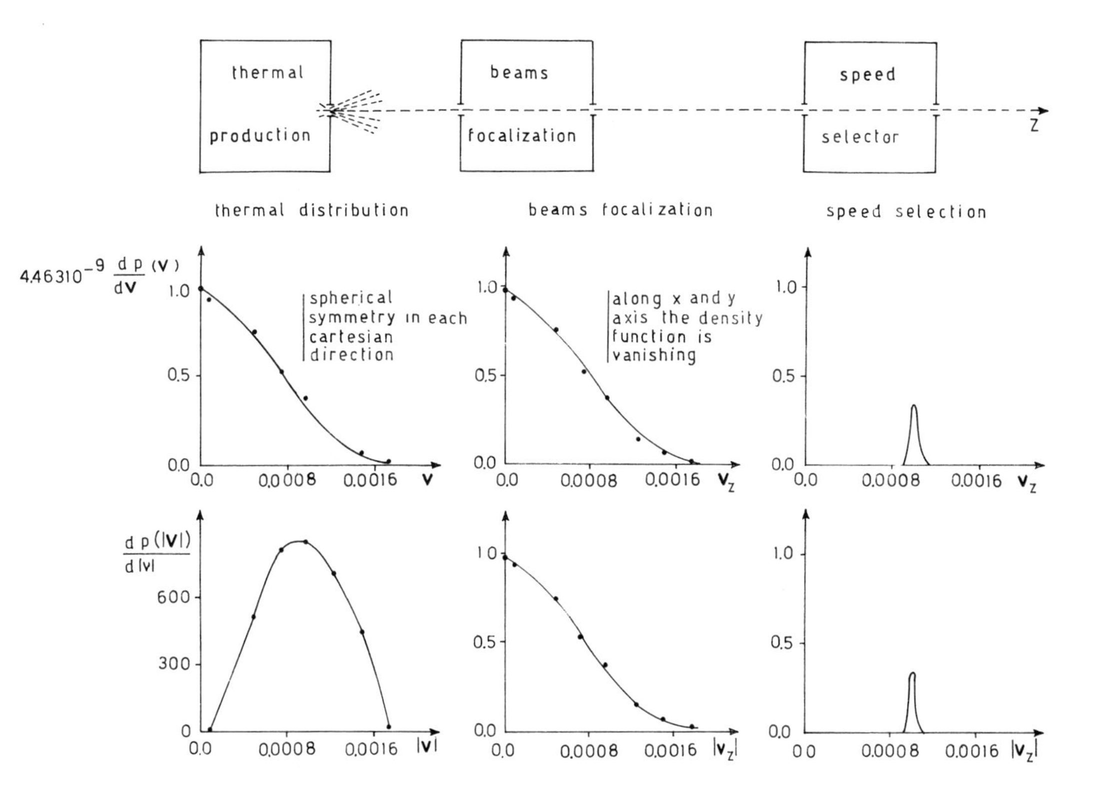

Figure 8.14 Getting localized and speed-selected beam for the D atom at 500 K (all distributions have not been normalized and speed is given in atomic units)

we can write:

$$\mathrm{d}p(\mathbf{v}_A\mathbf{v}_B) = \left(\frac{\mu M}{4\pi^2 k^2 T_A T_B}\right)^{3/2} \exp\left\{-\frac{1}{2k}\left[c^2\left(\frac{m_A}{T_A}+\frac{m_B}{T_B}\right)\right.\right.$$

$$\left.\left. + 2\mu\mathbf{v}_r\mathbf{c}\left(\frac{1}{T_A}-\frac{1}{T_B}\right)+\mu\mathbf{v}_r^2\left(\frac{m_B/T_A+m_A/T_B}{M}\right)\right]\right\}\mathrm{d}\mathbf{v}_r\,\mathrm{d}\mathbf{c} \quad (8.108)$$

If we now imagine a *thermal mixture* where $T_A = T_B = T$, the former equation becomes:

$$\mathrm{d}p(\mathbf{v}_A \cdot \mathbf{v}_B) = \frac{(\mu M)^{3/2}}{(2\pi kT)^3}\exp\left(-\frac{M\mathbf{c}^2+\mu\mathbf{v}_r^2}{2kT}\right)\mathrm{d}\mathbf{v}_r\,\mathrm{d}\mathbf{c} \quad (8.109)$$

As the collision frequency depends only on the norm of the vector $\mathbf{v}_r$ we shall integrate such a function over all the orientations of the relative velocity and over the centre of mass velocity. As:

$$\mathrm{d}\mathbf{v}_r\,\mathrm{d}\mathbf{c} = v_r^2 \sin\theta\,\mathrm{d}v_r\,\mathrm{d}\theta\,\mathrm{d}\varphi\,\mathrm{d}c_x\,\mathrm{d}c_y\,\mathrm{d}c_z \quad (8.110)$$

then:

$$\mathrm{d}p(v_r) = \underset{\theta}{\int_0^{\pi}}\underset{\varphi}{\int_0^{2\pi}}\underset{c_x}{\int_{-\infty}^{+\infty}}\underset{c_y}{\int_{-\infty}^{+\infty}}\underset{c_z}{\int_{-\infty}^{+\infty}}\mathrm{d}p(\mathbf{v}_r\mathbf{c}) \quad (8.111)$$

gives:

$$\mathrm{d}p(v_r) = \frac{4}{\sqrt{\pi}}\left(\frac{\mu}{2kT}\right)^{3/2} v_r^2 \exp\left(-\frac{\mu v_r^2}{2kT}\right)\mathrm{d}v_r \quad (8.112)$$

This is nothing more than the Maxwell distribution over the relative velocity. Replacing $\mu/2kT$ by α we can also write:

$$\mathrm{d}p(v_r) = \frac{4\alpha^{3/2}}{\sqrt{\pi}} v_r^2 \exp(-\alpha v_r^2)\,\mathrm{d}v_r \quad (8.113)$$

If A and B are not isolated atoms but molecules the same expressions may be obtained. However, the reduced mass μ must be generalized as:

$$\mu = \frac{M_A M_B}{M_A + M_B} \quad (8.114)$$

where M_A and M_B stand for the total masses of the A and B systems respectively. In Figure 8.15 we illustrate these distributions for some small molecules and atom–diatom collisions. The characteristics are as follows (see Table 8.1):

(a) The most probable speed, which corresponds to the maximum along the curve distribution, is given by

$$v_{mp} = \sqrt{(1/\alpha)}$$

where: (8.115)

$$\left(\frac{\mathrm{d}f(v)}{\mathrm{d}v}\right)_{v_{mp}} = 0$$

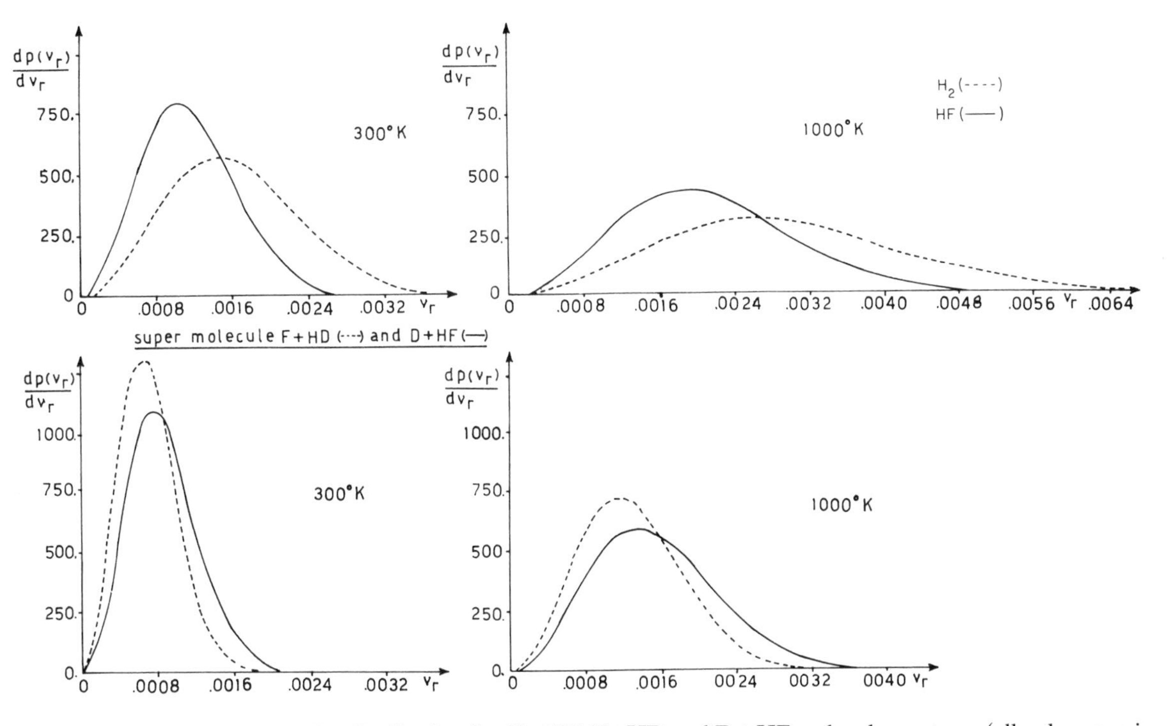

Figure 8.15 Relative velocity distribution for H_2, HF, F + HD and D + HF molecular systems (all values are in atomic units)

Table 8.1 Characteristic values for the normalized relative velocity distribution of Maxwell (1 a.u. $= 2.187653 \times 10^{-8}$ cm s^{-1})

	T(K)	$10^{-5}\,\alpha$	$1000\,v_{mp}$	$1000\,v_-$	$1000\,v_+$	$1000\,\langle v \rangle$	$10^6\,\langle v^2 \rangle$
HH	300	4.831	1.439	0.674	2.173	1.623	3.105
	1000	1.449	2.628	1.230	3.967	2.964	10.349
HF	300	9.176	1.043	0.488	1.576	1.177	1.635
	1000	2.753	1.906	0.892	2.878	2.151	5.449
F+HD	300	24.500	0.633	0.296	0.955	0.714	0.600
	1000	7.499	1.155	0.541	1.744	1.303	2.000
D+HF	300	17.544	0.755	0.353	1.140	0.852	0.855
	1000	5.263	1.378	0.645	2.081	1.555	2.850

(b) The two inflection points of the distribution curve occur at:

$$v_{\pm} = \sqrt{\left[(1/\alpha)\left(\frac{5 \pm \sqrt{17}}{4}\right)\right]}$$

with: (8.116)

$$\left(\frac{d^2 f(v)}{dv^2}\right)_{v_{\pm}} = 0$$

(c) The mean velocity is given by:

$$\langle v \rangle = \sqrt{(4/\pi\alpha)}$$

with: (8.117)

$$\frac{\int_0^{\langle v \rangle} f(v)\,dv}{\int_0^{\infty} f(v)\,dv} = \frac{\int_{\langle v \rangle}^{\infty} f(v)\,dv}{\int_0^{\infty} f(v)\,dv} = \tfrac{1}{2}$$

(d) The mean square velocity (related to the mean energy) corresponds to:

$$\langle v^2 \rangle = \frac{3}{2\alpha} \qquad \text{and} \qquad \langle \varepsilon \rangle = \frac{\mu \langle v^2 \rangle}{2} = \frac{3}{2}kT \tag{8.118}$$

These quantities are proportional to $\alpha^{-1/2}$ and may be classified in increasing order such as:

$$v_- < v_{mp} < \langle v \rangle < \sqrt{(<v^2>)} < v_+ \tag{8.119}$$

The associated coefficients are:

$$0.468 < 1.000 < 1.128 < 1.225 < 1.510$$

For the two last types of experiment given in Figure 8.14 similar deductions are feasible (in the case of the beam to thermal medium as well as in the case of the thermal diffusion). In the last experiment, integration over θ and

φ must be performed in a limited range depending on the experimental apparatus (at most $0 \leqslant \varphi, \theta \leqslant \pi$). Furthermore, we shall keep a maximum interest for the reactions in a thermal mixture at the equilibrium.

8.6.1.3 The normalized rotational–vibrational density function

The collision frequency also depends on the internal states of rotation or vibration when A and/or B are not single atoms. If E_{VJ} is the corresponding internal energy and g_{VJ} the associated weighting factor, the Boltzmann statistics enable us to write:

$$f(VJ) = \frac{g_{VJ}}{Q_{rv}} \exp\left(-\frac{E_{VJ}}{kT}\right) \tag{8.120}$$

where Q_{rv} is the normalizing factor also called the partition function and is equal to:

$$Q_{rv} = \sum_{\text{all } VJ \text{ states}} g_{VJ} \exp\left(-\frac{E_{VJ}}{kT}\right) \tag{8.121}$$

In a quasi-classical model the numbers J and V which characterize the rotational–vibrational level are restricted to integer values corresponding to the rotational and vibrational quantum numbers. Direct evaluation of such a function requires knowledge of the first internal energy levels (in practice we request that $(E_{VJ} - \text{ZPE})/kT$ at least equals 9 if ZPE stands for the zero point energy). In most computations one accepts the separation of the rotational and vibrational motions:

$$f(VJ) = f(V) \cdot f(J) \tag{8.122}$$

For a rigid rotator the available $f(J)$ functions are reported in Table 8.2.

Table 8.2 Rotational energy levels and statistical weights for some molecular types[29]

Molecule	Main inertia components†			Rotational energy‡	Statistical weight§
Linear	α	α	0	$AJ(J+1)$	$2J+1$
Oblate symmetric top	α	α	β	$AJ(J+1)+(B-A)K^2$	$2J+1$ for $K=0$ $2(2J+1)$ for $K>0$
Prolate symmetric top	α	β	β	$BJ(J+1)+(A-B)K^2$	$2J+1$ for $K=0$ $2(2J+1)$ for $K>0$
Spherical top	α	α	α	$AJ(J+1)$	$(2J+1)^2$
Asymmetric top	α	β	γ	—¶	$2J+1$

† Main inertia components are ordered in decreasing order.
‡ Where $A = \hbar^2/2\alpha$ and $B = \hbar^2/2\beta$.
§ Statistical weighting factor due to the nuclear spin must be added to the given value.
¶ Between oblate and prolate symmetric top molecule without corresponding degeneracy.

Moreover, when we are only interested in the lowest vibrational energy levels the harmonic approximation may be considered to be accurate enough. In this case, we find:

$$f(V) = \prod_i f(v_i) = \prod_i \frac{1}{Q_i} \exp\left[-\frac{\hbar\omega_i(v_i + \frac{1}{2})}{kT}\right] \tag{8.123}$$

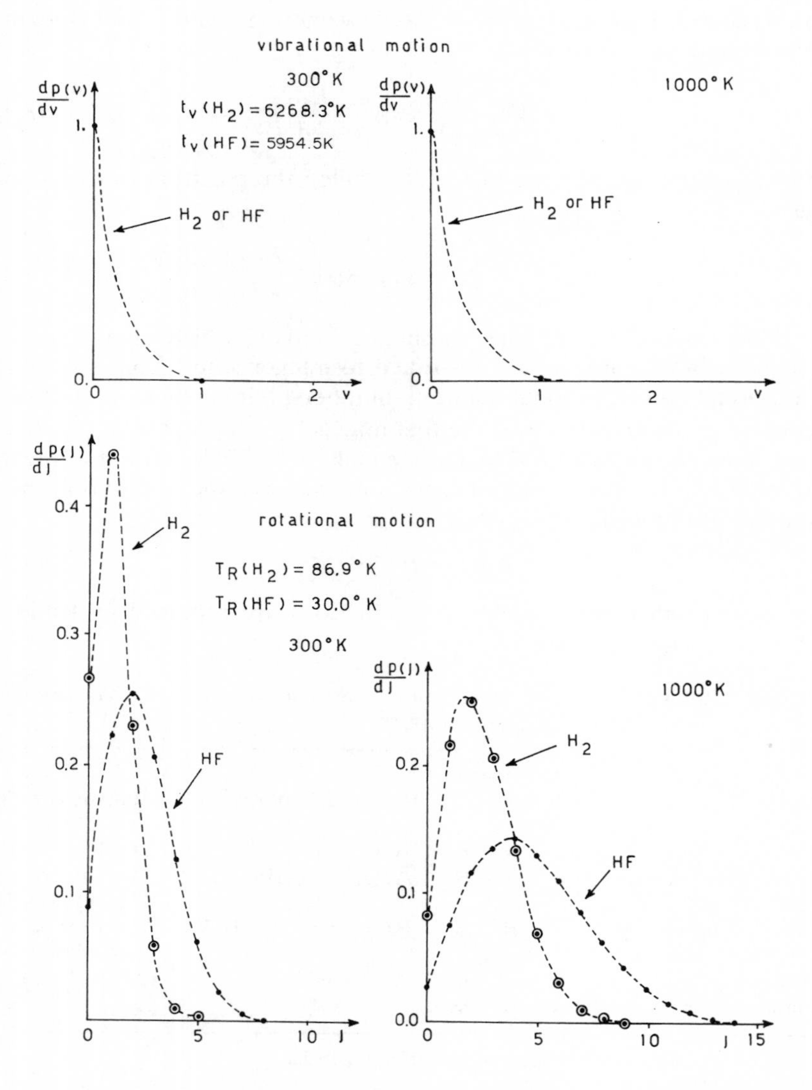

Figure 8.16 Rotational-vibrational distribution function in H_2 (⊙) and HF (·) molecules

where:

$$Q_i = \sum_{j=0}^{\infty} \exp\left[\frac{\hbar\omega_i(j+\frac{1}{2})}{kT}\right] \quad (8.124)$$

These $f(VJ)$ functions are shown in Figure 8.16 for the H_2 and HF molecules. It appears that the fundamental state of vibration is the only probable state, even at 1000 K. The most probable rotational states are:

	At 300 K	At 1000 K
For H_2:	$J=1$	$J=2$
For HF:	$J=2$	$J=4$

8.6.1.4 The density function for the internal coordinates and molecular orientations

Finally, the collision frequency depends on the nuclear parameters when A and/or B are not single atoms. Usually the internal coordinates (the bond lengths and bond angles) are restricted in the range of their outer and inner classical turning points. The phase distribution function for the bonding vibrational states of lowest energies may be found using the harmonic approximation. Then in the normal coordinate frame we write:

$$\frac{dp(Q)}{dQ} = f(\{Q\}) = \prod_i f(Q_i) = \prod_i \frac{N_i}{(Q_i^{(\pm)2} - Q_i^2)^{1/2}} \quad (8.125)$$

where $Q_i^{(\pm)}$ stands for the classical turning points associated with the ith vibrational mode as given previously (equation 7.277):

$$Q_i^{(\pm)} = \pm\sqrt{\left[\frac{\hbar(2v_i+1)}{\omega_i}\right]}$$

and N_i is a normalization factor defined such that:

$$\int_{Q^{(-)}}^{Q^{(+)}} f(Q)\,dQ = 1$$

Then we find:

$$N_i = \frac{\pi}{2}\sin(Q_i^{(\pm)^2}) \quad (8.126)$$

Another approach for the density function will be given in a later section.

The distribution function for the relative orientations of the A and/or B molecules with respect to a space-fixed coordinate frame is easy to write. When B is a diatomic compound two coordinates are required:

$$\theta \quad \text{with} \quad 0 \leqslant \theta \leqslant \pi$$

and

$$\varphi \quad \text{with} \quad 0 \leqslant \varphi \leqslant 2\pi$$

For a molecule which contains more than two atoms a third coordinate is needed:

$$\gamma \qquad \text{with} \qquad 0 \leqslant \gamma \leqslant 2\pi$$

The corresponding density functions have the form:

$$dp(\theta) = f(\theta)\, d\theta = \tfrac{1}{2} \sin\theta\, d\theta \tag{8.127a}$$

$$dp(\varphi) = f(\varphi)\, d\varphi = \frac{1}{2\pi}\, d\varphi \tag{8.127b}$$

$$dp(\gamma) = f(\gamma)\, d\gamma = \frac{1}{2\pi}\, d\gamma \tag{8.127c}$$

The usual definitions for those three angles are reported in Figure 8.17.

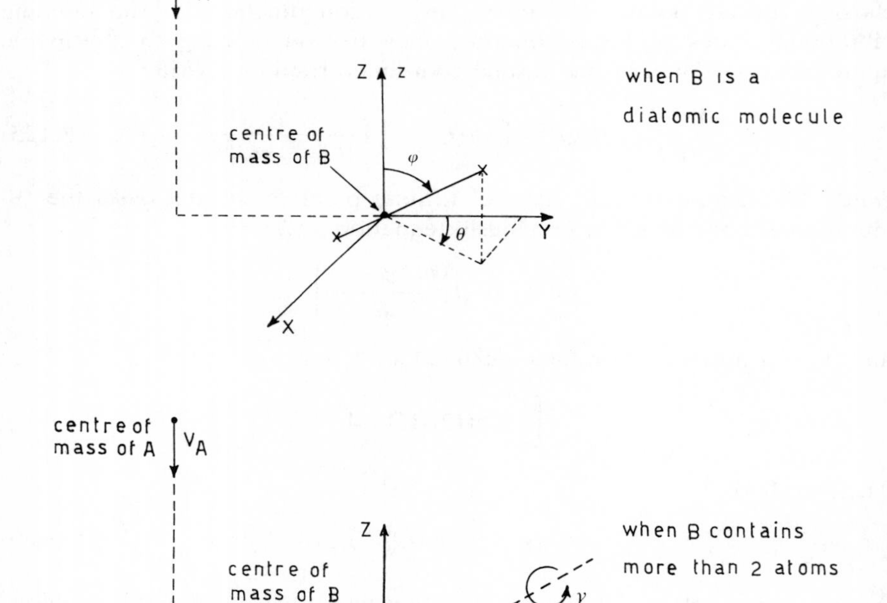

Figure 8.17 Relative molecular parameters θ, φ and γ

One remaining parameter must still be considered: the orientation of the rotational angular momentum. If η measures the angle between the Z space direction and the rotational axis, then:

$$\mathrm{d}p(\eta) = f(\eta)\,\mathrm{d}\eta = \frac{1}{2\pi}\,\mathrm{d}\eta \tag{8.128}$$

Let us now summarize the density functions introduced in this section. If we call ρ the set of variables relative to the internal coordinates and to the molecular orientation, then:

$$\rho_{\mathrm{A}} = \{Q_1 \cdots Q_{m_{\mathrm{A}}}, \theta, \varphi, \gamma, \eta\}_{\in \mathrm{A}} \tag{8.129a}$$

$$\rho_{\mathrm{B}} = \{Q_1 \cdots Q_{m_{\mathrm{B}}}, \theta, \varphi, \gamma, \eta\}_{\in \mathrm{B}} \tag{8.129b}$$

and

$$\rho = \{\rho_{\mathrm{A}}, \rho_{\mathrm{B}}\}$$

is a vector of $3N-6$ coordinates. The density expression will be given by:

$$\mathrm{d}p(\rho) = f(\rho)\,\mathrm{d}\rho = f(\rho_{\mathrm{A}})f(\rho_{\mathrm{B}})\,\mathrm{d}\rho_{\mathrm{A}}\,\mathrm{d}\rho_{\mathrm{B}} \tag{8.130}$$

8.6.1.5 The total volumic collision frequency

From previously defined relations (8.98 to 8.103, 8.113, 8.122 and 8.130) we write for the total volumic collision frequency the next general expression:

$$\mathrm{d}Z_{\mathrm{AB}} = N_{\mathrm{A}}N_{\mathrm{B}} \cdot v_{\mathrm{r}} \cdot 2\pi b\,\mathrm{d}b \cdot f(v_{\mathrm{r}})\,\mathrm{d}v_{\mathrm{r}} \cdot f(V_{\mathrm{A}}V_{\mathrm{B}}J_{\mathrm{A}}J_{\mathrm{B}}) \cdot f(\rho)\,\mathrm{d}\rho \tag{8.131}$$

We give now a more explicit formulation for some A + B kinds of experiment:

(a) The atom–atom cross-beams collision:

$$\mathrm{d}Z_{\mathrm{AB}} = N_{\mathrm{A}}N_{\mathrm{B}} \cdot 2\pi b\,\mathrm{d}b(v_{\mathrm{A}}^{02} + v_{\mathrm{B}}^{02} - 2v_{\mathrm{A}}^{0}v_{\mathrm{B}}^{0}\cos\gamma)^{1/2} \tag{8.132}$$

(b) The atom–atom thermal medium collision:

$$\mathrm{d}Z_{\mathrm{AB}} = N_{\mathrm{A}}N_{\mathrm{B}} \cdot 2\pi b\,\mathrm{d}b\frac{4\alpha^{1/2}}{\sqrt{\pi}}\,v_{\mathrm{r}}^{3}\exp(-\alpha v_{\mathrm{r}}^{2})\,\mathrm{d}v_{\mathrm{r}} \tag{8.133}$$

(c) The atom–diatom cross-beams collision:

$$\begin{aligned}\mathrm{d}Z_{\mathrm{AB}} = {} & N_{\mathrm{A}}N_{\mathrm{B}} \cdot 2\pi b\,\mathrm{d}b(v_{\mathrm{A}}^{02} + v_{\mathrm{B}}^{02} - 2v_{\mathrm{A}}^{0}v_{\mathrm{B}}^{0}\cos\gamma)^{1/2}f(V_{\mathrm{B}}J_{\mathrm{B}}) \\ & \times f(r_{\mathrm{B}})\,\mathrm{d}r_{\mathrm{B}} \cdot \tfrac{1}{2}\sin\theta_{\mathrm{B}}\,\mathrm{d}\theta_{\mathrm{B}}\frac{1}{2\pi}\,\mathrm{d}\varphi_{\mathrm{B}}\frac{1}{2\pi}\,\mathrm{d}\eta_{\mathrm{B}}\end{aligned} \tag{8.134}$$

(d) The atom–diatom thermal medium collision:

$$\mathrm{d}Z_{\mathrm{AB}} = N_{\mathrm{A}}N_{\mathrm{B}} \cdot 2\pi b\,\mathrm{d}b \frac{4\alpha^{3/2}}{\sqrt{\pi}} v_{\mathrm{r}}^{3} \exp\left(-\alpha v_{\mathrm{r}}^{2}\right) \mathrm{d}v_{\mathrm{r}} \mathrm{f}(V_{\mathrm{B}}J_{\mathrm{B}})$$

$$\times f(r_{\mathrm{B}})\,\mathrm{d}r_{\mathrm{B}} \cdot \frac{1}{2} \sin\theta_{\mathrm{B}}\,\mathrm{d}\theta_{\mathrm{b}} \frac{1}{2\pi}\,\mathrm{d}\varphi_{\mathrm{B}} \frac{1}{2\pi}\,\mathrm{d}\eta_{\mathrm{B}} \tag{8.135}$$

(e) The diatom–diatom thermal medium collision:

$$\mathrm{d}Z_{\mathrm{AB}} = N_{\mathrm{A}}N_{\mathrm{B}} \cdot 2\pi b\,\mathrm{d}b \frac{4\alpha^{3/2}}{\sqrt{\pi}} v_{\mathrm{r}}^{3} \exp\left(-\alpha v_{\mathrm{r}}^{2}\right) \mathrm{d}v_{\mathrm{r}}\, f(V_{\mathrm{A}}J_{\mathrm{A}})$$

$$\times f(V_{\mathrm{B}}J_{\mathrm{B}}) f(r_{\mathrm{A}})\,\mathrm{d}r_{\mathrm{A}} f(r_{\mathrm{B}})\,\mathrm{d}r_{\mathrm{B}} \cdot \tfrac{1}{2} \sin\theta_{\mathrm{A}}\,\mathrm{d}\theta_{\mathrm{A}} \cdot \tfrac{1}{2} \sin\theta_{\mathrm{B}}\,\mathrm{d}\theta_{\mathrm{B}}$$

$$\times \frac{1}{2\pi}\,\mathrm{d}\varphi_{\mathrm{A}} \frac{1}{2\pi}\,\mathrm{d}\varphi_{\mathrm{B}} \frac{1}{2\pi}\,\mathrm{d}\eta_{\mathrm{A}} \frac{1}{2\pi}\,\mathrm{d}\eta_{\mathrm{B}} \tag{8.136}$$

(f) The atom–triatom thermal medium collision:

$$\mathrm{d}Z_{\mathrm{AB}} = N_{\mathrm{A}}N_{\mathrm{B}} \cdot 2\pi b\,\mathrm{d}b \frac{4\alpha^{3/2}}{\sqrt{\pi}} v_{\mathrm{r}}^{3} \exp\left(-\alpha v_{\mathrm{r}}^{2}\right) \mathrm{d}v_{\mathrm{r}} f(V_{\mathrm{B}}J_{\mathrm{B}})$$

$$\times f(Q_1(\mathrm{B}), Q_2(\mathrm{B}), Q_3(\mathrm{B}))\,\mathrm{d}Q_1\,\mathrm{d}Q_2\,\mathrm{d}Q_3$$

$$\times \tfrac{1}{2} \sin\theta_{\mathrm{B}}\,\mathrm{d}\theta_{\mathrm{B}} \frac{1}{2\pi}\,\mathrm{d}\varphi_{\mathrm{B}} \frac{1}{2\pi}\,\mathrm{d}\gamma_{\mathrm{B}} \frac{1}{2\pi}\,\mathrm{d}\eta_{\mathrm{B}} \tag{8.137}$$

8.6.2 The Reaction Rate Constant Expression

The reaction velocity is classically expressed as the variation of the number of reactant molecules per unit of time and volume. Such a quantity is considered as proportional to the number of particles per unit of volume for each reactant partners (N_{A} and N_{B} respectively):

$$-\frac{\mathrm{d}N_{\mathrm{AB}}}{\mathrm{d}t} = kN_{\mathrm{A}}N_{\mathrm{B}} \tag{8.138}$$

where k is the reaction rate constant.

According to the previous section, we find that the reaction velocity is equal to the number of reactive collisions per unit of time and volume. This means that it is equal to the product of the volumic collision frequency Z_{AB} and the collision efficiency P_{AB}:

$$-\frac{\mathrm{d}N_{\mathrm{AB}}}{\mathrm{d}t} = \int P_{\mathrm{AB}}\,\mathrm{d}Z_{\mathrm{AB}} \tag{8.139}$$

The collision efficiency P_{AB} is an adimensional function depending on the same factors as $\mathrm{d}Z_{\mathrm{AB}}$ and equals unity when the collision is reactive and zero otherwise.

The integration reported in the last formula must include all the available experimental situations. Such integration may be done in two parts. First, we want to average all molecular internal and orientation coordinates ($\{\rho\}$) and the impact parameter b. These factors cannot be modified by changing the kind of the experiment we do. Then we define the total reaction cross-section $S(v_r, V, J)$ which depends only on the remaining factors: the relative velocity and the rotational–vibrational quantum numbers:

$$S(v_r VJ) = \int \mathrm{d}b \cdot 2\pi b \int \mathrm{d}\rho f(\rho) P_{AB}(v_r VJb\rho) \tag{8.140}$$

i.e. for the atom–diatom thermal collision we have:

$$S(v_r VJ) = \int_0^\infty \mathrm{d}b \cdot 2\pi b \int_{Q^{(-)}}^{Q^{(+)}} \mathrm{d}Q_B f(Q_B) \int_0^\pi \mathrm{d}\theta_B \cdot \tfrac{1}{2} \sin\theta_B$$
$$\times \int_0^{2\pi} \mathrm{d}\varphi_B \frac{1}{2\pi} \int_0^{2\pi} \mathrm{d}\eta_B \cdot \frac{1}{2\pi} P_{AB}(v_r V_B J_B b Q_B \theta_B \varphi_B \eta_B) \tag{8.141}$$

The expression of the reaction velocity becomes now:

$$-\frac{\mathrm{d}N_{AB}}{\mathrm{d}t} = N_A N_B \int_0^\infty \mathrm{d}v_r v_r f(v_r) \sum_{VJ} f(VJ) S(v_r VJ) \tag{8.142}$$

where the term $\sum_{VJ} f(VJ)$ includes summation over all the rotation–vibrational levels (it becomes $\sum_{VJK} f(VJK)$ when the K quantum number is to be included) for the two A and B partners if they are not single atoms.

The corresponding expansion for an atom–diatom thermal collision is:

$$-\frac{\mathrm{d}N_{AB}}{\mathrm{d}t} = N_A N_B \int_0^\infty \mathrm{d}v_r v_r^3 \frac{4\alpha^{3/2}}{\sqrt{\pi}} \exp(-\alpha v_r^2) \sum_{VJ} \frac{g_{VJ}}{Q} \exp\left(-\frac{E_{VJ}}{kT}\right) S(v_r VJ) \tag{8.143}$$

Comparing the two expressions to the reaction velocity (8.138 and 8.142) it is easy to obtain the rate constant expression:

$$k(v_r VJ) = v_r S(v_r VJ) \tag{8.144}$$

If we include in this expression, the translational–rotational–vibrational distribution for the particular experimental design of interest, we obtain:

$$k = -\frac{1}{N_A N_B} \frac{\mathrm{d}N_{AB}}{\mathrm{d}t} = \int_0^\infty \mathrm{d}v_r f(v_r) \sum_{VJ} f(VJ) k(v_r VJ) \tag{8.145}$$

8.6.3 The Reaction Rate Constant Calculation

8.6.3.1 The Monte Carlo integration method

The multidimensional integral reported above for $k(T)$ is insoluble in an analytical way. Nevertheless, we may use some numerical procedures. The simplest one is the Monte Carlo method.[30] By this method the integral evaluation is replaced by an estimator of it which is the average of the function to be integrated.

Let us imagine a function $\varphi(x_1 \cdots x_n)$ to be integrated in the range [0, 1]:

$$I = \int_0^1 \varphi(x_1 \cdots x_n)\, dx_1 \cdots dx_n \tag{8.146}$$

If $\{\xi_{11} \cdots \xi_{1p}, \ldots, \xi_{n1} \cdots \xi_{np}\}$ are p sets of n random numbers uniformly distributed in the range [0, 1], then:

$$\hat{\theta} = \frac{1}{p} \sum_{i=1}^{p} \varphi(x_1 = \xi_{1i} \cdots x_n = \xi_{ni}) \tag{8.147}$$

is an unbiased estimator of I ($\lim_{p\to\infty} \theta = I$) and its estimated variance ($\hat{\sigma}^2$) is:

$$\hat{\sigma}^2 = \frac{1}{p-1} \sum_{i=1}^{p} [\varphi(x_1 = \xi_{1i} \cdots x_n = \xi_{ni}) - \hat{\theta}]^2 \tag{8.148}$$

Note that instead of random numbers non-random series may be used (Diophantine integration). Then equation (8.147) must be replaced by $\hat{\theta} = 1/p \sum_i^{p-1} \varphi(\boldsymbol{\xi})$.[31]

Equation (8.148) shows that the standard deviation σ goes down as $p^{-1/2}$. To reduce σ by a factor two, the total number of functional evaluations p must be multiplied by four. Furthermore, the method converges at a rate approximately independent of the dimensionality of the integral. In practice $\{\xi_1 \cdots \xi_p\}$ are pseudorandom numbers where ξ_{i+1} depends on ξ_i in such a way that the set of ξ are uniformly distributed between 0 and 1 for all values of p which are not too small.

If we want to integrate a given $\varphi(\mathbf{x})$ function weighted by a normalized distribution function $w(\mathbf{x})$ it appears more useful to transform our integral in such a way that:

$$I = \int_0^1 \varphi(\mathbf{x}) w(\mathbf{x})\, d\mathbf{x} \tag{8.149}$$

becomes:

$$I = \int_0^1 \varphi(\boldsymbol{\xi})\, d\boldsymbol{\xi} \tag{8.150}$$

with:

$$\boldsymbol{\xi}_i = \int_0^{\mathbf{x}_i} w(\mathbf{x})\, d\mathbf{x} \tag{8.151}$$

for a normalized function w, where we choose $\mathbf{x}_i$ in order to reproduce each corresponding $\boldsymbol{\xi}_i$.

In general, when the integration range is $[\mathbf{a}, \mathbf{b}]$ instead of $[\mathbf{0}, \mathbf{1}]$ and if the function $w(\mathbf{x})$ is unnormalized, the equations become:

$$I = \int_{\mathbf{a}}^{\mathbf{b}} w(\mathbf{x})\varphi(\mathbf{x})\, d\mathbf{x} = \int_{\mathbf{a}}^{\mathbf{b}} w(\mathbf{x})\, d\mathbf{x} \int_{\mathbf{0}}^{\mathbf{1}} \varphi(\boldsymbol{\xi})\, d\boldsymbol{\xi} \tag{8.152}$$

$$I \simeq \hat{\theta} = \frac{\int_{\mathbf{a}}^{\mathbf{b}} w(\mathbf{x})\, d\mathbf{x}}{p} \sum_{i=1}^{p} \varphi(\mathbf{x}_i) \tag{8.153}$$

with:

$$\boldsymbol{\xi}_i = \frac{\int_{\mathbf{a}}^{\mathbf{x}_i} w(\mathbf{x})\, d\mathbf{x}}{\int_{\mathbf{a}}^{\mathbf{b}} w(\mathbf{x})\, d\mathbf{x}} \tag{8.154}$$

The last equation means that we choose for $\mathbf{x}_i$ a value such as $100 \times \boldsymbol{\xi}_i$ per cent. of the $\mathbf{x}$ values are less than or equal to $\mathbf{x}_i$ according to the distribution function $w(\mathbf{x})$. Equation (8.154) may be solved by standard techniques such as the Newton–Raphson method.

However, as the above expression has only one root, it is simpler to divide the interval $[a, b]$ into q parts (e.g. $q = 10$); for each x' selected value we compute the corresponding ξ' and look to see how close ξ'_0 is to ξ_i. We go over our procedure in a shorter interval $[x'_0 - (b-a)/q, x'_0 + (b-a)/q]$ until the difference $|\xi'_0 - \xi_i|$ becomes lower than a given convergence threshold. Such a method is particularly useful when $w(x)$ is the velocity distribution $v_r f(v_r)$, for which the Newton–Raphson method provides too slow a convergence for small ξ_i values.

Summarizing this section, we find that the multiple integral evaluation can be replaced by the average of the function to be integrated over a set of randomly chosen points. Further, we discuss the randomization process specially for atom–diatom thermal collisions. Generalization to any other reaction or experimental design is easy.

8.6.3.2 The total cross-section calculation

Randomization over θ, φ and η

These factors have as the weighting functions respectively:

$$w(\varphi) = w(\eta) = \frac{1}{2\pi}$$

$$w(\theta) = \tfrac{1}{2} \sin \theta$$

Using three random numbers ξ_{i1}, ξ_{i2} and ξ_{i3} we select the initial values for the trajectory i such that:

$$\varphi_i = 2\pi\xi_{i1} \tag{8.155a}$$

$$\eta_i = 2\pi\xi_{i2} \tag{8.155b}$$

$$\theta_i = \arc\cos\,(2\xi_{i3} - 1) \tag{8.155c}$$

Randomization over b:

Over such space direction the weighting function is:

$$w(b) = b$$

As the potential between an atom and a molecule has a limited range, we may assume that for an impact parameter $b \geqslant b_{\max}$ the trajectory becomes unreactive:

$$P_{AB} = 0 \qquad | \, b \geqslant b_{\max} \tag{8.156}$$

We only consider integration over the range $[0, b_{\max}]$; the normalized density function is:

$$w(b) = \frac{2b}{b_{\max}^2}$$

The initial values of b will be taken as:

$$b_i = b_{\max}\sqrt{\xi_{i4}} \tag{8.157}$$

Nevertheless, such a process is not statistically efficient because it overweights the large b values whereas the reactive trajectories tend to occur at small values. A better expression remains:

$$b_i = b_{\max}\xi_{i4} \tag{8.158}$$

which corresponds to a uniform distribution.

Randomization over the normal coordinates or over related other internal ones

The randomization of the initial values over the normal coordinates may be obtained using equation (8.125):

$$\xi_{i5} = \frac{\displaystyle\int_{Q^{(-)}}^{Q_i} f(QV)\,\mathrm{d}Q}{\displaystyle\int_{Q^{(-)}}^{Q^{(+)}} f(QV)\,\mathrm{d}Q} \tag{8.159a}$$

where $Q^{(+)}$ and $Q^{(-)}$ stand for the classical turning point values (equation 7.277). We may want to go beyond the harmonical approximation and include the rotational effect. The f function which appears in (8.159a) depends on the intramolecular potential as well as on the rotational and vibrational quantum numbers. In practice its analytical form is not known.

In the particular case of diatomic molecules we can replace (8.159a) by:

$$\xi_{i5} = \frac{\int_{R_-}^{R} G(RVJ)\,\mathrm{d}R}{\int_{R_-}^{R_+} G(RVJ)\,\mathrm{d}R} \tag{8.159b}$$

where R stands for the interatomic distance. To override this difficulty, we replace the integral over the vibrational phase by some equivalent one over the vibrational period:

$$G(R)\,\mathrm{d}R = \frac{\mathrm{d}t}{\tau_{1/2}} \tag{8.160}$$

where $\tau_{1/2}$ is the vibrational half-period.

Such a procedure is available as long as the time distribution is flat, regardless of the form of the radial distribution function G. Then:

$$\xi_{i5} = \frac{t_i}{\tau_{1/2}} \tag{8.161}$$

If we start at an initial time t_0 equal to $-t_i$, at an initial reaction radius ρ_0'

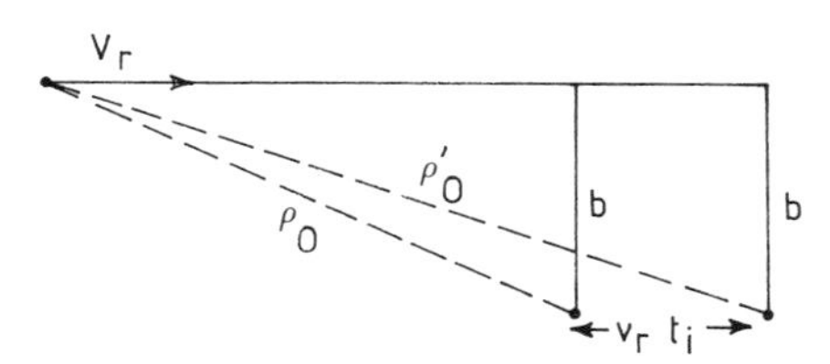

Figure 8.18 Back-starting on the reaction radius ρ

(see Figure 8.18) such as:

$$|\rho_0'|_i = \{[\sqrt{(\rho_0^2 - b^2)} + |v_r|\, t_i]^2 + b^2\}^{1/2} \tag{8.162}$$

and with an initial internuclear separation which alternates between R_+ and R_-, then at the zero value on the time scale we reach the correct value for the vibrational phase.

The turning point values calculation

The turning point values ($R_\pm$) are classically the roots of the following equation for a diatomic molecule:

$$E_{VJ} = \frac{J(J+1)\hbar^2}{2\mu R_\pm^2} + V(R_\pm) - V(R_e) \tag{8.163}$$

where $V(R)$ stands for the internuclear potential energy, R_e stands for the equilibrium internuclear distance and E_{VJ} is the vibrational–rotational energy which is characterized by the quantum numbers J and V. We assume for E_{VJ} the following form:

$$E_{VJ} = \omega_e(v+\tfrac{1}{2}) - \omega_e x_e(v+\tfrac{1}{2})^2 + B_e J(J+1) - D_e J^2(J+1)^2 - \alpha_e(v+\tfrac{1}{2})J(J+1) \tag{8.164}$$

which is related to the potential energy curve:

$$V(R-R_e) = \frac{k}{2}(R-R_e)^2 + \frac{g}{6}(R-R_e)^3 + \frac{j}{24}(R-R_e)^4 \tag{8.165}$$

in such a way that:

$$\begin{aligned} \omega_e &= \sqrt{(k/\mu)} \\ B_e &= (2\mu R_e^2)^{-1} \\ \omega_e x_e &= \frac{3}{2\omega_e^2\mu^2}\left(\frac{5g^2}{72\omega_e^2\mu} - \frac{j}{24}\right) \\ \alpha_e &= -\frac{4B_e^3 R_e^3 g}{\omega_e} - 6B_e^2\omega_e \\ D_e &= \frac{4B_e^3}{\omega_e} \end{aligned} \tag{8.166}$$

Equation (8.163) may be solved by the Newton–Raphson method. If:

$$G(R_\pm) = AR_\pm^{-2} + V(R_\pm) - B = 0 \tag{8.167}$$

with:

$$A = \frac{J(J+1)\hbar^2}{2\mu}$$

and

$$B = V(R_e) + E_{VJ}$$

then:

$$G(R_\pm + h) \simeq G(R_\pm) + \left[\frac{dG(R)}{dR}\right]_{R_\pm} h \tag{8.168}$$

For a short step size h we write:

$$h \simeq \frac{G(R_\pm + h)}{[dG(R)/dR]_{R_\pm + h}} \tag{8.169}$$

$$h \simeq \frac{AR^{-2} + V(R) - B}{[dV(R)/dR]_R - 2A/R^3} \tag{8.170}$$

The next trial value for R will be $R - h$ until h becomes as close as possible to zero. Table 8.3 provides some values for the turning point values R_+ and R_- in HD and HF molecules.

Table 8.3 Vibrational turning points and vibrational half-periods in HD and HF molecules (in atomic units)†

V	J		HD‡	HF§
0	0	$R_{(-)}$	1.21	1.64
		$R_{(+)}$	1.66	2.02
		$\tau_{1/2}$	194.98	201.42
0	1	$R_{(-)}$	1.21	1.64
		$R_{(+)}$	1.66	2.02
		$\tau_{1/2}$	195.26	201.51
1	0	$R_{(-)}$	1.10	1.54
		$R_{(+)}$	1.89	2.21
		$\tau_{1/2}$	197.78	207.32
1	1	$R_{(-)}$	1.10	1.54
		$R_{(+)}$	1.89	2.21
		$\tau_{1/2}$	197.96	207.41

† For length : 1 a.u. = 0.529167 Å;
for time : 1 a.u. = 2.41888×10^{-17} s;
for speed : 1 a.u. = 2.187653×10^{8} cm s^{-1}.
‡ $R_e(\text{HD}) = 1.40$ a.u.
§ $R_e(\text{HF}) = 1.81$ a.u.

The half-period computation

In order to reach the value of $\tau_{1/2}$ (for equation 8.161) which is the time associated with the nuclear motion from R_- to R_+, we must integrate the expression:

$$\tau_{1/2} = \int_{R_-}^{R_+} \frac{\mathrm{d}R}{v(R)} \tag{8.171}$$

where $v(R)$ stands for the nuclear velocity. We obtain $v(R)$ from the expression of the total energy associated with the vibrational motion:

$$E_{VJ} = \frac{J(J+1)\hbar^2}{2\mu R^2} + V(R) - V(R_e) + \frac{\mu v(R)^2}{2} \tag{8.172}$$

In Figure 8.19 we show the function $v(R)$ for the HF molecule. The integration of $\mathrm{d}R/v(R)$ may be simply obtained by a trapezoidal method. The values computed for HF and HD half-periods are reported in Table 8.3.

Cross-section calculation

When the initial conditions have been selected according to the previously described randomization process we can replace the multiple integral over

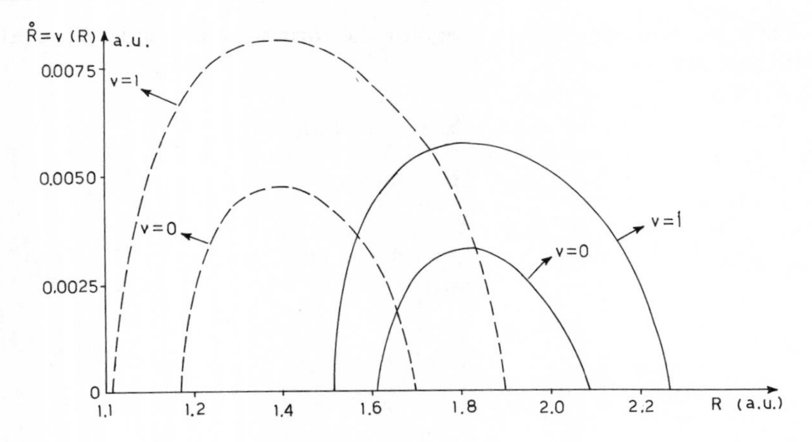

Figure 8.19 The velocity expression for H_2 (---) and HF (——) vibrator in the vibrational state $v = 0, 1$ (with $J = 0$)

φ, θ, η, b and R (or Q) by a simple summation over all trajectories. Then:

$$S(v_r, J, V) = \frac{2\pi b_M^2}{p} \sum_{i=1}^{p} \xi_{i4} P_r(i) \tag{8.173}$$

where p stands for the total number of trajectories and $P_r(i)$ for the efficacity of the ith trajectory. Note that P_r equals 1 if the trajectory is reactive and zero otherwise.

8.6.3.3 The total rate constant calculation

There remain three factors: v_r, V and J, to be selected suitably according to use experimental situation which is to be simulated.

Randomization over v_r

As the total energy (E_{tot}: the sum of the collision energy E_{col}, or $\mu_{AB}v_r^2/2$, and the rotation–vibration energy E_{VJ}) must be greater or equal to the potential barrier ($E^{\neq}$) between reactants and products it is not useful to compute trajectories which have too small a relative velocity. The lowest useful value (v_m) on v_r will be:

$$v_m = \left[\frac{2(E^{\neq} - E_{VJ})}{\mu_{AB}}\right]^{1/2} \tag{8.174}$$

where μ_{AB} stands for the total reduced mass of the reactants:

$$\mu_{AB} = \frac{M_A M_B}{M_A + M_B} \qquad \text{(equation 8.114)}$$

The selected relative velocities (v_i) will be the roots of:

$$\xi_{i6} = \frac{\int_{v_m}^{v_i} v_r f(v_r)\, dv_r}{\int_{v_m}^{\infty} v_r f(v_r)\, dv_r} \tag{8.175}$$

(we only need to compute such an expression in the case of a thermal medium), where:

$$f(v_r) = \frac{4\alpha^{3/2}}{\sqrt{\pi}} \exp(-\alpha v_r^2) v_r^2\, dv_r \qquad \text{(equation 8.113)}$$

with:

$$\alpha = \frac{\mu}{2kT}$$

It becomes:

$$\xi_{i6} = \frac{\int_{v_m}^{v_i} v_r^3 \exp(-\alpha v_r^2)\, dv_r}{\int_{v_m}^{\infty} v_r^3 \exp(-\alpha v_r^2)\, dv_r} = 1 - \frac{(\alpha v_i^2 + 1)\exp(-\alpha v_i^2)}{(\alpha v_m^2 + 1)\exp(-\alpha v_m^2)} \tag{8.176}$$

Such an equation may be easily solved by the numerical method described earlier (see the Monte Carlo averaging method). We finally note that the $\int_0^\infty v_r f(v_r)\, dv_r$ is not normalized.

Randomization over J and V

If separation between rotational–vibrational motions is assumed we shall write:

$$\xi_{i7} = \frac{1}{Q_J} \sum_{j=0}^{J_i} g_j \exp\left(-\frac{E_j}{kT}\right)\dagger \tag{8.177}$$

$$\xi_{i8} = \frac{1}{Q_V} \sum_{v=0}^{V_i} \exp\left(-\frac{E_v}{kT}\right) \tag{8.178}$$

where the rotational and vibrational partition functions are:

$$Q_J = \sum_{j=0}^{\infty} g_j \exp\left(-\frac{E_j}{kT}\right) \tag{8.179}$$

$$Q_V = \sum_{v=0}^{\infty} \exp\left(-\frac{E_v}{kT}\right) \tag{8.180}$$

† In equation (8.177) g_j stands for the rotational weighting factor (see Table 8.2); the summation over j must be extended over k for symmetric top molecules.

As in the quasi-classical approach, the numbers v and j are restricted to integral values, and it is required that V_i and J_i be the nearest integral solutions to the preceding equations (8.177) and 8.178).

The computations of E_j and E_v may be done assuming the harmonic approximation:

$$E_j = \frac{j(j+1)\hbar^2}{2\mu R_e^2} \tag{8.181}$$

$$E_v = (v + \tfrac{1}{2})\omega_e \tag{8.182}$$

or adding the anharmonic contribution (assuming $\alpha_e \simeq 0$).

If the separability of motions cannot be assumed, the first equations may be solved iteratively until:

$$\xi_{i7} = \frac{1}{Q_{VJ}} \sum_{j=0}^{J_i} g_j \exp\left(-\frac{E_{V_ij}}{kT}\right) \tag{8.183}$$

$$\xi_{i8} = \frac{1}{Q_{VJ}} \sum_{v=0}^{V_i} g_i \exp\left(-\frac{E_{vJ_i}}{kT}\right) \tag{8.184}$$

with:

$$Q_{VJ} = \sum_{j=0}^{\infty} \sum_{v=0}^{\infty} g_j \exp\left(-\frac{E_{vj}}{kT}\right) \tag{8.185}$$

Rate constant calculation

Assuming a correct randomization over V, J and v_r the rate constant expression becomes, in terms of Monte Carlo averaging:

$$k(T) = \frac{A}{q} \sum_{i=1}^{q} S(V_i J_i v_{r,i}) \tag{8.186}$$

where:

$$A = \frac{2}{\sqrt{(\mu\alpha)}} (\alpha v_m^2 + 1) \exp(-\alpha v_m^2)$$

The A coefficient corresponds to the normalization factor of the integral over the relative velocity:

$$A = \int_0^{\infty} v_r f(v_r)\, dv_r \tag{8.187}$$

as evidenced by variable changing from v_r to the random number ξ. We need for the atom–diatom thermal medium collision eight sets of random numbers (ξ_{ij} for $i = 1, p$ and $j = 1, 8$). We may reach such a requirement using only one vector of randomly distributed numbers (ξ_k for $k = 1, 8p$); the eight first values ξ_1 to ξ_8 are used as ξ_{11} to ξ_{18}, the eight following values ξ_9 to ξ_{16} become ξ_{21} to ξ_{28}, and so on, until ξ_k replaces ξ_{p8}.

8.7 THE THERMAL DEPENDENCE OF THE REACTION RATE CONSTANT

This last defined expression shows that the reaction rate constant depends on the temperature as long as the $f(v_r)$ and $f(VJ)$ distribution functions are temperature dependent.[32] Then we can write:

$$k(T) = \int \mathrm{d}v_r f(v_r, T) \sum_{VJ} f(VJ, T) k(v_r VJ) \tag{8.188}$$

As we have limited our discussion to a thermal mixture the general rate constant is:

$$k(T) = \int \mathrm{d}v_r \frac{4v_r^2}{\sqrt{\pi}} \left(\frac{\mu}{2kT}\right)^{3/2} \exp\left(-\frac{\mu v_r^2}{2kT}\right) \sum_{VJ} \frac{g_{VJ}}{Q_{rv}} \times \exp\left(-\frac{E_{VJ}}{kT}\right) k(v_r VJ) \tag{8.189}$$

8.7.1 The Arrhenius-like expression[33]

Equation 8.189 may also be written in terms of collision energy. Remembering that:

$$E_c = \frac{\mu v_r^2}{2} \tag{8.190}$$

then:

$$k(T) = \frac{(2/kT)^{3/2}}{(\pi\mu)^{1/2} Q_{rv}(T)} \sum_{VJ} g_{VJ} \int E_c \exp\left(-\frac{E_c + E_{VJ}}{kT}\right) S(E_c VJ)\, \mathrm{d}E_c \tag{8.191}$$

In order to simplify the writing of this last expression let us introduce the function φ_* defined as:

$$\varphi_*(E_c VJ, T) = E_c g_{VJ} \exp\left(-\frac{E_c + E_{VJ}}{kT}\right) S(E_c VJ) \tag{8.192}$$

Then:

$$k(T) = \frac{(2/kT)^{3/2}}{(\pi\mu)^{1/2} Q_{rv}(T)} \sum_{VJ} \int_0^\infty \varphi_*(E_c VJ, T)\, \mathrm{d}E_c \tag{8.193}$$

and

$$\ln k(T) = \ln \frac{(2/k)^{3/2}}{(\pi\mu)^{1/2}} - \frac{3}{2} \ln T - \ln Q_{rv}(T) + \ln\left(\sum_{VJ} \int \varphi_*\, \mathrm{d}E_c\right) \tag{8.194}$$

as:

$$\frac{\mathrm{d}\varphi_*}{\mathrm{d}T} = \frac{E_c + E_{VJ}}{kT^2} \varphi_* \tag{8.195}$$

From the derivative of $\ln k(T)$ (8.194) versus T we find that:

$$\frac{d \ln k(T)}{dT} = \frac{1}{kT^2} \left[\frac{\sum_{VJ} \int_0^{\infty} (E_c + E_{VJ}) \varphi_*(E_c VJ) \, dE_c}{\sum_{VJ} \int^{\infty} \varphi_*(E_c VJ) \, dE_c} - \frac{3}{2} kT - kT^2 \frac{d \ln Q_{rv}(T)}{dT} \right] \tag{8.196}$$

where the first term is simply the mean value of $E_c + E_{VJ}$ (which is the total internal energy) through function φ_*. Let us call this term $\langle E_* \rangle$. The second term represents the relative translational energy for the reactants and the last one is the corresponding rotational–vibrational mean energy. Let us replace these last two terms by $\langle E_R \rangle$. We can now write:

$$\frac{d \ln k(T)}{dT} = \frac{\langle E_* \rangle - \langle E_R \rangle}{kT^2} = \frac{E_a}{kT^2} \tag{8.197}$$

This expression is quite close to the Arrhenius expression but no assumption has been made about the independence of $\langle E_* \rangle - \langle E_R \rangle$ with respect to the temperature. We can also call $\langle E_* \rangle - \langle E_R \rangle$ the Arrhenius-like activation energy (E_a). This thermal independence occurs when we meet the next requirement:

$$\frac{d}{dT} (\langle E_* \rangle - \langle E_R \rangle) = 0 \tag{8.198}$$

where:

$$\frac{d\langle E_* \rangle}{dT} = \frac{\langle E_*^2 \rangle - (\langle E_* \rangle)^2}{kT^2} = C_* \tag{8.199}$$

and

$$\frac{d\langle E \rangle}{dT} = \frac{3}{2} k + \frac{\langle E_{VJ}^2 \rangle - (\langle E_{VJ} \rangle)^2}{kT^2} = C_v - \frac{3}{2} k \tag{8.200}$$

where C_v stands for the heat capacity of the reactants; subtraction of $\frac{3}{2}k$ arises when we have only to consider the relative motion ($\frac{3}{2}k$ instead of $\frac{3}{2}k + \frac{3}{2}k$). We can write the previous condition in a new form:

$$C_* \simeq C_v - \tfrac{3}{2} k \tag{8.201}$$

Let us now suppose that the cross-section ($S(E_c VJ)$) vanishes for a collision energy lower than a given E_c^0 value and increases for higher collision energies up to an asymptotic S_0 value. We can thus imagine a simple analytical form such as:

$$\begin{aligned} S(v_r VJ) &= 0 && \text{for } E_c \leqslant E_c^0 \\ S(v_r VJ) &= S_0\{1-\exp(-\alpha[E_c - E_c^0])\} && \text{for } E_c \geqslant E_c^0 \end{aligned} \tag{8.202}$$

We can further discuss the simple example of the rate constant k_{JV} for given values of J and V.

$$k_{JV} = \frac{(2/kT)^{3/2}}{\sqrt{(\pi\mu)}} S_0 \int_{E_c^0}^{\infty} E_c \exp\left(-\frac{E_c}{kT}\right)\{1-\exp(-\alpha[E_c - E_c^0])\}\, dE_c \tag{8.203}$$

where the J and V dependence of the cross-section has been suppressed for simplicity. As we know that:

$$I(a) = \int_a^{\infty} y \exp(-y)\, dy = \exp(-a)[1+a] \tag{8.204}$$

we can write:

$$k_{VJ}(T) = 4S_0\sqrt{\left(\frac{kT}{2\pi\mu}\right)}\left\{I\left(\frac{E_0}{kT}\right) - \frac{\exp(\alpha E_c^0)}{(1+\alpha kT)^2} I\left(E_c^0\left[\frac{1}{kT}+\alpha\right]\right)\right\}$$

or (8.205)

$$k_{VJ}(T) = 4S_0\sqrt{\left(\frac{kT}{2\pi\mu}\right)} \exp\left(-\frac{E_c^0}{kT}\right)\left[1+\frac{E_c^0}{kT} - \frac{1+E_c^0/kT+E_c^0\alpha}{(1+\alpha kT)^2}\right]$$

This rate constant vanishes when α tends to zero and when α increases up to an infinite value one finds:

$$\lim_{\alpha\to\infty} k_{VJ}(T) = 4S_0\sqrt{\left(\frac{kT}{2\mu\pi}\right)} \exp\left(-\frac{E_c^0}{kT}\right)\left[1+\frac{E_c^0}{kT}\right] \tag{8.206}$$

This situation corresponds to a cross-section like the one shown in Figure 8.20. From equation (8.205) we find the following Arrhenius-like activation energy:

$$\begin{aligned} E_a &= \langle E_* \rangle - \langle E_R \rangle = kT^2 \frac{d \ln k(T)}{dT} \\ &= \frac{kT}{2} + E_c^0 + \frac{2\alpha k^2T^2 \dfrac{1+E_c^0/kT+E_c^0\alpha}{(1+\alpha kT)^3} - E_c^0\left[1-\dfrac{1}{(1+\alpha kT)^2}\right]}{1+\dfrac{E_c^0}{kT} - \dfrac{1+E_c^0/kT+E_c^0\alpha}{(1+\alpha kT)^2}} \\ &= \frac{kT}{2} + E_c^0 + \frac{k^2T^2(2-\alpha E_c^0 - \alpha^2 E_c^0 kT)}{(1+\alpha kT)(E_c^0 + 2kT + \alpha E_c^0 kT + \alpha k^2T^2)} \end{aligned} \tag{8.207}$$

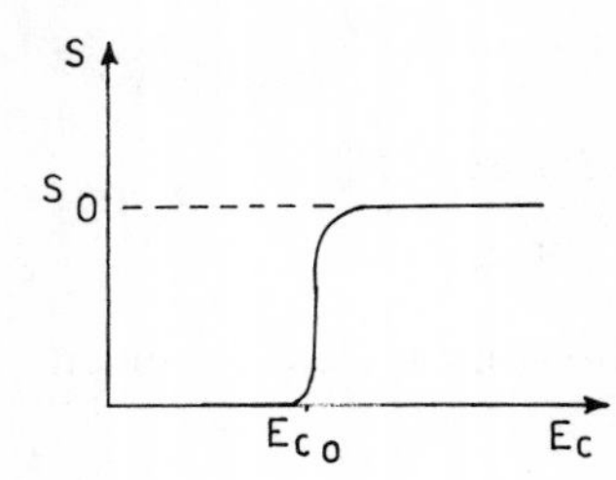

Figure 8.20 Profile of the total cross-section corresponding to $\alpha = \infty$

with:

$$\lim_{\alpha \to 0} (\langle E_{*} \rangle - \langle E_{R} \rangle) = \infty \tag{8.208}$$

and

$$\lim_{\alpha \to \infty} (\langle E_{*} \rangle - \langle E_{R} \rangle) = \frac{kT}{2} + \frac{E_c^0}{1 + kT/E_c^0} \tag{8.209}$$

This shows that even for E_c^0 equal to zero the Arrhenius-like activation energy for a given V and J value must be positive and in the range $[kT/2, \infty]$. The lowest available value, $kT/2$, corresponds in magnitude to the energy associated with the translation along the reaction pathway.

8.7.2 The Transition State Theory Expression

Let us now write the $k(T)$ expression in such a way that we find explicitly the relative translational partition function. Usually this function has the form:

$$Q_{tr} = V\left(\frac{2\pi\mu kT}{h^2}\right)^{3/2} \tag{8.210}$$

as:

$$k(T) = \frac{4(\mu/2kT)^{3/2}}{\sqrt{\pi} Q_{VJ}} \sum_{VJ} g_{VJ} \int v_r^3 \exp\left(-\frac{E_c + E_{VJ}}{kT}\right) S(v_r VJ) \, dv_r$$

then:

$$k(T) = \frac{4\pi V \mu^3}{h^3 Q_{tr} Q_{VJ}} \sum g_{VJ} \int v_r^3 \exp\left(-\frac{E_c + E_{VJ}}{kT}\right) S(v_r VJ) \, dv_r \tag{8.211}$$

According to Heisenberg's principle in each cartesian direction we have:

$$\Delta p_x \cdot \Delta x = h \tag{8.212a}$$

or

$$\Delta v_x \cdot \Delta x = \frac{h}{\mu} \tag{8.212b}$$

If we call τ_1 and τ_2 the elementary volume of the position and velocity spaces in which two situations are not physically distinct then:

$$\tau_1 \cdot \tau_2 = \frac{h^3}{\mu^3} \tag{8.213}$$

The reaction rate constant becomes, by introducing (8.213) into (8.211):

$$k(T) = \frac{kT}{h} \frac{V}{Q_{VJ}Q_{tr}} \sum_{VJ} g_{VJ} \int \frac{4\pi v_r^2 \, dv_r}{\tau_2} \times \int \frac{(h/kT)v_r \cdot 2\pi b \, db}{\tau_1} \exp\left(-\frac{E_c + E_{VJ}}{kT}\right) S(bv_r VJ) \tag{8.214}$$

where $4\pi v_r^2 \, dv_r/\tau_2$ corresponds to the number of distinct ways one finds v_r between v_r and $v_r + dv_r$ in the velocity space and $(h/kT)v_r \cdot 2\pi b \, db/\tau_1$ corresponds to the distinct positions for a given relative velocity and an impact parameter in the range $[b, b + db]$. The remaining adimensional cross-section $S(b, v_r, VJ)$ stands for an efficiency factor which only selects the reactive conditions on b, v_r, V and J.

For all these reasons we introduce the partition function Q_* defined as:

$$Q_* = \frac{8\pi\mu}{h^2 kT} \sum_{VJ} g_{VJ} \int E_c \exp\left(-\frac{E_c + E_{VJ}}{kT}\right) S(E_c VJ) \, dE_c \tag{8.215}$$

The reaction rate constant finds a simple form:

$$k(T) = \frac{kT}{h} V \frac{Q_*(T)}{Q_{tr}(T)Q(T)} \tag{8.216}$$

or replacing $Q_{tr}Q_{rv}$ by Q_R, which corresponds to the relative partition function in the reactants:

$$k(T) = \frac{kT}{h} V \frac{Q_*(T)}{Q_R(T)} \tag{8.217}$$

Let us now take the logarithm of the last expression:

$$\ln k(T) = \ln \frac{kV}{h} + \ln T + \frac{kT \ln Q_* - kT \ln Q_R}{kT} \tag{8.218}$$

Remembering the statistical thermodynamics relation:

$$F = -kT \ln Q \tag{8.219}$$

we obtain:

$$\begin{aligned} \ln k(T) &= \ln \left(\frac{kV}{h}\right) + \ln T - \frac{F_* - F_r}{kT} \\ &= \ln \left(\frac{kV}{h}\right) + \ln T - \frac{\Delta F_*}{kT} \end{aligned} \tag{8.220}$$

This equation is quite close to the one deduced from the transition state theory, also called the absolute rate theory. Nevertheless, in the present case no assumption is made concerning the properties of the transition state. If we now derive the former expression of $\ln k(T)$ with respect to the temperature we find:

$$\frac{\mathrm{d}\ln k(T)}{\mathrm{d}T}=\frac{1}{T}+\frac{\mathrm{d}\ln Q_{*}}{\mathrm{d}T}-\frac{\mathrm{d}\ln Q_{\mathrm{R}}}{\mathrm{d}T} \tag{8.221}$$

In statistical thermodynamics we know that:

$$E=kT^{2}\frac{\mathrm{d}\ln Q}{\mathrm{d}T} \tag{8.222}$$

Our preceding expression (8.221) becomes:

$$\begin{aligned}\frac{\mathrm{d}\ln k(T)}{\mathrm{d}T}&=\frac{1}{kT^{2}}(kT+E_{*}-E_{\mathrm{R}})\\&=\frac{1}{kT^{2}}(kT+\Delta E_{*})\end{aligned} \tag{8.223}$$

Therefore a term kT exists between ΔE_{*} and E_{a} which we have previously deduced:

$$\Delta E_{*}=E_{\mathrm{a}}-kT \tag{8.224}$$

Such a result shows that the lower limit for ΔE_{*} (according to the previous discussion of equation 8.209) is:

$$\Delta E_{*}=-\frac{kT}{2}$$

However, as:

$$F=E-TS$$

we find that:

$$S_{*}=\frac{\Delta E_{*}-\Delta F_{*}}{T} \tag{8.225}$$

8.7.3 Summarizing

The rate constant may be written as:

$$k(T)=\frac{kTV}{h}\exp\left(\frac{\Delta S_{*}}{k}\right)\exp\left(-\frac{\Delta E_{*}}{kT}\right) \tag{8.226}$$

or

$$k(T)=A\exp\left(-\frac{E_{\mathrm{a}}}{kT}\right) \tag{8.227}$$

where:

$$A = \frac{kTVe}{h} \exp\left(\frac{\Delta S_*}{k}\right)$$

These expressions are formally identical to the Arrhenius equation and to the relation obtained by the transition state theory, but they do not justify in any way the different assumptions made to obtain them. In particular, we do not introduce the concept of transition state and we do not assume thermal independence of parameters such as E_a, A, ΔS_* and ΔE_*. In the present case we only consider the reactions by setting some mean value differences according to whether we count all the initial situations or only those which correspond to a reactive process.

8.8 AN ADVANCED EXAMPLE: THE DHF SUPERSYSTEM[34]

8.8.1 The Potential Energy Surface and the Reaction Pathway

On the DHF potential energy hypersurface we can find three absolute minima which are respectively the F+HD, D+HF and H+DF atom–diatom supersystems (the last two have the same energy and differ only by the D and H exchange). We have also three first-order minimax which correspond to the collinear F...H...D, F...D...H and D...F...H supermolecule. The last stationary point is an absolute maximum corresponding to the completely dissociated D+H+F system. We summarize these features in Figure 8.21.

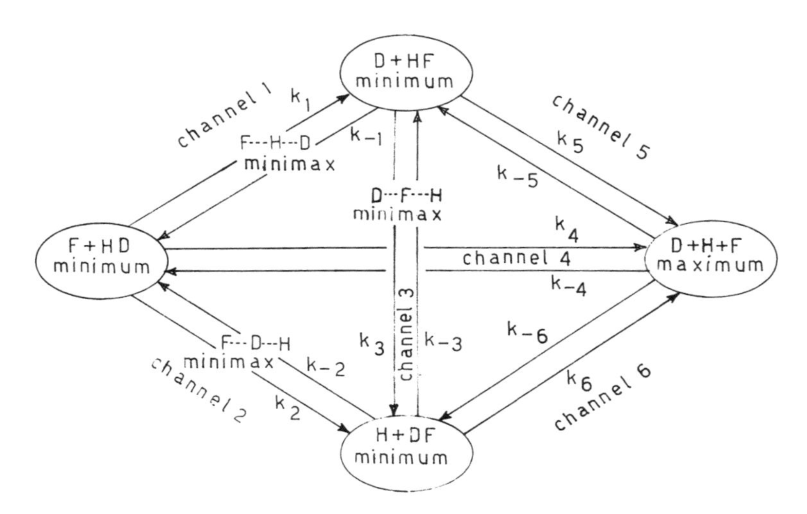

Figure 8.21 Stationary points on the DHF potential energy hypersurface

A LEPS-type augmented function has been used to represent analytically the potential energy hypersurface. The parameters were adjusted to fit as close as possible a set of 450 CI calculations for different nuclear configurations. A linear regression enables us to dispose of the analytical form for the H_2 and HF dissociation curves:

$$^1E(r) = \sum_i C_i \frac{1}{r^i}$$

The unknown C_i coefficients found are given in Table 8.4. The remaining

Table 8.4 The expansion parameters (C_i) for the H_2 and HF dissociation curve (in atomic units)

i	H_2	HF
1	0.173509	0.125438
2	−2.794195	−1.226192
3	17.286903	0.291981
4	−52.486069	46.496138
5	86.598199	−242.441525
6	−84.411485	578.067148
7	49.948094	−796.842818
8	−17.112864	686.410466
9	2.604342	−376.192049
10	0.210075	127.661719
11	−0.133916	−24.468423
12	0.013929	2.024410

parameters ^3D_e, $^3\alpha$ and ^3r_e of the purely LEPS function (equation 8.10) have been first chosen by a random search and optimized using the three-dimensional simplex method. The results obtained are:

$$^3D(\text{HD}) = 0.13226 \qquad ^3D(\text{HF}) = 0.15976 \text{ (a.u. of energy)}$$
$$^3\alpha(\text{HD}) = 0.9635 \qquad ^3\alpha(\text{HF}) = 1.5620 \text{ (a.u. of distance)}^{-1}$$
$$^3r_e(\text{HD}) = 0.9163 \qquad ^3r_e(\text{HF}) = 1.6028 \text{ (a.u. of distance)}$$

The resulting LEPS reproduces CI calculations with a standard error of 10 kcal mol^{-1}. The analytical representation has been improved up to a standard error of 2 kcal mol^{-1} by using a residual function ($f(r_1r_2r_3)$) of the form (8.19):

$$f(r_1r_2r_3) = \sum C_{lmn} \frac{1}{r_{\text{HD}}^m}\left(\frac{1}{r_{\text{HF}}^l}\frac{1}{r_{\text{DF}}^n} + \frac{1}{r_{\text{DF}}^l}\frac{1}{r_{\text{HF}}^n}\right)$$

A stepwise linear regression method enables us to select an appropriate set of C_{lmn} coefficients such that l, m and n are in the range [0, 7] (see Table 8.5). The relative energies found in this way are reported in Table 8.6.

Table 8.5 The expansion coefficients (C_{lmn}) for the residual function ($f(r_1r_2r_3)$) used for the DHF supermolecule (in atomic units)

l	m	n	C_{lmn}
7	0	1	0.428963
6	0	1	47.385889
5	0	1	−75.622582
4	0	1	45.661974
3	0	1	−13.851910
7	1	1	263.090285
4	1	1	−597.018842
1	4	1	−0.564983
3	1	1	174.774462
2	3	0	−0.517921
2	0	1	1.670824
6	1	1	−821.618821
5	1	1	1,037.644539
2	1	1	−25.915646
1	4	0	0.394701
4	3	0	3.265107
4	2	0	2.432668
1	2	0	0.407943
7	0	3	33.895584
7	1	0	−2.824472
5	0	4	1.929291
2	0	2	3.546453
7	0	2	−43.924898
6	2	0	−26.783891
7	2	0	15.380292
5	1	5	−14.782299
1	2	1	2.401366
3	1	2	−11.394305
1	3	0	−0.803263
7	3	7	3.179478

Table 8.6 Relative energy of the different stationary points on the DHF potential energy hypersurface (in kilocalories per mole)

Stationary point	Geometry (a.u.)	Relative energy
F+HD	r(HD) = 1.401	17.05
H+DF or D+HF	r(DF) = 1.806	0.00
H . . . D . . . F or D . . . H . . . F	r(HD) = 1.615, r(DF) = 2.535	17.68
H . . . F . . . D	r(HF) = 2.372, r(FD) = 2.372	44.95
H+D+F		111.66

Activation energies in channels 1 or 2 are sufficiently accurate compared to the best CI calculations[35] but results in channel 3 are somewhat underestimated. In channels 4 to 6 the energy variations are such that the corresponding chemical process remains unlikely.

According to our previous discussion we have already determined the reaction pathway (see Section 8.4.2 as well as Section 7.2.5). The step by step method we have explained enables us to obtain a tabulated reaction pathway. Effort has been made to generate a continuous representation of this reference curve on the potential energy hypersurface. If σ stands for the reaction coordinate, any point of the reaction pathway (RPW) may be expressed as a function of the previously defined z and Z coordinate frame:

$$\sigma = f(z, Z)$$

or as inverse functions:

$$z(\text{RPW}) = f(\sigma)$$

$$Z(\text{RPW}) = f(\sigma)$$

(Only two coordinates are requested, in fact, because the reaction pathway appears to occur for only the collinear approaches.) As seen in Figure 8.22 it

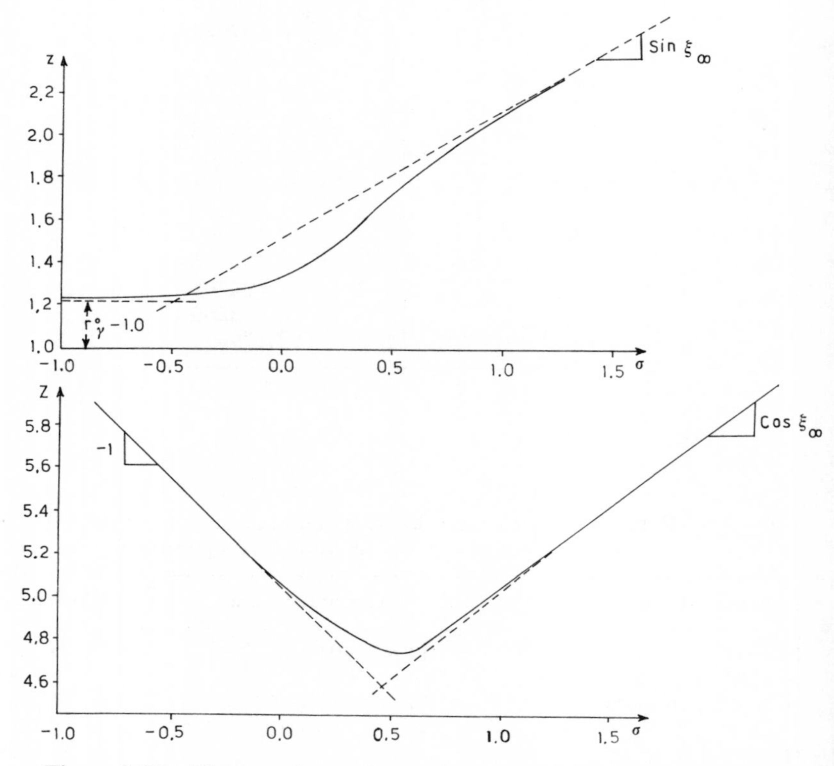

Figure 8.22 The reaction pathways for the F+HD→D+HF reaction

appears that the last expressions correspond to hyperboles. If we choose:

$$Z = A\sigma + B + \frac{\sqrt{(D\sigma^2 + E\sigma + F)}}{2C}$$

$$z = a\sigma + b + \frac{\sqrt{(d\sigma^2 + e\sigma + f)}}{2c} + \alpha \exp[-\beta(\sigma - \gamma)^2]$$

for numerical convenience we have augmented the z function by a gaussian function.

In determining the coefficients we ensure (to preserve distances along σ) the asymptotic behaviour. The results are reported in Table 8.7.

Table 8.7 Parameters of the reaction pathway for the F+HD → H+FD and F+HD → D+HF chemical processes

F+HD → H+FD		F+HD → D+HF	
$A = -0.2254$	$a = 0.4178$	$A = -0.1021$	$a = 0.3027$
$B = 3.7399$	$b = 1.0235$	$B = 4.6396$	$b = 1.1573$
$C = 1.0000$	$c = 1.0000$	$C = 1.0000$	$c = 1.0000$
$D = 2.3999$	$d = 0.6984$	$D = 3.2249$	$d = 0.3666$
$E = -3.3686$	$e = 0.0986$	$E = -2.7764$	$e = 0.3956$
$F = 1.3172$	$f = 0.0979$	$F = -0.7366$	$f = 0.1335$
	$\alpha = -0.102$		$\alpha = -0.197$
	$\beta = 0.590$		$\beta = 3.564$
	$\gamma = 2.887$		$\gamma = 0.075$

8.8.2 The Classical Trajectories for the F+HD Reaction

Some reactive and non-reactive trajectories for the F+HD collisions have already been shown in Figures 8.9 to 8.11. In these cases the interatomic distance frame (r_{HF}, r_{DH}, r_{DF}) was used. Let us now use another coordinate frame which explicitly takes into account the reaction pathway reference curve previously defined.
We choose a set of six coordinates named seminatural coordinates (SNC):

σ : the reaction coordinate
n : defined in cuts perpendicular to the reaction pathway
m : distance characterizing the non-collinear arrangement
$\boldsymbol{\theta} = \{\theta\varphi\chi\}$: three eulerian angles bonded to the rotation of the DHF plane

Details have been given elsewhere.[36] By using such a reference frame we can visualize the kinetic energy evolution along any classical trajectories (Figure 8.23). Partitioning the total kinetic energy leads to:

T(dissociation)
T(rotation)
T(bending–stretching) $= T(X) + T(Y) + f(m, n)$
with: $T = T$(dissociation) $+ T$(rotation) $+ T$(bending–stretching)

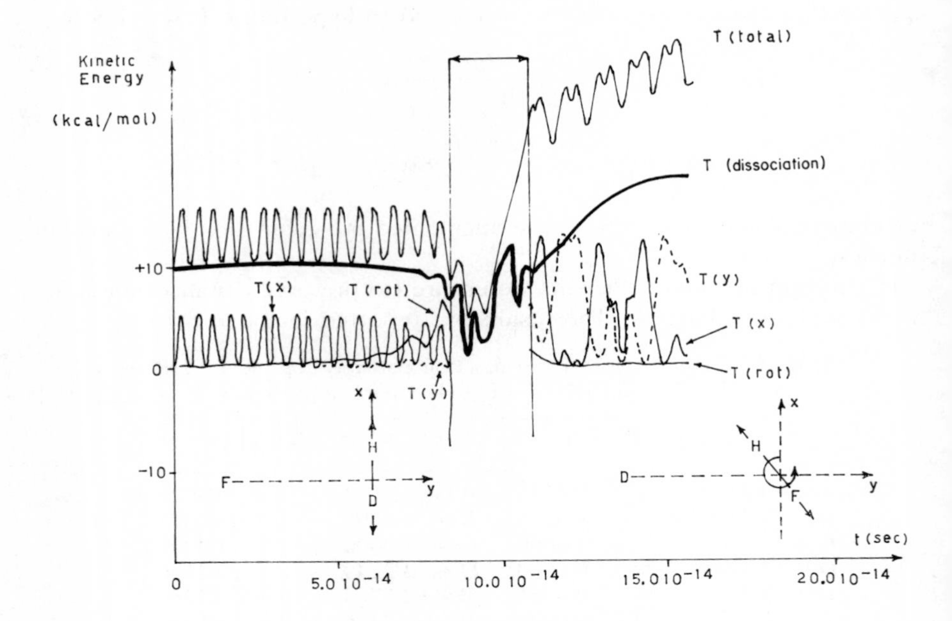

Figure 8.23 Kinetic energy along a classical trajectory in terms of the SNC frame. The initial conditions are: $V = 0$, $J = 0$, $E_c = 10.16$ kcal mol^{-1}

From Figure 8.23 it appears that the kinetic dissociation energy first decreases. After crossing the transition point ($\sigma = 0$) it goes up again. This corresponds to an ascent on the potential surface followed by a descent to the product.

When $J = 0$ in the reactant region we have only a stretching contribution to the kinetic energy; when the transition point approaches, rotational motion becomes effective—the outcoming diatomic molecule on the product side has simultaneous bending and stretching contributions to the total kinetic energy (for reasons of clarity all curves are not given in the transition region).

Plotting the evolution of σ along the trajectory, a possible measurement of the time needed by the energy reorganization (Figure 8.24) appears. During 2.31×10^{-14} s, evolution of the system along the reaction coordinate vanishes; in fact, $\dot{\sigma}$ velocity becomes approximately zero. We observe that this transition time increases as the initial relative velocity goes down. For a large set of trajectories the values effectively observed are in the range $[0.1 \times 10^{-14}, 2.5 \times 10^{-14}]$ s; this lies among a fourth and is 2.2 times a vibrational period.

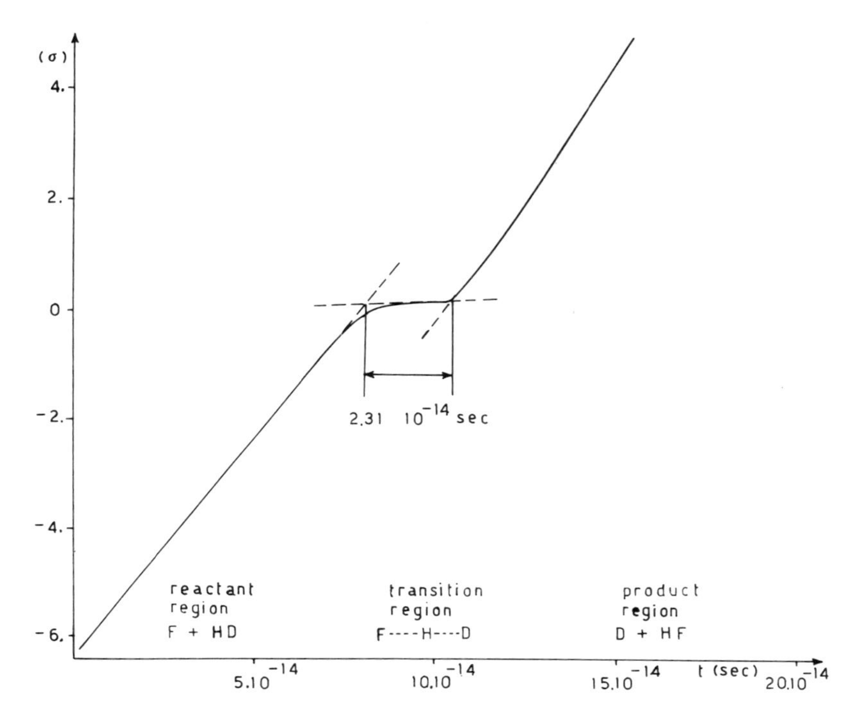

Figure 8.24 Function $\sigma(t)$ along the trajectory defined in Figure 8.23

8.8.3 Rate Constant and Its Thermal Dependence

We report here the reaction rate constant at two temperatures (300 and 1000 K) for the F+HD thermal collision. The following assumptions have been used for the impact parameter and for the minimum collision relative speed:

$$b_{\text{max}} = 4.0.\ \text{a.u.} = 2.12\ \text{Å}$$

$$v_{\text{m}} = 0.0005\ \text{a.u.} = 1.094 \times 10^5\ \text{cm s}^{-1}$$

Even at 1000 K the occupation probability of the first vibrational excited state in the HD molecule ($V = 1$) remains to small to generate collisions in this quantum state (see Figure 8.15). For the rotational quantum number and for the relative velocity the range scanned effectively by the random generator is given in Table 8.8. Details about reactive and non-reactive collisions are reported in Table 8.9.

Table 8.8 The range over J and v_r scanned by the random generator

	300 K	1000 K
$[J_{min}, J_{max}]$	[0, 6]	[0, 8]
$[v_m, v_{max}]$	[0.0005, 0.0024]	[0.005, 0.0043]

Table 8.9 Generated trajectories in each channel for F+HD collisions; the D+HF results are also provided for comparison

Channel	Reaction	Number of trajectories (and estimate of the integrate standard error) 300 K	1000 K
0	F+HD→F+HD	247 (3%)	219 (4%)
1	F+HD→D+HF	55 (12%) } (8%)	68 (11%) } (7%)
2	F+HD→H+DF	48 (13%)	63 (11%)
	Total number	350	350
0	D+HF→D+HF	—	470 (5%)
−1	D+HF→F+HD	—	30 (18%)
3	D+HF→H+DF	—	0
	Total number	—	500

Table 8.9 shows that the Monte Carlo integrator variance remains at a level of 10 to 20 per cent. according to the relatively small number of trajectories computed. From these results we may compute the reaction rate constants and their thermal dependence (see Table 8.10).

Table 8.10 Reaction rate constant and associated thermal dependence for the F+HD thermal collision

$$k(T) = A \exp\left(-\frac{E_a}{kT}\right) = \frac{kT}{h} V \exp\left(-\frac{\Delta S_*}{k}\right) \exp\left(-\frac{\Delta E_*}{kT}\right)$$

	F+HD→H+FD 300 K	F+HD→H+FD 1000 K	F+HD→D+HF 300 K	F+HD→D+HF 1000 K	D+HF→F+HD 1000 K
$k(T)$ (cm^3 mol^{-1} s^{-1})	1.065×10^{13}	4.560×10^{13}	1.699×10^{13}	3.308×10^{13}	5.461×10^{10}
E_a(kcal mol^{-1})	0.582	1.101	0.466	1.484	12.955
A (cm^3 mol^{-1} s^{-1})	2.830×10^{13}	7.933×10^{13}	3.713×10^{13}	6.973×10^{13}	—
ΔF_* (kcal mol^{-1})	−1.758	−6.358	−2.037	−5.720	7.01
ΔE_*(kcal mol^{-1})	−0.014	−0.886	−0.130	−0.503	11.00
ΔS_* (kcal mol^{-1} K^{-1})	5.814	5.472	6.354	5.217	−18.0

Comparison with experimental results is not possible; nevertheless, the F+H_2 collision reported by Fettis[37] and by Homann *et al.*[38] leads to a total reaction rate constant of the same order as this one:

$$k(T)=k_1+k_2=2.764\times10^{13}\text{ (at 300 K) and }7.868\times10^{13}\text{ (at 1000 K)}$$

to be compared with 2.65×10^{12} (computed from the formula $k=4.68\times 10^{13}\exp(-7{,}155\text{ J mol}^{-1}/RT)$[39]) at 300 K and 1.98×10^{13}[37] or 7.08×10^{13}[38] at 1000 K. If our results seem to be slightly overestimated it is due to the somewhat smaller transition barrier we obtained compared to the one usually accepted (0.62 kcal mol^{-1} versus 1.6 kcal mol^{-1}[40]).

Table 8.10 also shows a very low Arrhenius-like activation energy (E_a). Nevertheless, this term always remains larger than $kT/2$ leading to a positive value for ΔE_*. Considering Figure 8.25 it appears that any reactant supersystem contains enough energy to bypass the transition point when we add to the potential energy the ZPE correction due to the residual vibrational terms. The fact that some trajectories are non-reactive may only be understood from a dynamical point of view. The ZPE calculations lead to the following vibrational frequencies for the transition point:

(a) Vibrational stretching $r_{HF}+r_{DH}$: 1,823.45 cm^{-1}
(b) Bending (two perpendicular modes): 460.46 cm^{-1}
(c) Reactive stretching $r_{HF}-r_{DH}$: i 731.81 cm^{-1} (imaginary frequency)

These values induce a negative ZPE correction of -2.4 kcal mol^{-1} to the transition barrier (Figure 8.26).

From another point of view we may discuss the activation energies in terms of cross-sections (S). From the available trajectories we have tried to estimate the dependence of the cross-section with respect to the parameters v_r and J (Figure 8.27 and Table 8.11). The computational design oriented to

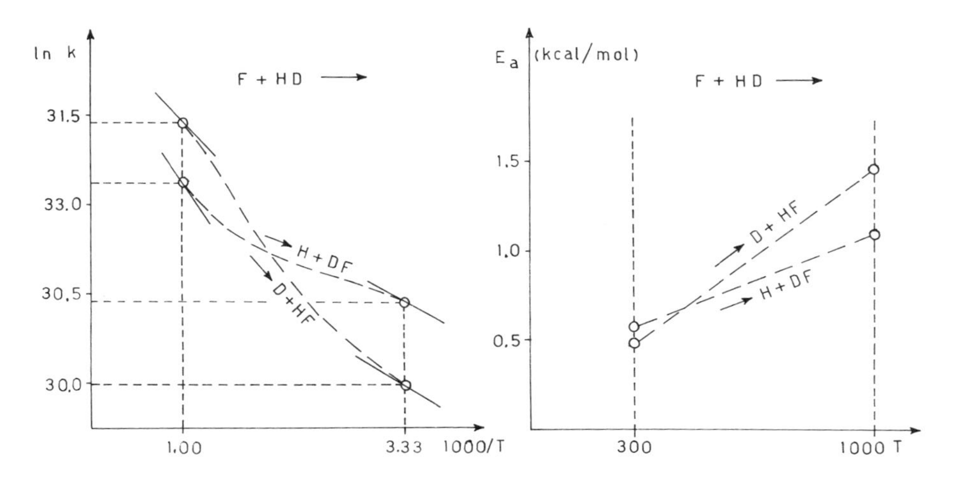

Figure 8.25 Usual plot of ln $k(T)$ versus $1/T$ and the thermal evolution of E_a with respect to T

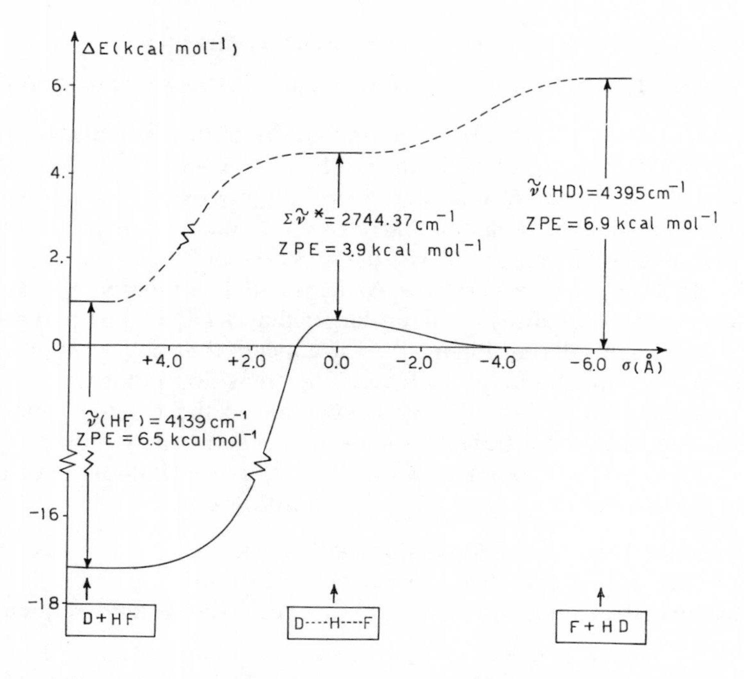

Figure 8.26 Relative energy and zero point energy (ZPE) evolution along the reaction pathway (σ)

the direct evaluation of $k(T)$ as well as the small number of trajectories enables us to give only qualitative results. Nevertheless; we observe a fast increase of the cross-section with respect to the relative velocity; modelization of the curves $S(E_c)$ is of the type (equation 8.202):

$$S(E_c) = 0 \qquad \text{for } E_c \leqslant E_c^0$$
$$S(E_c) = S^0\{1 - \exp[-\alpha(E_c - E_c^0)]\} \qquad \text{for } E_c \geqslant E_c^0$$

with a quasi-infinite α value. In Table 8.11, we report the effectively observed values for E_c^0 and S^0 and those obtained assuming $\alpha = \infty$ where we

Table 8.11 Lowest reactive collision energy and highest cross-section (observed value and model values)

Reactions	Temperatures	Observed values		Value assuming $\alpha = \infty$	
		E_c^0(kcal mol^{-1})	S^0(e.u.)	E_c^0(kcal mol^{-1})	S^0(e.u.)
F+HD→D+HF	300 K	0.470	7.93	0.571	8.61
	1000 K	0.464	7.05	0.520	7.09
F+HD→H+DF	300 K	0.420	4.79	0.414	4.77
	1000 K	0.765	10.08	1.261	10.95
D+HF→F+HD	1000 K	—	—	—	—

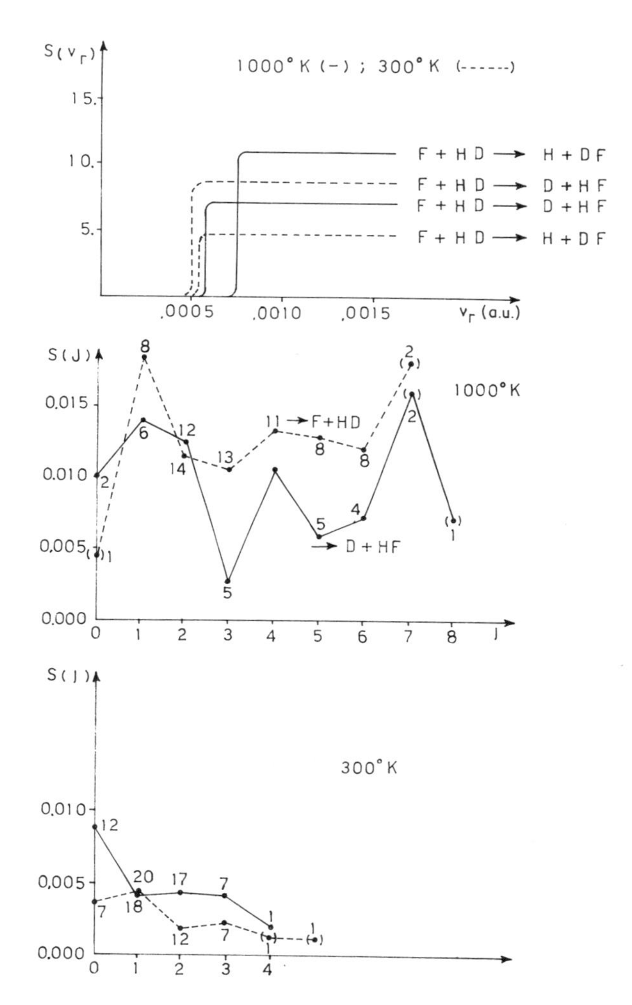

Figure 8.27 Dependence of the cross-section on the relative collision velocity (v_r) and the rotational quantum number (J)

want to reproduce the values previously given (Table 8.10) for E_a and $k(T)$. The agreement between observed and computed values shows that the small Arrhenius-like activation energy corresponds to a rapidly increasing cross-section between zero and a constant S^0 at a given E_c^0 collision energy.

REFERENCES

1. N. Rosen and P. M. Morse, *Phys. Rev.*, **42,** 210 (1932).
2. S. N. Mohammad, *Internat. J. Quantum Chem.*, **13,** 429 (1978).

3. A. Ralston and H. S. Will, *Mathematical Methods for Digital Computers*, Vol. II, Wiley, 1968, p. 156.
4. M. Sana, *Rapport d'Activité Scientifique du CECAM* CECAM, 1978, p. 176.
5. J. L. Dunham, *Phys. Rev.*, **41,** 721 (1932).
6. C. L. Beckel and R. Engelke, *J. Chem. Phys.*, **49,** 5199 (1968).
7. G. Simons, R. G. Parr and J. M. Finlan, *J. Chem. Phys.*, **59,** 3229 (1973).
8. K. D. Jordan, J. L. Kinsey and R. Silbey, *J. Chem. Phys.*, **61,** 911 (1974).
9. T. F. George and K. Morokuma, *J. Chem. Phys.*, **59,** 1959 (1973); *Chem. Phys.*, **2,** 129 (1973).
10. K. D. Jordan, *Internat. J. Quantum Chem. Symp.*, **9,** 325 (1975).
11. F. London, *Z. Elektrochem.*, **35,** 552 (1929); P. J. Kuntz, in *Atom–Molecule Collision Theory* (Ed. R. B. Bernstein), Plenum, 1979, Chap. 3, p. 55.
12. J. Hirschfelder, H. Eyring and B. Topley, *J. Chem. Phys.*, **4,** 170 (1936); J. T. Muckerman, *J. Chem. Phys.*, **54,** 1135 (1971); **56,** 2997 (1972); P. J. Kuntz, *Chem. Phys. lett.*, **16,** 581 (1972).
13. R. N. Porter and M. Karplus, *J. Chem. Phys.*, **40,** 1105 (1964); D. G. Thruhlar and C. J. Horowitz, *J. Chem. Phys.*, **68,** 2466 (1978).
14. H. Conroy, *J. Chem. Phys.*, **51,** 3979 (1969).
15. D. R. McLaughlin and D. L. Thomson, *J. Chem. Phys.*, **59,** 4393 (1973); X. Chapuisat and Y. Jean, *Topics in Current Chemistry*, Vol. 8, Springer-Verlag, 1976; S. K. Gray and J. S. Wright, *J. Chem. Phys.*, **66,** 2867 (1977); S. K. Gray, J. S. Wright and X. Chapuisat, *Chem. Phys. Lett.*, **48,** 155 (1977).
16. K. T. Tang and J. P. Toennies, *J. Chem. Phys.*, **68,** 5501 (1978).
17. P. J. Kuntz, in *Dynamics of Molecular Collisions* (Ed. W. H. Miller), Plenum Press, 1976, Part B, Chap. 2, p. 68.
18. S. W. Harrison, S. Swaminathan, D. L. Beveridge and R. Ditchfield, *Internat. J. Quantum Chem.*, **14,** 319 (1978).
19. W. L. J. Jorgensen and M. E. Cournoyer, *J. Amer. Chem. Soc.*, **100,** 4942 (1978); see also D. R. Yarkony, S. V. O'Neil and H. F. Schaefer III, *J. Chem. Phys.*, **60,** 855 (1974).
20. G. C. Lie and E. Clementi, *J. Chem. Phys.*, **62,** 2195 (1975); O. M. Matsuoka, E. Clementi and M. Yoshimine, *J. Chem. Phys.*, **64,** 1351 (1976).
21. W. L. Jorgensen, *J. Amer. Chem. Soc.*, **100,** 7824 (1978).
22. K. Fukui, S. Kato and H. Fujimoto, *J. Amer. Chem. Soc.*, **97,** 1 (1975); Z. Ishida, K. Morokuma and A. Komornicki, *J. Chem. Phys.*, **66,** 2153 (1977); A. Tachibana and K. Fukui, *Theoret. Chim. Acta*, **49,** 321 (1978), **51,** 189 (1979).
23. M. Sana, G. Reckinger and G. Leroy, *Theoret. Chim. Acta*, **58,** 145 (1981).
24. K. Fukui, S. Kato and H. Fujimoto, *J. Amer. Chem. Soc.*, **97,** 1 (1975).
25. W. H. Miller, N. C. Handly and J. E. Adams, *J. Chem. Phys.*, **72,** 99 (1980).
26. J. T. Hougen, P. R. Bunker and J. W. C. Johns, *J. Mol. Spect.*, **34,** 136 (1970);
27. F. S. Acton, *Numerical Methods that Work*, Harper, 1970, Chap. 5, p. 129.
28. M. Karplus, R. N. Porter and R. D. Sharma, *J. Chem. Phys.*, **43,** 3259 (1965); R. N. Porter and L. Ralf, in *Dynamics of Molecular Collisions* (Ed. M. H. Miller), Plenum Press, 1976, Chap. 1, p. 1; D. G. Truhlar and J. T. Muckerman, in *Atom–Molecule Collision Theory* (Ed. R. B. Nernstein), Plenum Press, 1979, Chap. 16, p. 505.
29. G. Herzberg, in *Molecular Spectra and Molecular Structure*, Vol. 2, Van Nostrand-Reinhold, 1945, Chap. I, p. 13; for calculations which go beyond the rigid rotation approximation see D. M. Dennison, *Rev. Mod. Phys.*, **3,** 280 (1931), **12,** 175 (1940).
30. J. M. Hammersley and D. C. Handscomb, in *Les Méthodes de Monte Carlo*, Dunod, 1967.
31. C. B. Haselgrove, *Math. Comput.*, **15,** 323 (1961); or for applications see H.

Conroy, *J. Chem. Phys.*, **47,** 5307 (1967); H. H. Suzukawa, D. L. Thompson, V. B. Cheny and M. Wolfsberg, *J. Chem. Phys.*, **59,** 3992, 4000 (1973).
32. S. F. Christov, *Collision Theory and Statistical Theory of Chemical Reactions*, Lecture Notes in Chemistry no. 18, Springer-Verlag, 1980, Chap. III, p. 122.
33. R. C. Tolman, in *Statistical Mechanics*, Chemical Catalog Company, 1927, Chap. 21; R. H. Fowler and E. A. Guggenheim, in *Statistical Thermodynamics*, Cambridge Press, 1939, Chap. 12; M. Karplus, R. N. Porter and R. D. Sharma, *J. Chem. Phys.*, **43,** 3259 (1965).
34. G. Reckinger, Ph.D. Thesis, Université Catholique de Louvain (1983).
35. J. C. Polanyi and J. L. Schreiber, *Faraday Disc. Chem. Soc.*, **62,** 267 (1977).
36. A. Nauts and X. Chapuisat, *Chem. Phys.*, **46,** 333 (1980).
37. G. C. Fettis, *Prog. React. Kinet.*, **2,** 3 (1964).
38. K. H. Homann, H. Schweinfurth and J. Warnatz, *Ber. Bunsenges Phys. Chem.*, **81,** 724 (1977).
39. P. Carsky and R. Zahradnik, *Internat. J. Quantum Chem.*, **16,** 243 (1979).
40. R. Foon and M. Kaufman, *Progr. Reaction Kinetics*, **8,** 81 (1975).

Author Index

Bold type numbers refer to pages on which the references are fully given.

Subject Index

Chemical Compound Index

Bold type numbers refer to chemical reactions

Commonly used Abbreviations

AO, atomic orbital

BDE, bond dissociation energy
BF, body fixed
BO, Born–Oppenheimer

CF, configuration function
CGTO, contracted gaussian type orbital
CI, configuration interaction
CIPSI, configuration interaction by perturbation second-order iterative
CMO, canonical molecular orbital
CNDO, complete neglect differential overlap
CSF, configuration state function

DIM, diatom in molecule
DM, density matrix

EA, electron affinity
EH, extended Hückel

FSGO, floating spherical gaussian orbital

GTO, gaussian type orbital

HF, Hartree–Fock
HFF, Hellmann–Feynman force
HFR, Hartree–Fock–Roothaan
HOMO, highest occupied molecular orbital

INDO, intermediate neglect differential overlap
INO, iterative natural orbital
IP, ionization potential
IR, infra-red
IRP, intrinsic reaction pathway

LCAO, linear combination of atomic orbital
LCBO, linear combination of bond orbital
LCSD, linear combination of Slater determinant
LCVO, linear combination of valence orbital
LEPS, London–Eyring–Polanyi–Sato
LMO, localized molecular orbital
LUMO, lowest unoccupied molecular orbital

MINDO, modified intermediate neglect differential overlap
MO, molecular orbital

NC, natural coordinate
NDDO, neglect of diatomic differential overlap
NO, natural orbital

PCILO, perturbative configuration interaction on localized orbitals
PCSO, pseudo canonical spin orbital
PES, potential energy surface
PPP, Pariser–Parr–Pople

QCPE, Quantum Chemistry Program Exchange

RHF, restricted Hartree–Fock
RPW, reaction pathway

SCF, self-consistent field
SE, stabilization energy
SF, space fixed
SNC, semi-natural coordinate
SO, spin orbital
SOMO, single occupied molecular orbital
STO, Slater type orbital

TST, transition state theory

UHF, unrestricted Hartree–Fock

WF, wave-function
WFF, wave-function force

ZDO, zero differential overlap
ZPE, zero point energy